SPACE-TIME WIRELESS SYSTEMS

One of the most promising technologies to resolve the bottlenecks in traffic capacity of future wireless networks is multiple-input multiple-output (MIMO) communications and space-time processing. MIMO wireless technology has progressed from the stage of fundamental research to commercially available products within a decade. With over sixty contributors from the field, this book provides an extensive overview of the state-of-the-art in MIMO communications, ranging from its roots in antenna array processing to advanced cellular communication systems. A balanced treatment of three key areas – information theory, algorithms and systems studies, and implementation issues – has been assembled by four editors with a broad range of academic and industry experience. This comprehensive reference will appeal to practitioners, researchers, and graduate students in wireless communications.

HELMUT BÖLCSKEI is an assistant professor of Communication Theory at ETH Zurich, Switzerland.

DAVID GESBERT is an associate professor with the Department of Mobile Communications, Eurecom Institute, France.

CONSTANTINOS B. PAPADIAS is a technical manager in Lucent Technologies' Bell Labs Wireless and Broadband Access Networks research center. He is currently on leave from Lucent as an Associate Professor at Athens Information Technology (AIT) in Athens, Greece.

ALLE-JAN VAN DER VEEN is a professor in the Circuits and Systems Group at TU Delft, the Netherlands.

T0155731

SPACE-TIME WIRELESS SYSTEMS

From Array Processing to MIMO Communications

Edited by

H. BÖLCSKEI, D. GESBERT, C. B. PAPADIAS,
AND A.-J. VAN DER VEEN

CAMBRIDGE UNIVERSITY PRESS
Cambridge, New York, Melbourne, Madrid, Cape Town, Singapore, São Paulo

Cambridge University Press
The Edinburgh Building, Cambridge CB2 8RU, UK

Published in the United States of America by Cambridge University Press, New York

www.cambridge.org
Information on this title: www.cambridge.org/9780521851053

© Cambridge University Press 2006

First published 2006
Reprinted 2006
This digitally printed version 2008

A catalogue record for this publication is available from the British Library

ISBN 978-0-521-85105-3 hardback
ISBN 978-0-521-07120-8 paperback

Professor Arogyaswami J. Paulraj, to whom this book is dedicated.

Arogyaswami J. Paulraj

It is a great pleasure to write a brief summary of the career of Professor Arogyaswami J. Paulraj (Paul as everyone calls him), whose many contributions we are honoring with this book. Paul has had an unusually varied and brilliant career, spanning several countries, technologies and roles: leading major system development projects, founding new industrial research laboratories, guiding a large university research group, starting innovative companies, and advising several large and small enterprises. This has given him a range of experiences and a versatility that is indeed rare.

His talents were evident very early. Throughout his studies, Paul was always and easily the top-ranking student. Finishing high school at age fifteen, Paul was selected through a nationwide competitive examination for the elite Indian National Defense Academy, a joint services institution, after which he chose a Naval career. The focus of the Naval engineering cadre was mostly on practical skills for maintaining weapons systems. Paul was interested in more, and self-tutored himself in subjects such as linear algebra, control theory and signal processing. These theoretical inclinations were increasingly evident and in 1969 the Navy deputed him to the Indian Institute of Technology (IIT) at Delhi, for an MTech. program. There, he quickly caught the eye of an influential EE Professor, PV Indiresan, who urged the Navy to allow Paul to enroll in the Ph.D. program. After opposition both from the IIT Senate (which worried about Paul's academic preparation) and from the Navy (which did not see the value of a Ph.D. degree), Indiresan finally got his way. That was when I first met Paul, having been invited by Indiresan to give a few lectures at IIT Delhi on my current research on nonlinear estimation theory. To be frank, the lectures were too mathematical for most of the audience, but the challenges caught Paul's fancy. And despite some mild discouragement from Indiresan and from me, Paul persisted. And a couple of years later I was very happy to serve as an external examiner for his PhD thesis on the estimation of Markov processes using advanced results and tools from stochastic diffusion theory.

The first pay-back for the Navy from Paul's advanced studies came sooner than they expected. In 1971, a brief war with Pakistan exposed the short-comings of the Navy's (British origin) sonar units, leading to the loss of an Indian ship. The Navy flew Paul to a postmortem the next morning, which led to his being put in charge of a sonar-improvement project. Paul did this very successfully, working in a new laboratory set up at IIT Delhi and two years later the redesigned sonar units entered fleet service. Paul's reward was a year as a research fellow at Loughborough University in the U.K. On his return he was charged to lead a very much more ambitious project to develop a large surface-ship sonar unit, which was not available to India because of military export restrictions. Paul and his team worked under very difficult circumstances and overcame many technical (and bu-reaucratic) challenges to induct into fleet service in 1983 a world-class sonar system (called APSOH for Advanced Panoramic Sonar, Hull mounted). This landmark military electronics achievement in India won him several major service awards and commendations. The reward this time for Paul, by now a Naval Captain, was a two-year fully-paid sabbatical anywhere he wished in the world. I had lost track of Paul soon after his Ph.D. and did not know of his sonar work till I got a letter from him in early 1983 asking to spend the time at Stanford. After much skepticism, given Paul's more than ten years in system development, that he could fit into our pretty the-oretical research group, I did agree to his request. Of course, I should have known better. Paul settled very quickly into our academic research envi-ronment and significantly helped to advance our group's recently initiated research in the area of subspace methods (launched by Ralph Schmidt's invention of the MUSIC algorithm) for direction finding using antenna ar-rays. In fact, just before returning to India in 1986, Paul came up with a completely new idea for direction finding called ESPRIT (Estimation of Sig-nal Parameters by Rotational Invariance Techniques). Paul urged Richard Roy to adopt this topic for his thesis research, and their work led to a mini-revolution in subspace methods. It has spawned more than 700 pa-pers and several doctoral dissertations; its applications go beyond array signal processing to spectral estimation and to model-based system identifi-cation.

Back in India, Paul was charged to start a new R&D center in Artificial Intelligence and Robotics in 1986. He focused it on developing command post systems for the armed forces and in parallel (!) worked on initiating an avionics development group. In 1988, he was invited to join Bharat Electronics (India's largest military electronics company) as chief scientist to set up a new Central Research Laboratory (CRL). Paul determined to introduce the structure and work culture that he had seen in the U.S. and CRL soon attracted a very bright group of researchers who developed several leading-edge products for these communication and radar systems. Again, in parallel, he set up and ran a systems software group in Bangalore for the Center for the Development of Advanced Computing, whose charter was to develop a massively parallel computer, again a technology that India was unable to acquire from elsewhere. However, administrative duties and bureaucratic battles began to take their toll and in 1991 Paul inquired about the possibility of returning to Stanford. At the time, our major funding was in a different area: a DARPA-sponsored project on the application of control and signal processing techniques to problems in semiconductor manufacturing. This time I was wiser and quickly welcomed Paul's assistance, and he and his family returned to Stanford in late 1991. By then our research in the antenna array area, which Paul had helped to accelerate before he left in 1986, had reached a stage of maturity. So with Paul and Richard Roy, now a research associate, we began to explore its applications to the re-emerging field of wireless communications. The challenges here were very different from those in direction finding, because the communication channel was much less well-defined and the earlier central concept of array manifolds became much more tenuous. Fortunately our research sponsors (DARPA) were quite open-minded and the success of our efforts in the manufacturing area allowed us to devote some resources to exploring the wireless area. This support was augmented by gift funds obtained through friends at Qualcomm, AT&T Bell Laboratories, GE, and Rockwell. The work progressed well enough that after some effort a faculty appointment as Professor (Research) came through in early 1993 and Paul began to lead an independent research program, dubbed the Smart Antennas Research Group (SARG), specifically to explore the brand new field of using antenna arrays in mobile wireless communication.

Paul attracted brilliant researchers and visitors to his group and together they developed many key ideas in smart antennas. An early Ph.D. thesis studied algorithms and system performance for Space Division Multiple Access (SDMA), which after some ups and downs is now a candidate for 4G mobile networks. In a 1993 patent application (issued in 1994), Paul proposed the use of spatial multiplexing for multiple-input multiple-output (MIMO) antenna systems and noted that it could dramatically increase link capacity. However, this patent first emphasized applications to HDTV digital video broadcasting, which in 1993 was facing a major capacity challenge; mobile wireless was still largely in the analog AMPS era. By now MIMO is a major focus of the wireless community, and it is already entering many wireless LAN and mobile standards. Regrettably, in 1994, Paul's applications for research grants to study MIMO were unsuccessful and so he and his group focused on more immediate problems related to the use of multiple antennas only at base stations. Several theses followed studying algorithms and usage concepts in CDMA, TDMA, and OFDMA systems. Paul returned to the MIMO area in 1998 after researchers at Bell Laboratories had by then laid its theoretical foundations. New theses studied MIMO capacity, space-frequency coding, space-time spreading, space-time equalization, precoding, and interference cancellation. SARG also initiated an annual Stanford workshop on Smart Antennas to bring together participants in academia, industry and government. These workshops became very popular and grew into the focal event for this new field. SARG, over the past dozen years, has graduated twenty Ph.D. students and hosted over twelve postdoctoral students and twenty long-term visitors. Many SARG alumni have gone on to become important leaders in their own right. The group has produced over 300 archival research publications, an introductory book on smart antennas, and 23 U.S. patents. Paul has received a number of awards in India to reflect his long service there. These include the Jain Gold Medal (1974), Distinguished Service Medal (1974), Most Distinguished Service Medal (1983), VASVIK Award (1984), and Scientist of the Year (1985). His other honors include Best Paper Award of the IEEE Signal Processing Society (with Alle-Jan van der Veen as the lead young author) (1997), Distinguished Alumnus, Indian Institute of Technology, Delhi (1999), and IEEE Technical Achievement Award, Signal Processing Society, 2003. He is also a Fellow of the IE (India), IETE (India), IEEE, and the Indian National Academy of Engineering. Notably, his dozen years in the US also led to election to the U.S. National Academy of Engineering in 2006. More honors are certainly on the horizon.

Besides being a widely-sought plenary and keynote speaker at numerous academic and industry venues and conferences, Paul is active on the advisory boards of several research consortia and centers in Europe, India, and the U.S. He has also worked with industry on smart antenna development via major consulting and advising roles. In 1999, he himself founded Gigabit (later called Iospan) Wireless, which successfully developed a MIMO-OFDMA based fixed-wireless system. Iospan, the first company to commercialize MIMO technology, was acquired by Intel Corp. in 2003. In 2004, Paul was invited to cofound Beceem Communications to develop chip sets for the mobile wireless Internet market. Smart antenna technology, which has now grown to dominate research and applications in wireless, owes much to Paul's contributions. He asked many of the right questions about this emerging technology and helped evolve a balanced picture of its potential and challenges.

Paul has been at different points in his career a scientist, engineer, teacher, manager, and advisor—sometimes many of these at the same time. And to everything he has brought dedication, vision, and humanity, which, combined with his scientific and technological abilities, has transformed and enhanced everything he has worked on. We are all very indebted to the editors of this book, themselves alumni of SARG, for organizing a conference and putting together this book to honor, on the occasion of his 60th birthday, the many contributions of Professor Arogyaswami Joseph Paulraj.

Thomas Kailath, Stanford, CA

Contents

Part III

Receiver algorithms and parameter estimation

Part IV

System-level issues of multiantenna systems

Contributors

Jørgen Bach Andersen
Aalborg University, Denmark

Gwen Barriac
Qualcomm, Inc., San Diego, CA, U.S.

Jean-Claude Belfiore
École Nationale Supérieure des Télécommunications, Paris, France

Ezio Biglieri
Universitat Pompeu Fabra, Barcelona, Spain

Ernst Bonek
Technische Universität Wien, Austria

Mark Brady
Stanford University, CA, U.S.

David Browne
University of California Los Angeles, CA, U.S.

R. Michael Buehrer
Virginia Tech, Blacksburg, VA, U.S.

Andreas Burg
ETH Zurich, Switzerland

Giuseppe Caire
University of Southern California, Los Angeles, CA, U.S.

A. Robert Calderbank
Princeton University, NJ, U.S.

John Cioffi
Stanford University, CA, U.S.

Antonio Maria Cipriano
École Nationale Supérieure des Télécommunications, Paris, France

Marc de Courville
Motorola Labs, Paris, France

Michael P. Fitz
University of California Los Angeles, CA, U.S.

David Garrett
Beceem Communications, Inc., Santa Clara, CA, U.S.

Alex B. Gershman
Darmstadt University of Technology, Germany

Georgios B. Giannakis
University of Minnesota, MN, U.S.

Karine Gosse
Motorola Labs, Paris, France

Babak Hassibi
California Institute of Technology, Pasadena, CA, U.S.

Noah Jacobsen
University of California, Santa Barbara, CA, U.S.

Magnus Jansson
Royal Institute of Technology (KTH), Stockholm, Sweden

Thomas Kailath
Stanford University, CA, U.S.

Heechoon Lee
University of California Los Angeles, CA, U.S.

Daniel Liu
University of California Los Angeles, CA, U.S.

Angel Lozano
Bell Laboratories, Lucent Technologies, Crawford Hill, NJ, U.S.

Xiaoli Ma
Georgia Institute of Technology, Atlanta, GA, U.S.

Upamanyu Madhow
University of California, Santa Barbara, CA, U.S.

Ninoslav Marina
École Polytechnique Fédérale de Lausanne (EPFL), Switzerland

Abdelkader Medles
Bell Laboratories, Lucent Technologies, Swindon, U.K.

Mehdi Mohseni
Stanford University, CA, U.S.

Markus Muck
Motorola Labs, Paris, France

Ayman F. Naguib
Qualcomm, Inc., Campbell, CA, U.S.

Björn Ottersten
Royal Institute of Technology (KTH), Stockholm, Sweden

Pirjo Pasanen
Nokia Research Center, Helsinki, Finland

H. Vincent Poor
Princeton University, NJ, U.S.

Daryl Reynolds
West Virginia University, Morgantown, WV, U.S.

Stéphanie Rouquette
Motorola Labs, Paris, France

Brian M. Sadler
U.S. Army Research Laboratory, Adelphi, MD, U.S.

Anna Scaglione
Cornell University, Ithaca, NY, U.S.

Shlomo Shamai (Shitz)
Technion, Haifa, Israel

Sébastien Simoens
Motorola Labs, Paris, France

Dirk Slock
Institut Eurécom, Sophia Antipolis, France

Robert A. Soni
Lucent Technologies, Whippany, NJ, U.S.

Yossef Steinberg
Technion, Haifa, Israel

Petre Stoica
Uppsala University, Sweden

Tae Eung Sung
Cornell University, Ithaca, NY, U.S.

Youngchul Sung
Cornell University, Ithaca, NY, U.S.

A. Lee Swindlehurst
Brigham Young University, Provo, UT, U.S.

Giorgio Taricco
Politecnico di Torino, Italy

Stephan ten Brink
Realtek Semiconductors, Irvine, CA, U.S.

Olav Tirkkonen
Nokia Research Center, Helsinki, Finland

Lang Tong
Cornell University, Ithaca, NY, U.S.

Antonia M. Tulino
Universitá degli Studi di Napoli, Naples, Italy

Sergio Verdú
Princeton University, NJ, U.S.

Mats Viberg
Chalmers University of Technology, Göteborg, Sweden

Haris Vikalo
California Institute of Technology, Pasadena, CA, U.S.

Pramod Viswanath
University of Illinois, Urbana-Champaign, IL, U.S.

Xiaodong Wang
Columbia University, New York, NY, U.S.

Werner Weichselberger
woolf solutions IT consulting and development, Vienna, Austria

Hanan Weingarten
Technion, Haifa, Israel

Jack Winters
Motia, Inc., Middletown, NJ, U.S.

Gregory W. Wornell
Massachusetts Institute of Technology, Cambridge, MA, U.S.

Huan Yao
Massachusetts Institute of Technology, Cambridge, MA, U.S.

Lizhong Zheng
Massachusetts Institute of Technology, Cambridge, MA, U.S.

Weijun Zhu
University of California Los Angeles, CA, U.S.

Acknowledgments

We, all four editors, are disciples of Prof. Arogyaswami J. Paulraj. We owe much of what we know about MIMO wireless and space-time processing to Prof. Paulraj, whose style of thinking and way of approaching research problems have influenced us profoundly. We dedicate this book in deep admiration and respect to Prof. Paulraj.

We would like to thank Moritz Borgmann for his tremendous help in typesetting the book and thereby contributing significantly to making the entire project happen. We are furthermore grateful to U. Schuster, A. Kapur, and all the reviewers for their assistance. Finally, we are especially grateful to all the contributors for their efforts and their smooth and highly professional collaboration.

H. Bölcskei, Zurich
D. Gesbert, Sophia-Antipolis
C. Papadias, Holmdel, NJ
A.-J. van der Veen, Delft

Introduction

1 The MIMO wireless success story

Wireless products in all shapes and sizes penetrate our offices, homes, and pockets. Despite the bursting of the dot-com bubble, the wireless communications industry has seen significant growth in the past decades and has become an essential part of our lives. Among the technologies with a chance to resolve the bottlenecks in traffic capacity likely to occur in many of the future wireless networks, few are as promising as multiple-input multiple-output (MIMO) technology and space-time processing.

This chapter serves as a brief introduction to MIMO wireless and space-time processing, with the aim of providing the reader with a perspective on historical and recent progress in this rapidly evolving field of communications research. A taxonomy of the various branches of study in MIMO wireless is provided, pointing to the remaining chapters in this book, which address the corresponding areas in greater detail.

The popularity of MIMO wireless systems is, by all measures, staggering. MIMO technology constitutes one of the rare cases in wireless communications where the full cycle of developing ideas on the blackboard, experimental verification, integrated circuits, standards, and, finally, commercially available products has been completed within a decade. This success is further accentuated by the synergy brought about between researchers in information and coding theory, signal processing, propagation theory and antenna design, VLSI, physics, and applied mathematics. The reasons behind this success story are manifold; we may cite a few: to begin with, the know-how in multiple-antenna theory and practice, starting with work conducted in the days of Marconi and later boosted by military and civilian applications in the second half of the 20th century, has helped tremendously to advance the field once the MIMO concept had emerged. In addition, more than fifty years of research in information and coding theory have resulted in a significant set of accessible tools that allowed the potential of MIMO systems to be established early on. Furthermore, the abundance of fascinating problems with solutions having almost immediate practical implications attracted a wide community of talented researchers. Finally, the ever improving computational capabilities of communications devices have resulted in making the prospect of widespread MIMO wireless applications a realistic one.

2 MIMO and space-time processing

What is a MIMO system? The definition of MIMO is a challenging task given the multiplicity of terms in use to denote various aspects of multiple-antenna communications. In this introduction, we refer to a wireless link as MIMO when multiple antenna elements are used on *both sides* of the link. (We similarly denote a wireless link as MISO or SIMO[†] when multiple antennas are used only for transmission or only for reception, respectively.) The more general notion of a MIMO system or a MIMO network refers to an entire wireless system or network that contains a single or (in the network case) multiple MIMO links. The terminology "smart antenna" is used for an antenna array in conjunction with signal processing or combining capabilities. Finally, we adopt the general notion of space-time processing, embracing all aspects of signal processing using multiple antennas or sensors including, for example, parameter estimation, such as direction-of-arrival estimation, and transmission techniques, such as space-time coding. We conclude by noting that a MIMO system employs space-time processing to realize the spectral efficiency and link reliability gains offered by MIMO channels.

2.1 Core principles of space-time processing

At the core of space-time processing lies the concept of *spatial sampling* of a propagating wave field (e.g., radio or acoustic) through the use of spatially distributed sensing devices. The processing of the resulting vector signals is referred to as space-time processing.

Consider a source emitting in the far field, as shown in Fig. 1. The response of an antenna (sensor) array to the unobstructed propagating wave field created by this source can be represented by a vector[‡] $\mathbf{h} = \begin{bmatrix} h_1 & h_2 & \cdots & h_N \end{bmatrix}^T$, where N is the number of antenna elements. The vector $\mathbf{h}$ is often called

---

[†]The abbreviations MISO and SIMO stand for multiple-input single-output and single-input multiple-output, respectively.

[‡]The superscript T stands for transposition.

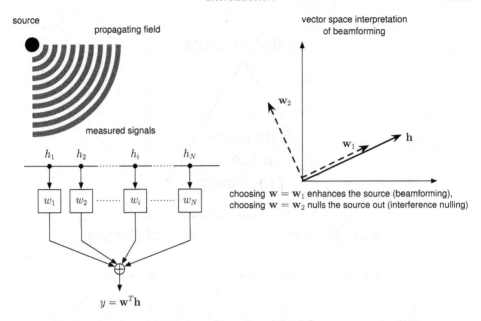

Fig. 1. A vector-space interpretation of multiple-antenna processing: the response of an array to the propagating wave field emitted by a source can be represented by a signature vector. Knowledge of this signature helps to identify the position of the source and can be used to amplify the source signal or attenuate it in case it is interference.

the "spatial signature" of the source. In practice, scattering objects in the propagation environment result in a spatial signature consisting of a super-position of several contributions corresponding to the individual scattering objects and causing waves to impinge from various angles and with various delays. The resulting space-time signature will in general exhibit multipath fading and may cause intersymbol interference (ISI).

Research in space-time processing is mostly concerned with how to *estimate space-time signatures or, equivalently, space-time channels* and how to use these estimates to *extract information (in the presence of noise and interference) on certain parameters* pertaining to the source (e.g., angle-of-arrival (AOA), propagation delay, or Doppler frequency) or to *facilitate the detection of the signal or message* transmitted by the source. The key concepts in space-time processing are illustrated in Fig. 2 and can briefly be summarized as follows:

(i) *Spatial structure.* Spatial sampling is in many ways analogous to temporal sampling. For example, in the context of AOA estimation, similar to the Nyquist condition in temporal sampling, there is a con-

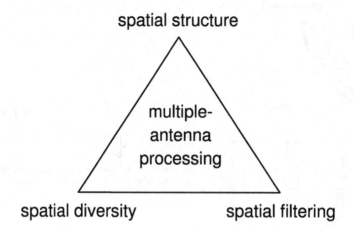

Fig. 2. Key concepts in multiple-antenna processing.

dition on antenna spacing that needs to be satisfied to ensure that the spatial signature relates to the source's AOA in a one-to-one fashion. Exploiting a-priori known structure of the signature vector is crucial in performing source parameter estimation as well as estimating the signature vector itself. Fundamental progress on AOA estimation exploiting such structural properties was realized in the late 1970s and the 1980s through the development of so-called subspace methods known under the names of MUSIC and ESPRIT. Chapter 13 discusses subspace methods in detail. A general treatment of basic concepts in array signal processing is provided in Chapter 12.

(ii) *Spatial filtering.* Once the signature (channel) is estimated, spatial filtering can be applied to extract the information signal in the presence of various impairments such as thermal noise, cochannel interference and attenuation due to shadowing and multipath fading.

Spatial filtering consists of applying a vector of antenna (sensor) combining coefficients $\mathbf{w} = \begin{bmatrix} w_1 & w_2 & \cdots & w_N \end{bmatrix}^T$ to the received signal vector (see Fig. 1), typically with the goal of maximizing the received signal to noise-plus-interference ratio. In the absence of cochannel interference, the signal-to-noise ratio (SNR) becomes the relevant quantity; the SNR-maximizing solution is[†] $\mathbf{w} = \begin{bmatrix} h_1^* & h_2^* & \cdots & h_N^* \end{bmatrix}^T$, resulting in what is known as coherent combining or, equivalently, matched

---

[†]The superscript * stands for complex conjugation.

filtering. Coherent combining yields an SNR gain that scales linearly in the number of antennas and is often denoted as array gain or beamforming gain.

In the presence of cochannel interference, spatial filtering can be employed to perform interference reduction. The basic idea is to choose the weight vector **w** such that it is "as orthogonal as possible" to the signature(s) of the interfering signal(s). With N sensors, in principle, it is possible to suppress $N-1$ interfering sources while forming a beam toward a single desired source. A detailed treatment of beamforming techniques is provided in Chapter 2.

(iii) *Spatial diversity.* The concept of spatial diversity is closely tied to a probabilistic description of the signature vector. In a multipath propagation environment with rich scattering, the signature coefficients are adequately modeled as (typically independent) random variables. The basic principle of spatial diversity is that if several replicas of the information signal (corresponding to a multitude of antenna elements) are received through independently fading links (branches), then with high probability at least one or more of these links will not be in a fade at any given instant. The number of independent diversity branches is the diversity order offered by the channel (N in our case). The corresponding gains on the link level are reduced error probability for a given SNR or, equivalently, SNR savings for a given target error probability. For example, in an uncoded Rayleigh fading link with QPSK modulation and coherent combining, assuming a target bit error rate of 10^{-3}, we obtain an SNR saving of 10 dB due to second-order receive diversity (in addition to a 3 dB array gain). However, SNR savings due to diversity alone saturate quickly with increasing number of antennas.

Spatial diversity can be realized at the receiver (typically through coherent combining) or the transmitter (or both) of a wireless system. In the case where the channel is not known at the multiple-antenna transmitter, spatial transmit diversity gain can be obtained through space-time coding, discussed in Chapter 7, which effectively spreads information bits cleverly across antennas (space) and time. We finally note that besides spatial diversity there are two additional forms of diversity traditionally exploited in wireless communications systems, namely time diversity due to Doppler spread and frequency diversity due to delay spread. A detailed treatment of the concept of diversity can be found in Chapter 3.

2.2 *The MIMO paradigm*

Fundamentally, space-time processing as described in Section 2.1 requires multiple antennas at one end of the communication link only. In cellular networks, it is typically the base station that is equipped with an antenna array. The underlying reason is that the extra cost incurred and space needed by antenna arrays is more affordable at the base station than at the user terminal. As we are, however, witnessing the convergence of mobile communications and IP-based applications, handheld phones are turning into sophisticated bandwidth-hungry data-processing devices and the need for higher data rates is increasing. Combined with restrictions on form factors and computational complexity getting more relaxed, this provides increasing motivation for considering the use of multiple antennas at the user end as well. Swift progress in theoretical MIMO wireless research (as detailed below and illustrated in the rest of this book) in combination with the practical motivation outlined above has dramatically fueled the interest of the IT engineering community in the area of MIMO wireless systems.

Employing multiple antennas at both the transmitter and the receiver yields higher diversity gains but also, and more importantly, increased capacity through "spatial bandwidth". The resulting capacity gains are realized through *spatial multiplexing*.

Spatial multiplexing

The principle of spatial multiplexing stipulates that multiple independent data streams can be sent simultaneously and in the same frequency band over the MIMO channel and recovered at the receiver if appropriate signal processing is employed. Using an algebraic analogy, the input signals are the unknowns in a linear system of equations with the observations given by the measurements made at the receive array. Provided that enough observations are available (i.e., the number of independent data streams is smaller than or equal to the number of receive antennas), the input signals can be identified uniquely. The transmitters can be distributed or colocated on a single device. An early (1994) proposal of a MIMO wireless system was a capacity-enhanced HDTV network, where TV channels were broadcast simultaneously from independent relay sites (Paulraj and Kailath, 1994), corresponding to the case of distributed transmitters. For colocated transmit antennas, it is multipath propagation that provides the richness of the scattering environment required for spatial multiplexing gain. We note that the basic principle of spatial multiplexing is in sharp contrast to frequency, time or code-division based multiplexing schemes, which realize orthogonality of the multiplexed data streams at the cost of bandwidth. In the MIMO case, one relies on the channel to induce this separation in the spatial domain, albeit the spatial signatures corresponding to the different data streams will, in general, not be strictly orthogonal (but they will be diverse enough so that they can be separated through receiver processing). The spatial multiplexing effect was quantified rigorously (Foschini and Gans, 1998; Telatar, 1999) and shown to result in a theoretical capacity gain (over a single-input single-output system) of $\min(M, N)$, where M is the number of transmit antennas and N is the number of receive antennas. This gain has first been demonstrated in practice in 1998 using the BLAST[†] indoor prototype developed at Lucent Bell Labs (Foschini and Gans, 1998). Fundamentals of MIMO channel capacity and spatial multiplexing are discussed in Chapters 4 and 5.

2.3 Commercial applications of MIMO and space-time processing

Constrained by cost and deployment restrictions, the integration of MIMO technology into cellular networks and wireless local area networks (WLANs) has experienced a slow start but is now accelerating. This is partly thanks to a better understanding of the benefits of the technology, progress on algorithm development and associated hardware implementations and a wider acceptance in standards. The application of MIMO and space-time processing in commercial wireless networks aims at improving system performance along one or more of the following directions:

- Increased quality of service (data rates, error rates, latency)
- Increased battery life
- Increased coverage
- Increased spectrum efficiency (bits/s/Hz/cell)

MIMO technology and space-time processing affect nearly all aspects of transceiver design. Historically, the diversity concept (dating back to the early days of cellular communications) has proved the most robust and easiest to put into practice. Today, the base stations of second-generation cellular systems such as GSM (Global System for Mobile Communications) routinely use antenna diversity on the uplink to combat fading in non–line-of-sight (NLOS) scenarios. The performance of downlink beamforming is typically quite limited due to imperfect channel state information (CSI) at the base station, which results from insufficient feedback or the fact that the channel is not fully reciprocal, even in time division duplex (TDD) systems.

[†]The acronym BLAST stands for Bell labs LAyered Space-Time architecture.

Solutions limited to "dominant-path beamforming" are used, however, in 2G (GSM and IS95) and 3G systems. Transmit diversity on the downlink based on the Alamouti scheme (Alamouti, 1998) is employed in 3G (both UMTS and CDMA2000) systems; see Chapter 25 for details.

More recently, the development of advanced wireless data-centric networks has opened the doors to widespread adoption of MIMO and space-time processing concepts. Still evolving standards like IEEE 802.16 (WiMax), IEEE 802.11 (WiFi), discussed in Chapter 26, and IEEE 802.20 (Mobile Broadband Wireless Access) focus on spectrally efficient Internet access and, thus, constitute active playgrounds for real-life testing of even the most recent MIMO research advances.

3 Today's MIMO and space-time processing research

Depicting the state-of-the-art in the rapidly evolving field of MIMO and space-time processing is a difficult task due to the multifaceted and very dynamic nature of the area.

In the following, we propose one possible (certainly incomplete) taxonomy of ongoing MIMO and space-time processing research. Using a classification into "information-theoretic studies", "algorithmic studies", and "system studies and implementation", Fig. 3 summarizes the different areas of research. Although represented separately in Fig. 3, many of these branches of study are often overlapping and benefit from cross-fertilization.

3.1 Information-theoretic studies

Information theory has been successful at establishing the ultimate performance limits of MIMO systems (Telatar, 1999; Foschini and Gans, 1998), see Chapter 4. This success has triggered significant effort in several directions, summarized below.

First, a clear need for more refined statistical MIMO channel models, taking into account real-world indoor and outdoor propagation conditions, was established. Basics of MIMO channel models and insights based on measurements are presented in Chapters 1 and 23, respectively. The impact of propagation conditions on MIMO channel capacity is analyzed in Chapter 5.

Spatial multiplexing and space-time coding are capable of realizing capacity and diversity gains without requiring CSI at the transmitter. Additional performance improvements obtainable from transmit CSI are discussed in Chapter 6.

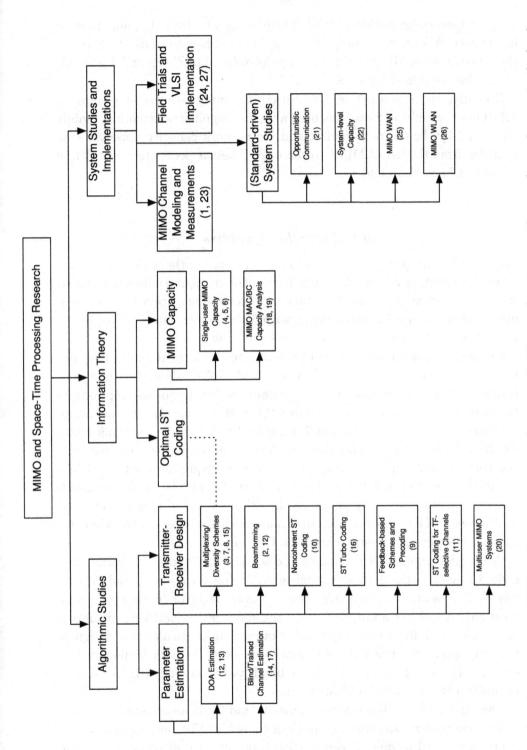

Fig. 3. A taxonomy of MIMO and space-time processing research. The numbers in parentheses refer to chapters in the book.

Space-time codes exploit spatial multiplexing and diversity gains to varying degrees. A framework for quantifying the fundamental trade-offs between these two forms of MIMO gains has been introduced in Zheng and Tse (2003) and is discussed in Chapter 8.

Recently, the extension of single-user information-theoretic results on MIMO to the multiuser case has been a topic of significant attention, mainly motivated by the desire to understand the network capacity improvements resulting from the use of MIMO technology. These aspects are dealt with in detail in Chapters 18 and 19.

3.2 Algorithmic studies

Triggered by and partly in parallel to the information-theoretic efforts summarized in Section 3.1, progress has been made on algorithmically oriented research aimed at designing transmission schemes that approach the information-theoretic performance limits within the power, latency, and implementation-related constraints of wireless products.

Probably the most active area of research in the algorithmic context are the design of space-time codes (Tarokh *et al.*, 1998; Alamouti, 1998) and corresponding receiver (decoding) algorithms under various assumptions on the channel model and CSI availability. Space-time codes for channel-aware receivers are discussed in Chapter 7. Chapter 10 deals with space-time codes for the noncoherent case, where neither the transmitter nor the receiver know the channel. Space-time coding for time- and frequency-selective MIMO channels is treated in Chapter 11. The issue of how CSI at the transmitter affects transceiver design from an algorithmic point-of-view is addressed in Chapter 9. The basics of computationally efficient receive algorithms for spatial multiplexing are summarized in Chapter 15.

With the exception of the material discussed in Chapter 10, space-time code design, as well as most spatial multiplexing techniques, traditionally rely on the assumption of a fully channel-aware receiver (coherent communication). Acquiring accurate CSI at the receiver is, therefore, an important topic. Training-based CSI acquisition and, in particular, the design of training signals motivated by information-theoretic criteria is discussed in Chapter 17. The general problem of blind and semiblind MIMO channel estimation is dealt with in Chapter 14.

The gap between MIMO channel capacity and the rates achieved by actual space-time transmit-receive schemes can be reduced by borrowing independently developed advanced transmission concepts such as turbo codes. The application of turbo codes to the multiple-antenna case is discussed in Chapter 16.

To realize MIMO gains in the network context, corresponding algorithms are required. A survey of multiuser MIMO transmit and receive algorithms is provided in Chapter 20.

3.3 System studies and implementations

Understanding the impact of MIMO on wireless network performance is key to assessing the value of this new technology in practical systems. An analysis along these lines is provided in Chapter 22. Chapter 21 demonstrates how multiple ("dumb") antennas can be used to realize opportunistic beamforming gains in a cellular network.

On the way from progress in information-theoretic and algorithmic aspects to a final product, basic concept validation is usually performed through experimental platforms. Chapter 24 reports the results obtained from experiments conducted on such a MIMO platform.

One of the most crucial factors determining the ultimate value of MIMO technology in practical systems is the extent to which MIMO transmit and receive algorithms can be implemented in VLSI. Chapter 27 provides a discussion of general VLSI implementation aspects arising in MIMO detection algorithms.

3.4 Emerging topics

The list of topics covered in this book is certainly far from exhaustive. At the time of writing, several new areas, expected to push the frontiers of MIMO and space-time processing research even further, are emerging. Some of these areas are summarized below. Among all these evolutions, a general trend that is clearly visible is the convergence of distinct disciplines: signal processing with VLSI, information theory with networking, information theory with signal processing, and networking with wireless system engineering.

The information-theoretic performance limits of multiuser MIMO systems are reasonably well understood. However, the search for signal processing and coding algorithms that realize a satisfactory portion of these ultimate limits at a reasonable complexity and latency is still going on. The role played by multiple-antenna technology in improving the performance of ad-hoc networks and, in particular, "cooperative communication schemes" is currently attracting significant attention. Finally, understanding how the MIMO-enabled physical layer and higher layers in the OSI model can be jointly optimized to maximize overall system performance offers numerous new and challenging problems (IEEE Signal Process. Mag. special issue, 2004).

References

Alamouti, S. M. (1998). A simple transmit diversity technique for wireless communications. *IEEE J. Sel. Areas Commun.*, **16** (8), 1451–1458.

Foschini, G. J. and Gans, M. J. (1998). On limits of wireless communications in a fading environment when using multiple antennas. *Wireless Personal Commun.*, **6** (3), 311–335.

IEEE Signal Process. Mag. special issue (2004). *Special issue on cross-layer design. IEEE Signal Process. Mag.*, **21** (5).

Paulraj, A. J. and Kailath, T. (1994). Increasing capacity in wireless broadcast systems using distributed transmission/directional reception. U.S. Patent 5,345,599.

Tarokh, V., Seshadri, N., and Calderbank, A. R. (1998). Space-time codes for high data rate wireless communication: Performance criterion and code construction. *IEEE Trans. Inf. Theory*, **44** (2), 744–765.

Telatar, I. E. (1999). Capacity of multi-antenna Gaussian channels. *Eur. Trans. Telecommun.*, **10** (6), 585–595.

Zheng, L. and Tse, D. N. C. (2003). Diversity and multiplexing: A fundamental tradeoff in multiple antenna channels. *IEEE Trans. Inf. Theory*, **49** (5), 1073–1096.

Part I

Multiantenna basics

1

Propagation aspects of MIMO channel modeling

Jørgen Bach Andersen

Aalborg University

1.1 Introduction

The MIMO link consists of two antennas, one at each end of a link, where each antenna consists of several elements, which can be treated independently. In addition, the environment should allow several paths to exist from each antenna element at one end to each element at the other end. If these paths are "different" in a way to be explained, the simple two-antenna case turns into a multichannel case with high order of diversity and the possibility of enhanced spectral efficiency due to the parallel nature of the channels. We thus have a matrix channel in contrast to the usual scalar channel. The environment may also be modeled, but we shall restrict the analysis to rather simple scattering situations, which hopefully will have enough generality to cover most real cases. To define the words used, a path (or ray) is defined as a transmission between an antenna element and a scatterer or between two scatterers, while by transmission coefficient we mean the sum of paths connecting an element on one side to an element on the other side.

The modeling of MIMO channels is a multistep procedure. If realistic modeling is wanted, it must be based on measurements from which the relevant features are extracted. Significant determinants are the correlations between the transmission coefficients and the number of scatterers. Assuming more general randomness, a simulation model may be required to satisfy the experimental conditions only approximately in order to simplify matters. It is a common approximation to assume random phases from scatterers, in which case summation from a few scatterers leads to the well-known complex Gaussian distributions with Rayleigh distributions for the amplitudes. Another lesson learned from experiments is the clustering of rays in delay and angle. Such features have also been utilized in models.

3

The general description of a MIMO link is

$$y = Hx + n,$$

where x is the signal vector for the transmit side and y for the receive side, and H is the transmission matrix connecting the M transmit antennas to the N receive antennas. The noise is described as a random noise vector n with Gaussian statistics. This is the narrowband case valid for one frequency. For a wideband situation we would have a matrix at each delay, but in the present chapter we shall be limited to the narrowband case.

In a MIMO situation we are interested in the capacity and its distribution and in creating models that give the same distribution as measured. The capacity equation used in the following is the standard one for no knowledge of the channel at the transmitter with power divided equally between the antennas,

$$C = \log_2 \left(\det \left(I + \frac{P}{M} HH^H \right) \right),$$

where $(\cdot)^H$ indicates conjugate transpose and P is the per-receive antenna SNR. For high values of SNR we can separate the power term from the matrix term leading to a measure of the multipath richness depending only on the eigenvalues of the rectangular HH^H matrix. The definition of richness is

$$R(k) = \sum_{i=1}^{k} \log_2(\lambda_i), \tag{1.1}$$

i.e., the cumulative sum of the log of eigenvalues. The variable k may range from 1 to N for a full rank matrix where it has been assumed that $N \leq M$. Occasionally we will denote the last term $R(N)$ as the richness, while we reserve the name richness curve to the N numbers. Essentially the capacity is equal to the richness plus a term depending on the power level. It is convenient for comparing capacities in various situations without referring to a specific SNR, since a limited SNR has the effect of suppressing the weaker eigenvalues. By using the richness function we are sure to include all eigenvalues correctly. The H-matrix is normalized to unit average power in each path, which corresponds to the sum of eigenvalues being equal to NM.

The chapter is a tutorial introduction to MIMO channels, so we start with some simple deterministic cases, while adding the randomness later. Geometrical models of the interaction with scatterers are very useful, since we can model single-bounce and double-bounce models without having to specify concrete environments. Correlation between the antennas and between the

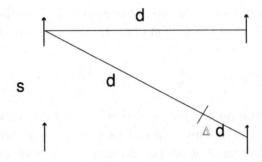

Fig. 1.1. Two linear arrays with element spacing s and distance d are facing each other.

paths in general is important. Other important parameters to be studied are the influence of the number of scatterers and the impact of a line-of-sight path, the Ricean channel. The chapter concludes with an example of experimental results for an indoor environment. For further reading on MIMO channels the reader is referred to Molisch (2004), Yu and Ottersten (2002), and Wallace and Jensen (2002).

1.2 Simple deterministic cases

It is instructive to study a few deterministic cases, first of all to indicate that randomness is not required, and secondly to show how we can obtain maximum capacity with equal eigenvalues. The situation is the simplified symmetric case with two antenna elements on each side with equal spacing s.

1.2.1 *Two arrays in line-of-sight (LOS), no scatterers*

The capacity or richness is determined by the channel matrix $\boldsymbol{H}$ connecting the transmit side with the receive side. In this very simple situation, only considering the phases, it is given by

$$\boldsymbol{H} = \begin{bmatrix} 1 & e^{-j2\pi\Delta d} \\ e^{-j2\pi\Delta d} & 1 \end{bmatrix}, \qquad \boldsymbol{H}\boldsymbol{H}^H = 2\begin{bmatrix} 1 & \cos(2\pi\Delta d) \\ \cos(2\pi\Delta d) & 1 \end{bmatrix},$$

where the lengths are measured in wavelengths. The two nonzero singular values of the matrix depend on Δd, the phase difference between the straight path and the skew path (see Fig. 1.1). The sum of the eigenvalues (the square of the singular values) equals 4 when the matrix is normalized to mean power of 1. When the arrays are far from each other, Δd tends to zero, the matrix becomes singular, and there is only one nonzero eigenvalue of magnitude 4.

This is to be expected since the gain of one array is 2, so the total link gain
is 4. To determine when the far field occurs, we can expand Δd for large d
as

$$\Delta d = d\sqrt{1 + \frac{s^2}{d^2}} - d \cong d(1 + \frac{1}{2}\frac{s^2}{d^2}) - d = \frac{s^2}{2d},$$

so for $s^2/2d \ll \lambda$ the rank is low, and there is only one nonzero eigenvalue,
or in other words the antennas appear as point sources with zero angular
spread. On the other hand, when the antennas are in each others "radiating
near field", MIMO with more than one nonzero eigenvalue is possible. The
maximum richness is 2 bits/s/Hz occurring when the two eigenvalues are
equal, each equal to 2. This occurs for $\Delta d = 0.25\lambda$ where

$$\boldsymbol{H} = \begin{bmatrix} 1 & -j \\ -j & 1 \end{bmatrix}, \qquad \boldsymbol{H}\boldsymbol{H}^H = \begin{bmatrix} 2 & 0 \\ 0 & 2 \end{bmatrix}.$$

For $\Delta d = 0.5\lambda$, the matrix is singular and has only one nonzero eigenvalue.
The situation is illustrated in Fig. 1.2, showing the richness as a function of
distance between arrays for an element spacing s of 2 wavelengths.

It is clear that two LOS arrays is an impractical MIMO structure because
they must be fairly close to offer sufficient angular width. The discussion here
is similar to the treatment in Gesbert *et al.* (2002), where it is generalized
to larger arrays. There is an important exception for the 2 by 2 array. The
above discussion was for a scalar wave function, but by using two orthogonal
polarizations in the electromagnetic case, it is possible to have a perfect LOS
MIMO independent of distance.

1.2.2 Two arrays and two scatterers in between

The situation can be largely improved by introducing widely spread scatterers (real or artificial) between the arrays. Each scatterer is in the far field
from the arrays. We can think of the scatterers as being part of the antennas,
making them look large. As an example, we change the situation above with
two scatterers of equal strength at $+45°$ and $-45°$ (see Fig. 1.3). Again, the
situation is rather idealized, but serves a tutorial purpose. The $\boldsymbol{H}$ matrix
may be easily calculated, again considering only the phases. It is a single-
bounce situation where each antenna couples via both scatterers. The result
is

$$\boldsymbol{H} = \begin{bmatrix} \cos(2\pi s \sin(\pi/4)) & 1 \\ 1 & \cos(2\pi s \sin(\pi/4)) \end{bmatrix}.$$

The richness is shown in Fig. 1.4 as a function of s for the normalized $\boldsymbol{H}$.

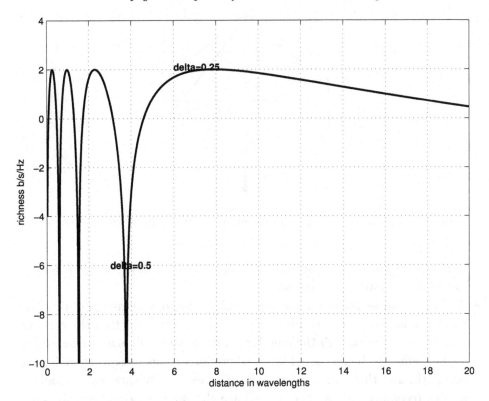

Fig. 1.2. Richness as a function of distance between two two-element arrays for a spacing of two wavelengths.

It is clear that the parameter of importance F is

$$F = \frac{s \sin(\theta)}{\lambda} \cong \frac{sw}{\lambda d},$$

where w is the total width of the scatterer and s the total length of the array. F should be above 0.1 in order to have a system with maximum richness. In Vaughan and Bach Andersen (2003), F is denoted F_{cor} because it is related to correlation in the random cases to be discussed later. F may also be related to the angular extent of the scatterer (in this case the 90° to the two scatterers) relative to the beam width of the array. The beam width of the array should be so small that it can resolve some details of the scatterer in order to have more than one significant eigenvalue.

Several important observations can be derived from this simple case:

- Maximum richness (2 bits/s/Hz) is obtained for several values of spacing.
- Richness goes to minus infinity (second eigenvalue goes to zero) for spacing

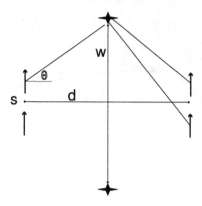

Fig. 1.3. Two scatterers at $\pm\pi/4$ of equal magnitude between two arrays.

tending to zero. Even though the angular spread is large, some spacing is required to separate the signals.

- The low richness around a spacing of 0.7 is due to a singular matrix. The null is similar to the deep nulls in Fig. 1.2. It is interesting that the richness may be low even though the angular spread is large. This is similar to the keyhole effect to be discussed later, where for various reasons there is only one path, and thus only one nonzero eigenvalue. Another way to look at it is to recognize that all the paths have the same electrical length, since there is an extra phase shift of π on each half of the paths due to the large spacing. In ordinary array terms grating lobes are created when the element spacing is too large. In MIMO applications, the capacity suffers due to equality of electrical path lengths, even though the physical path lengths differ.

From the simple, idealized situations treated in this section we can conclude that LOS MIMO systems will work satisfactorily for a short range of distances. When scatterers are present with large angular spreads the capacity is satisfactory (in a loose sense) when the spacing between elements exceeds a certain minimum value depending on the angular spread. Too large spacing may in special cases lead to situations of low rank and low capacity, even with large angular spreads.

1.3 Random channel models

The random channel models contain the previous deterministic models as special cases; one realization of a stochastic model can be considered deterministic. There are essentially two different aspects of randomness, which

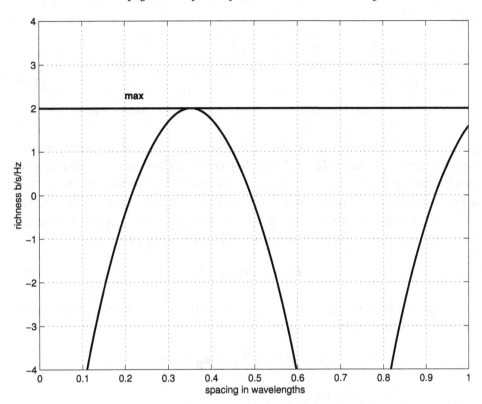

Fig. 1.4. Richness for the 2,2 case with two identical scatterers in the far field at angles of ±45 degrees as a function of the spacing between the elements.

are apparent when we consider movement. If the environment is static, movement of one (or both) terminals has the effect of varying path lengths to the scatterers, and phase mixing of the various contributions will occur, resulting in a Doppler spread. Since the resulting addition of a large number of complex signals with arbitrary phase leads to a normal distribution of the real and imaginary parts, we obtain a Rayleigh distribution in magnitude. An alternative interpretation is to consider random environments and a static terminal. The complex scattering amplitudes are then chosen to belong to a certain family, most often complex Gaussian again.

The first case to consider is the single bounce model of Fig. 1.3, where the whole space between $+\theta$ and $-\theta$ is filled with random scatterers, in our case chosen as complex Gaussian, with Rayleigh envelope. Thus, the array response from the scatterers to the receive array can be written as

$$V = \Phi_{R}\, G_{S}\,,$$

where

$$\boldsymbol{\Phi}_{\mathrm{R}} = \begin{bmatrix} e^{-j2\pi s \sin\theta_1} & e^{-j2\pi s \sin\theta_2} & \cdots & e^{-j2\pi s \sin\theta_S} \\ e^{-j2\pi 2s \sin\theta_1} & e^{-j2\pi 2s \sin\theta_2} & \cdots & e^{-j2\pi 2s \sin\theta_S} \\ \vdots & & \ddots & \vdots \\ e^{-j2\pi Ns \sin\theta_1} & e^{-j2\pi Ns \sin\theta_2} & \cdots & e^{-j2\pi Ns \sin\theta_S} \end{bmatrix} \tag{1.2}$$

is a matrix defining the electrical path lengths from the S scatterers to the N antenna elements, and $\boldsymbol{G}_{\mathrm{S}} = \mathcal{CN}(0, \boldsymbol{I}_{\mathrm{S}})$ the complex normally distributed amplitudes of the S scatterers with $\boldsymbol{I}_{\mathrm{S}}$ as the $S \times S$ identity matrix. The array is as before a uniform array with spacing s. Let $\boldsymbol{\Phi}_{\mathrm{T}}$ be a similar matrix $(M \times S)$ for the transmit case to the same scatterers, then the channel transfer matrix $(N \times M)$ for this single-bounce case is given by

$$\boldsymbol{H} = \boldsymbol{\Phi}_{\mathrm{R}} \, \boldsymbol{G}_{\mathrm{S}} \, \boldsymbol{\Phi}_{\mathrm{T}}{}^{T}. \tag{1.3}$$

It is apparent that $\boldsymbol{\Phi}_{\mathrm{R}}$ is related to the correlation between the elements on the receive side; in fact,

$$\boldsymbol{R}_{\mathrm{R}} = \mathrm{E}\{\boldsymbol{V}\boldsymbol{V}^{H}\} = \boldsymbol{\Phi}_{\mathrm{R}} \, \boldsymbol{\Phi}_{\mathrm{R}}{}^{H}$$

so (1.3) may be written as

$$\boldsymbol{H} = \boldsymbol{R}_{\mathrm{R}}{}^{1/2} \boldsymbol{G}_{\mathrm{S}} \, \boldsymbol{R}_{\mathrm{T}}{}^{1/2}.$$

This is an exact formulation for the single-bounce case.

Simulations based on (1.3) are shown in Fig. 1.5. The angles of arrival and departure are random and uncorrelated. An i.i.d. uncorrelated Rayleigh matrix would have given a richness of 2.9 bits/s/Hz and the ideal maximum with equal eigenvalues is 8 bits/s/Hz. It is interesting to compare with the deterministic case of Fig. 1.4. The low richness at small spacing is still valid due to the high antenna correlations, but the low values around 0.7 do not exist in the many-scatterer case—they were a result of the particular case with only two scatterers. Even at large spacing it is apparent that the richness (or capacity) is less than for the Rayleigh case due to the small, but nonzero correlations. It should also be emphasized that these results neglect mutual coupling, which in general will have the effect of reducing correlations and increasing capacity, bringing them closer to the Rayleigh case, independent of spacing. By proper matching it is possible to compensate completely for the mutual coupling even at close spacing (Dossche *et al.*, 2004), the cost being reduced bandwidth.

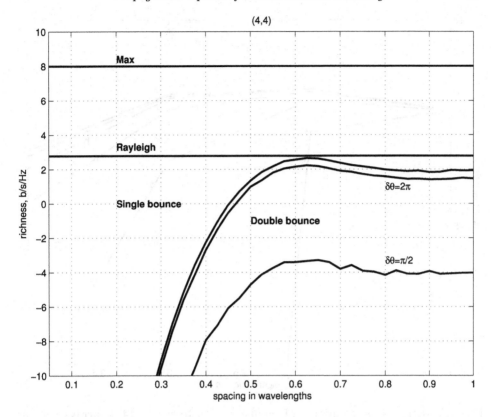

Fig. 1.5. Mean richness for single-bounce and double-bounce case for two symmetric 4-element arrays as a function of element spacing. The scatterers are uniformly distributed between +50 and -50 degrees with complex Gaussian amplitudes. The line "Rayleigh" is for an uncorrelated i.i.d matrix, and "maximum" is the ideal case of equal eigenvalues. The number of realizations is 2000.

1.3.1 Number of paths or clusters

The mean value and statistical distribution of the capacity depend not only on the angular distribution and the number of antenna elements, but also on the number of scatterers N_S. This is obvious when the number of scatterers is smaller than the minimum of N and M, in which case N_S determines the number of eigenvalues or subchannels. It is not so clear what the dependency is when N_S is larger. It can be noted that a completely random, full channel matrix requires NM independent coefficients, but if N_S is less than NM, this condition is not fulfilled. Some simulations are shown for a $(4,4)$ case with a spacing of 0.5 wavelength in Fig. 1.6. Note that the curves converge when N_S approaches $NM = 16$. It is seen that a small number of scatterers significantly reduces the richness, and that after 20 scatterers the values

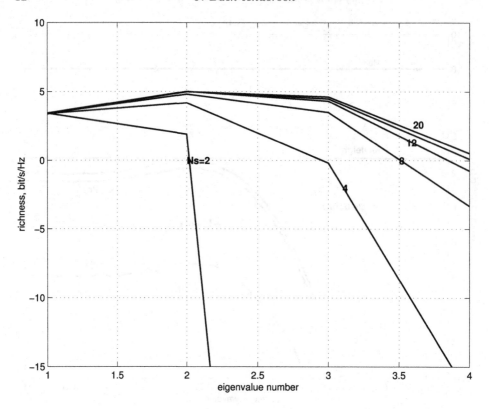

Fig. 1.6. Richness as a function of number of scatterers for $(4,4)$ arrays with a spacing of 0.5 wavelengths.

approach the asymptotic values determined by the correlations. Since the statistical distribution is also of interest with the random variation of the scatterers, the capacity is plotted for an SNR of 20 dB in Fig. 1.7. The severe degradation of the mean capacity is noted for $N_S = 4$, and the outage capacity at 1% level is degraded even further, indicating a reduction of the diversity. By comparison with the richness, we can conclude that the difference would be even greater for larger values of SNR.

It has been found experimentally that often the energy is distributed among clusters of paths where the angular spread inside the cluster is so small that the details are unresolvable. By approximation, such a cluster can then be perceived as a point source from an array point of view, which means that signals from one cluster are similar to those from one path or scatterer. We can thus expect that a low number of such clusters will have the same impact on the capacity as a low number of scatterers as in Figs. 1.6 and 1.7.

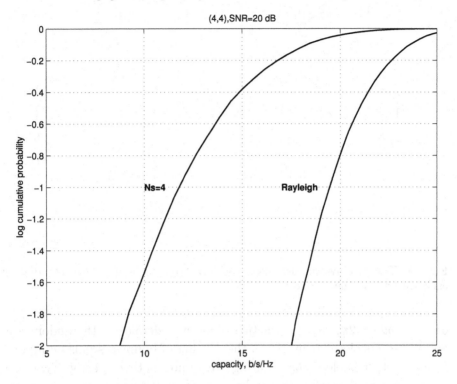

Fig. 1.7. Cumulative distribution of capacity for $N_S = 4$ and the Rayleigh case for $(4, 4)$ arrays with a spacing of 0.5 wavelengths.

1.3.2 Double-bounce channel

It is unlikely that all the rays from one antenna only interact with one scatterer before propagating to the receiving antenna. The previous single-bounce model may be simply extended by adding an extra layer (or ring) of scatterers with random propagation from each scatterer in one layer to each scatterer in the other layer. It is also known as the double-ring model. By an obvious modification of (1.3) we get

$$H = \boldsymbol{\Phi}_\mathrm{R} \, \boldsymbol{G}_\mathrm{R} \, \boldsymbol{X}_\mathrm{RT} \, \boldsymbol{G}_\mathrm{T} \, \boldsymbol{\Phi}_\mathrm{T}{}^T, \qquad (1.4)$$

where $\boldsymbol{X}_\mathrm{RT}$ is the matrix connecting the two layers. In the present work we choose $x_{ij} = e^{-j\theta}$, where θ is uniformly distributed between 0 and $\delta\theta$. The underlying idea is that the waves may follow some common paths limiting the phase excursions. This could be a simple separation in free space as discussed in Gesbert *et al.* (2002) or a more general guidance effect. The resulting richness for $\delta\theta = 2\pi$ and $\pi/2$ are shown in Fig. 1.5 with the rather surprising result that the richness is slightly less than for the single-bounce

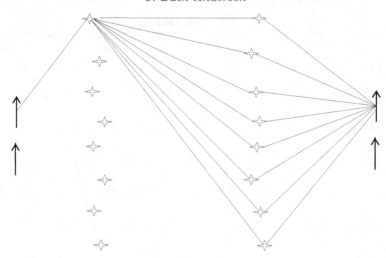

Fig. 1.8. The two-layer model where each scatterer in the first layer is radiating to each scatterer in the second layer.

case for $\delta\theta = 2\pi$, but in practice indistinguishable—with random angles and scatterers closely connected, the number of bounces seems unimportant. In contrast, it is clear that sharing a common path has a significant effect as shown for $\delta\theta = \pi/2$. In the limit of $\delta\theta = 0$ there is only one nonzero eigenvalue (an example of a pinhole or keyhole, cf. Chizhik *et al.*, 2000; Gesbert *et al.*, 2002)

It is interesting to observe that the double-bounce model is fundamentally different from the single bounce. Even if the number of scatterers goes to infinity, the channel coefficients may not be zero-mean complex Gaussian according to Yu and Ottersten (2002). Burr (2003) gives some examples of the dependency on the number of scatterers.

1.3.3 Correlations and the Kronecker model

In the channel model characterized as Rayleigh it was assumed that all the matrix elements, the transmission coefficients, are identically distributed and uncorrelated. In practice this is not the case as seen in Fig. 1.5 where the richness decreases as the spacing decreases. It is therefore important to understand the correlations between the transmission coefficients, and we take the simple (2,2) case with two elements at each side (Fig. 1.9). For brevity

$$H = \begin{bmatrix} a & b \\ c & d \end{bmatrix},$$

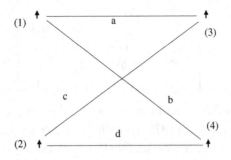

Fig. 1.9. Correlated transmission coefficients for a $(2, 2)$ case.

and we are interested in the full correlation matrix, a 4×4 matrix, between all transmission coefficients. It is customary to consider the antenna correlations, i.e., in our case the correlations between antennas 1 and 2, and between antennas 3 and 4, since they are often known from SIMO or MISO cases, where there is only one antenna at one of the ends. Considering the correlation between 1 and 2, this could be achieved in two different ways. By transmitting from antenna 3, move the antennas around to get sufficient independent samples, and find

$$\langle a, c \rangle_3 = \rho_3,$$

where $\langle x, y \rangle$ denotes the correlation between two transmission coefficients. Instead of using antenna 3 we could have chosen antenna 4,

$$\langle b, d \rangle_4 = \rho_4.$$

Considering the double-bounce situation it seems intuitively plausible that the two correlations must be the same, independent of which transmitter antenna we use, assuming that the antennas are identical and not placed in different global environments. Thus, we assume

$$\langle a, c \rangle = \langle b, d \rangle = \rho.$$

Similarly, we have

$$\langle a, b \rangle = \langle c, d \rangle = \mu.$$

Filling in this assumption in the full correlation matrix (1.5), it is noted that the elements along the antidiagonal, the 'internal' coefficients, are still not determined. This tells us that what can be correlated from the antenna sides is not the whole story.

$$\boldsymbol{R_H} = \begin{bmatrix} 1 & \rho & \mu & \langle a, d \rangle \\ \rho^* & 1 & \langle c, b \rangle & \mu \\ \mu^* & \langle c, b \rangle^* & 1 & \rho \\ \langle a, d \rangle^* & \mu^* & \rho^* & 1 \end{bmatrix}. \tag{1.5}$$

Nevertheless, it is possible to assume that the full matrix can be determined from the antenna correlations, the so-called Kronecker model (Shiu *et al.*, 2000; Kermoal *et al.*, 2002; Yu *et al.*, 2004) where the missing terms in (1.5) are forced to be products of the other correlations. As an example, we quote the $(2, 2)$ case (Kermoal *et al.*, 2002)

$$\boldsymbol{R_T} = \begin{bmatrix} 1 & \mu \\ \mu^* & 1 \end{bmatrix}, \qquad \boldsymbol{R_R} = \begin{bmatrix} 1 & \rho \\ \rho^* & 1 \end{bmatrix}$$

$$\boldsymbol{R}_{\text{MIMO}} = \boldsymbol{R_T} \otimes \boldsymbol{R_R} = \begin{bmatrix} 1 & \rho & \mu & \mu\rho \\ \rho^* & 1 & \mu\rho^* & \mu \\ \mu^* & \mu^*\rho & 1 & \rho \\ (\mu\rho)^* & \mu^* & \rho^* & 1 \end{bmatrix}$$

The Kronecker model is an approximation, and it is shown in Yu *et al.* (2004) that the error grows with the number of antenna elements. We can immediately see also that any internal correlations in the double-bounce model (lower rank $\boldsymbol{X}_{\text{RT}}$ in (1.4)) will not be covered by the Kronecker model. We shall see now that the single-bounce case in (1.2) is definitely not covered when receive and transmit angles are correlated. An improved version of the Kronecker model is the Weichselberger (Weichselberger *et al.*, 2006) channel model, which models correlation properties at the receiver and transmitter jointly.

1.3.4 Correlations in the single-bounce case

The two transmission coefficients in Fig. 1.10 have the same path length when the scatterer is moved up or down, so we will expect a high correlation instead of the low correlation predicted by the Kronecker rule. A simulation result with 20 random scatterers in the center plane is shown in Fig. 1.11 for a 4 by 4 case. The model is the same as was used for Fig. 1.5. The white squares indicated a correlation of 1. Below the diagonal there are 14 correlations, which are equal to 1. It may be shown that this does not have a detrimental effect on the richness, on the contrary. We can expect, however, that the diversity order may have decreased, since the number of

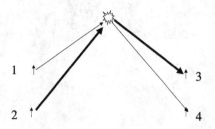

Fig. 1.10. High correlation between transmission coefficients in the single-bounce case. The thick 2–3 transmission coefficient is correlated with the thin 1–4 transmission coefficient when the scatterer is moved up and down.

independent paths has been reduced. It should be recalled that the case shown is symmetric, and as such an idealized case.

1.3.5 Ricean channel

It is often the case in practice that there will be a LOS path between the transmitter and receiver, allowing for a constant signal amidst all the random parts. Due to the well-known statistical properties of such a signal the channel is called a Ricean channel or a LOS channel. It is customary to introduce a K-factor, which is the ratio between the constant power and the random power. It may be introduced in the modeling as

$$ \boldsymbol{H}_{\text{Rice}} = \frac{\boldsymbol{H}_{\text{random}} + \sqrt{K}\boldsymbol{J}}{\sqrt{1+K}}, $$

where $\boldsymbol{J}$ is a matrix of all ones.

For K tending to infinity we will approach the case in Section 1.2 with two arrays facing each other and the normalized channel matrix having a 1 for each entry, leading to only one nonzero eigenvalue and one channel. For $K = 20$ dB, there is a loss of richness of about 20 bits/s/Hz compared with the purely random case (see Fig. 1.12). However, since it is capacity and not richness that is of primary interest, we must in this case take the power into account. It is very likely (McNamara *et al.*, 2000) that the power increases when the LOS situation occurs, since there are no obstructions, so there are two competing forces; the richness decreases but the SNR increases. If we assume that all eigenvalues are active and that the SNR increases proportionally with K, then the power contribution to the capacity is $\Delta C = 0.33\,N\,\text{SNR}(dB) = 0.33 \cdot 4 \cdot 20 = 26.4$ bits/s/Hz, so there is a net gain of about 6 bits/s/Hz in this example.

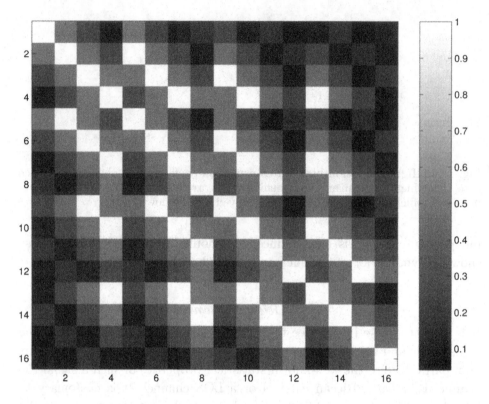

Fig. 1.11. Correlation (absolute value of complex correlations) between the 16 channel coefficients of the single-bounce symmetric case with four elements on each side with a spacing of 0.4 wavelengths.

1.4 An experimental example

Only comparisons with experiments can determine whether a channel model is accurate or not. The perfect channel model should be able to reproduce the eigenvalues and their statistical distributions. This will ensure that the model will give the correct mean and outage capacity for any outage level and for any SNR. In this section, we shall briefly discuss one indoor case, which requires most of the elements discussed earlier. A multichannel sounder was used for measuring MIMO channels in indoor environments at Aalborg University. The sounder had a bandwidth of 100 MHz at 5.8 GHz and a capability of 16 transmit antennas and 32 receive antennas. For the reported measurements, two planar arrays of vertical monopoles over a ground plane were used, and a multitude of measurements were obtained by moving the receiver array. The sounder is explained in greater detail in Nielsen *et al.* (2004). The environment consists of a multistory building with laboratories

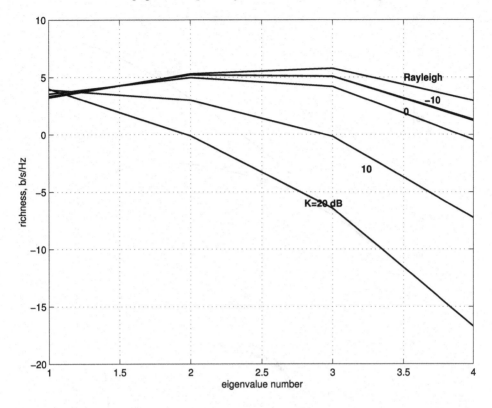

Fig. 1.12. The richness curve as a function of Ricean K-factor.

and offices, and the measurement shown here is for a situation with the transmitter in an office and the receiver in another office at the same floor. The double-layer model (1.4) is matched to the mean richness curve by selecting appropriate values of angular spreads (obtained independently) and values of N_S and $\delta\theta$. The resulting fit is shown in Figs. 1.13 a) and b), where a) shows the total mean richness $E\{R(k)\}$, and b) the cumulative distribution of $R(N)$. It is worth emphasizing that it was not possible to obtain a matching fit without taking both N_S and $\delta\theta$ into account. The optimum values were $N_S = 100$ and $\delta\theta = 0.6\pi$.

1.5 Discussion

In this chapter, we have highlighted some aspects of channel modeling and hopefully given an overview to the reader. The classical MIMO models have focused on the parameters that can be directly measured from the antennas, such as angular parameters and antenna element correlations. However, it is

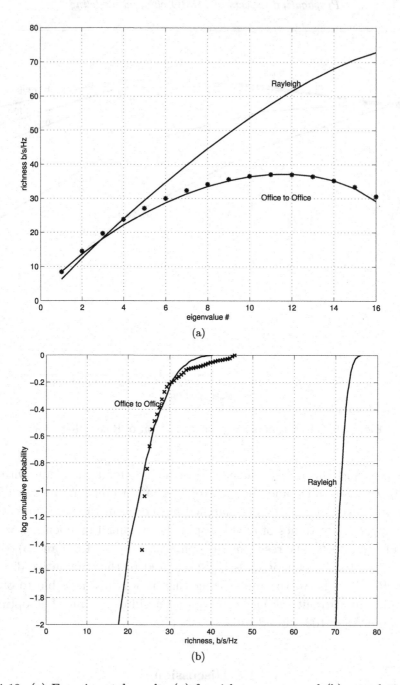

Fig. 1.13. (a) Experimental results (x) for richness curve and (b) cumulative distribution of richness for an environment consisting of offices at two different floors. The full curves are fitted to the experimental results by choosing $N_S = 100$ and $\delta\theta = 0.6\pi$. The antennas are $(16, 32)$ planar monopole arrays. The Rayleigh curve is for independent complex Gaussian matrix elements.

clear that the radio waves may experience additional phenomena on the way from transmitter to receiver that are not directly accessible at the antennas. Such phenomena may be the effective number of scatterers, N_S, which should be seen relative to the total number of matrix elements, MN, and the phase variation in the coupling matrix in the double-bounce model. Some of the conclusions obtained were:

- Multipath richness, defined as the cumulative sum of log of eigenvalues is a sensitive measure for comparing environments and for fitting to channel models.
- A single bounce model with uncorrelated arrival and departure angles is sufficient for rich environments.
- Environments with correlated arrival and departure angles will lead to decreased diversity and internal transmission coefficient correlations.
- The simple Kronecker model is reasonably accurate for small N and M, but fails to capture the internal phenomena.
- The Ricean channel reduces the richness, but may still increase capacity.
- A double-bounce model has the additional capability of modeling internal transmission coefficient correlations, including keyhole effects.
- The number of effective scatterers is an important parameter for both single-bounce and double-bounce models.
- Using only phase variations is a simple way of modeling the coupling matrix in double-bounce situations.

Acknowledgment

The work reported here has been supported by DoCoMo-Eurolabs. The author also acknowledges useful comments from anonymous reviewers.

References

Burr, A. G. (2003). Capacity bounds and estimates for the finite scatterers MIMO wireless channel. *IEEE J. Sel. Areas Commun.*, **21** (5), 812–818.

Chizhik, D., Foschini, G. J., and Valenzuela, R. A. (2000). Capacities of multi-element transmit and receive antennas. *Electron. Lett.*, **36** (13), 1099–1100.

Dossche, S., Blanch, S., and Romeu, J. (2004). Optimum antenna matching to minimise signal correlation on a two-port antenna diversity system. *Electron. Lett.*, **40** (19), 1164–1165.

Gesbert, D., Bölcskei, H., Gore, D. A., and Paulraj, A. (2002). Outdoor MIMO wireless channels: Models and performance prediction. *IEEE Trans. Commun.*, **50** (12), 1926–1934.

Kermoal, J. P., Schumacher, L., I.Pedersen, K., and Morgensen, P. E. (2002). A stochastic MIMO radio channel model with experimental validation. *IEEE J. Sel. Areas Commun.*, **20** (6), 1211–1226.

McNamara, D. P., Beach, M. A., Fletcher, P. N., and Karlsson, P. (2000). Capacity variation of indoor multiple-input multiple-output channels. *Electron. Lett.*, **36** (24), 2037–2038.

Molisch, A. F. (2004). A generic model for MIMO wireless propagation channels in macro- and microcells. *IEEE Trans. Signal Process.*, **52** (1), 62–71.

Nielsen, J. Ø., Bach Andersen, J., Eggers, P. C. F., Pedersen, G. F., Olesen, K., Sørensen, E. H., and Suda, H. (2004). Measurements of indoor 16×32 wideband MIMO channels at 5.8 GHz. In *Proc. IEEE Int. Symp. Spread Spectrum Tech. and Applications*, 864–868.

Shiu, D.-S., Foschini, G. J., Gans, M. J., and Kahn, J. M. (2000). Fading correlation and its effect on the capacity of multielement antenna systems. *IEEE Trans. Commun.*, **48** (3), 502–513.

Vaughan, R. and Bach Andersen, J. (2003). *Channels, Propagagation and Antennas for Mobile Communications* (IEE Press, London).

Wallace, J. W. and Jensen, M. A. (2002). Modeling the indoor MIMO wireless channel. *IEEE Trans. Antennas Propagat.*, **50** (5), 591–599.

Weichselberger, W., Herdin, M., Özcelik, H., and Bonek, E. (2006). A stochastic MIMO channel model with joint correlation on both link ends. *IEEE Trans. Wireless Commun.* In press.

Yu, K., Bengstsson, M., Ottersten, B., McNamara, D., Karlsson, P., and Beach, M. (2004). Modeling of wideband MIMO radio channels based on NLOS indoor measurements. *IEEE Trans. Veh. Technol.*, **53** (3), 655–665.

Yu, K. and Ottersten, B. (2002). Models for MIMO propagations channels, a review. *Wireless Commun. and Mobile Comput.*, **2** (7), 653–666.

2

Beamforming techniques

Jack Winters

Motia, Inc.

2.1 Introduction

With receiver beamforming, the signals received from multiple antenna elements in an antenna array are weighted and combined to produce an output signal, as shown in Fig. 2.1. This figure shows M antennas with weights $\boldsymbol{W} = \begin{bmatrix} w_1 w_2 \cdots w_M \end{bmatrix}$. If the input signal is $\boldsymbol{X} = \begin{bmatrix} x_1 x_2 \cdots x_M \end{bmatrix}$, then the output signal is $y = \boldsymbol{W}\boldsymbol{X}^T$, where the superscript T denotes transpose. These weights can be complex, with both phase and amplitude. On transmission, the beamformer applies the complex weights to the transmit signal for each antenna element of the array.

Although there are M weights, we note that only the relative weights matter, i.e., the phase and amplitude differences between antennas. Thus, there are only $M-1$ degrees of freedom with M weights, with these degrees of freedom used to increase the array gain, mitigate multipath fading, suppress interference, etc., as discussed in this chapter (Jakes, 1974; Monzingo and Miller, 1980; Stutzman and Thiele, 1981; Litva and Lo, 1996; Winters, 1998b). The simplest antenna array is a linear array as shown in Fig. 2.2.

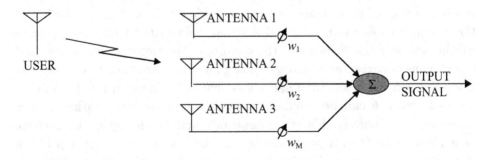

Fig. 2.1. A beamformer.

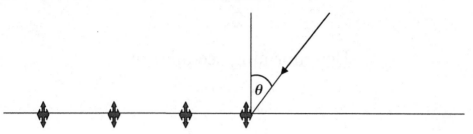

Fig. 2.2. A linear antenna array.

If the antenna elements are spaced half a wavelength apart and the weights are all equal to one, then the beamformer forms an antenna pattern with gain

$$\frac{\sin\left(M\frac{\pi}{2}\sin(\theta)\right)}{\sin\left(\frac{\pi}{2}\sin(\theta)\right)},$$

where θ is the angle relative to broadside. Thus, with M antenna elements, the array has a gain of M and a beamwidth given by π/M. If the antenna elements are spaced closer together, the array gain decreases proportionally to the array spacing, as the gain is degraded by mutual coupling. If the elements are farther apart, the gain remains the same, but the beamwidth decreases proportional to the spacing, and the sidelobes increase in strength. At a spacing of a wavelength or greater, grating lobes appear in the pattern as the array pattern repeats itself.

When a plane wave (as in a line-of-sight system) arrives at angle θ at a linear array with half-wavelength spacing, the difference in phase between the antenna elements is $\pi\sin(\theta)$. Thus, if the beamformer weights are set to $w_i = e^{-ji\pi\sin(\theta)}$, $i = 1, 2, \ldots, M$, the antenna pattern points the main beam at angle θ, i.e., a signal received at angle θ will be weighted such that the signals combine in phase, and a signal transmitted with these weights will be sent at angle θ. Because the spacing of the elements in the received signal plane decreases as θ increases, the gain of the array decreases below M when this spacing is less than half a wavelength. This variation of gain with angle-of-arrival θ can be avoided by using a circular array rather than a linear array. However, this is an issue primarily with signals that arrive as a plane wave from a point source, and this is generally not a problem in most wireless communication environments with multipath, as discussed next.

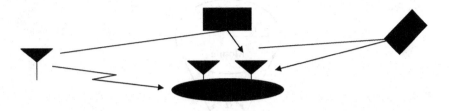

Fig. 2.3. A multipath environment.

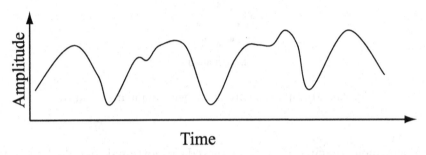

Fig. 2.4. Variation in received signal amplitude with time in a multipath environment.

2.2 Multipath environments

In most wireless environments, the signals arrive at the receiver via a number of different paths rather than as a plane wave, as shown in Fig. 2.3. The signals from the various paths add in and out of phase at each antenna and create signal fading. With enough paths, the channel gain between the transmitter and the receiver becomes complex Gaussian distributed, with the amplitude having a Rayleigh distribution (Jakes, 1974). Thus, as shown in Fig. 2.4, the amplitudes of the received signals vary with time as objects and users move. The frequency of that variation, called the Doppler frequency, depends on the movement speed. Specifically, it is given by the velocity divided by the wavelength. For example, at 100 km/h and 2 GHz, the Doppler shift is about 200 Hz. The fading also varies with the location of the antenna elements, although this depends on the angular spread α, which is defined as the range in angles over which the received signal arrives at the receiver. The angular spread is shown in Fig. 2.5. As shown in this figure, the signal arrives at the base station (which uses a circular array in the figure) over a range of angles α, due to reflections within a radius r around the mobile. If the angular spread is 360 degrees, the antenna elements only need to be spaced about a quarter wavelength apart for the fading to be statistically uncorrelated at each antenna. For α less than 360 degrees (2π radians), the

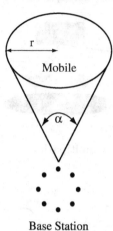

Fig. 2.5. Multipath environment with angular spread α.

antenna element spacing must be increased proportional to $2\pi/\alpha$ for the fading to be uncorrelated. However, uncorrelated fading can also be obtained by pointing different directional antennas in different directions, or by using orthogonal polarizations on the antennas. Of course, the effective use of directional and orthogonally polarized antennas requires that the angular spread is large enough, and that the received signals have multiple polarizations.

In a digital communication system, the fading leads to a variation in the bit error rate (BER) at the receiver, which can be quantified by an outage probability P_o, i.e., the probability that the BER is greater than some threshold BER_0.

Multipath fading has a profound effect on beamforming. Since the phase and amplitude of the signal at each antenna element is due to the sum of multiple paths of the arriving signal, the antenna pattern itself is no longer of interest. Indeed, such an antenna pattern is meaningless in a multipath environment, since the relative gain of the antenna as a function of angle θ is not constant with distance. Instead, weights now need to be complex, not just a phase shift based on antenna spacing and angle-of-arrival as in the line-of-sight system. Note that the antenna geometry then plays a less important role in multipath environments. For example, spacing linear array elements several wavelengths apart results in multiple grating lobes in the antenna pattern, which is undesirable in a line-of-sight system. But in a multipath environment, such spacing results in less correlation of the fading and thus better performance, as discussed below; the grating lobes have no effect and are meaningless.

2.3 Array and diversity gains

There are two basic types of gain that can be obtained with beamforming (Litva and Lo, 1996; Winters, 1998b). They can be explained by the fact that the fading channel is statistical in nature with a given distribution, with the two types of gains manifesting in the mean and variance of this distribution. The first one is array gain, which is defined as the increase in the average output SNR. With multipath fading, this average is taken over the multipath fading. As discussed above, with M antennas the array gain is a maximum of M. This is because, with proper beamforming, the desired signal adds coherently at the receiver (giving an increase of M in voltage and M^2 in power), whereas the noise, when independent on each antenna element, adds incoherently for an increase of M in power. Thus, the SNR increases by a factor of M. This, of course, requires sufficient spacing of the antennas, so that the mutual coupling is negligible. For a linear array with half-wavelength spacing in a line-of-sight environment, this gain of M corresponds to a narrower main beam out of the array. Note that this array gain of M occurs independently of the environment, i.e., in both line-of-sight and multipath environments.

The second type of gain is diversity gain. This is defined as the decrease in the average received SNR for a given BER, averaged over the fading. Note that this diversity gain only occurs in multipath environments when using digital communications, and results because the BER is a nonlinear function of the SNR. That is, for a given average SNR, deep fades cause a disproportionate number of bit errors, requiring a higher average SNR to obtain the same average BER. The diversity gain then varies with the modulation and detection technique, i.e., the BER as a function of the SNR, and the BER itself. For example, for coherent detection of a binary phase-shift-keyed signal (BPSK) in Rayleigh fading, the diversity gain at a BER of 10^{-2} is 5.2 dB with two antennas, but increases to 14.7 dB at a BER of 10^{-4}. The diversity gain also increases less when more antennas are added. For example, the 5.2 dB diversity gain with two antennas increases to 7.6 dB with four antennas, and it increases to a maximum of 9.5 dB with an infinite number of antennas. The reason is that, as more antennas are used, the fading becomes less and less, and after a point there is no significant fading, and the performance is the same as that in additive white Gaussian noise. Of course, this gain decreases with fading correlation, and it also will be less if other types of diversity are used in conjunction with antenna diversity, such as when a RAKE receiver is used in CDMA when there is significant delay spread.

There are three main types of antenna diversity: spatial, polarization, and pattern diversity. Spatial diversity is the most widely known, whereby the antennas are spaced sufficiently far apart to obtain low fading correlation. A rule of thumb is that the spacing should be greater than that required to obtain a beamwidth equal to the angular spread. For example, if the angular spread is 360 degrees, then a spacing of about a quarter wavelength is sufficient. However, if the angular spread is only a few degrees, then the spacing needs to be on the order of ten wavelengths. This is fortunate, because a mobile device, where only spacings less than a wavelength may be practical, is usually located in the clutter on the ground, where the angular spread is close to 360 degrees, while a base station located high on a tower, where spacings of ten wavelengths are practical, generally operates with a line-of-sight to the vicinity of the mobile, resulting in a few degrees of angular spread. Note that horizontal separation is generally used (although some European countries also use vertical separation to obtain a smaller profile antenna on a tower), because the angular spread in the horizontal plane is generally greater than that in the vertical plane as seen at the base station.

A second type of antenna diversity is polarization diversity. For cellular systems, both horizontal and vertical polarizations can be used. When cellular systems were first deployed in the early 1980s, the base station antennas were high up on a tower (40 m typically) and the mobile antennas were located on top of vehicle roofs. In that case, the received signals were mainly vertically polarized, with the horizontal polarization 6 to 10 dB lower in power. In this case, polarization diversity was not an effective form of antenna diversity. However, today handsets are held in different positions, so that the polarization of the antenna varies, and the base station antenna may be lower, resulting in horizontal and vertical polarizations having about the same signal strength. Thus, polarization diversity is effective. Note that polarization diversity allows twice the order of diversity in the same form factor as a single antenna, since a single antenna can be used with two different feeds to provide both polarizations. In indoor systems, there may also be a significant signal component from above and below the antennas, and thus a third polarization can also be used.

The third form of diversity is pattern diversity. The most common method for pattern diversity is the use of a multibeam antenna. In this case, sometimes referred to as angle diversity, each beam will see a different set of paths for the signal. Thus, the fading will be uncorrelated in each beam, and if the angular spread is larger than the width of each beam, the power of the signals can be similar in different beams, and effective diversity gain can be achieved. But pattern diversity can also be achieved by each antenna just having a different pattern. Again, different antennas will see different paths, and the fading can be uncorrelated. An interesting case for this is when similar antennas are placed close together. As the antennas become closer, the spatial diversity will decrease. However, as the antennas are more closely spaced, mutual coupling between the antennas will change the antenna pattern of each antenna, so that they are different; the net effect, as has been measured, is that the diversity gain remains about the same, no matter how close the antennas are placed. Of course, due to mutual coupling, the array gain decreases and is nearly zero when the antennas are very close. However, as noted above, the diversity gain can dominate over the array gain, such that in some cases substantial improvement can be obtained by placing multiple antennas very close together. Indeed, by using a combination of spatial, polarization, and pattern diversity, nearly full diversity gain can be obtained in very small form factors, e.g., four or more antennas on a PCMCIA card, sixteen antennas on a handset, and dozens on a laptop. The key point is that the size of the device generally does not limit the number of antennas that can be used - it is usually the cost and power consumption of the multiple radio frequency (RF) chains that may be required with these antennas.

2.4 A brief history

Adaptive beamforming had its roots in classic papers in the 1960s and 1970s, such as Widrow *et al.* (1967) and Applebaum (1976). Widrow *et al.* (1967) describes the LMS adaptive array, which is a technique to adaptively determine the weights that are derived from the received signal to minimize the least mean squared error between the received signal and a reference signal, and thus allow the adaptive array to track a signal with known properties while suppressing interference, with application in communication systems. Applebaum (1976) describes the Applebaum array, which adaptively suppresses sidelobe energy when the desired signal angle-of-arrival is known, such as in a radar system. Monzingo and Miller (1980) provide an overview of the numerous adaptive beamforming techniques during this period, when the primary application was expensive military communication and radar systems, operating in a line-of-sight environment, often with intentional jamming. Starting in the 1980s, there was increased interest in commercial systems, particularly mobile and wireless systems, where multipath and unintentional interference (from other users) was the main concern, e.g., Winters (1984). However, it was not until the cost of signal processing was dramatically reduced, and commercial wireless systems matured in the late 1990s that adaptive beamforming became commercially feasible, and interest ballooned.

2.5 Basic receiver combining techniques

The simplest combining technique is selection diversity. With selection diversity, the antenna element with the strongest signal is selected for the output signal. As a result, the outage probability with M antennas is just the outage probability with one antenna raised to the Mth power, thus dramatically reducing the outage probability in many cases. Variations of this include switched diversity, where the antenna element with the strongest signal is selected and used until the received SNR falls below a given threshold, at which time the next antenna is selected.

Selection diversity, however, does not use all the received signal energy. A better technique, therefore, is maximum ratio combining (MRC), whereby the received signals from all the antennas are weighted and combined to maximize the output SNR. It can easily be shown that this is done when the beamforming weights are just the complex conjugates of the elements of the received signal channel vector, i.e., the maximal ratio combiner is a spatial matched filter. That is, the received signals are cophased and weighted according to their signal strength before combining. Referring to Fig. 2.1, if $c_1, c_2, \ldots, c_M$ are the complex channel gains, the weights w_i are $c_1^*, c_2^*, \ldots, c_M^*$, where the superscript * denotes complex conjugation. This is the optimum beamforming technique in terms of minimizing the BER, when only the desired signal and noise are present at the receiver. In this case, with independent (uncorrelated) Rayleigh fading at each antenna, the BER with M antennas is approximately given by the BER with a single antenna raised to the Mth power. This is what is referred to as an M-fold diversity gain. Note that in a linear array without multipath, the MRC weights are phases only and form a narrow beam as described above. That is, in a noise-limited environment, MRC is the optimum technique independent of the multipath environment.

2.6 Interference suppression

The antenna array can also be used to suppress interference. When interference is present, the received signals can be combined to maximize the output signal-to-interference-plus-noise ratio, rather than just the SNR. The weights that do this are given by (Monzingo and Miller, 1980) $\boldsymbol{W} = \boldsymbol{R}_{\mathrm{xx}}^{-1} \boldsymbol{C}^*$, where $\boldsymbol{R}_{\mathrm{xx}} = E[\boldsymbol{X}^* \boldsymbol{X}^T]$ is the spatial correlation matrix of the received signal vector $\boldsymbol{X}$, and $\boldsymbol{C} = [c_1 c_2 \cdots c_M]$ is the array response vector for the desired signal as above. What these weights do in general is to minimize the mean squared error in the output signal, and thus this is referred to as minimum mean squared error (MMSE) combining. What is being done is to

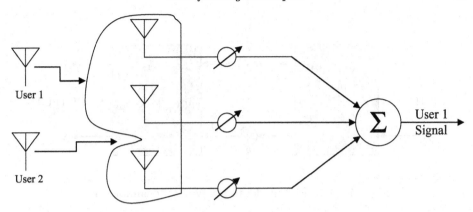

Fig. 2.6. Interference suppression in a line-of-sight system.

utilize the correlation of the interference at the receive antennas to reduce interference power and increase the desired SNR. Note that this technique is the same as MRC when interference is not present, i.e., when $\boldsymbol{R}_{xx}$ is just the identity matrix (times a constant), so that $\boldsymbol{W} = \boldsymbol{C}^*$.

To visualize interference suppression, first consider the case of no multipath. Fig. 2.6 shows a line-of-sight system with a desired user 1 and an interfering user 2 arriving from different angles. In this case, the weights generate an antenna pattern with the main beam in the direction of the desired signal and a null in the direction of the interference. Thus, the spatial dimension is utilized to increase the signal-to-interference-plus-noise ratio. Now, with M antennas, as noted above, the array has $M - 1$ degrees of freedom. These degrees can be used to increase the gain of the desired signal or null interference. Specifically, with $M-1$ degrees of freedom, the beamformer can completely null up to $M - 1$ interfering signals while providing gain toward the desired signal. The type of beamformer that completely nulls out interference is referred to as a zero-forcing (ZF) combiner (Monzingo and Miller, 1980). The MMSE and ZF combiners are described further below (see also Paulraj and Papadias, 1997). The amount of gain toward the desired signal depends on the angular separation of the desired and interfering signals, with a maximum SNR gain of M for the desired signal.

Next, consider the beamformer in a multipath environment. In this case, the signals from the desired user 1 and the interfering user 2 arrive from multiple paths at the receiver. At first it might seem impossible to do effective beamforming in this case, as the beamformer does not have enough degrees of freedom to put a main beam in the direction of all the desired signal paths, and nulls in the direction of all the interfering signal paths.

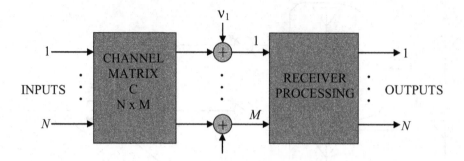

Fig. 2.7. A MIMO model to study a ZF beamformer.

However, as with MRC in a multipath environment, the antenna pattern is meaningless. That is, the signals from user 1 and user 2 may arrive at the receive antennas via a number of different paths, but the signals from user 1 add to give a single phase and amplitude for this signal at each antenna, and similarly for user 2. Thus, there is an array response vector for user 1 and one for user 2, as in the line-of-sight case; but in the multipath case these response vectors do not correspond to a particular geometry or angle-of-arrival. However, the equations for the MMSE and ZF weights remain the same, and thus a beamformer with M antennas can still suppress up to $M-1$ interfering signals. In particular, a ZF beamformer in a Rayleigh fading environment that suppresses N interferers has the same performance as a beamformer that has $M-N$ antennas with no interference. This is shown in detail below.

2.7 Performance of a ZF beamformer

Consider a ZF beamformer in a multipath environment. We consider a multiple-input multiple-output (MIMO) model with N signal inputs and N signal outputs, as shown in Fig. 2.7. At each output, there is one desired signal and $N-1$ interfering signals. We consider N transmit and M receive antennas, with channel matrix C, where $c_{ij} = [C]_{ij}$ is the channel coefficient between transmit antenna i and receive antenna j. We will assume that there is independent Rayleigh fading at each antenna, so that the columns of C are linearly independent and the c_{ij} are complex independent and identically-distributed (i.i.d.) zero-mean Gaussian random variables. The noise is assumed to be additive zero-mean i.i.d. Gaussian as well. With the ZF beamformer, the goal is to null out the $N-1$ interfering signals and maximize the SNR of the desired signal at each of the outputs.

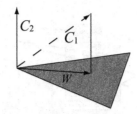

Fig. 2.8. The ZF weights with one interferer.

Now, if $N = 1$ (no interferers), then the weights are just the MRC weights, $W = C_1^*$, where $C_1 = [c_{1j}]$, $j = 1, 2, \ldots, M$. The BER is then bounded by (Winters *et al.*, 1994)

$$\text{BER} \leq E_c \left[\exp \left(-\rho \sum_{j=1}^{M} |c_{ij}|^2 \right) \right] = (1 + \rho)^{-M},$$

where ρ is the SNR. To cancel the interferers for $N \geq 2$ ($N - 1$ interferers), the weights must be orthogonal to the interference $C_2, \ldots, C_N$. Thus, the weight vector is just the projection of C_1^* onto the $M - N + 1$ dimensional space orthogonal to $C_2, \ldots, C_N$, as shown in Fig. 2.8 for the case of $N = 2$. But since the elements of C_1^* are i.i.d. Gaussian random variables, W has $M - N + 1$ dimensions with the same statistics as C_1, independent of $C_2, \ldots, C_N$. Thus, the BER is given by

$$\text{BER} \leq E_z \left[\exp \left(-\rho \sum_{i=1}^{M-N+1} |z_i|^2 \right) \right] = (1 + \rho)^{-(M-N+1)},$$

where the z_i are complex i.i.d. Gaussian random variables. The result is thus that a receiver using ZF combining with M antennas and $N - 1$ interferers has the same performance as a receiver with $M - N + 1$ antennas and no interference. Thus, the ZF beamformer can null $N - 1$ interferers with an $(M - N + 1)$-fold diversity improvement.

2.8 MIMO

Note that the $N - 1$ interferers can actually be coming from the same device, as shown in Fig. 2.9 (see, e.g., Winters (1987)), which is referred to as MIMO with spatial multiplexing of N. In this figure, a laptop with four antennas is communicating with a base station with four antennas. The laptop can transmit a different signal out of each of the antennas—each at the same time and in the same frequency band, but with different data. At the receiver,

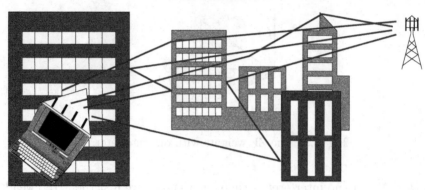

Fig. 2.9. Example of the use of MIMO with spatial multiplexing.

the received signals at each of the four antennas are combined using one set of weights to generate the first signal, another set of weights to generate the second signal, etc. Note that only four receive antennas are required to do this, as the four received signals can be digitized and processed in four different ways. By using this technique with M transmit and M receive antennas, M spatial channels can be created to increase the data rate M-fold in the same signal bandwidth. It has been shown that this can be done without any increase in the total transmit signal power. The only requirement is that the fading is nearly uncorrelated at each of the transmit and receive antennas. Specifically, this requirement means that the channel correlation matrix $\mathbf{C}$ must have a rank of M, i.e., the multipath environment must be rich enough, which, in general, means that there must be at least M sufficiently strong signal paths/reflectors between the transmit and receive antennas (Paulraj *et al.*, 2003). Experimental results (Ling *et al.*, 2001) have shown that outdoors the multipath environment can generally support up to a ten-fold MIMO capacity increase, while modeling results for indoors (Driessen and Foschini, 1999), with its richer multipath environment, show that at least a 100-fold MIMO capacity increase is feasible. However, most systems today consider up to four antennas, which results have shown to give a four-fold capacity increase in almost all environments (even line-of-sight environments, since some multipath usually exists even with such a line-of-sight). This allows four antennas to achieve data rates in excess of 1.5 Mbps in EDGE, 19 Mbps in WCDMA, and 216 Mbps in 802.11a/g ($4{\times}54$ Mbps).

Thus, a ZF combiner with M antennas can provide up to an M-fold diversity gain with an array gain of M, suppress up to $M-1$ interferers, and provide up to an M-fold capacity increase (with M antennas on the other side as well). However, since the degrees of freedom are limited to $M-1$, the total gain of these three effects is limited to M. However, these three gains can be traded off, e.g., with M antennas providing higher diversity and array gain to users at the edge of the coverage area, higher interference suppression to those users with interference problems, and higher MIMO capacity to those without range/interference issues. Such adaptive MIMO techniques can maximize each user's performance, or be used to maximize overall system performance.

2.9 MMSE combining

In the above discussion, we considered a zero-forcing beamformer that completely nulled any interference in order to illustrate the capabilities of adaptive arrays. In practice, though, we rarely want to completely suppress the interference - it is generally better to merely suppress the interference into the noise. The usual criterion to be optimized is the BER, which is minimized by minimum mean squared error (MMSE) combining (under the assumption of Gaussian noise and interference). With MMSE, the weights are given by $W = R_{xx}^{-1}C^*$.

Since typically there may be many more interferers than antennas, but only one or two dominant interferers, MMSE combining permits a substantial increase in performance even with as few as two antennas. Indeed, in 1999, AT&T Wireless changed its combining algorithm from MRC to MMSE at its base stations, resulting in the ability to decrease the frequency reuse factor from 7 to 4 with only a change in software (two receive antennas were already in use on most base stations). Note that MMSE combining has the same performance as MRC when interference is not present, but has substantially better performance than MRC when strong interference is present, as in interference-limited cellular systems. Also note that MMSE will yield the optimum performance under any type of fading conditions, whether it is line-of-sight (in which case a narrow beam would be formed), Rayleigh fading, Ricean fading, etc.

2.10 Smart antennas

A smart antenna is a multielement antenna where the signals received at each antenna element are intelligently combined to improve the performance of the wireless system. The reverse is performed on transmission. As discussed above, smart antennas can a) increase signal range, b) suppress interfering signals, c) combat signal fading, and d) increase the capacity of wireless systems. Smart antenna technologies can be used to improve most wireless ap-

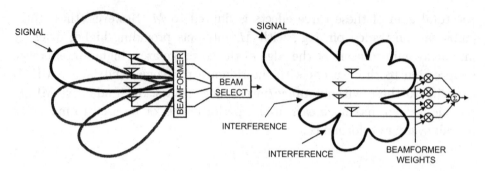

Fig. 2.10. Multibeam (left) and adaptive array (right) smart antennas.

plications, including not just the Wi-Fi systems IEEE 802.11a/b/g (Alastalo
and Kahola, 2003), but also cellular systems (Wong and Cox, 1998) such as
3G, and in-vehicle mobile DBS entertainment systems (Wang and Winters,
2004) such as mobile video, internet, and gaming, as well as satellite (and
digital) radio systems. Other applications include GPS, RFIDs, and WiMax
systems (IEEE, 2004).

There are two basic types of smart antennas: the switched multibeam
antenna and the adaptive antenna array, as shown in Fig. 2.10. With the
multibeam antenna, multiple nonoverlapping beams are used to cover a given
angle. For example, Fig. 2.10 shows four 30 degree beams that would cover
a 120 degree sector. The receiver measures the received signal strength in
each of the four beams and selects for the output signal the beam with the
strongest signal. The same beam is then used for transmission. Note that this
assumes that the uplink and downlink channels are reciprocal, see, e.g., Dias
et al. (2004). Since the beam with the strongest signal would generally not
change over a period of seconds or even minutes, depending on the coverage
area, the control circuitry for this system is very simple.

The adaptive array, on the other hand, uses multiple antenna elements,
with each element having generally the same antenna pattern (unless pattern
diversity is used). In this case, the antenna weights must track the phase and
amplitude of the received signals, which in a multipath fading environment
can change at a Doppler rate up to several hundred hertz, depending on
the carrier frequency and the speed of the user, and the weights must be
calculated at a rate that is a couple of orders of magnitude faster than the
Doppler frequency to accurately combine the received signals. Note that
the environments that most wireless systems, including wireless local area
networks (WLANs), operate in are multipath environments. For example, at
a carrier frequency of 2 GHz and a vehicle speed of 100 km/h, the weights

(which are complex) need to be calculated every 50 μs. Thus, the complexity of the control circuitry in the adaptive array is substantially higher than that of the multibeam antenna.

However, the adaptive array can have substantially better performance than the switched multibeam antenna. Although both types of smart antennas have a gain of M with M beams or antenna elements in a line-of-sight environment, the adaptive array has other advantages in a line-of-sight environment, and significant advantages in a multipath environment. Specifically, in either a line-of-sight or a multipath system, the adaptive array with M antennas can eliminate up to $M - 1$ interferers, while the switched multibeam antenna only reduces interference that falls outside of the selected beam.

Furthermore, in a multipath environment, the adaptive array still has a gain of M with M antennas, as well as providing an M-fold diversity gain against multipath fading, while the switched multibeam antenna has reduced gain (which becomes smaller as more of the signal energy falls out of the main beam), and very limited multipath diversity. Finally, as discussed above, only the adaptive array can be used with MIMO for spatial multiplexing in multipath environments (although a multibeam antenna could also be used for spatial reuse in an environment with small angular spread).

2.11 Transmit techniques

So far, we have concentrated on receive techniques. Let us now consider transmit beamforming techniques. If a multibeam antenna has been used, then the transmitter can simply use the same beam for transmission as was used for reception and achieve the same gain in both directions. However, with the adaptive array it is not always so simple. Specifically, since the fading changes with frequency, if the transmit frequency differs from the receive frequency by more than the coherence bandwidth, which is the bandwidth over which the fading remains correlated (the reciprocal of the delay spread), then the receive weights cannot be used for transmission to obtain the same gain in both directions.

However, if the wireless system uses the same frequency for transmission as for reception (transmitting and receiving at different times), as in a time-division duplex (TDD) system, then the receive weights can be used for transmission to obtain the same gain in both directions (this is true in principle, although in practice differences in transmit and receive filters, drift between clocks on each end of the link, etc., can reduce the gain on transmission, see, e.g., Dias *et al.* (2004)). For example, WLAN systems use

TDD. In this case, adaptive arrays need only be used at the access point or at the clients, but not both, to get gains on the uplink and downlink.

In frequency-division duplex (FDD) systems, such as cellular systems, where a different frequency is used for transmission and reception, the receive weights cannot be used for transmission (unless the frequencies are within the coherence bandwidth of the channel, which is typically not the case). One solution, which has been used in cellular base stations, is to use an adaptive array for reception, but a multibeam antenna on transmission (Winters, 1998b). The system will then be unbalanced in terms of the gain on the uplink and the downlink, but part of this difference can be mitigated through transmit power control and the fact that the base station transmit power is usually much higher than that of the mobile. A second method is to use feedback from the receiver to change the transmit antenna pattern. For example, switched diversity with feedback (Winters, 1983) can be used to send a single bit feedback from the receiver to have the transmit antenna changed if the signal level falls below a threshold. Similarly, the power control feedback information in a CDMA system can be used to adjust the phase of a second antenna relative to the first to maximize performance for MRC on transmit (Wong and Cox, 1998). With more transmit antennas, feedback of channel state information can be used, although this can become a prohibitive amount of information to transmit (methods with reduced state information have been used to try to overcome this limitation) (Blum *et al.*, 2002).

If feedback is not possible, or if the signal is being broadcast to multiple users, then transmit beamforming cannot achieve an array gain. However, diversity gain is still possible by the use of multiple transmit antennas with some form of temporal diversity. One method is as follows. With fast moving users, interleaving with coding can achieve a coding gain that averages out the fading, but slow moving users can be stuck in a fade. In this case, a continuously changing phase shift can be used between transmit antennas, e.g., with a slight frequency offset between antennas corresponding to a desired Doppler frequency, such that artificial fast fading is created at the receiver. Then, with interleaving and coding the performance of a fast fading system can be achieved.

A second method is to introduce delay spread into a flat fading channel. Although delay spread generally degrades the performance of a wireless system, when equalization is used at the receiver a diversity gain can be achieved, as discussed below. Specifically, a symbol delay can be inserted between each transmit antenna, resulting in an M-fold diversity gain with M transmit antennas and appropriate equalization at the receiver (Winters,

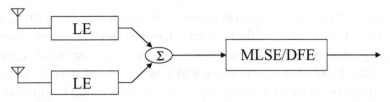

Fig. 2.11. A spatial temporal matched filter for a delay spread environment.

1998a). Furthermore, more sophisticated spatial-temporal encoding schemes at the transmitter can permit simplified receive techniques. This encoding includes the Alamouti code (Alamouti, 1998), along with the entire field of space-time coding, which is discussed elsewhere in this book.

2.12 Issues—delay spread

So far, we have considered only frequency-flat channels. However, when the time delay difference between the signal paths exceeds about 10% of the symbol period, the frequency selective fading due to delay spread can degrade performance. One method to avoid this problem is the use of orthogonal frequency-division multiplexing (OFDM), whereby the signal is transmitted using multiple carriers, called tones, each with a frequency narrow enough so that delay spread is not an issue. The weights are then calculated for each tone. This method is being used in many wireless systems, including IEEE802.11a/g, and is a primary candidate for fourth generation cellular systems.

Another method is to use the degrees of freedom in the adaptive array to suppress the delayed signals. The M-element adaptive array can suppress up to $M-1$ interfering signals, independent of their delay, or can suppress any number of delayed signals over $(M-1)/2$ symbol periods (Falconer *et al.*, 1993; Paulraj *et al.*, 1998). However, this suppression of the delayed signals is not optimum in that the energy in the delayed signals is lost. Furthermore, it is generally not wise to use the degrees of freedom of the array to mitigate temporal distortions - these distortions can be handled at much lower cost and power by digital signal processing rather than the hardware of the adaptive array. Thus, typically temporal processing is used in combination with spatial processing (the adaptive array) to provide performance improvement in a delay-spread environment. The optimum receiver is a spatio-temporal whitened matched filter (Li *et al.*, 1998; Paulraj *et al.*, 1998; Tidestav and Lindskog, 1998) as shown in Fig. 2.11. In this case, a linear equalizer is used with each antenna, with signals combined to maximize the output signal-to-noise-plus-interference ratio. That is, the correlation matrix of the received signal plus interference is whitened, which suppresses the interference and increases the desired signal output power. However, the combined signal may still contain intersymbol interference, which is then mitigated by a matched filter (in practice, with a nonlinear equalizer, such as a maximum-likelihood sequence estimator or decision-feedback equalizer).

A simpler structure is the use of the adaptive array followed by temporal equalization of the combined signal. Although this is suboptimal and has degradation relative to the optimum receiver in Fig. 2.11, in many cases the performance degradation is small.

2.13 Issues—implementation

Smart antennas can be implemented either in the digital or the analog domain. In the digital implementation, the signal from each antenna element is downconverted to baseband and A/D converted, and digital signal processing is used to determine the weights and combine the received signals. The disadvantage of this method is that with M antennas, M complete RF chains are needed, including M A/D converters, which are costly and consume significant power. This can be avoided by analog processing, where the received signals are weighted and combined at RF, and then the output signal requires only one downconversion and A/D conversion, i.e., one complete RF chain. Thus, for the same number of antennas, the cost of the analog smart antenna system can be lower, and it consumes less power. However, the digital processor can provide more accurate weights and have better performance when there is delay spread or frequency-selective fading, since analog RF combining generally uses only a frequency-independent complex weight across the band, i.e., does spatial processing only. Digital processing is also needed when using MIMO techniques with spatial multiplexing for increased data rates. But analog signal processing can also enable the use of an appliqué architecture, since the combiner output signal is at RF. In this architecture, a smart antenna system can be added to an existing system by just replacing the antenna with the appliqué, which combines the received signals from multiple antennas to form an RF output signal for input to the receiver, and divides and weights the RF signal from the transmitter for each of the transmit antennas, i.e., without modification of an existing transmitter or receiver (Martin *et al.*, 2001).

Another implementation issue is the generation of the combining weights. To generate these weights properly, the smart antenna must be able to distinguish the desired signal from noise and interference. Two basic methods can be used: nonblind and blind. In the nonblind method (Winters, 1993), the received signal is demodulated to determine the transmitted data, and this data is used as a reference signal to distinguish between the desired and interfering signals (which would not have the training sequence of the desired signal). For example, in MIMO with spatial multiplexing, nonblind methods are generally required as the different data streams differ only in their data and training sequences. Blind techniques do not demodulate the data, but require some other method to distinguish the desired signal from noise and interference. For example, the weights can be set to maximize the output power for a given normalized gain for the weights. In this case, the weights will basically cophase the received signals, adjusting the gain to be proportional to the received signal strength on each antenna element. Other methods include: a) power inversion (Monzingo and Miller, 1980), where the strongest signal is assumed to be interference and is suppressed, and b) the constant modulus algorithm (Treichler and Agee, 1983; Touzni *et al.*, 2001), where the weights are adjusted to minimize envelope fluctuations in the output signal, typically assuming that the desired signal has constant envelope, whereas the interference, particularly if it is offset in frequency from the desired signal, does not (constant modulus types of algorithms have also been used for MIMO with spatial multiplexing, see, e.g., Touzni *et al.*, 2001).

Finally, in most of the above discussion we have assumed that the fading correlation between antennas is zero. In practice, this correlation is of course nonzero, and well-designed antennas in most environments will have correlations less than 0.5. However, even in the worst case scenarios for antennas and environment, the correlation rarely exceeds 0.7. This is fortunate because the degradation is generally less than 1 dB with MRC for a fading correlation less than 0.7 (Jakes, 1974), and less than 1 dB with MMSE combining and interference with correlation less than 0.5 (Salz and Winters, 1994). Thus, fading correlation, although an impairment that needs to be considered, is generally not a significant impairment.

Conclusions

Beamforming has evolved from a technique to generate a desired antenna pattern in line-of-sight environments to one providing gain, diversity, interference suppression, and capacity increase in multipath environments. As wireless systems mature, and the cost of the required signal processing dramatically falls, adaptive beamforming is now seen as a key technique to provide dramatic performance and capacity improvement at reasonable cost and will see widespread deployment in nearly all wireless systems.

References

Alamouti, S. (1998). A simple transmit diversity technique for wireless communications. *IEEE J. Sel. Areas Commun.*, **16** (8), 1451–1458.

Alastalo, A. T. and Kahola, M. (2003). Smart-antenna operation for indoor wireless local area networks using ODFM. *IEEE Trans. Wireless Commun.*, **2** (2), 392–399.

Applebaum, S. R. (1976). Adaptive arrays. *IEEE Trans. Antennas Propagat.*, **24** (5), 585.

Blum, R. S., Winters, J. H., and Sollenberger, N. R. (2002). On the capacity of cellular systems with MIMO. *IEEE Commun. Lett.*, **6** (8), 322–324.

Dias, A. R., Bateman, D., and Gosse, K. (2004). Impact of RF front-end impairments and mobility on channel reciprocity for closed-loop multiple antenna techniques. In *Proc. IEEE Int. Symp. Personal, Indoor, Mobile Radio Commun.*, vol. 2, 1434–1438.

Driessen, P. F. and Foschini, G. J. (1999). On the capacity formula for multiple-input multiple-output wireless channels: A geometric interpretation. *IEEE Trans. Commun.*, **47** (2), 173–176.

Falconer, D. D., Abdulrahman, M., Lo, N. K. K., Petersen, B. R., and Sheikh, A. U. H. (1993). Advances in equalization and diversity for portable wireless systems. *Digital Signal Process.: A Rev. J.*, **3** (3), 148–162.

IEEE (2004). *IEEE standard for local and metropolitan area networks—Part 16: Air interface for fixed broadband wireless access systems.* IEEE Std 802.16-2004.

Jakes, W. C. (1974). *Microwave Mobile Communications* (AT&T, reprinted by IEEE Press, Piscataway, NJ, U.S.A.).

Li, Y., Winters, J. H., and Sollenberger, N. R. (1998). Optimal spatial-temporal equalization for diversity receiving systems with co-channel interference. In *Proc. IEEE Int. Conf. Commun.*, vol. 3, 1355–1359.

Ling, J., Chizhik, D., Wolniansky, P., Valenzuela, R., Costa, N., and Huber, K. (2001). Multiple transmitter multiple receiver capacity survey in Manhattan. *Electron. Lett.*, **37** (16), 1041–1042.

Litva, J. and Lo, T. K.-Y. (1996). *Digital Beamforming in Wireless Communications* (Artech House, Boston, MA, U.S.A.).

Martin, C. C., Winters, J. H., and Sollenberger, N. R. (2001). MIMO radio channel measurements: Performance comparison of antenna configurations. In *Proc. IEEE Veh. Technol. Conf. (VTC Fall)*, vol. 2, 1225–1229.

Monzingo, R. A. and Miller, T. W. (1980). *Introduction to Adaptive Arrays* (Wiley, New York, NY, U.S.A.).

Paulraj, A., Nabar, R., and Gore, D. (2003). *Introduction to Space-Time Wireless Communications* (Cambridge Univ. Press, Cambridge, U.K.).

Paulraj, A. and Papadias, C. B. (1997). Space-time processing for wireless communications. *IEEE Signal Process. Mag.*, **14** (6), 49–83.

Paulraj, A., Papadias, C. B., Reddy, V. U., and van der Veen, A. (1998). Space-time blind signal processing for wireless communication systems: Recent advances and practical considerations. In Wornell, G. and Poor, H. V. (eds.), *Wireless Communications: a Signal Processing Perspective* (Prentice Hall, Englewood Cliffs, NJ, U.S.A.).

Salz, J. and Winters, J. H. (1994). Effects of fading correlation on adaptive arrays in digital mobile radio. *IEEE Trans. Veh. Technol.*, **43** (4), 1049–1057.

Stutzman, W. L. and Thiele, G. A. (1981). *Antenna Theory and Design* (John Wiley & Sons, New York, NY, U.S.A.).

Tidestav, C. and Lindskog, E. (1998). Bootstrap equalization. In *Proc. IEEE Int. Conf. Universal Personal Commun.*, vol. 2, 1221–1225.

Touzni, A., Fijalkow, I., Larimore, M. G., and Treichler, J. R. (2001). A globally convergent approach for blind MIMO adaptive deconvolution. *Signal Process.*, **49** (6), 1166–1178.

Treichler, J. R. and Agee, B. (1983). A new approach to multipath correction of constant modulus signals. *IEEE Trans. Acoust., Speech, Signal Process.*, **31** (2), 459–472.

Wang, J. and Winters, J. H. (2004). An embedded antenna for mobile DBS. In *Proc. IEEE Veh. Technol. Conf. (VTC Fall)*, vol. 6, 4092–4095.

Widrow, B., Mantey, P. E., Griffiths, L. J., and Goode, B. B. (1967). Adaptive antenna systems. *Proc. IEEE*, **55** (12), 2143.

Winters, J. H. (1983). Switched diversity with feedback for DPSK mobile radio systems. *IEEE Trans. Veh. Technol.*, **32** (1), 134–150.

—— (1984). Optimum combining in digital mobile radio with cochannel interference. *IEEE J. Sel. Areas Commun.*, **2** (4), 528–539.

—— (1987). On the capacity of radio communications systems with diversity in a Rayleigh fading environment. *IEEE J. Sel. Areas Commun.*, **5** (5), 871–878.

—— (1993). Signal acquisition and tracking with adaptive arrays in the digital mobile radio system IS-54 with flat fading. *IEEE Trans. Veh. Technol.*, **42** (4), 377–384.

—— (1998a). The diversity gain of transmit diversity in wireless systems with Rayleigh fading. *IEEE Trans. Veh. Technol.*, **47** (1), 119–123.

—— (1998b). Smart antennas for wireless systems. *IEEE Personal Commun. Mag.*, **5** (1), 23–27.

Winters, J. H., Salz, J., and Gitlin, R. D. (1994). The impact of antenna diversity on the capacity of wireless communication systems. *IEEE Trans. Commun.*, **42** (234), 1740–1751.

Wong, P. B. and Cox, D. C. (1998). Low-complexity interference cancelation and macroscopic diversity for high capacity PCS. *IEEE Trans. Veh. Technol.*, **47** (1), 124–132.

3

Diversity in wireless systems

Ayman F. Naguib

Qualcomm, Inc.

A. Robert Calderbank

Princeton University

3.1 Introduction

The main impairment in wireless channels is fading or random fluctuation of the signal level. This signal fluctuation happens across time, frequency, and space. Diversity techniques provide the receiver with multiple independent looks at the signal to improve reception. Each one of those independent looks is considered a diversity branch. The probability that all diversity branches will fade at the same time goes down as the number of branches increases. Hence, with a high probability, there will be at least one branch or link with a good signal such that the transmitted data can be detected reliably.

Wireless channels are, in general, characterized by frequency-selective multipath propagation, Doppler-induced time-selective fading, and space-selective fading. An emitted signal propagating through the wireless channel is reflected and scattered from a large number of scatterers, thereby arriving at the receiver through different paths and hence arriving at different times. This results in the *time dispersion* of the transmitted signal. A measure of this dispersion is called the channel *delay spread* $\tau_{\max}$. The *coherence bandwidth* of the channel $B_c \approx 1/\tau_{\max}$ measures the frequency bandwidth over which the propagation channel remains correlated. Therefore, a propagation channel with a small delay spread will have a large coherence bandwidth, i.e., the channel frequency response will remain correlated over a large bandwidth, and vice versa.

In addition, transmitter and receiver mobility as well as changes in the propagation medium induce time variations in the propagation channel. This time variation is characterized by the *Doppler spread* F_d, which measures the width of the received signal spectrum when a single sinusoid is transmitted. The *coherence time* $T_c \approx 1/F_d$ measures the amount of time over which the propagation channel remains correlated. A propagation channel with slow time variation will remain correlated for a long period of time, and vice versa.

In addition to time and frequency selectivity, wireless channels are also characterized by spatial selectivity. As we mentioned earlier, an emitted signal will arrive at the receiver through a large number of different paths. Signal energy in each arriving path will be different and each path will have a different angle of arrival at the receiving antenna. Hence, these paths will add up differently at different points in space, thereby giving rise to *space-selective fading*. Space-selective fading depends (mainly) on the *angle spread* σ_θ of the arriving paths. When different paths arrive within a very narrow angle, a large separation in space is needed such that the propagation channel becomes uncorrelated, and vice versa.

Wireless channels in general exhibit all or a combination of space-time-frequency-selective fading. The actual diversity captured by the receiver depends on the inherent diversity that is available in the channel, the coding and modulation scheme used for transmission, and the receiver design itself. This inherent diversity depends on the size of the codeword used, the number of transmit antennas M_T, the number of receive antennas M_R, the coherence bandwidth B_c, and the coherence time T_c.

There are many ways to obtain diversity. In the next few sections, we will discuss several diversity schemes in time, frequency, and space. In Section 3.2, we will discuss time diversity techniques. In Section 3.3, several ways to obtain frequency diversity are discussed. Space-diversity techniques at the receiver and transmitter are discussed in Sections 3.4 and 3.5, respectively. Finally, Section 3.6 includes a few notes and remarks.

3.2 Time diversity

Time diversity exploits the time-varying nature of wireless channels. This will in general be the case when the transmitter and/or the receiver is moving. Time diversity can be obtained by using channel coding and interleaving. Here, the information symbols are first encoded. These encoded symbols are dispersed over time in different coherence periods so that different symbols in the codewords experience independent fades. This is done by using an appropriately designed interleaver, such that any two consecutive symbols at the output of the channel code are separated by more than the coherence time of the channel at the output of the interleaver.

The simplest way to obtain time diversity via coding and interleaving is by using repetition code. Let us consider a time-varying flat channel. In this case, the symbol duration T is much larger than the delay spread of the

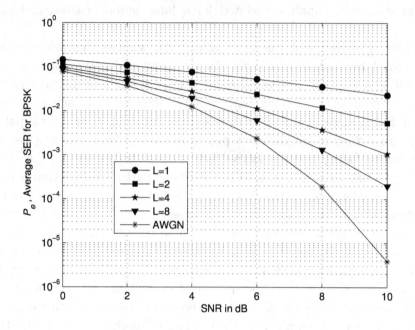

Fig. 3.1. Average symbol error probability as a function of L.

channel $\tau_{\max}$ (the delay spread of the channel is equal to the inverse of the channel coherence bandwidth $1/B_c$). Assume that a symbol s, drawn from a scalar constellation with unit energy, is to be transmitted. Every symbol s is repeated L times and then interleaved before it is transmitted over the wireless channel. The received signal corresponding to the lth transmission is given by

$$r_l = \sqrt{\frac{E_s}{L}} h_l \cdot s + n_l, \qquad l = 1, 2, \ldots, L, \qquad (3.1)$$

where $\sqrt{E_s}$ is the total symbol energy for transmitting a single symbol, n_l is the complex Gaussian additive white noise with zero mean and variance N_o, and h_l is the complex channel gain, which is assumed to be Gaussian with zero mean and unit variance. We assume that we are using an ideal interleaver such that channel gains h_l are independent. Proakis (2001) shows that, given the received signals in (3.1), the sufficient statistic for detecting s is given by

$$z = \sum_{l=1}^{L} h_l^* r_l = \alpha \cdot s + \tilde{n},$$

where

$$\alpha = \sqrt{\frac{E_s}{L} \sum_{l=1}^{L} |h_l|^2} \quad \text{and} \quad \tilde{n} = \sum_{l=1}^{L} h_l^* n_l.$$

The signal-to-noise ratio (SNR) is given by

$$\Gamma = \frac{E_s}{N_o} \cdot \frac{1}{L} \sum_{l=1}^{L} |h_l|^2 = \gamma \cdot \frac{1}{L} \sum_{l=1}^{L} |h_l|^2,$$

where $\gamma = E_s/N_o$ is the SNR without the repetition (i.e., when $L = 1$). Assuming maximum-likelihood (ML) detection is used at the receiver, it was shown by Proakis (2001), Cioffi (2002) that the probability of symbol error (SER) is given by

$$P_e \approx \bar{N}_e Q \left(\sqrt{\frac{\Gamma \cdot d_{\min}^2}{2}} \right), \tag{3.2}$$

where $\bar{N}_e$ and $d_{\min}$ are the average number of nearest neighbors and the minimum distance of the constellation used, respectively. Using the fact that $Q(x) \le e^{-x^2/2}$, $x \ge 0$, the probability of error P_e may be upper-bounded as

$$P_e \le \bar{N}_e \cdot \exp\left\{ -\frac{\Gamma d_{\min}^2}{4} \right\} = \bar{N}_e \cdot \exp\left\{ -\frac{\gamma d_{\min}^2 \|\mathbf{h}\|^2}{4L} \right\}, \quad \|\mathbf{h}\|^2 = \sum_{l=1}^{L} |h_l|^2. \tag{3.3}$$

Since we assumed h_l to be i.i.d. complex Gaussian with zero mean and unit variance, by averaging over the distribution of $\|\mathbf{h}\|^2$, the average SER $\bar{P}_e$ can be upper-bounded by (Proakis, 2001)

$$\bar{P}_e \le \bar{N}_e \left(\frac{1}{1 + \gamma d_{\min}^2/4L} \right)^L \le \bar{N}_e \left(\frac{\gamma d_{\min}^2}{4L} \right)^{-L}. \tag{3.4}$$

The last inequality follows when the SNR is high.

In Fig. 3.1, we plot the average SER as a function of the code length or the diversity order L for BPSK modulation. The length of the repetition code L affects the slope of the average SER as function of SNR on a log-log scale. For comparison, we also plot the average SER for an AWGN channel. As we can see, as the length of the repetition code, which is also called the *diversity order*, is increased, the SER approaches the SER of an AWGN link.

In the above example, we considered repetition coding to achieve time diversity. There are two problems with this approach. First, this comes at the expense of spectral efficiency, since L symbol times are used to transmit only one symbol. Secondly, a repetition code does not effectively exploit all the degrees of freedom in the fading channel. By using a more sophisticated channel code, a coding gain in addition to the diversity gain may be obtained (Seshadri et al., 1995). For example, we may be able to use linear block codes or convolutional codes along with interleaving to exploit the time diversity in the wireless link. In addition to the coding gain provided by the specific code used, the diversity order obtained by these codes is equal to the *minimum Hamming distance* of the code in the case of the linear block code, and to the *free distance* of the code in the case of convolutional codes (Seshadri et al., 1995; Proakis, 2001). Also, we may note that the above analysis assumed that we have an ideal interleaver and hence independent channel gains over the code symbols. When practical interleavers are used, the channel gains may be correlated. However, the above analysis can be generalized to cover this case.

3.3 Frequency diversity

In our analysis of time diversity, we assumed that the wireless channel is flat, i.e., it is represented by a complex gain (a single tap filter). However, when the symbol duration T is less than the delay spread of the channel, the transmitted signal is received over multiple symbol durations and the different multipath components can be resolved at the receiver (in this case the transmission bandwidth W is much larger than the channel coherence bandwidth B_c). In this case, the channel is no longer flat, i.e., the channel frequency response changes over the transmission bandwidth, and the channel cannot be modeled by a single-tap filter. In a similar fashion to time-diversity techniques, frequency-diversity techniques exploit this variation of the wireless channel to extract the diversity gain in the channel.

Let us assume that the wireless channel in this case can be modeled by an L-tap discrete-time filter, whose tap gains are h_l, $l = 0, 1, \ldots, L - 1$. In this case, the channel impulse response $h(k)$ is

$$h(k) = \sum_{l=0}^{L} h_l \delta(k - l), \tag{3.5}$$

and the received signal at time k corresponding to the transmission of symbol $s(k)$ is given by

$$r(k) = \sum_{l=0}^{L} h_l s(k - l) + n(k). \tag{3.6}$$

There will be *intersymbol interference* or *ISI*. One approach to exploit the diversity here without having to worry about this ISI is to send one information symbol every L symbol durations. This will be very similar to the time diversity with repetition coding described above, except that the multiple receptions are due to the channel response instead of the transmitter itself. This, however, is an inefficient way of transmission since, again, only one symbol is transmitted every L symbol durations. There are techniques to mitigate the ISI without a loss in spectral efficiency that would allow us to exploit the frequency diversity in the channel. These techniques can be broadly classified into three different categories discussed in the following.

3.3.1 Single-carrier systems with equalization

In single-carrier systems, the receiver can use linear and nonlinear processing to *equalize* the channel and mitigate the ISI. These techniques include linear minimum mean squared error (MMSE) equalizers (Lee and Messerschmitt, 1994; Proakis, 2001; Cioffi, 2002), where the equalizer taps $w = [w_1 \; w_2 \cdots w_L]^T$ are chosen such that the error between the equalizer output and the desired symbol is minimized. That is,

$$w = \arg\min_{w} \mathbb{E}\left\{|w * r(n) - s(n)|^2\right\}.$$

The output of the linear equalizer is then fed into a symbol-by-symbol detector. The performance of the linear MMSE equalizer can be improved by using a *decision-feedback equalizer* (DFE) (Lee and Messerschmitt, 1994; Proakis, 2001; Cioffi, 2002), where the previous decisions of the symbol-by-symbol detector are fed back and filtered to cancel some of the residual ISI remaining at the output of the linear equalizer. DFE will provide a significant improvement over an MMSE linear equalizer with the same number of taps (See Proakis (2001), Cioffi (2002) for a detailed comparison). In this case, the feedforward filter w and the feedback filter b are chosen such that the error between the input to the symbol-by-symbol detector is minimized,

$$\{w, b\} = \arg\min_{w,b} \mathbb{E}\left\{|w * r(n) - b^* s(n - \Delta)|^2\right\}.$$

Another approach that can be used here is the *maximum-likelihood sequence estimation (MLSE)* implemented using the Viterbi algorithm. MLSE will in general give a better performance as compared with DFE (Forney, 1972, 1973; Lee and Messerschmitt, 1994; Bottomley and Chennakeshu, 98; Proakis, 2001). However, the problem with MLSE is that its complexity increases exponentially with the number of taps in the channel. Hence, its practical use is limited in general to channels with a small number of taps.

Hybrid techniques that combine DFE equalizers and MLSE can be used (Eyuboglu and Qureshi, 1988; Falconer, 1993; Lee and Messerschmitt, 1994; Ariyavisitakul *et al.*, 2000). Here, a feedforward front-end filter w is used such that the equivalent digital channel at the output of w has a number of taps b that can be handled by the MLSE with a reasonable complexity.

The equalization techniques discussed above are called *time-domain equalizers* since the equalization step is performed in the time domain. An alternative approach is to use *frequency-domain equalizers* (FDE) (Sari *et al.*, 1995; Clark, 1998; Falconer *et al.*, 2002). Let us consider a burst of N information symbols at time k, $x_k = \begin{bmatrix} x_k(0) & x_k(1) & \cdots & x_k(N-1) \end{bmatrix}^T$ that is transmitted over an additive white Gaussian noise (AWGN) and frequency-selective channel $h(k)$. We assume that each burst is appended with a cyclic prefix of length L. This is done to eliminate the interburst interference by ignoring the first L received symbols in every burst that correspond to the cyclic prefix. In this case, we can write the remaining received samples as

$$r_k = H \cdot x_k + n_k, \tag{3.7}$$

where $x_k = \begin{bmatrix} x_k(N-1) & x_k(N-2) & \cdots & x_k(0) \end{bmatrix}^T$ is the vector of input symbols, which is assumed to be zero mean with covariance $E_s \cdot I$ (i.e., the input symbols are assumed to be i.i.d.), $r_k = \begin{bmatrix} r_k(N-1) & r_k(N-2) & \cdots & r_k(0) \end{bmatrix}^T$ is the received signal vector, $n_k = \begin{bmatrix} n_k(N-1) & n_k(N-2) & \cdots & n_k(0) \end{bmatrix}^T$ is the additive Gaussian noise vector, which is assumed to be zero mean with covariance $N_o \cdot I$, and H is the $N \times N$ channel matrix. The matrix H is a *circulant matrix*, i.e., each row is a cyclic shift of the first row $\begin{bmatrix} h_1 & h_2 & \cdots & h_L & 0 & \cdots & 0 \end{bmatrix}$. A basic result from matrix theory is that a circulant matrix has the eigenvalue decomposition

$$H = Q^* \Lambda_h Q, \tag{3.8}$$

where Q is the *discrete Fourier transform matrix* (DFT), whose (i, l)th element is

$$Q(i, l) = \frac{1}{\sqrt{N}} e^{-j2\pi i l / N}, \qquad 0 \leq i, l \leq N-1,$$

and Λ_h is the diagonal eigenvalue matrix whose diagonal elements $\lambda_{1,1}^h$, $\lambda_{2,2}^h, \ldots, \lambda_{N,N}^h$ are the N-point DFT of $h_0, h_1, \ldots, h_L$ (Horn and Johnson, 1985; Golub and Loan, 1989; Cioffi, 2002). Let us now consider the DFT of the received signal vector r_k

$$R_k = \text{DFT}(r_k) = Q r_k = \Lambda_h Q x_k + Q n_k = \Lambda_h X_k + N_k, \tag{3.9}$$

where X_k is the DFT of the input symbols vector and N_k is the DFT of the noise vector. Here, we have used the fact that $QQ^* = I$ since Q is orthonormal. The *single-carrier frequency-domain equalizer* (SC-FDE) is the $N \times N$ matrix filter W that minimizes the mean squared error (MSE)

$$e^2 = \mathbb{E}\left\{||W^*Y_k - X_k||^2\right\}.$$

It can be shown that W is given by (Proakis, 2001; Cioffi, 2002)

$$W = \left(\Lambda_h \Lambda_h^* + \frac{1}{\rho} \cdot I\right)^{-1} \Lambda_h,$$

where $\rho = E_s/N_o$ is the signal-to-noise ratio (SNR). We immediately notice that the matrix filter W is a diagonal matrix whose (i, i)th element is given by $w_{i,i} = \lambda_{i,i}^h / (|\lambda_{i,i}^h|^2 + 1/\rho)$. Hence, the equalizer in this case is made of N single-tap equalizers. The output of the MMSE SC-FDE, defined as $Z_k = W^* R_k$, is transformed back to the time domain (by applying the inverse DFT matrix Q^*) to yield the *soft decisions* $\hat{x}_k$ for the input symbols vector x_k

$$\hat{x}_k = Q^* R_k = Q^* W^* \Lambda_h Q x_k + v_k,$$

where $v_k = QW^* N_k$ is the output noise vector. Final decisions on the transmitted symbols are made by feeding the soft decisions $\hat{x}_k$ into a slicer in the case of uncoded transmission, otherwise they are fed into a channel decoder in the case of coded transmission. Note that when N is a power of 2, the DFT is efficiently implemented using the *fast Fourier transform* (FFT).

3.3.2 Multicarrier systems

Another way to exploit the frequency diversity in the channel is to *precode* the transmitted symbols such that the frequency-selective channel is converted into a set of orthogonal subcarriers, each being affected by a narrowband flat-fading channel. This technique is called *orthogonal frequency-division multiplexing* or OFDM (Proakis, 2001). Again, we consider a burst of N information symbols at time k, $X_k = \begin{bmatrix} X_k(0) & X_k(1) & \cdots & X_k(N-1) \end{bmatrix}$ that is transmitted over a frequency-selective AWGN channel $h(k)$. Let H_k be the N-point DFT of the channel impulse response. The transmitter performs an inverse discrete Fourier transform (if N is a power of 2, then the FFT is used) on X_k. The output of the inverse DFT is

$$x_k = Q \cdot X_k,$$

where Q is the DFT matrix defined above. A new sequence $\tilde{x}_k$ is constructed by appending to $x(k)$ a cyclic prefix of length L, which consisrs of the last L

symbols of $x(k)$. Again, this is done to eliminate the interburst interference by ignoring the first L received symbols in every burst that correspond to the cyclic prefix. The elements of $\tilde{x}_k$ are then transmitted serially over the channel. These $N+L$ samples constitute one OFDM symbol. Note that every block of N information symbols is now transmitted over $N+L$ transmission periods. As before, the receiver strips off the samples corresponding to the cyclic prefix to obtain N samples $r_k = \begin{bmatrix} r_k(0) & r_k(1) & \cdots & r_k(N-1) \end{bmatrix}^T$ such that

$$r_k = H \cdot x_k + n_k, \tag{3.10}$$

where the channel matrix H is circulant (Cioffi, 2002). Note that this is similar to the model in (3.7) except that there, the vector x_k represents the actual information symbols, whereas in the case at hand it is the *inverse fast Fourier transform (IFFT)* of the information symbol. The receiver applies the DFT matrix to the received signal vector to get

$$R_k = Q \cdot r_k.$$

Using again the fact that H is circulant, equation (3.10) can be rewritten as

$$R_{k,l} = H_{k,l} \cdot X_{k,l} + N_{k,l}, \qquad l = 1, 2, \ldots, N.$$

This looks similar to (3.1). The receiver multiplies $R_{k,l}$ by $H_{k,l}^*$ to get $\hat{X}_{k,l}$, a soft decision or sufficient statistic for $X_{k,l}$. Note that contrary to the FDE scheme described earlier, in the case of OFDM both channel equalization and detection are performed in the frequency domain. Hence, by using a channel code across the "*tones*", we can exploit frequency diversity. Another way of exploiting frequency diversity when the transmitter is only using a fraction of the bandwidth, i.e., it is using a subset of the *tones*, as in the case of *orthogonal frequency-division multiple access (OFDMA)* (Xia et al., 2003), is to use coding and frequency hopping. In this case, a codeword is transmitted over a number of OFDM symbols, and in each OFDM symbol, a different subset of tones is used for transmission, hence providing frequency diversity.

3.3.3 Direct-sequence spread spectrum

In direct-sequence spread spectrum systems, the multipath structure in the channel is exploited by modulating the information symbols with a pseudonoise sequence and transmitting the resulting waveform over a bandwidth W that is much larger than the information rate. In this case, the

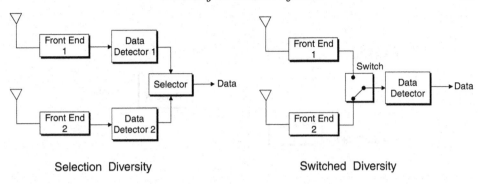

Fig. 3.2. Selection and switched diversity.

effect of the ISI is reduced, which leads to a much simpler receiver structure called *rake receiver* (Price and Green, 1958; Turin, 1980). The rake receiver can be viewed as an equalizer whose taps are sparsely spaced. See Proakis (2001) for full details.

3.4 Space diversity at the receiver

Receive diversity relies on the availability of $M \geq 2$ receive antennas at the receiver that are spaced far enough apart such that the channels from the transmitter to each of the receive antennas can be assumed uncorrelated. Decorrelation between receive antennas can also be obtained with polarized antennas for $M = 2$ (Paulraj *et al.*, 2003). The receiver processes the M received signals r_i, $i = 1, 2, \ldots, M$, according to some criteria to improve the overall performance. There are several methods to combine signals from different receive antennas to obtain diversity. These methods can, in general, be classified into two groups of combining, namely (1) selection combining, and (2) gain combining.

3.4.1 Selection combining

In selection diversity, the receiver chooses the best received signal according to some criteria for detection. In this case, the chosen signal is

$$r = \mathcal{C} \{r_1, r_2, \ldots, r_m\},$$

where $\mathcal{C}$ represents the selection criteria. In theory, this signal can be chosen based on a number of quality metrics, including total received signal power/strength, SNR, etc. In practice, however, these metrics are difficult

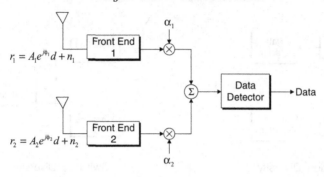

Fig. 3.3. Gain-combining diversity.

to obtain since they would require that each receive antenna has its own receiver chain to estimate the selection metric. Moreover, this technique also
requires that this selection metric be computed or monitored at a rate that
is faster than the fading rate. A block diagram of the selection diversity technique is shown in Fig. 3.2. Another form of selection diversity is switched
diversity, in which an alternate antenna is chosen for reception if the received signal level falls blow a certain threshold, while above the threshold,
the selected antenna is used for detection. The threshold may be fixed or
variable. Setting the threshold level requires knowledge of the mean signal
level in a given geographical area. A block diagram for switched diversity is
shown in Fig. 3.2.

3.4.2 Gain combining

In gain combining, the signal used for detection is a linear combination of
all received signals $r_i, i = 1, 2, \ldots, m$, i.e.,

$$r = \sum_{i=1}^{M} \alpha_i r_i,$$

where the α_i are the combining gains for each branch. There are different
combining methods based on the choice of the combining weights. Let us
assume that s is the data symbol, which belongs to some constellation with
unit energy, that is received by the M received antennas. The channel to
each receive antenna is characterized by a complex gain $h_i, i = 1, 2, \ldots, M$,
which is assumed to be distributed as a zero-mean complex Gaussian random variable with unit variance, and the gains are independent across the

antennas. The received signal at each receive antenna is then

$$r_i = \sqrt{E_s} h_i s + n_i, \qquad i = 1, 2, \ldots, M,$$

where r_i is the received signal at the ith receive antenna, E_s is the average symbol energy in each link, and n_i is the additive white Gaussian noise in each link, which is assumed to have zero mean and variance N_o. We assume that the noise is independent across all receive antennas. Note that the received signal model in (3.4.2) looks exactly the same as that used in (3.1), except that the multiple received signals are now collected over *multiple receive antennas* instead of *multiple symbol periods*. Hence, we should expect that the diversity performance here would be similar to that of time diversity with repetition coding except that there is no loss in spectral efficiency and that the postprocessing SNR will increase with the number of antennas. A block diagram for the gain combining diversity schemes is shown in Fig. 3.3. Let $h_i = A_i e^{j\phi_i}$. The weighting coefficients α_i can be chosen in several ways.

In *equal gain combining* (EGC), the weights are chosen as $\alpha_i = e^{-j\phi_i}$. In this way, the antennas are cophased and added together. This approach requires perfect knowledge of the complex channel gain phases. Assuming that the complex channel gains are perfectly known at the receiver (both phase and magnitude), a second approach that will maximize the postcombining SNR is *maximal ratio combining (MRC)* (Yacoub, 1993). In this case, the weighting coefficients are chosen as $\alpha_i = h_i^*$. Let $\boldsymbol{r} = [r_1 \; r_2 \; \cdots \; r_M]^T$, $\boldsymbol{h} = [h_1 \; h_2 \; \cdots \; h_M]^T$, and $\boldsymbol{n} = [n_1 \; n_2 \; \cdots \; n_M]^T$. In this case, the output of the MR combiner is given by

$$z = \sqrt{E_s} \boldsymbol{h}^* \boldsymbol{h} s + \boldsymbol{h}^* \boldsymbol{n} = \sqrt{E_s} \|\boldsymbol{h}\|^2 s + \tilde{n}.$$

The post-MRC SNR is then given by

$$\Gamma = \gamma \sum_{i=1}^{M} |h_i|^2, \tag{3.11}$$

where $\gamma = E_s/N_o$ is the average SNR per receive antenna. Assuming that the receiver employs an ML detector, it is shown in Cioffi (2002) that the SER is given by

$$P_e \approx \bar{N}_e Q \left(\sqrt{\frac{\Gamma \cdot d_{\min}^2}{2}} \right), \tag{3.12}$$

where $\bar{N}_e$ and $d_{\min}$ are the number of nearest neighbors and the minimum distance of the constellation used. Again, combining (3.11) and (3.12), it is shown in Proakis (2001) that the average SER $\bar{P}_e$ is given by

$$\bar{P}_e \leq \bar{N}_e \left(\frac{1}{1 + \gamma d_{min}^2/4} \right)^M \leq \bar{N}_e \left(\frac{\gamma d_{min}^2}{4} \right)^{-M}. \tag{3.13}$$

The last inequality again follows when the SNR is high. Hence, a diversity order equal to the number of receive antennas M is achieved. Moreover, the average postprocessing SNR is given by $\bar{\Gamma} = M \cdot \gamma$. Hence, in addition to the increased diversity order, the average received SNR is also enhanced by a factor of M over the single-receive-antenna case due to the array gain.

A third approach is the *minimum mean squared error (MMSE) combining*, where the weighting coefficients $\boldsymbol{\alpha} = (\alpha_1 \ \alpha_2 \ \cdots \ \alpha_M)$ are chosen such that the difference between the combiner output and the transmitted symbol s is minimized. That is,

$$\boldsymbol{\alpha} = \arg \min_{\boldsymbol{\alpha}} \mathbb{E} \left\{ |\boldsymbol{\alpha}^* \boldsymbol{r} - s|^2 \right\}. \tag{3.14}$$

It can be easily shown that the optimal choice of $\boldsymbol{\alpha}$ that will minimize the MMSE in (3.14) is given by

$$\boldsymbol{\alpha} = \left(\boldsymbol{h}\boldsymbol{h}^* + \frac{1}{E_s} \boldsymbol{R}_{nn} \right)^{-1} \boldsymbol{h}, \tag{3.15}$$

where $\boldsymbol{R}_{nn}$ is the noise covariance. When the noise across antennas is uncorrelated, i.e., $\boldsymbol{R}_{nn} = N_o \cdot \boldsymbol{I}$, it is easy to see that the weights of MMSE combining reduce to those of MRC up to a scaling factor. The impact of receive diversity on the SER performance can be seen, in general, in two distinct ways: (1) the slope of the SER curve as a function of SNR, and (2) the average SNR at the output of the combiner. It is shown in Yacoub (1993) that for MRC, equal gain, and selection diversity, the slope of the SER as a function of SNR per antenna increases (i.e., the curve becomes much steeper) with the number of receiver antenna. In the case of switched diversity, the slope of the SER is smaller (which indicates a lower diversity order in this case). It was also shown that the average SNR at the output of the combiner for each of the above receive diversity schemes is given by

$$\bar{\Gamma} = \begin{cases} \gamma \cdot \sum_{i=1}^{M} \frac{1}{i} & \text{pure selection} \\ \gamma \cdot M & \text{maximal ratio combining} \\ \gamma \cdot \left(1 + (M-1)\frac{\pi}{4} \right) & \text{equal gain.} \end{cases}$$

The above discussion for receive diversity assumed that the channel to each of the receive antennas is flat. Extensions to the case when the channel to each receive antenna is frequency selective are straightforward. For example, for MMSE combining, the antenna weights in (3.15) become filters instead of scalars. That is, the received signal at each receive antenna is filtered by an MMSE filter and the output of all filters is then combined for detection. Moreover, DFE architectures are also possible. See Balaban and Salz (1991, 1992), Paulraj *et al.* (2003) for a detailed treatment of this subject.

3.5 Space diversity at the transmitter

From the above discussion, we can easily see that receive-diversity techniques are capable of exploiting full diversity and array gains. The performance improvement due to receive diversity is almost proportional to the number of antennas. However, deploying multiple receive antennas at terminal devices might be difficult due to cost and/or space limitations. Instead, transmit-diversity techniques with multiple transmit antennas can be used. The information-theoretic aspects of transmit diversity were addressed by Telatar (1995), Foschini (1996), Foschini and Gans (1998), Narula *et al.* (1999). The impact of transmit diversity on the channel capacity can be summarized as follows (Paulraj *et al.*, 2003). Over a nonfading channel, transmit diversity does not improve the channel capacity. However, over fading channels, transmit diversity will achieve a higher capacity. This capacity, however, is still lower than the capacity of receive diversity with the same number of antennas. This is due to the inability of the transmitter to exploit the transmit array gain when the channel is not known (in the receiver diversity case, the channel can be estimated at the receiver). When the channel is known at the transmitter (i.e., through some sort of feedback from the receiver), all transmit power can be allocated to a single spatial mode in order to achieve a capacity equal to that of receive diversity. Transmit-diversity schemes can be classified into three broad categories—schemes using feedback, schemes with preprocessing and feedforward or training information but no feedback, and blind schemes based on channel coding.

3.5.1 Transmit diversity with feedback

This transmit-diversity scheme uses feedback, either explicitly or implicitly, from the receiver to the transmitter to train the transmitter. Fig. 3.4 shows a conceptual block diagram for transmit diversity with feedback. A signal is weighted differently and transmitted from two different antennas. The weights w_1 and w_2 are varied such that the received signal power $|r(t)|^2$ is maximized. The weights are adapted based on feedback information from

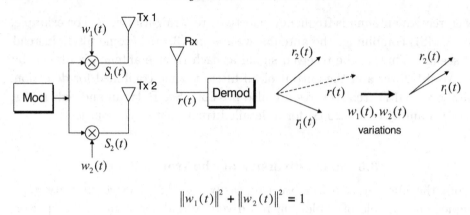

$$\left\| w_1(t) \right\|^2 + \left\| w_2(t) \right\|^2 = 1$$

Fig. 3.4. Transmit diversity with feedback.

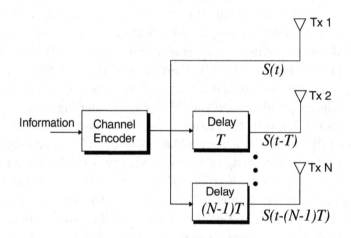

Fig. 3.5. Transmit delay diversity.

the receiver. For instance, in time division duplex (TDD) systems (Henry and Glance, 1972), the same antenna weights are used for reception and transmission, so that feedback is implicit in the exploitation of channel symmetry. These weights are chosen during reception to maximize the receive SNR, and during transmission to weight the amplitudes of the transmitted signals, and will therefore also maximize the SNR at the receiver. Explicit feedback includes switched-diversity systems with feedback (Winters, 1983). However, in practice, movement by either the transmitter or the receiver (or the surroundings such as cars) and interference dynamics cause a mismatch between the channel perceived by the transmitter and the one perceived by the receiver.

3.5.2 *Transmit diversity with preprocessing at transmitter*

Transmit-diversity schemes mentioned in the second category use linear and nonlinear preprocessing at the transmitter to spread the information across antennas. At the receiver, information is recovered by an optimal receiver. Training information is required to estimate the channel from the different transmit antennas to the receiver. These channel estimates are used to compensate for the channel response at the receiver.

Delay diversity

The first scheme of this type is the *delay-diversity* (DD) scheme (see Fig. 3.5) proposed by Wittneben (1991) and it includes the delay-diversity scheme of Seshadri and Winters (1994) as a special case. Similar linear processing techniques were also studied in Wittneben (1993), Guey *et al.* (1996). It was shown in Winters (1994, 1998) that delay diversity schemes are indeed optimal in providing diversity in the sense that the diversity gain experienced at the receiver (which is assumed to be optimal) is equal to the diversity gain obtained with receive diversity. The delay-diversity scheme can be viewed as intentionally creating multipath propagation or frequency selectivity in the channel, which can be exploited at the receiver by using an equalizer. The linear filtering used (to create delay diversity) at the transmitter can also be viewed as a channel code that takes binary/integer inputs and creates real-valued outputs. Let $h_i(k)$ denote the equivalent discrete-time channel impulse response from the ith transmit antenna to the receive antenna. The overall channel response and its corresponding frequency-domain response seen at the receiver are

$$h(k) = \sum_{i=1}^{M} h_i(k - \Delta_i) \ \text{ and } \ h(z) = \sum_{i=1}^{M} z^{-\Delta_i} h_i(z),$$

where Δ_i is the antenna-specific delay. Note that these delays can be arbitrary. However, to ensure that the overall channel response will experience frequency selectivity, these time delays should be chosen such that (Seshadri and Winters, 1994; Kaiser, 2000)

$$\Delta_i \geq \frac{1}{W} = \frac{1}{\text{Transmission Bandwidth}}, \qquad i = 2, 3, \ldots, M.$$

Note that we assume that the delay for the first antenna is zero. In general, delay diversity does not require any changes in the receiver. In fact, the receiver does not even need to know that delay diversity is used. However, one has to be careful in the choice of the time delays given the complexity of the required equalizer at the receiver. For example, let us assume that delay diversity is used in a single-carrier system and that the channel impulse responses from each of the two transmit antennas can be modeled as a linear filter with two taps. That is, $h_1(z) = \alpha_0 + \alpha_1 z^{-1}$ and $h_1(z) = \beta_0 + \beta_1 z^{-1}$. With delay diversity, the overall channel response will in general have three taps, $h(z) = \alpha_0 + (\alpha_1 + \beta_0)z^{-1} + \beta_2 z^{-2}$. As long as the equalizer used at the receiver can handle such a channel response, delay diversity will provide the full diversity gain without any loss. As another example, let us consider the case when delay diversity is used along with the OFDM scheme described above, where T_s is the OFDM symbol duration and N is the number of information symbols transmitted in a single OFDM symbol. In this case, (Kaiser, 2000),

$$\Delta_i \geq \frac{1}{W} = \frac{T_s}{N}, \qquad i = 2, 3, \ldots, M.$$

Recall that a cyclic prefix is added at the beginning of the OFDM symbol to avoid any interblock interference. Let ν_g be the length of the cyclic prefix and L be the maximum delay spread for the digital multipath channel (from any of the transmit antennas to the receiver antenna). In this case, to have no interblock interference, the antenna-specific delays need to satisfy

$$\Delta_i \leq \nu_g - L, \qquad i = 2, 3, \ldots, M.$$

This last constraint on the delay parameters Δ_i can be seen in two (similar) ways. First, it can be seen as a key drawback of the delay-diversity scheme in the sense that the additional delays will increase the total delay spread as seen at the receiver, and hence the length of the cyclic prefix ν_g needs to be increased by $\max\{\Delta_i\}$ in order to maintain zero interblock interference between consecutive OFDM symbols. This translates into a reduction in the bandwidth efficiency of the system. On the other hand, in a system where the guard time is tightly designed to be only slightly more than the delay spread of the channel, the choice of the delay parameter Δ_i will be restricted. This problem of increased total delay spread due to DD is fixed by the use of *cyclic delay diversity*, which is discussed next. Fig. 3.6 shows a simplified block diagram of an OFDM system with cyclic delay diversity. Again, the same OFDM signal is transmitted over M antennas, whereas the different antenna signals differ only in their antenna-specific cyclic shift. In other words, let $\boldsymbol{X}_k = \begin{bmatrix} X_k(0) & X_k(1) & \cdots & X_k(N-1) \end{bmatrix}$ be the input symbols to the IFFT at time k and $\boldsymbol{x}_k = \begin{bmatrix} x_k(0) & x_k(1) & \cdots & x_k(N-1) \end{bmatrix}$ be the corresponding output (time-domain samples). The transmitter applies a cyclic shift Δ_i on the time-domain samples before transmission such that $\boldsymbol{x}_{i,k} = \boldsymbol{x}_k((n-\Delta_i) \bmod N) = \begin{bmatrix} x_k(N-\Delta_i) & \cdots & x_k(N-1) & x_k(0) & x_k(1) & \cdots & x_k(N-\Delta_i-1) \end{bmatrix}$ is transmitted from antenna i. Hence, using the property that (Oppenheim and Schafer,

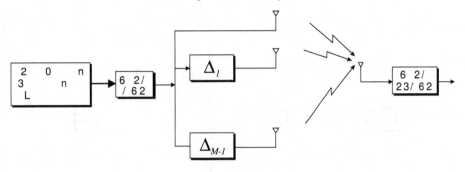

Fig. 3.6. OFDM with cyclic delay diversity.

1989) $\mathrm{DFT}\left(x_k\left((n-\Delta_i)\bmod N\right)\right) = \mathrm{DFT}\left(x_k\right) \cdot e^{-j(2\pi/N)n\Delta_i}$, the signal transmitted from antenna $i = 1, 2, \ldots, M$ is given by

$$x_k\left((l-\Delta_i)\bmod N\right) = \frac{1}{\sqrt{N}}\sum_{l=0}^{N-1}X_k(n)e^{-j(2\pi/N)n\Delta_i}\cdot e^{-j(2\pi/N)nl}.$$

In this case, the overall channel frequency response $H(n)$ is given by

$$H(n) = \sum_{i=1}^{M}H_i(n)\cdot e^{-j(2\pi/N)n\Delta_i}.$$

Note that the cyclic shift Δ_i can take any integer value, but since $(\Delta_i + l \cdot N)\bmod N = \Delta_i$ and $e^{j2\pi(\Delta_i+l\cdot N)/N} = e^{j2\pi\Delta_i/N}$, then the range of values that Δ_i can take is limited to $\Delta_i = 0, 1, \ldots, N-1$.

Another form of transmit diversity is *space-time block coding* (STBC), originally proposed by Alamouti (1998) for flat channels and later extended to frequency-selective channels in Al-Dhahir (2001), Lindskog and Paulraj (2000). With two antennas, it was shown (Naguib, 2001) that the Alamouti scheme will have the same diversity order as delay diversity. It was also shown that with more than two antennas, it will have a higher diversity order and a larger matched-filter bound (Mazo, 1991) than delay diversity. The STBC approach is discussed in details in Chapter 7 of this book.

3.5.3 Channel coding-based transmit diversity

In these schemes, multiple transmit antennas are combined with channel coding to provide diversity. An example of this approach is the use of channel coding along with phase-sweeping (Hiroike *et al.*, 92) or of frequency offset (Hattori and Hirade, 1978) with multiple transmit antennas, to simulate fast fading, as shown in Fig. 3.8. An appropriately designed pair of channel code

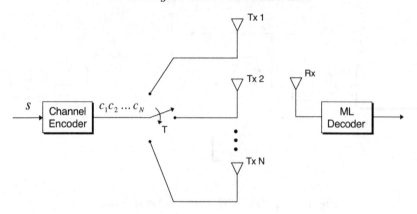

Fig. 3.7. Transmit diversity with channel coding.

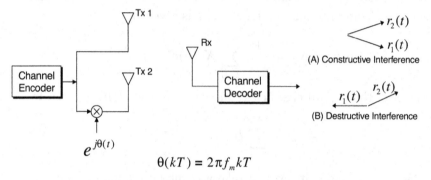

$$\theta(kT) = 2\pi f_m kT$$

Fig. 3.8. Transmit diversity with phase sweeping.

and interleaver is used to provide the diversity benefit. This scheme relies exactly on the same concept as time diversity with repetition coding that we discussed earlier. The difference here is that the wireless channel does not have to be time varying as in the case of time diversity with repetition coding. Here, *virtual* time-variations are created by using the phase-sweeping on the second antenna. Hence, this scheme can be used over static channels. Let $h_1(t)$ and $h_2(t)$ be the channel impulse response from the first and second antennas, respectively. Then, the overall channel response at the receiver is $h(t) = h_1 + h_2 e^{j\theta(t)}$. As the phase $\theta(t)$ changes over time, the overall channel response will undergo peaks and valleys simulating fast fading, which is exploited by the use of the channel code.

Another approach in this category is to encode information by a channel code (Fig. 3.7) and to transmit the code symbols using different antennas in an orthogonal manner. This can be done by either time multiplexing (Hiroike *et al.*, 92), or by using orthogonal spreading sequences for different antennas (Weerackody, 1993). The disadvantage of these schemes is the loss in bandwidth efficiency. Using coding techniques that are designed specifically for use with multiple transmit antennas can yield an overall larger diversity and coding gains with no or very little loss in bandwidth efficiency (see Chapter 7).

3.6 Notes

There are other forms of diversity that can be used to improve system performance. For example, in cellular systems, *macrodiversity* can be exploited by the fact that the signal from any given terminal can be received at two base stations. Improved reception for that specific terminal can be obtained, if signals from these two base stations are combined. This can be seen as a system-level diversity. Another form of system-level diversity that exploit the nature of the fading channel is *multiuser diversity* (Viswanath *et al.*, 2002). When there are many users in the system whose channels fade independently, then with high probability there will be at least one user whose channel is much stronger than the channels of other users. By allocating all the system resources to that user to transmit and/or receive, the channel is used efficiently, and hence the system throughput is maximized. Also, in our treatment of transmit diversity, we assumed M transmit antennas and one receive antenna. The basic idea remains the same when using multiple receive antennas. However, in receive diversity where the transmitter is using multiple transmit antennas to send different data streams, the treatment is different. These streams will be interfering with each other. Some of the available degrees of freedom at the receiver will be used to suppress interference from the other streams while decoding the stream of interest (Wolniansky *et al.*, 1998; Naguib *et al.*, 2000; Paulraj *et al.*, 2003).

References

Al-Dhahir, N. (2001). Single-carrier frequency-domain equalization for space-time block-coded transmission over frequency-selective fading channels. *IEEE Commun. Lett.*, **5** (7), 304–306.

Alamouti, S. (1998). A simple transmit diversity technique for wireless communications. *IEEE J. Sel. Areas Commun.*, **16** (8), 1451–1458.

Ariyavisitakul, L., Winters, J. H., and Sollenberger, N. R. (2000). Joint equalization and interference suppression for high data rate wireless systems. *IEEE J. Sel. Areas Commun.*, **18** (7), 1214–1220.

Balaban, P. and Salz, J. (1991). Dual diversity combining and equalization in digital cellular mobile radio. *IEEE Trans. Veh. Technol.*, **40** (2), 342–354.

—— (1992). Optimum diversity combining and equalization in data transmission with application to cellular mobile radio—part I: Theoretical considerations. *IEEE Trans. Commun.*, **40** (5), 885–894.

Bottomley, G. and Chennakeshu, S. (98). Unification of MLSE receivers and extension to time-varying channels. *IEEE Trans. Commun.*, **46** (4), 464–472.

Cioffi, J. (2002). Class reader for EE379a–digital communication. URL `http://www.stanford.edu/class/ee379a`. Stanford University, Stanford, CA.

Clark, M. V. (1998). Adaptive frequency-domain equalization and diversity combining for broadband wireless communications. *IEEE J. Sel. Areas Commun.*, **16** (8), 1385–1395.

Eyuboglu, M. V. and Qureshi, S. U. (1988). Reduced-state sequence estimation with set partioning and decision feedback. *IEEE Trans. Commun.*, **36** (1), 12–20.

Falconer, D. (1993). *Advances in Equalization and Diversity for Portable Wireless Systems*, vol. 3, 148–162 (Academic, New York).

Falconer, D., Ariyavisitakul, L., Benyamin-Seeyar, A., and Eidson, B. (2002). Frequency domain equalization for single-carrier broadband wireless systems. *IEEE Commun. Mag.*, **40** (4), 58–66.

Forney, G. D. (1972). Maximum likelihood sequence estimation of digital sequencess in the presence of intersymbol interference. *IEEE Trans. Inf. Theory*, **18** (3), 363–378.

—— (1973). The Viterbi algorithm. *Proc. IEEE*, **61** (3), 268–278.

Foschini, G. J. (1996). Layered space-time architecture for wireless communication in a fading environment when using multi-element antennas. *Bell Labs Tech. J.*, **1** (2), 41–59.

Foschini, G. J. and Gans, M. J. (1998). On limits of wireless communications in a fading environment when using multiple antennas. *Wireless Commun. Mag.*, **6** (3), 311–335.

Golub, G. H. and Loan, C. F. V. (1989). *Matrix Computations* (The John Hopkins Univ. Press, Baltimore and London), second edn.

Guey, J.-C., Fitz, M. P., Bell, M. R., and Kuo, W.-Y. (1996). Signal design for transmitter diversity wireless communication systems over Rayleigh fading channels. In *Proc. IEEE Veh. Technol. Conf.*, vol. 1, 136–140.

Hattori, T. and Hirade, K. (1978). Multitransmitter Simulcast digital signal transmission by using frequency offset strategy in land mobile radio-telephone. *IEEE Trans. Veh. Technol.*, **27** (4), 231–238.

Henry, P. S. and Glance, B. S. (1972). A new approach to high capacity digital mobile radio. *Bell Syst. Tech. J.*, **51**, 1611–1630.

Hiroike, A., Adachi, F., and Nakajima, N. (92). Combined effects of phase sweeping transmitter diversity and channel coding. *IEEE Trans. Veh. Technol.*, **41** (2), 170–176.

Horn, R. A. and Johnson, C. R. (1985). *Matrix Analysis* (Cambridge Univ. Press, Cambridge, MA).

Kaiser, S. (2000). Spatial transmit diversity techniques for broadband OFDM systems. In *Proc. IEEE Global Telecommun. Conf.*, vol. 3, 1824–1828.

Lee, E. A. and Messerschmitt, D. G. (1994). *Digital Communications* (Kluwer Academic Publishing, AH Dordrecht, The Netherlands), second edn.

Lindskog, E. and Paulraj, A. (2000). A transmit diversity scheme for channels with intersymbol interference. In *Proc. IEEE Int. Conf. Commun.*, vol. 1, 307–311.

Mazo, J. E. (1991). Exact matched filter bound for two-beam Rayleigh fading. *IEEE Trans. Commun.*, **39** (7), 1027–1030.

Naguib, A. F. (2001). On the matched filter bound of transmit diversity techniques. In *Proc. IEEE Int. Conf. Commun.*, vol. 2, 596–603.

Naguib, A. F., Seshadri, N., and Calderbank, R. (2000). Increasing data rates over wireless channels: Space-time coding and signal processing for high data rate wireless communications. *IEEE Signal Process. Mag.*, **17** (3), 76–92.

Narula, A., Trott, M., and Wornell, G. (1999). Performance limits of coded diversity methods for transmitter antenna arrays. *IEEE Trans. Veh. Technol.*, **45** (7), 2418–2433.

Oppenheim, A. V. and Schafer, R. W. (1989). *Discrete-Time Signal Processing* (Prentice Hall, Englewood Cliffs, NJ), first edn.

Paulraj, A., Nabar, R., and Gore, D. (2003). *Introduction to Space-Time Wireless Communications* (Cambridge Univ. Press, Cambridge, U.K.), first edn.

Price, R. and Green, P. E., Jr. (1958). A communication technique for multipath channels. *Proc. IRE*, **46** (3), 555–570.

Proakis, J. G. (2001). *Digital Communications* (McGraw-Hill, New York, NY), fourth edn.

Sari, H., Karam, G., and Jeanclaude, I. (1995). Transmission techniques for digital terrestrial TV broadcasting. *IEEE Commun. Mag.*, **33** (2), 100–109.

Seshadri, N., Calderbank, A. R., and Pottie, G. (1995). Channel coding for cochannel interference suppression in wireless communication systems. In *Proc. IEEE Veh. Technol. Conf.*, vol. 2, 629–633.

Seshadri, N. and Winters, J. H. (1994). Two schemes for improving the performance of frequency-division duplex (FDD) transmission systems using transmitter antenna diversity. *Int. J. Wireless Inf. Netw.*, **1** (1), 49–60.

Telatar, E. (1995). Capacity of multi-antenna Gaussian channels. Technical memorandum, AT&T Bell Laboratories.

Turin, G. L. (1980). Introduction to spread-spectrum antimultipath techniques and their application to urban digital radio. *Proc. IEEE*, **68** (3), 328–353.

Viswanath, P., Tse, D., and Laroia, R. (2002). Opportunistic beamforming using dumb antennas. *IEEE Trans. Inf. Theory*, **48** (6), 1277–1294.

Weerackody, V. (1993). Diversity for the direct-sequence spread spectrum system using multiple transmit antennas. In *Proc. IEEE Int. Conf. Commun.*, vol. 3, 1503–1506.

Winters, J. H. (1983). Switched diversity with feedback for DPSK mobile radio systems. *IEEE Trans. Veh. Technol.*, **32** (1), 134–150.

—— (1994). The diversity gain of transmit diversity in wireless systems with Rayleigh fading. In *Proc. IEEE Int. Conf. Commun.*, vol. 2, 1121–1125.

—— (1998). Diversity gain of transmit diversity in wireless systems with Rayleigh fading. *IEEE Trans. Veh. Technol.*, **47** (1), 119–123.

Wittneben, A. (1991). Base station modulation diversity for digital SIMULCAST. In *Proc. IEEE Veh. Technol. Conf.*, vol. 1, 848–853.

—— (1993). A new bandwidth efficient transmit antenna modulation diversity scheme for linear digital modulation. In *Proc. IEEE Int. Conf. Commun.*, vol. 3, 1630–1634.

Wolniansky, P. W., Foschini, G. J., Golden, G. D., and Valenzuela, R. A. (1998). V-BLAST: An architecture for realizing very high data rates over rich scattering wireless channels. In *Proc. URSI Int. Symp. Signals, Syst., Electron.*, 295–300.

Xia, P., Zhou, S., and Giannakis, G. B. (2003). Bandwidth and power efficient multicarrier multiple access. *IEEE Trans. Commun.*, **51** (11), 1828–1837.

Yacoub, M. D. (1993). *Foundation of Mobile Radio Engineering* (CRC Press, Boca Raton, FL), first edn.

4
Fundamentals of MIMO channel capacity

Ezio Biglieri

Universitat Pompeu Fabra

Giorgio Taricco

Politecnico di Torino

4.1 Introduction and channel models

We consider a radio system with t antennas simultaneously transmitting
one signal each, and with r antennas receiving these signals at the output
of a channel affected by fading. Assuming two-dimensional constellations
throughout, the channel model becomes

$$\mathbf{y} = \mathbf{Hx} + \mathbf{z},\tag{4.1}$$

where $\mathbf{x} \in \mathbb{C}^t$, $\mathbf{y} \in \mathbb{C}^r$, $\mathbf{H} \in \mathbb{C}^{r \times t}$ (i.e., $\mathbf{H}$ is an $r \times t$ complex, possibly
random, matrix, whose entries h_{ij} describe the gains of each transmission
path from a transmit to a receive antenna), and $\mathbf{z}$ is a circularly-symmetric,
complex Gaussian noise vector. The component x_i, $i = 1, \ldots, t$, of vector
$\mathbf{x}$ is the elementary signal transmitted from antenna i; the component y_j,
$j = 1, \ldots, r$, of vector $\mathbf{y}$ is the signal received by antenna j. We also assume
that the noise components affecting the different receivers are independent
with variance N_0, i.e.,

$$\mathbb{E}[\mathbf{zz}^\dagger] = N_0 \mathbf{I}_r$$

where $\mathbf{I}_r$ is the $r \times r$ identity matrix, and the signal energy is constrained
by $\mathbb{E}[\mathbf{x}^\dagger \mathbf{x}] = t\mathcal{E}$, where $\mathcal{E}$ denotes the average energy per elementary signal.
The additional assumption that $\mathbb{E}[|h_{ij}|^2] = 1$ for all i, j yields the average
signal-to-noise ratio (SNR) at the receiver:

$$\rho = t\frac{\mathcal{E}}{N_0}.$$

Then, rather than assuming a power or energy constraint, we may refer to
an SNR constraint, i.e.,

$$\mathbb{E}[\mathbf{x}^\dagger \mathbf{x}] \le \rho N_0$$

For later reference, we define

$$m \triangleq \min\{t, r\}, \qquad n \triangleq \max\{t, r\}.$$

Explicitly, we have from (4.1):

$$y_i = \sum_{j=1}^{t} h_{ij} x_j + z_i, \qquad i = 1, \ldots, r,$$

which shows how every component of the received signal consists of a linear combination of the signals emitted by each antenna. We say that **y** is affected by *spatial interference*, generated by the signals transmitted by the various antennas.

For frequency-nonselective fast-fading channels, we have, with n denoting discrete time,

$$\mathbf{y}_n = \mathbf{H}_n \mathbf{x}_n + \mathbf{z}_n, \tag{4.2}$$

with $\mathbf{H}_n$, $-\infty < n < \infty$, an ergodic random process (Biglieri *et al.*, 1998).

For frequency-nonselective slow-fading channels, the model becomes

$$\mathbf{y}_n = \mathbf{H} \mathbf{x}_n + \mathbf{z}_n,$$

and each code word, however long, experiences only one channel state. This fading model is nonergodic (Biglieri *et al.*, 1998).

4.1.1 Channel state information

A crucial factor in determining the performance of transmission over a channel affected by fading is the availability, at the transmitting or at the receiving terminal, of *channel-state information* (CSI), that is, the value taken on by the fading gains (the entries of $\mathbf{H}_n$) in a transmission path. In a fixed wireless environment, the fading gains can be expected to vary slowly, so their estimate can be obtained by the receiver with a reasonable accuracy and possibly fed back to the transmitter, even in a system with a large number of antennas. In some cases, we may assume that partial knowledge of the CSI is available. One way of obtaining this estimate is by periodically sending pilot signals on the same channel also used for data transmission (these pilot signals are used in wireless systems also for acquisition, synchronization, etc.). We shall address this issue in Section 4.4.

4.1.2 Narrowband multiple-antenna channel models

Assume that the $r \times t$ channel matrix $\mathbf{H}$ remains constant during the transmission of an entire code word. Analysis of this channel requires the joint probability density function (pdf) of the rt entries of $\mathbf{H}$. A number of relatively simple models for this pdf have been proposed in the technical literature, based on experimental results and analyses. Among these, we consider the following:

Rich scattering The entries of $\mathbf{H}$ are independent, circularly symmetric, zero-mean complex Gaussian random variables.

Completely correlated The entries of $\mathbf{H}$ are correlated, circularly symmetric, zero-mean complex Gaussian random variables. To specify this model, the correlation coefficients of all pairs of elements are required.

Separately correlated ("Kronecker model") The entries of $\mathbf{H}$ are correlated, circularly symmetric, zero-mean, complex Gaussian random variables, with the correlation between two entries of $\mathbf{H}$ separated in two factors accounting for the receive and transmit correlation (Chuah *et al.*, 2002):

$$\mathbb{E}[(\mathbf{H})_{i,j}(\mathbf{H})^*_{i',j'}] = (\mathbf{R})_{i,i'}(\mathbf{T})_{j,j'},$$

for two Hermitian, positive definite matrices $\mathbf{R}$ $(r \times r)$ and $\mathbf{T}$ $(t \times t)$. This model is justified because only the objects surrounding the receiver and the transmitter cause the correlation of the local antenna elements, while they have no impact on the correlation at the other end of the link. The channel matrix can be expressed in the form

$$\mathbf{H} = \mathbf{R}^{1/2}\mathbf{H}_u\mathbf{T}^{1/2}, \tag{4.3}$$

where $\mathbf{H}_u$ is a matrix of uncorrelated, circularly symmetric, zero-mean, complex Gaussian random variables with unit variance, and $(\cdot)^{1/2}$ denotes matrix square root.[†] For a fair comparison of different correlation cases, we assume that the total average received power

---

[†]The square root of matrix $\mathbf{A} \geq \mathbf{0}$ whose singular-value decomposition is $\mathbf{A} = \mathbf{UDV}^\dagger$ is defined as $\mathbf{A}^{1/2} \triangleq \mathbf{UD}^{1/2}\mathbf{V}^\dagger$.

is constant, i.e.,

$$\mathbb{E}[\|\mathbf{H}\|^2] = \mathbb{E}[\mathrm{tr}(\mathbf{R}\mathbf{H}_u\mathbf{T}\mathbf{H}_u^\dagger)]$$

$$= \sum_{i,j,k,\ell} \mathbb{E}[(\mathbf{R})_{ij}(\mathbf{H}_u)_{jk}(\mathbf{T})_{k\ell}(\mathbf{H}_u)_{i\ell}^*]$$

$$= \sum_{i,k}(\mathbf{R})_{ii}(\mathbf{T})_{kk}$$

$$= \mathrm{tr}(\mathbf{R})\,\mathrm{tr}(\mathbf{T}) = rt.$$

Since $\mathbf{H}$ is not affected if $\mathbf{R}$ is scaled by a factor $\alpha \neq 0$ and $\mathbf{T}$ by a factor α^{-1}, we can assume, without loss of generality, that

$$\mathrm{tr}(\mathbf{R}) = r \qquad \text{and} \qquad \mathrm{tr}(\mathbf{T}) = t. \tag{4.4}$$

Uncorrelated keyhole The rank of $\mathbf{H}$ may be smaller than $\min\{t,r\}$. A special case occurs when $\mathbf{H}$ has rank one ("keyhole" channel) (Chizhik *et al.*, 2002; Gesbert *et al.*, 2002). Assume $\mathbf{H} = \mathbf{h}_r\mathbf{h}_t^\dagger$, with the entries of the vectors $\mathbf{h}_r$ and $\mathbf{h}_t$ being independent, circularly symmetric, zero-mean, complex Gaussian random variables. This model applies in the presence of walls, where the propagating signal passes through a small aperture, such as a keyhole. In this way, the incident electric field is a linear combination of the electric fields arriving from the transmit antennas, and radiates from the hole after scalar multiplication by the scattering cross-section of the keyhole. As a result, the channel matrix can be written as the product of a column vector by a row vector. Similar phenomena arise in indoor propagation through hallways or tunnels.

Rice channel The channel models listed above are zero-mean. However, for certain applications where a line-of-sight signal component is present, the channel matrix $\mathbf{H}$ should be modeled as having a nonzero mean.

4.2 Channel capacity

In this section we evaluate the capacity of the MIMO transmission system described by (4.1). Several models for the matrix $\mathbf{H}$ can be considered (Telatar, 1999):

(a) $\mathbf{H}$ is deterministic.

(b) $\mathbf{H}$ is a random matrix with a given probability distribution. Each channel use (i.e., the transmission of one symbol from each of the t

transmit antennas) corresponds to an *independent* realization of $\mathbf{H}$ (ergodic channel).

(c) $\mathbf{H}$ is a random matrix, which remains fixed for the whole transmission after it is chosen (nonergodic channel).

When $\mathbf{H}$ is random (cases (b) and (c) above), we assume that its entries are $\sim \mathcal{N}_c(0, 1)$, i.e., i.i.d. complex Gaussian with zero-mean, independent real and imaginary parts, each with variance $1/2$. Equivalently, each entry of $\mathbf{H}$ has uniform phase and Rayleigh magnitude. This choice models Rayleigh fading with enough separation between antennas, so that the fades for each TX/RX antenna pair are independent. We also assume, unless otherwise stated, that the CSI (that is, the realization of $\mathbf{H}$) is perfectly known at the receiver, while only the probability distribution of $\mathbf{H}$ is known at the transmitter (the latter assumption is necessary for capacity computations, since the transmitter must choose an optimum code for that specific channel).

4.2.1 Deterministic channel

Assume first that the nonrandom value of $\mathbf{H}$ is known at both transmitter and receiver. We derive the capacity by maximizing the average mutual information $I(\mathbf{x}; \mathbf{y})$ between input and output of the channel over the choice of distributions of $\mathbf{x}$. Singular-value decomposition of the matrix $\mathbf{H}$ yields

$$\mathbf{H} = \mathbf{U}\mathbf{D}\mathbf{V}^\dagger,$$

where $\mathbf{U} \in \mathbb{C}^{r \times r}$ and $\mathbf{V} \in \mathbb{C}^{t \times t}$ are unitary, and $\mathbf{D} \in \mathbb{R}^{r \times t}$ is diagonal. We can write

$$\mathbf{y} = \mathbf{U}\mathbf{D}\mathbf{V}^\dagger\mathbf{x} + \mathbf{z}. \tag{4.5}$$

Premultiplication of (4.5) by $\mathbf{U}^\dagger$ shows that the original channel is equivalent to the channel described by the input-output relationship

$$\tilde{\mathbf{y}} = \mathbf{D}\tilde{\mathbf{x}} + \tilde{\mathbf{z}}, \tag{4.6}$$

where $\tilde{\mathbf{y}} \triangleq \mathbf{U}^\dagger\mathbf{y}$, $\tilde{\mathbf{x}} \triangleq \mathbf{V}^\dagger\mathbf{x}$ (so that $\mathbb{E}[\tilde{\mathbf{x}}^\dagger\tilde{\mathbf{x}}] = \mathbb{E}[\mathbf{x}^\dagger\mathbf{x}]$), and $\tilde{\mathbf{z}} \triangleq \mathbf{U}^\dagger\mathbf{z} \sim \mathcal{N}_c(0, N_0\mathbf{I}_r)$. Now, the rank of $\mathbf{H}$ is at most $m \triangleq \min\{t, r\}$; hence, at most m of its singular values are nonzero. Denote them by $\sqrt{\lambda_i}$, $i = 1, \ldots, m$, and rewrite (4.6) componentwise in the form

$$\tilde{y}_i = \begin{cases} \sqrt{\lambda_i}\tilde{x}_i + \tilde{z}_i, & i = 1, \ldots, m \\ 0, & i = m+1, \ldots, r, \end{cases} \tag{4.7}$$

which shows how this channel is equivalent to a set of m parallel independent channels. In addition, we see that, for $i > m$, $\tilde{y}_i$ is *independent of the transmitted signal*, and $\tilde{x}_i$ plays no role. Notice that the second line of the RHS of (4.7) is not present when $r \leq m$.

Maximization of the mutual information requires independent $\tilde{x}_i$, $i = 1, \ldots, m$, each with independent Gaussian, zero-mean real and imaginary parts. Their variances should be chosen via "waterfilling:"

$$\mathbb{E}[\Re(\tilde{x}_i)]^2 = \mathbb{E}[\Im(\tilde{x}_i)]^2 = \frac{N_0}{2}\left(\mu - \lambda_i^{-1}\right)_+, \qquad (4.8)$$

where $(\cdot)_+ \triangleq \max(0, \cdot)$. This comes from a result concerning parallel channels (Gallager, 1968). With μ chosen so as to meet the SNR constraint, we see that the SNR, as parametrized by μ, is

$$\rho(\mu) = \sum_i \left(\mu - \lambda_i^{-1}\right)_+,$$

and the capacity takes on the value (in bits per complex dimension)

$$C(\mu) = \sum_i \left(\log_2(\mu\lambda_i)\right)_+.$$

Example 4.1 Take $t = r = m$, and $\mathbf{H} = \mathbf{I}_m$. Due to the structure of $\mathbf{H}$, there is no spatial interference here, and transmission occurs over m parallel additive white Gaussian noise (AWGN) channels, each with SNR ρ/m, and therefore with capacity $\log_2(1 + \rho/m)$ bit/complex dimension. Thus,

$$C = m\log_2(1 + \rho/m).$$

We see here that the capacity is proportional to the number of transmit antennas. Notice also that, as $m \to \infty$, the capacity tends to the limiting value $C = \rho\log_2 e$.

4.2.2 Independent Rayleigh fading channel

We assume now that $\mathbf{H}$ is independent of both $\mathbf{x}$ and $\mathbf{z}$, with entries $\sim \mathcal{N}_c(0, 1)$, and that for each channel use an independent realization of $\mathbf{H}$ is drawn, so that the channel is ergodic. If the receiver has perfect CSI, the mutual information between the channel input (the vector $\mathbf{x}$) and its output (the pair $\mathbf{y}, \mathbf{H}$), is:

$$I(\mathbf{x}; \mathbf{y}, \mathbf{H}) = I(\mathbf{x}; \mathbf{H}) + I(\mathbf{x}; \mathbf{y} \mid \mathbf{H}).$$

Since $\mathbf{H}$ and $\mathbf{x}$ are independent, then $I(\mathbf{x}; \mathbf{H}) = 0$, and hence

$$I(\mathbf{x}; \mathbf{y}, \mathbf{H}) = I(\mathbf{x}; \mathbf{y} \mid \mathbf{H}) = \mathbb{E}_{\tilde{\mathbf{H}}}[I(\mathbf{x}; \mathbf{y} \mid \mathbf{H} = \tilde{\mathbf{H}})],$$

where $\tilde{\mathbf{H}}$ denotes a realization of the random matrix $\mathbf{H}$. The maximum of $I(\mathbf{x}; \mathbf{y}, \mathbf{H})$ with respect to the distribution of $\mathbf{x}$ is the channel capacity C. If the entries of $\mathbf{H}$ are uncorrelated, the capacity is achieved by a transmitted signal $\mathbf{x} \sim \mathcal{N}_c(0, (\rho/t)\mathbf{I}_t)$, and is equal to (Telatar, 1999)

$$C = \mathbb{E}\left[I(\mathbf{H})\right], \tag{4.9}$$

where

$$I(\mathbf{H}) \triangleq \log_2 \det\left(\mathbf{I}_r + \frac{\rho}{t}\mathbf{H}\mathbf{H}^\dagger\right) \tag{4.10}$$

is a random variable sometimes called *instantaneous mutual information*.

The exact computation of (4.9) will be examined soon. For the moment, note that, if r is fixed and $t \to \infty$, the strong law of large numbers yields

$$\frac{1}{t}\mathbf{H}\mathbf{H}^\dagger \to \mathbf{I}_r \qquad \text{almost surely (a.s.).}$$

Thus, as $t \to \infty$ the capacity tends to

$$\log_2 \det(\mathbf{I}_r + \rho\mathbf{I}_r) = \log_2(1 + \rho)^r = r \log_2(1 + \rho)$$

and hence increases *linearly* with r.

A simplistic interpretation of the above would qualify fading as *beneficial* to MIMO transmission, as independent path gains generate r independent spatial channels. Actually, high capacity is generated by a multiplicity of nonzero singular values in $\mathbf{H}$, which is typically achieved if $\mathbf{H}$ is a random matrix, but not if it is deterministic.

Exact computation of C

Exact calculation of (4.9) (Telatar, 1999; Shin and Lee, 2003; Biglieri and Taricco, 2004) yields

$$C = \log_2(e)\frac{m!}{(n-1)!}\sum_{\ell=0}^{m-1}\sum_{\mu=0}^{m}\sum_{p=0}^{\ell+\mu+n-m}\frac{(-1)^{\ell+\mu}(\ell + \mu + n - m)!}{\ell!\mu!}e^{t/\rho}E_{p+1}(t/\rho)$$

$$\cdot\left[\binom{n-1}{m-1-\ell}\binom{n}{m-1-\mu} - \binom{n-1}{m-2-\ell}\binom{n}{m-\mu}\right], \tag{4.11}$$

where

$$E_n(x) \triangleq \int_1^\infty e^{-xy}y^{-n}\,dy$$

is the exponential integral function of order n.

Some special cases as well as asymptotic approximations to the values of C are examined in the examples that follow.

Example 4.2 $(r \gg t)$ Consider first $t = 1$, so that $m = 1$ and $n = r$. Application of (4.11) yields

$$C = \log_2(e) \sum_{k=1}^{r} e^{1/\rho} E_k(1/\rho). \tag{4.12}$$

An asymptotic expression of C, valid as $r \to \infty$, can be obtained as follows. Using in (4.12) the approximation, valid for large k,

$$e^x E_k(x) \approx \frac{1}{x + k}, \tag{4.13}$$

we obtain

$$C \approx \log_2(e) \sum_{k=1}^{r} \frac{1}{1/\rho + k} \approx \log_2(e) \int_0^r \frac{1}{1/\rho + x}\, dx = \log_2(1 + r\rho).$$

We see here that, if $t = 1$, the capacity increases only logarithmically as the number of receive antennas is increased—hardly an efficient way of boosting capacity.

For finite $t > 1$ (and $r \to \infty$), we set $\mathbf{W} = \mathbf{H}^\dagger \mathbf{H} \to r\mathbf{I}_t$ a.s. Hence, the following asymptotic expression holds:

$$C = \log_2 \det(\mathbf{I}_t + (\rho/t)\mathbf{W}) \approx t \log_2(1 + (\rho/t)r). \tag{4.14}$$

Example 4.3 $(t \gg r)$ Consider first $r = 1$, so that $m = 1$ and $n = t$. Application of (4.11) yields

$$C = \log_2(e) \sum_{k=1}^{t} e^{t/\rho} E_k(t/\rho).$$

Proceeding as in Example 4.2, an asymptotic expression of C as $t \to \infty$ can be obtained, yielding $C \approx \log_2(1 + \rho)$. For finite $r > 1$ (and $t \to \infty$), we set $\mathbf{W} = \mathbf{H}\mathbf{H}^\dagger \to t\mathbf{I}_r$ a.s. Hence, the following asymptotic expression holds:

$$C = \log_2 \det(\mathbf{I}_r + (\rho/t)\mathbf{W}) \approx r \log_2(1 + \rho). \tag{4.15}$$

Example 4.4 $(r = t)$ With $r = t$, we have $m = n = r$, and application of (4.11) yields the capacity plotted in Fig. 4.1.

The results of Fig. 4.1 show that capacity increases almost linearly with $m \triangleq \min\{t, r\}$. This fact can be analyzed in a general setting by showing that, when t and r both grow to infinity, the capacity per antenna tends

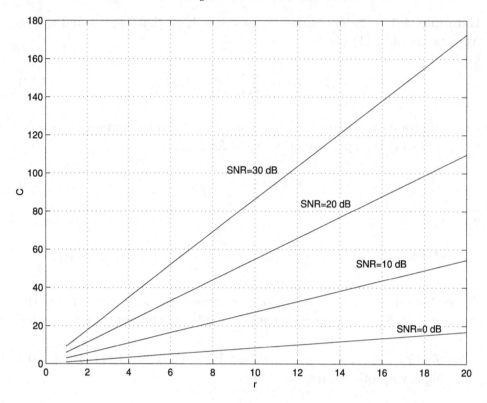

Fig. 4.1. Capacity with independent Rayleigh fading and $t = r$ antennas.

to a constant. The following can be proved (see the paper by Biglieri and Taricco (2004) and references therein):

$$\frac{C}{m} \approx (\log_2(w_+\rho) + (1 - \alpha) \log_2(1 - w_-) - (w_-\alpha) \log_2 e) \cdot \max\{1, 1/\alpha\},$$

(4.16)

where

$$w_\pm \triangleq (w \pm \sqrt{w^2 - 4/\alpha})/2$$

and

$$w \triangleq 1 + \frac{1}{\alpha} + \frac{1}{\rho}.$$

This asymptotic result can be used to approximate the value of C for finite r and t by setting $\alpha = t/r$. The approximation provides values very close to the true capacity even for small r and t, as shown in Fig. 4.2. The figure shows the asymptotic value of C/m (for $t, r \to \infty$ with $t/r \to \alpha$)

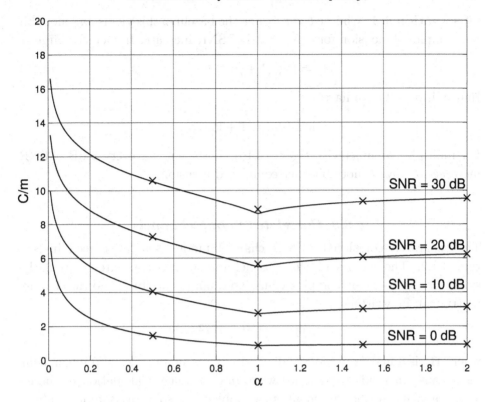

Fig. 4.2. Asymptotic ergodic capacity per antenna (C/m) with independent Rayleigh fading as $t, r \to \infty$ and $t/r \to \alpha$ (solid curves). The exact ergodic capacity per antenna for $r = 2$ is also shown for comparison ($\times$).

versus α and the nonasymptotic values of C/m corresponding to $r = 2$ and 4, respectively.

Observation 4.1 It can be observed, from (4.16) and a modicum of algebra, that, for large SNR, i.e., for $\rho \to \infty$, the ergodic capacity is asymptotically equal to $m \log_2 \rho$: comparing this result with the asymptotic capacity of the single-input, single-output channel $C \approx \log_2 \rho$, we see that use of multiple antennas increases the capacity by a factor m. That is, multiple antennas generate m independent parallel channels; hence, they provide a *rate* (or *multiplexing*) *gain* m.

Observation 4.2 For the validity of (4.16), it is not necessary to assume that the entries of **H** are Gaussian, as needed for the preceding nonasymptotic results: a sufficient condition is that **H** has i.i.d. entries with unit variance (see, e.g., Biglieri and Taricco, 2004).

Observation 4.3 Choose $t = r = 1$ as the baseline; this yields *one* more bit per complex dimension for every 3 dB of SNR increase. In fact, for large ρ,

$$C = \log_2(1 + \rho) \approx \log_2 \rho. \tag{4.17}$$

Hence, if $\rho \rightarrow 2\rho$, we have

$$\log_2(2\rho) = 1 + \log_2 \rho.$$

For multiple antennas with $t = r$, (4.16) shows that for every 3 dB of SNR increase we have t more bits per complex dimension.

4.2.3 Correlated fading channels

The separately-correlated MIMO channel model was introduced in Section 4.1.2. The entries of the channel matrix are correlated, circularly-symmetric, zero-mean, complex Gaussian random variables, and the channel matrix can be written as

$$\mathbf{H} = \mathbf{R}^{1/2}\mathbf{H}_u\mathbf{T}^{1/2},$$

where $\mathbf{H}_u$ is a matrix of independent, circularly-symmetric, zero-mean, complex Gaussian random variables with unit variance. Calculation of the ergodic capacity in this case is an open problem; its solution is known only in some special cases (Goldsmith *et al.*, 2003). A lower bound C^* to capacity can be obtained under the constraint $\mathbf{x} \sim \mathcal{N}_c(0, (\rho/t)\mathbf{I}_t)$ (i.i.d. signals across transmit antennas: this is the sensible choice if the correlations among the entries of $\mathbf{H}$ are unknown at the transmitter, in which case C^* yields the actual capacity). We have

$$C^* = \mathbb{E}\left[\log_2 \det\left(\mathbf{I}_r + \frac{\rho}{t}\mathbf{H}_u\mathbf{T}\mathbf{H}_u^\dagger\mathbf{R}\right)\right].$$

In the special case $t = r$, for high SNR the following asymptotic approximation can be derived (Shin and Lee, 2003):

$$C^* \approx m \log_2(\rho/t) + \log_2(m!) + \log_2 \det(\mathbf{TR}).$$

This result is tantamount to saying that, when $t = r = m$, the asymptotic loss in C^* due to correlation is $\log_2 \det(\mathbf{TR})/m$ bit/s/Hz. To prove that correlation actually causes a loss, let $t_i, i = 1, \ldots, m$, denote the positive eigenvalues of $\mathbf{T}$, and recall the trace constraint (4.4). We obtain

$$\det(\mathbf{T})^{1/m} = \prod_i t_i^{1/m} \leq \frac{1}{m}\sum_i t_i = 1.$$

Since a similar result applies to $\mathbf{R}$, we obtain

$$-\log_2 \det(\mathbf{TR})/m \geq 0,$$

with equality if and only if $\mathbf{T} = \mathbf{R} = \mathbf{I}_m$. This confirms that, under the "fair comparison" conditions dictated by (4.4), the asymptotic power loss due to separate correlation is always nonnegative, and zero only in the uncorrelated case. This proves the following asymptotic (in the SNR) statements:

- (Separate) correlation degrades system performance.
- The linear growth of C^* with respect to the minimum number of transmit/receive antennas is preserved.

The above can be extended, with the help of some algebra, to the case $t \neq r$ (Shin and Lee, 2003).

4.3 Nonergodic Rayleigh fading channel

When $\mathbf{H}$ is chosen randomly at the beginning of the transmission and is held fixed for all channel uses, average capacity has no meaning, as the channel is nonergodic. In this case, the quantity to be evaluated is outage probability rather than capacity, that is, the probability that the transmission rate ρ exceeds the mutual information of the channel. The maximum rate that can be supported by the channel with a given outage probability is referred to as the *outage capacity*.

Under these conditions, the outage probability is defined as

$$P_{\text{out}}(R) \triangleq \mathbb{P}(I(\mathbf{H}) < \rho), \tag{4.18}$$

where $I(\mathbf{H})$ is the "instantaneous mutual information" defined in (4.10). The evaluation of (4.18) should be done by Monte Carlo simulation. However, one can profitably use an asymptotic result stating that, if the entries of $\mathbf{H}$ are i.i.d. zero-mean Gaussian, and t and r grow to infinity, the instantaneous mutual information $I(\mathbf{H})$ tends to a Gaussian random variable in distribution. Thus, by computing its asymptotic mean μ_I and variance σ_I^2, one can characterize its asymptotic behavior. The value of this asymptotic result is strongly enhanced by the fact that $I(\mathbf{H})$ is very well approximated by a Gaussian random variable even for small numbers of antennas. Thus, the outage probability for any pair (t, r) is given by

$$P_{\text{out}}(R) \approx Q\left(\frac{\mu_I - R}{\sigma_I}\right), \tag{4.19}$$

where

$$\mu_I \triangleq -t\Big\{(1+\beta)\log_2 w + q_0 r_0 \log_2 e + \log_2 r_0 + \beta\log_2(q_0/\beta)\Big\}, \quad (4.20)$$

$$\sigma_I^2 \triangleq -\log_2 e \cdot \log_2(1 - q_0^2 r_0^2/\beta) \qquad\qquad\qquad\qquad (4.21)$$

are expressed in bit/s/Hz and (bit/s/Hz)^2, respectively, with $w \triangleq \sqrt{1/\rho}$, $\beta \triangleq \alpha^{-1}$, and

$$q_0 = \frac{\beta - 1 - w^2 + \sqrt{(\beta - 1 - w^2)^2 + 4w^2\beta}}{2w},$$

$$r_0 = \frac{1 - \beta - w^2 + \sqrt{(1 - \beta - w^2)^2 + 4w^2}}{2w}. \qquad (4.22)$$

Based on these results, we can use (4.19) to approximate closely the outage probabilities as in Fig. 4.3. This figure shows the rate that can be supported by the channel for a given SNR and a given outage probability, that is, from (4.19):

$$\rho = \mu_I - \sigma_I Q^{-1}(P_{\text{out}}). \qquad (4.23)$$

As r, t increase, the outage probability curves come closer to each other: this fact can be interpreted by saying that, as r and t grow to infinity, the channel tends to an ergodic channel.

Fig. 4.4 shows the outage capacity (at $P_{\text{out}} = 0.01$) of a nonergodic Rayleigh fading MIMO channel.

4.3.1 Block-fading channel

Here we take the approach of choosing a block-fading channel model (Biglieri et al., 1998) as shown in Fig. 4.5. The channel is characterized by the F matrices $\mathbf{H}_k$, $k = 1, \ldots, F$, each describing the fading gains in a block. The channel input-output equation is

$$\mathbf{y}_k[n] = \mathbf{H}_k\mathbf{x}_k[n] + \mathbf{z}_k[n], \qquad (4.24)$$

for $k = 1, \ldots, F$ (block index) and $n = 1, \ldots, N$ (symbol index along a block), $\mathbf{y}_k, \mathbf{z}_k \in \mathbb{C}^r$, and $\mathbf{x}_k \in \mathbb{C}^t$. Moreover, the additive noise $\mathbf{z}_k[n]$ is a vector of circularly symmetric complex Gaussian RVs with zero mean and variance N_0. Hence,

$$\mathbb{E}[\mathbf{z}_k[n]\mathbf{z}_k[n]^\dagger] = N_0\mathbf{I}_r.$$

It is convenient to use the SVD

$$\mathbf{H}_k = \mathbf{U}_k\mathbf{D}_k\mathbf{V}_k^\dagger,$$

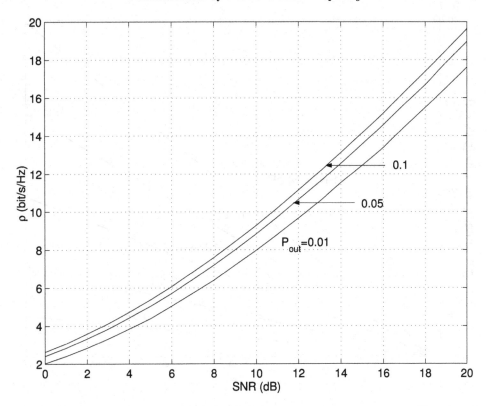

Fig. 4.3. Transmission rate that can be supported with $r = t = 4$ and a given outage probability by a nonergodic Rayleigh channel. The results are based on the Gaussian approximation.

where $\mathbf{D}_k$ is an $r \times t$ real matrix whose main-diagonal entries are the ordered singular values $\sqrt{\lambda_{k,1}} \geq \cdots \geq \sqrt{\lambda_{k,m}}$, with $\lambda_{k,i}$ the ith largest eigenvalue of the Hermitian matrix $\mathbf{H}_k\mathbf{H}_k^\dagger$, and $m \triangleq \min\{r, t\}$. Since $\mathbf{U}_k$ and $\mathbf{V}_k$ are unitary, by premultiplying $\mathbf{y}_k[n]$ by $\mathbf{U}_k^\dagger$ the input-output relation (4.24) can be rewritten in the form

$$\widetilde{\mathbf{y}}_k[n] = \mathbf{D}_k\widetilde{\mathbf{x}}_k[n] + \widetilde{\mathbf{z}}_k[n],$$

where $\widetilde{\mathbf{y}}_k[n] \triangleq \mathbf{U}_k^\dagger\mathbf{y}_k[n]$, $\widetilde{\mathbf{x}}_k[n] \triangleq \mathbf{V}_k^\dagger\mathbf{x}_k[n]$, $\widetilde{\mathbf{z}}_k[n] \triangleq \mathbf{U}_k^\dagger\mathbf{z}_k[n]$, and $\widetilde{\mathbf{z}}_k[n] \sim \mathcal{N}_c(\mathbf{0}, N_0\mathbf{I}_r)$ since

$$\mathbb{E}[\widetilde{\mathbf{z}}_k[n]\widetilde{\mathbf{z}}_k[n]^\dagger] = \mathbf{U}^\dagger\mathbb{E}[\widetilde{\mathbf{z}}_k[n]\widetilde{\mathbf{z}}_k[n]^\dagger]\mathbf{U} = N_0\mathbf{I}_r.$$

No delay constraints When the random matrix process $\{\mathbf{H}_k\}_{k=1}^F$ is i.i.d., the channel is ergodic as $F \to \infty$, and the average capacity is the relevant quantity. When the entries of the channel matrices are uncorrelated, and

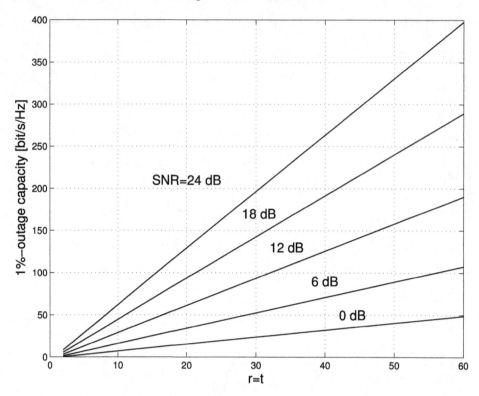

Fig. 4.4. Outage capacity (at $P_{\text{out}} = 0.01$) with independent Rayleigh fading and $r = t$ antennas.

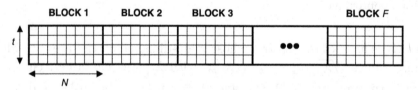

Fig. 4.5. One code word in an F-block fading channel.

perfect CSI is available to the receiver only, this is given by

$$C = \mathbb{E}\left[\sum_{i=1}^{m} \log_2\left(1 + \frac{\rho}{t}\lambda_i\right)\right]. \qquad (4.25)$$

If perfect CSI is available to transmitter and receiver,

$$C = \mathbb{E}\left[\sum_{i=1}^{m} \left(\log_2(\mu\lambda_i)\right)_+\right], \qquad (4.26)$$

where μ is the solution of the "waterfilling" equation

$$\mathbb{E}\left[\sum_{i=1}^{m}(\mu - 1/\lambda_i)_+\right] = \rho.$$

For all block lengths N, the capacities (4.25) and (4.26) are achieved by code sequences with length FNt with $F \to \infty$. Capacity (4.25) is achieved by random codes whose symbols are i.i.d. $\sim \mathcal{N}_c(0, \rho/t)$. Thus, all antennas transmit the same average energy per symbol. Capacity (4.26) can be achieved by generating a random code with i.i.d. components $\sim \mathcal{N}_c(0, 1)$ and having each code word split into F blocks of N vectors $\tilde{\mathbf{x}}_k[n]$ with t components each. For block k, the optimal linear transformation

$$\mathbf{W}_k = \mathbf{V}_k \ \mathrm{diag}(\sqrt{\rho_{k,1}}, \ldots, \sqrt{\rho_{k,m}}, \underbrace{0, \ldots, 0}_{t-m})$$

is computed, where $\rho_{k,i} \triangleq (\mu - 1/\lambda_{k,i})_+$. The vectors $\mathbf{x}_k[n] = \mathbf{W}_k\tilde{\mathbf{x}}_k[n]$ are transmitted from the t antennas. This optimal scheme can be viewed as the concatenation of an optimal encoder for the unfaded AWGN channel, followed by a linear transformation ("beamforming") described by the weighting matrix $\mathbf{W}_k$ varying from block to block (Biglieri *et al.*, 2001).

Delay constraints Consider now a delay constraint that forces F to take on a finite value. Define $\mathbf{\Lambda} \triangleq \{\lambda_{k,i}\}_{k=1,i=1}^{F,m}$, $\mathbf{\Gamma} \triangleq \{\rho_{k,i}\}_{k=1,i=1}^{F,m}$, the *instantaneous* mutual information

$$I(\mathbf{\Lambda}, \mathbf{\Gamma}) \triangleq \frac{1}{F}\sum_{k=1}^{F}\sum_{i=1}^{m}\log_2(1 + \lambda_{k,i}\rho_{k,i}),$$

and the *instantaneous* SNR per block

$$\rho_F \triangleq \frac{1}{F}\sum_{k=1}^{F}\sum_{i=1}^{m}\rho_{k,i}.$$

Assuming that the receiver has perfect knowledge of the CSI (and hence of $\mathbf{\Lambda}$), we can define a power allocation rule depending on $\mathbf{\Lambda}$, so that $\rho_{k,i}$ and ρ_F are functions of $\mathbf{\Lambda}$. We may consider two power constraints:

$$\rho_F(\mathbf{\Lambda}) \leq \rho \qquad \text{(short-term)}, \tag{4.27}$$

$$\mathbb{E}[\rho_F(\mathbf{\Lambda})] \leq \rho \qquad \text{(long-term)}. \tag{4.28}$$

The optimum power allocation rules minimizing the outage probability

$$P_{\text{out}}(R) \triangleq \mathbb{P}\left(I(\mathbf{\Lambda}, \mathbf{\Gamma}) < R\right)$$

under constraints (4.27) and (4.28) are derived by Biglieri *et al.* (2001).

Another important concept related to outage probability is given in the following.

Definition 4.1 *The* zero-outage capacity, *sometimes also referred to as* delay-limited capacity, *is the maximum rate for which the minimum outage probability is zero under a given power constraint (Tse and Hanly, 1998; Biglieri et al., 2001).*

It was shown by Biglieri *et al.* (2001) that, under a positive long-term power constraint, the zero-outage capacity of a block-fading channel is positive if the channel is *regular*. A regular channel is defined as follows.

Definition 4.2 *A block-fading channel is said to be* regular *if the fading distribution is continuous, and*

$$\mathbb{E}[1/\bar{\lambda}_F] < \infty,$$

where $\bar{\lambda}_F$ is the geometric mean of the $\lambda_{k,i}$:

$$\bar{\lambda}_F \triangleq \prod_{k,i} \lambda_{k,i}^{1/mF},$$

where, as usual, $m \triangleq \min\{t, r\}$.

Example 4.5 *The Rayleigh fading channel with $F = m = 1$ is not regular, and its zero-outage capacity is zero. The Rayleigh block fading channel is regular if $mF > 1$ (refer to the paper by Biglieri et al. (2001) for a proof). For example, if $F > 1$ and $m = 1$, we have*

$$\mathbb{E}[1/\bar{\lambda}_F] = (\mathbb{E}[\lambda_1^{-1/F}])^F = [\Gamma(1 - 1/F)]^F < \infty,$$

where $\Gamma(x) \triangleq \int_0^\infty u^{x-1} e^{-u} du$ is the standard Gamma function.

4.3.2 Asymptotics

Under a long-term power constraint, and with optimal transmit power allocation the zero-outage capacity of a regular block-fading channel, as $\rho \to \infty$, is given by (Biglieri *et al.*, 2001)

$$C_{\text{zero-outage}} \approx m \log_2 \left(\frac{\rho}{m\mathbb{E}\left[1/\bar{\lambda}_F\right]} \right).$$

As $m \to \infty$ and $\max\{t, r\}/m \to \alpha > 0$, the limiting value of the normalized zero-outage capacity per degree of freedom C/m coincides with the limiting normalized ergodic capacity (Biglieri *et al.*, 2001).

4.4 Influence of channel-state information

As we have seen, in a system with t transmit and r receive antennas and an ergodic Rayleigh fading channel modeled by a $t \times r$ matrix with random i.i.d. complex Gaussian entries, the average channel capacity with perfect CSI at the receiver is about $m \triangleq \min\{t, r\}$ times larger than that of a single-antenna system for the same transmitted power and bandwidth. The capacity increases by about m bit/s/Hz for every 3 dB increase in signal-to-noise ratio (SNR). Due to the assumption of perfect CSI available at the receiver, this result can be viewed as a fundamental limit for coherent multiple-antenna systems.

4.4.1 Perfect CSI at the receiver

The most commonly studied situation is that of perfect CSI available at the receiver, which is the ideal assumption under which we developed our study of multiple-antenna systems above.

4.4.2 Imperfect CSI at the receiver

If the assumption of a quasi-static fading channel is valid, system performance can be enhanced if the receiver is made aware of the realization of the random fading gains affecting the transmission paths from each transmit to each receive antenna. For this purpose, a fraction of the transmitted frame may consist of pilot symbols, whose composition is known to the receiver and is used by the latter to estimate the channel parameters. Due to noise and to the finite number of pilot symbols in a frame, the channel estimate is not perfect. Taricco and Biglieri (2005) investigate effects that this imperfect estimation has on system performance.

4.4.3 No CSI

Fundamental limits of noncoherent communication, where estimates of the fading coefficients are not available, will now be derived. Consider a block-fading channel model. To compute the capacity of this channel, we assume that coding is performed using blocks, each of them consisting of tN elementary symbols to be transmitted by t antennas in N time instants. One block is represented by the $t \times N$ matrix $\mathbf{X}$. We further assume that the $r \times N$ noise matrix $\mathbf{Z}$ has i.i.d. $\mathcal{N}_c(0, N_0)$ entries. The received signal is the $r \times N$ matrix

$$\mathbf{Y} = \mathbf{HX} + \mathbf{Z}. \tag{4.29}$$

The following can be proved (Marzetta and Hochwald, 1999):

(a) The pdf of $\mathbf{Y}$ depends on its argument only through the product $\mathbf{Y}^\dagger\mathbf{Y}$, which consequently plays the role of a sufficient statistic. If $N < r$, the $N \times N$ matrix $\mathbf{Y}^\dagger\mathbf{Y}$ provides a representation of the received signals that is more economical than the $r \times N$ matrix $\mathbf{Y}$.

(b) The pdf of $\mathbf{Y}$ depends on the transmitted signal $\mathbf{X}$ only through the $N \times N$ matrix $\mathbf{X}^\dagger\mathbf{X}$.

Observation (b) above is the basis of the following theorem, which, in its essence, says that there is no increase in capacity if we have $t > N$; hence, there is no point in making the number of transmit antennas greater than N if there is no CSI. In particular, if $N = 1$ (an independent fade occurs at each symbol period), only one transmit antenna is useful. Note how this result contrasts sharply with its counterpart of CSI known at the receiver, where the capacity grows linearly with $\min\{t, r\}$.

Theorem 4.1 *If the entries of $\mathbf{H}$ are i.i.d., then the channel capacity for $t > N$ equals the capacity for $t = N$.*

Marzetta and Hochwald (1999) also showed that the signal matrix that achieves capacity can be written in the form:

$$\mathbf{X} = \mathbf{D\Phi},$$

where $\mathbf{\Phi}$ is a $t \times N$ matrix such that $\mathbf{\Phi\Phi}^\dagger = \mathbf{I}_t$. Moreover, $\mathbf{\Phi}$ has a pdf that is unchanged when the matrix is multiplied by a deterministic unitary matrix (this is the matrix counterpart of a complex scalar having unit magnitude and uniformly-distributed phase). $\mathbf{D}$ is a $t \times t$ real, nonnegative, diagonal matrix independent of $\mathbf{\Phi}$, whose role is to scale $\mathbf{X}$ to meet the power constraint. In general, the optimizing $\mathbf{D}$ is unknown, as is the exact expression of capacity. However, for the high-SNR regime ($\rho \gg 1$), the following results are available (Marzetta and Hochwald, 1999; Zheng and Tse, 2002), but note that they depend critically on the assumed fading model (Lapidoth and Moser, 2003):

(a) If $N \gg t$ and $t \leq \min\{N/2, r\}$, then capacity is attained when $\mathbf{D} = \sqrt{\rho N N_0/t}\,\mathbf{I}_t$, so that $\mathbf{X} = \sqrt{\rho N N_0/t}\,\boldsymbol{\Phi}$.

(b) For every 3 dB increase of ρ, the capacity increase is $t^*(1 - t^*/N)$, where $t^* \triangleq \min\{t, r, \lfloor N/2 \rfloor\}$.

(c) If $N \geq 2r$, there is no capacity increase by using $r > t$.

An obvious upper bound to capacity can be obtained if we assume that the receiver is provided with perfect knowledge of the realization of $\mathbf{H}$. Hence, the bound to capacity per block of N symbols is

$$C \leq N \log_2 \det\left[\mathbf{I}_t + \frac{\rho}{t}\mathbf{H}^\dagger\mathbf{H}\right].$$

We can reasonably expect that the actual capacity tends to the RHS of the previous inequality, because a certain (small) fraction of the coherence time can be reserved for sending training data to be used by the receiver for its estimate of $\mathbf{H}$.

References

Biglieri, E., Caire, G., and Taricco, G. (2001). Limiting performance of block-fading channels with multiple antennas. *IEEE Trans. Inf. Theory*, **47** (4), 1273–1289.

Biglieri, E., Proakis, J., and Shamai (Shitz), S. (1998). Fading channels: Information-theoretic and communication aspects. *IEEE Trans. Inf. Theory*, **44** (6), 2619–2692.

Biglieri, E. and Taricco, G. (2004). Transmission and reception with multiple antennas: Theoretical foundations. *Foundations and Trends in Commun. and Inf. Theory*, **1** (2), 1–156.

Chizhik, D., Foschini, G. J., Gans, M. J., and Valenzuela, R. A. (2002). Keyholes, correlations, and capacities of multielement transmit and receive antennas. *IEEE Trans. Wireless Commun.*, **1** (2), 361–368.

Chuah, C.-N., Tse, D. N. C., Kahn, J. M., and Valenzuela, R. A. (2002). Capacity bounds via duality with applications to multiple-antenna systems on flat-fading channels. *IEEE Trans. Inf. Theory*, **48** (3), 637–650.

Gallager, R. G. (1968). *Information Theory and Reliable Communication* (J. Wiley & Sons, New York).

Gesbert, D., Bölcskei, H., Gore, D. A., and Paulraj, A. J. (2002). Outdoor MIMO wireless channels: Models and performance prediction. *IEEE Trans. Commun.*, **50** (12), 1926–1934.

Goldsmith, A., Jafar, S. A., Jindal, N., and Vishwanath, S. (2003). Capacity limits of MIMO channels. *IEEE J. Sel. Areas Commun.*, **21** (5), 684–702.

Lapidoth, A. and Moser, S. M. (2003). Capacity bounds via duality with applications to multiple-antenna systems on flat-fading channels. *IEEE Trans. Inf. Theory*, **49** (10), 2426–2467.

Marzetta, T. L. and Hochwald, B. M. (1999). Capacity of a mobile multiple-antenna communication link in Rayleigh flat fading. *IEEE Trans. Inf. Theory*, **45** (1), 139–157.

Shin, H. and Lee, J. H. (2003). Capacity of multiple-antenna fading channels: Spatial fading correlation, double scattering, and keyhole. *IEEE Trans. Inf. Theory*, **49** (10), 2636–2647.

Taricco, G. and Biglieri, E. (2005). Space–time coding with imperfect channel estimation. *IEEE Trans. Wireless Commun.*, **4** (4), 1874–1888.

Telatar, E. (1999). Capacity of multi-antenna Gaussian channels. *Eur. Trans. Telecommun.*, **10** (6), 585–595.

Tse, D. and Hanly, V. (1998). Multi-access fading channels—Part I: Polymatroid structure, optimal resource allocation and throughput capacities. *IEEE Trans. Inf. Theory*, **44** (7), 2796–2815.

Zheng, L. and Tse, D. N. C. (2002). Communication on the Grassman manifold: A geometric approach to the noncoherent multiple-antenna channel. *IEEE Trans. Inf. Theory*, **48** (2), 359–383.

5

Multiantenna capacity: myths and realities

Angel Lozano

Bell Laboratories, Lucent Technologies

Antonia M. Tulino

Universitá degli Studi di Napoli

Sergio Verdú

Princeton University

Over the last decade, the increases in capacity promised by multiantenna communication techniques have spurred many information-theoretic analyses. Furthermore, information theory has been used as a design tool to optimize the signals fed to the transmit array and to motivate signal processing strategies at the receiver. In this chapter, we catalog a number of misconceptions that have arisen in the multiantenna literature. The focus is on information-theoretic results and their interpretations, rather than on the validity of the various modeling assumptions. After some introductory material, we address several misconceptions about optimum signaling and about the impact of model features, such as antenna correlation, line-of-sight components, and intercell interference. Particular attention is given to the low- and high-power regions. We also briefly touch upon the relationship of the capacity results with some practical transmit and receive architectures. The chapter deals mostly, but not exclusively, with coherent communication.

5.1 Definitions

Denoting by n_T and n_R the number of transmit and receive antennas, respectively, we shall abide by the frequency-flat complex vector model[†]

$$y = \sqrt{g}\, \mathbf{H}\mathbf{x} + \mathbf{n}, \tag{5.1}$$

where g is a deterministic scalar that represents the average channel gain, while the random matrix $\mathbf{H}$ is normalized to satisfy

$$E[\text{Tr}\{\mathbf{H}\mathbf{H}^\dagger\}] = n_T n_R. \tag{5.2}$$

[†]If the channel is frequency-selective, it can be decomposed into parallel noninteracting channels, each of which conforms to (5.1).

While the distribution of $\mathbf{H}$ is known to both transmitter and receiver, we shall specify when its realization is known to either.

The additive noise is modeled as Gaussian with i.i.d. (independent identically distributed) entries, and[†]

$$N_0 = \frac{E\left[\|\mathbf{n}\|^2\right]}{n_{\mathsf{R}}}.$$

The input is constrained such that

$$E\left[\|\mathbf{x}\|^2\right] = P, \tag{5.3}$$

with P the power available for transmission. The normalized input covariance is denoted by

$$\boldsymbol{\Phi}_{\mathbf{x}} = \frac{E[\mathbf{x}\mathbf{x}^\dagger]}{P/n_{\mathsf{T}}},$$

where (5.3) translates into $\mathrm{Tr}\{\boldsymbol{\Phi}_{\mathbf{x}}\} = n_{\mathsf{T}}$. If the input is isotropic, $\boldsymbol{\Phi}_{\mathbf{x}} = \mathbf{I}$.

5.2 Multiantenna capacity: the coherent realm

With transmit power P and bandwidth B, the rate R [bits/s] achievable with arbitrary reliability must obey the fundamental limit $R/B \leq C(\mathsf{SNR})$, where $C(\mathsf{SNR})$ is the capacity [bits/s/Hz], and

$$\mathsf{SNR} = g\frac{P}{N_0}.$$

Most existing results on multiantenna capacity apply to the coherent regime, where the channel realization is available at the receiver. A glimpse of several issues concerning the capacity of noncoherent multiantenna channels is given in Section 5.11.

With coherent reception, the input-output mutual information is maximized if the input $\mathbf{x}$ in (5.1) is circularly symmetric complex Gaussian. Conditioned on $\mathbf{H}$, the mutual information [bits/s/Hz] is

$$\mathcal{I}(\mathsf{SNR}, \boldsymbol{\Phi}_{\mathbf{x}}) = \log_2 \det\left(\mathbf{I} + \frac{\mathsf{SNR}}{n_{\mathsf{T}}}\mathbf{H}\boldsymbol{\Phi}_{\mathbf{x}}\mathbf{H}^\dagger\right). \tag{5.4}$$

Formula (5.4) was given in the mid 1990s by Foschini (1996) and Telatar (1999). Earlier embodiments of this formula and its generalizations were found by Pinsker (1964), Tsybakov (1965), Root and Varaiya (1968), Brandenburg and Wyner (1974), Verdú (1986), and Cover and Thomas (1990).

[†]Any nonsingular noise covariance can be absorbed into the correlation of $\mathbf{H}$. Section 5.9 deals specifically with noise that is colored and subject to fading.

Without knowledge of the realizations of the channel matrices at the transmitter, the capacity is

$$C(\text{SNR}) = \max_{\mathbf{\Phi_x}:\text{Tr}\{\mathbf{\Phi_x}\}=n_T} E[\mathcal{I}(\text{SNR}, \mathbf{\Phi_x})], \tag{5.5}$$

as long as the sequence of random matrices $\{\mathbf{H}_k\}$ is stationary and ergodic.

When the variation of $\{\mathbf{H}_k\}$ within the horizon of each codeword is not fast enough for the time-averaged mutual information to approach its statistical average, the ergodic capacity (5.5) loses its operational meaning, and the distribution (not just the average) of the mutual information (the so-called outage capacity) has to be considered. In addition to fast-fading channels, the distribution of the mutual information concentrates around its mean in wideband frequency-selective fading channels, and in channels with hybrid ARQ. The ergodic capacity is further operationally meaningful, even if the channel is invariant over the codeword, with achievable-rate feedback.

A particularly attractive ergodic channel, from which plenty of insight can be drawn, is the *canonical* multiantenna channel, where the matrix has i.i.d. Rayleigh-fading entries. The evaluation of the capacity of the canonical channel is facilitated by several facts:

- The isotropy of the capacity-achieving input distribution (Telatar, 1999).
- The availability of an explicit expression for the marginal distribution of the squared singular values of random matrices with Gaussian i.i.d. entries (Bronk, 1965; Telatar, 1999).
- The availability of an explicit asymptotic expression for the squared singular value distribution of matrices with i.i.d. entries whose size goes to infinity with a constant aspect ratio (Marčenko and Pastur, 1967). This asymptotic expression, furthermore, does not depend on the marginal distribution of the matrix entries.

Using these expressions, closed-form formulas for the capacity of the canonical channel were derived, first asymptotically in the numbers of antennas (Verdú and Shamai, 1999; Rapajic and Popescu, 2000) and then for arbitrary numbers thereof (Shin and Lee, 2003).

5.3 Input optimization

5.3.1 Transmitter side information

If the channel realization is known at the transmitter, capacity is achieved with an input covariance whose eigenvectors coincide at time k with those of $\mathbf{H}_k^\dagger \mathbf{H}_k$, and whose eigenvalues are obtained via waterfilling on those of $\mathbf{H}_k^\dagger \mathbf{H}_k$. This solution was first derived by Tsybakov (1965) for deterministic matrix channels, inspired by Shannon's original frequency-domain waterfilling approach (Shannon, 1949). It was rederived by several authors in the 1990s, specifically for multiantenna communication, once the interest on the topic was sparked (Raleigh and Cioffi, 1998; Telatar, 1999).

Depending on the time horizon over which the power is allowed to be averaged and the temporal dynamics of the problem, several solutions have been obtained for the power assigned to each channel realization:

- Temporal waterfilling (Goldsmith and Varaiya, 1997; Biglieri *et al.*, 2001).
- Dynamic programming (Negi and Cioffi, 2002).

The same techniques can be used if the variability takes place along the frequency (rather than time) axis.

Note that, in practice, discrete constellations such as PSK or QAM are used in lieu of the capacity-achieving Gaussian inputs. Not only does this reduce the mutual information, but power allocation via waterfilling on the channel singular values is no longer optimal (Lozano *et al.*, 2005b).

5.3.2 No transmitter side information

The optimization of the input becomes more problematic, and in fact it has not been fully solved, when the channel realizations are unknown to the transmitter. For the canonical channel (i.i.d. Rayleigh-fading entries), the intuitively appealing isotropy of the optimal input was proved by Telatar (1999). It is, however, frequently forgotten that isotropic inputs are not necessarily optimal for noncanonical channels. A popular nonisotropic signaling, which is a natural counterpart to the solution with transmit side information, is *eigenbeamforming*: signaling on the eigenvectors of $E[\mathbf{H}^\dagger \mathbf{H}]$.

Myth 1   *Regardless of the distribution of* $\mathbf{H}$, *eigenbeamforming achieves capacity.*

The optimality of inputs whose eigenstructure is that of $E[\mathbf{H}^\dagger \mathbf{H}]$ holds true for several important classes of channels such as Rayleigh fading with certain correlation structures (Jafar *et al.*, 2004; Jorswieck and Boche, 2004a) and Ricean fading (Venkatesan *et al.*, 2003; Hoesli and Lapidoth, 2004a). However, this principle fails to hold in general, as the following counterexample shows.

Example 5.1 *Given* $n_T = n_R = 3$, *let* **H** *take three possible realizations with respective probabilities* $q_1 = 0.4$, $q_2 = 0.58$ *and* $q_3 = 0.02$. *Specifically,*

$$\mathbf{H}_k = \mathbf{\Lambda}_k^{1/2} \mathbf{V}_k^{\dagger},$$

where $\mathbf{V}_3 = \mathbf{I}$ *while* $\mathbf{V}_1 = \mathbf{V}_2 = \mathbf{V} \neq \mathbf{I}$ *is a unitary matrix and*

$$\mathbf{\Lambda}_1 = \mathrm{diag}\{2.5, 2.5, 1.75\},$$
$$\mathbf{\Lambda}_2 = \mathrm{diag}\{0, 0, 0.5172\},$$
$$\mathbf{\Lambda}_3 = \mathrm{diag}\{15.2, 10.3, 8.1\}.$$

The eigenvectors of $E[\mathbf{H}^{\dagger}\mathbf{H}] = \mathrm{diag}\{1.304, 1.206, 1.162\}$ *are the columns of the identity, but an input covariance* $\mathbf{\Phi_x}$ *whose eigenvectors coincide with those of* **V** *can achieve strictly higher* $E[\mathcal{I}(\mathsf{SNR}, \mathbf{\Phi_x})]$.

The optimization of the eigenvalues of $\mathbf{\Phi_x}$ has only been explicitly solved at low SNR (Verdú, 2002b) and, for $n_T \leq n_R$, at high SNR (Hoesli and Lapidoth, 2004b; Tulino *et al.*, 2004). Beyond these limiting scenarios, the eigenvalues have only been characterized through fixed-point conditions that must be solved for iteratively (Hoesli and Lapidoth, 2004b; Tulino *et al.*, 2006).

A low-complexity approach consists of determining the eigenvalues of the input covariance matrix by *statistical waterfilling*, namely, a waterfilling on the eigenvalues of $E[\mathbf{H}^{\dagger}\mathbf{H}]$ (e.g., Ivrlac *et al.* (2003), Simeone and Spagnolini (2003)). For $n_R = 1$, this approach minimizes the pairwise error probability at high SNR (Zhou and Giannakis, 2003).

Myth 2 *The eigenvalues of the capacity-achieving input covariance are determined by statistical waterfilling.*

The capacity-achieving eigenvalues are in general not given by a waterfilling on the eigenvalues of any channel covariance matrix.

At low SNR, it is inviting to solve the optimization of $\mathbf{\Phi_x}$ by simply maximizing the first term in the corresponding series expansion of $\mathcal{I}(\mathsf{SNR}, \mathbf{\Phi_x})$, i.e., by solving the much simpler problem

$$C(\mathsf{SNR}) = \max_{\mathbf{\Phi_x}:\mathrm{Tr}\{\mathbf{\Phi_x}\}=n_T} \frac{\mathsf{SNR}}{n_T} E\left[\mathrm{Tr}\left\{\mathbf{H}\mathbf{\Phi_x}\mathbf{H}^{\dagger}\right\}\right] \log_2 e + o(\mathsf{SNR}). \qquad (5.6)$$

The cost function in (5.6), however, only reflects the energy at the output of the channel. There is no notion of bandwidth in the optimization, which leads to the following misconception.

Myth 3   *Rank-1 beamforming in the direction of the largest-eigenvalue eigenvector of $E[\mathbf{H}^\dagger\mathbf{H}]$ is always optimal at low SNR. If such eigenvalue is multiple, beamforming in the direction of any associated eigenvector is optimal.*

As the following example shows unequivocally, unless bandwidth is a free commodity, beamforming is not optimal (regardless of how low the SNR is) if the largest eigenvalue of $E[\mathbf{H}^\dagger\mathbf{H}]$ has plural multiplicity. For second-order optimality, in particular, the power should be uniformly divided among the associated eigenvectors. An extreme case is that of the canonical channel, where the multiplicity is n_T and the power has to be isotropically radiated.

Example 5.2   *(Verdú, 2002a) Consider n_T transmit and one receive antennas with i.i.d. channel entries. If B_ℓ is the bandwidth required to sustain rate R with power P and rank-ℓ signaling, then for vanishing P (and R)*

$$\frac{B_1}{B_{n_T}} = \frac{2\,n_T}{1+n_T}.$$

For $n_T = 2$, beamforming requires 1/3 more bandwidth than an isotropic input. For large n_T, it needs twice the bandwidth.

The suboptimality of beamforming is largely upheld even if the dominating eigenvalues of $E[\mathbf{H}^\dagger\mathbf{H}]$ are distinct but only modestly dissimilar: except at very low SNR, beamforming remains suboptimal. The precise SNR below which beamforming is strictly optimal when the eigenvalues of $E[\mathbf{H}^\dagger\mathbf{H}]$ are distinct has been determined for certain Rayleigh-fading channels (Simon and Moustakas, 2003; Jafar *et al.*, 2004; Jorswieck and Boche, 2004a).

The fallacy in Myth 3, revealed in Verdú (2002b), is evidenced by a low-SNR expansion of the capacity as function of

$$\frac{E_b}{N_0} = \frac{P/R}{N_0}, \qquad (5.7)$$

which, letting $x|_{3\,\mathrm{dB}} = (10\log_{10} x)/(10\log_{10} 2)$, yields

$$\mathsf{C}\left(\frac{E_b}{N_0}\right) = \max_{\boldsymbol{\Phi}_\mathbf{x}:\mathrm{Tr}\{\boldsymbol{\Phi}_\mathbf{x}\}=n_T} \left(\frac{E_b}{N_0}\Big|_{3\,\mathrm{dB}} - \frac{E_b}{N_0}_{\min}\Big|_{3\,\mathrm{dB}}\right) S_0 + \epsilon, \qquad (5.8)$$

where $(E_b/N_0)_{\min}$ is the minimum required energy per bit, while S_0 is the capacity slope therein, in bits/s/Hz/(3 dB), and ϵ is a lower-order term. Both key measures, $(E_b/N_0)_{\min}$ and S_0, are functions of $\boldsymbol{\Phi}_\mathbf{x}$.

The first-order optimization in (5.6) is tantamount to the minimization of $(E_b/N_0)_{\min}$, with no regard for S_0, and it leads to wrong conclusions regardless of how low the SNR is.

At high SNR, the following is a common misconception:

Myth 4 *Isotropic inputs are optimal at high* SNR.

The rationale that leads to Myth 4 is to approximate $C(\text{SNR})$ at high SNR as

$$C(\text{SNR}) = \max_{\mathbf{\Phi_x}:\text{Tr}\{\mathbf{\Phi_x}\}=n_\text{T}} E\left[\log_2 \det\left(\frac{\text{SNR}}{n_\text{T}}\mathbf{H}^\dagger\mathbf{H}\right) + \log_2 \det(\mathbf{\Phi_x})\right] + o(1),$$

and then conclude that, because of the concavity of $\log_2 \det(\cdot)$, the input covariance $\mathbf{\Phi_x} = \mathbf{I}$ is optimal. Reinforced by Myth 2 and the limiting uniformity of the waterfilling solution at high SNR, the statement in Myth 4 fails to hold if $\mathbf{H}^\dagger\mathbf{H}$ is singular with positive probability, as for example when $n_\text{T} > n_\text{R}$ (Hoesli and Lapidoth, 2004b; Tulino *et al.*, 2004).

Example 5.3 *Consider a Rayleigh-fading channel with $n_\text{T} = 2$ and $n_\text{R} = 1$, and with λ_1 and λ_2 the (nonzero) eigenvalues of the transmit correlation matrix. Denote by p_1 and p_2 the eigenvalues of $\mathbf{\Phi_x}$ that achieve capacity. In the high-*SNR *limit (Tulino* et al.*, 2004),*

$$p_1 = \lambda_2 - \frac{\lambda_1\lambda_2}{2}\log_e\frac{p_2\lambda_2}{p_1\lambda_1} \quad \text{and} \quad p_2 = \lambda_1 + \frac{\lambda_1\lambda_2}{2}\log_e\frac{p_2\lambda_2}{p_1\lambda_1}.$$

If $\lambda_1 \neq \lambda_2$, then $p_1 \neq p_2$.

Scant progress has been made in the optimization of the input covariances that maximize the rate for a given allowed outage probability. Even for the canonical channel, Telatar's conjecture (Telatar, 1999) that the power is to be equally distributed among a subset of the transmit antennas (that decreases as the outage probability increases) remains open.

5.4 Low SNR

In addition to Myth 3, a number of misleading observations can be made on the basis of a first-order expansion of the coherent capacity at low SNR. For the canonical channel,

$$C(\text{SNR}) = \text{SNR}\, n_\text{R} \log_2 e + o(\text{SNR}). \tag{5.9}$$

Fix R, B, and n_R. From (5.7), (5.8), (5.11), and (5.12), the increase in transmit power with $n_T = 1$ relative to arbitrary n_T is, in 3 dB units,

$$\Delta P|_{3\,dB} = \frac{R}{2B}\left(1 - \frac{1}{n_T}\right). \tag{5.10}$$

The first-order series expansion in (5.9) and the fact that (5.10) vanishes for $R/B \to 0$ buttress the following widespread misconception.

Myth 5 *At low* SNR, *the capacity is unaffected by* n_T.

While the minimum transmit energy per bit is indeed unaffected by n_T, the following example characterizes its impact on the low-SNR power-bandwidth trade-off within the simple context of the canonical channel.

Example 5.4 *(Verdú, 2002a) Denoting by* B_{n_T} *the bandwidth required to sustain rate* R *with power* P *and* n_T *transmit antennas, for vanishing* P,

$$\frac{B_1}{B_{n_T}} = n_T \frac{1 + n_R}{n_T + n_R}.$$

With $n_T = 1$, *we require* $(n_R + 1)/2$ *times the bandwidth needed with* $n_T = n_R$, *and* $(n_R + 1)$ *times the bandwidth needed with large* n_T.

This result is easily obtained from (5.8) with the minimum energy per bit and slope of the canonical channel:

$$\frac{E_b}{N_0}_{\min} = \frac{\log_e 2}{g\,n_R} \tag{5.11}$$

$$S_0 = 2\frac{n_T n_R}{n_T + n_R}. \tag{5.12}$$

Myth 5 stems from the fact that, in terms of power at low SNR, the value of having multiple antennas resides exclusively at the receiver[†] but it fails to recognize that transmit and receive antennas are equally valuable in terms of bandwidth efficiency at a given power.

Consider now a channel with correlated entries, given by

$$\mathbf{H} = \mathbf{\Theta}_R^{1/2}\mathbf{H}_w\mathbf{\Theta}_T^{1/2}, \tag{5.13}$$

where $\mathbf{\Theta}_T$ and $\mathbf{\Theta}_R$ are deterministic transmit and receive correlation matrices, while $\mathbf{H}_w$ is a canonical channel matrix. The first-order expansion in (5.6) yields

$$C(\text{SNR}) = \text{SNR}\,n_R\lambda_{\max}(\mathbf{\Theta}_T)\log_2 e + o(\text{SNR}). \tag{5.14}$$

[†]This stems directly from the fact that the captured power grows with n_R but not with n_T.

Myth 6 *At low* SNR, *receive correlation has no impact on the capacity.*

This misconception stems again from the fact that the notion of bandwidth is lacking in (5.6) and (5.14). The impact of correlation is indeed small (and it vanishes with the SNR) in terms of power increase for constant R and B, but it may be sizeable in terms of bandwidth expansion with constant R and P (Lozano *et al.*, 2003).

Example 5.5 *The bandwidth B_c required to sustain a rate R with power P and with receive correlation Θ_R relative to the bandwidth B_u required with no receive correlation is, for vanishing P*

$$\frac{B_c}{B_u} = \frac{n_R + \mathrm{Tr}\{\Theta_R^2\}/n_R}{n_R + 1},$$

which approaches 2 if n_R is large and the correlations therein strong.

5.5 High SNR

At high SNR, the capacity with coherent reception expands as

$$C(\text{SNR}) = S_\infty \, \text{SNR}|_{3 \text{ dB}} + O(1), \tag{5.15}$$

where S_∞ in bits/s/Hz/(3 dB) is known variously as the (maximum) multiplexing gain, the pre-log, the high-SNR slope, or the number of degrees of freedom. For the canonical channel,

$$S_\infty = \min(n_T, n_R),$$

an observation that ignited the enthusiasm in multiantenna communication (Foschini, 1996; Telatar, 1999).

Myth 7 *At high* SNR, *the multiantenna capacity is determined by S_∞.*

The reality is that almost all coherent multiantenna channels[†] have $S_\infty = \min(n_T, n_R)$ regardless how they deviate from the canonical model. Yet, the power required to achieve a given capacity is highly sensitive to antenna correlation, line-of-sight components, noise color, the fading distribution, etc. Pragmatically, at any given SNR a certain number of nonzero singular values of the channel matrix are "dormant" (i.e., well below the noise), and

---

[†]Possible exceptions are channels with singular correlation and the keyhole channel (whose matrix is the outer product of two vectors (Chizhik *et al.*, 2000)), where $S_\infty \leq \min(n_T, n_R)$.

thus at such SNR it is not the rank of the channel matrix that matters but the number of nonnegligible singular values (Tse and Viswanath, 2005). This more pragmatic interpretation of S_∞, nonetheless, still fails to capture the impact on required power at high SNR of various channel features. An expansion, which unlike (5.15), is able to capture such impact is (Shamai and Verdú, 2001; Tulino et al., 2004)

$$C(\text{SNR}) = S_\infty(\text{SNR}|_{3 \text{ dB}} - \mathcal{L}_\infty) + o(1), \tag{5.16}$$

where $\mathcal{L}_\infty$ represents the power offset, in 3 dB units, with respect to a reference channel having the same S_∞ but with unfaded and orthogonal dimensions (i.e., such that $(1/n_\text{T})\mathbf{HH}^\dagger = \mathbf{I}$). The power offset plays a chief role at SNR values of operational interest in current high spectral efficiency applications.

Myth 8 *At high SNR, antenna correlation is immaterial.*

Nonsingular antenna correlations have no effect on S_∞ but they do shift the power offset.

Example 5.6 *Consider a correlated channel represented by (5.13) with nonsingular $\mathbf{\Theta}_\text{T}$ and $\mathbf{\Theta}_\text{R}$, and with $n_\text{T} = n_\text{R}$. Denoting the power offset in the absence of correlation by $\mathcal{L}_\infty^{\text{i.i.d.}}$, the power penalty in 3-dB units incurred with correlation is*

$$\mathcal{L}_\infty - \mathcal{L}_\infty^{\text{i.i.d.}} = -\frac{1}{n_\text{T}} \sum_{\ell=1}^{n_\text{T}} \log_2 \lambda_\ell(\mathbf{\Theta}_\text{T}) - \frac{1}{n_\text{R}} \sum_{\ell=1}^{n_\text{R}} \log_2 \lambda_\ell(\mathbf{\Theta}_\text{R}),$$

where $\lambda_\ell(\cdot)$ indicates the ℓth eigenvalue of a matrix. If some of the eigenvalues of $\mathbf{\Theta}_\text{T}$ or $\mathbf{\Theta}_\text{R}$ are small, this power penalty can be arbitrarily large.

In nonergodic channels, it is sometimes desirable to sacrifice some rate in exchange for spatial diversity, so as to ensure a faster decay of the error probability. (In the presence of other diversity mechanisms or when enough delay is tolerable, the need for spatial diversity is far less acute and approaching capacity is the primary goal.) In the high SNR region, the trade-off between degrees of freedom (or multiplexing gain) and diversity was established by Zheng and Tse (2003) for the canonical channel. Note that, at the point of zero diversity, the multiplexing gain attains its maximum value S_∞. For the reasons discussed above, the diversity-multiplexing trade-off does not completely capture the high SNR behavior for noncanonical channels as it neglects the power offset.

5.6 Antenna correlation

In addition to the low- and high-SNR regions, explicit expressions for the coherent capacity of Rayleigh-fading channels whose correlation conforms to (5.13) have appeared in the literature (Kiessling and Speidel, 2004). The capacity for large numbers of antennas has also been reported as the solution of a fixed-point equation, for the correlation structure in (5.13) by Moustakas *et al.* (2000), Mestre *et al.* (2003), and for more general correlations by Tulino *et al.* (2003). All these results, however, require a separate optimization of the eigenvalues of the input covariance.

For $n_R = 1$, with the channel realization known at the transmitter, it has been shown that antenna correlation lowers capacity at any SNR (Jorswieck and Boche, 2004b).[†] However, this is not true in general.

Myth 9 *Antenna correlation is detrimental at any* SNR.

The mutual information achievable with isotropic inputs is indeed lowered by antenna correlation (Jorswieck and Boche, 2004b). When the capacity is achieved by nonisotropic inputs, in contrast, correlation need not be detrimental. The effect is easily assessed when the correlations obey (5.13):

- Transmit correlation reduces the effective dimensionality of the transmitter, but it also enables power focusing. The net effect depends on the SNR and the ratio between n_T and n_R. Below a certain SNR, correlation is always advantageous; above, it is unfavorable if $n_T \leq n_R$, but it may be again advantageous if $n_T > n_R$.[‡]
- Receive correlation reduces the effective dimensionality of the receiver without increasing the captured power and thus it lowers the mutual information at any SNR.

5.7 Ricean fading

The computation of the capacity of Ricean channels is substantially more involved than that of Rayleigh-fading channels, and fewer results are available (Kang and Alouini, 2002; Lebrun *et al.*, 2004; Alfano *et al.*, 2004).

Consider the Ricean channel

$$\mathbf{H} = \sqrt{\frac{K}{K+1}}\bar{\mathbf{H}} + \sqrt{\frac{1}{K+1}}\mathbf{H}_w, \tag{5.17}$$

[†]This also holds if the transmitter not only does not know the channel realization but only knows that the channel distribution satisfies certain symmetry constraints.

[‡]Transmit correlation is also beneficial at every SNR in the case of noncoherent reception (Jafar and Goldsmith, 2003).

where $\bar{\mathbf{H}}$ is deterministic while $\mathbf{H}_w$ is a canonical channel matrix and K is the Ricean K-factor. If $n_\mathsf{T} = n_\mathsf{R} = 1$, the capacity increases monotonically with K. In the multiantenna arena, however, line-of-sight components are sometimes perceived as being detrimental to the capacity.

Myth 10 *The presence of a line-of-sight component reduces the capacity of a multiantenna channel.*

If $\mathbf{H}$ is normalized as per (5.2), then, depending on the singular values of $\bar{\mathbf{H}}$, we can find instances where the capacity is either improved or degraded by the line-of-sight component. The following examples illustrate these different behaviors:

Example 5.7 *Let $\bar{\mathbf{H}}$ be unit rank, and let $n_\mathsf{T} = n_\mathsf{R} = 2$. At high SNR, the capacity behaves as (5.16) with $S_\infty = 2$ and with (Lozano et al., 2005a)*

$$\mathcal{L}_\infty(K) = 1 + \log_2 \frac{K+1}{2\sqrt{K}} - \frac{\log_2 e}{2}\left(\mathrm{E}_1(4K) - \gamma + \frac{1 - e^{-4K}}{4K}\right), \quad (5.18)$$

where γ is Euler's constant, and $\mathrm{E}_1(z) = \int_1^\infty e^{-z\xi}/\xi\, d\xi$ is an exponential integral. The power offset in (5.18) increases monotonically with K and for large K it grows as $(1/2)\log_2 K$. With $K = 100$, for instance, the excess power offset due to the line-of-sight component amounts to 7.9 dB.

Example 5.8 *For the full-rank line-of-sight matrix*

$$\bar{\mathbf{H}} = \begin{bmatrix} 1 & 0.4 \\ 0.9 & 0.1 \\ 0.2 & -2 \end{bmatrix}$$

it can be verified, using the expressions in Lozano et al. (2005a), that

$$\mathcal{L}_\infty(5) = \mathcal{L}_\infty(0) - 0.33,$$

or about a 1 dB gain at high SNR thanks to a $K = 5$ line-of-sight component.

Notice that, by using (5.17), we have evaluated the impact of the line-of-sight component at a fixed SNR. This requires that, in the presence of a line-of-sight component, the power received over the fading portion of the channel declines. Alternatively, we can consider the modified Ricean channel

$$\mathbf{H} = \sqrt{K}\bar{\mathbf{H}} + \mathbf{H}_w, \quad (5.19)$$

which, in general, does not satisfy (5.2). The addition of a line-of-sight component does not alter the faded power, and the corresponding capacity increases monotonically with K (Hoesli and Lapidoth, 2004a). Therefore, the behavior is seen to depend critically on the channel model.

As both n_T and n_R go to infinity with constant ratio, the change in capacity caused by a rank-1 line-of-sight component takes a remarkably simple form. Making explicit the dependence of the mutual information on K, for the model in (5.17)

$$\lim_{n_T, n_R \to \infty} \frac{\mathcal{I}(\mathsf{SNR}, K)}{n_R} = \lim_{n_T, n_R \to \infty} \frac{\mathcal{I}(\frac{\mathsf{SNR}}{K+1}, 0)}{n_R}, \tag{5.20}$$

whereas, for the model in (5.19),

$$\lim_{n_T, n_R \to \infty} \frac{\mathcal{I}(\mathsf{SNR}, K)}{n_R} = \lim_{n_T, n_R \to \infty} \frac{\mathcal{I}(\mathsf{SNR}, 0)}{n_R}. \tag{5.21}$$

In either case, the mutual information behaves as if the line-of-sight component were absent. This is a direct manifestation of the fact that only a single eigenvalue of $\mathbf{H}\boldsymbol{\Phi}_\mathsf{x}\mathbf{H}^\dagger$ is perturbed by the line-of-sight component. Asymptotically, this perturbation is not reflected in the empirical eigenvalue distribution of $\mathbf{H}\boldsymbol{\Phi}_\mathsf{x}\mathbf{H}^\dagger$, which determines the mutual information.[†]

5.8 Asymptotic analysis

Asymptotic analyses as n_T and n_R go to infinity with a constant ratio are often feasible using results in random matrix theory (Tulino and Verdú, 2004) and provide valuable engineering insights. Furthermore, the number of antennas required for the asymptotics to be closely approached are often very small. However, this principle should not be taken for granted without numerical verification. For example, in the scenario considered in the last paragraph of Section 5.7 the limits therein are approached very slowly.

Myth 11 *Asymptotic capacity results are always closely approached for small numbers of antennas.*

If either n_T or n_R is held fixed while the other dimension grows without bound, application of the strong law of large numbers to a channel with zero-mean i.i.d. entries yields

$$\frac{1}{n_R}\mathbf{H}^\dagger\mathbf{H} \overset{a.s.}{\to} \mathbf{I} \qquad (n_R \to \infty)$$

$$\frac{1}{n_T}\mathbf{H}\mathbf{H}^\dagger \overset{a.s.}{\to} \mathbf{I} \qquad (n_T \to \infty).$$

[†]The behavior in (5.20) and (5.21) extends to channels whose line-of-sight component has rank $r > 1$ as long as $\lim_{n_R \to \infty} r/n_R = 0$.

In these asymptotes, the capacity of the canonical channel (and of channels that can be expressed as function thereof) becomes particularly simple. If both n_T and n_R go to infinity with constant ratio, the (i, j) entries of both $(1/n_R)\mathbf{H}^\dagger\mathbf{H}$ and $(1/n_T)\mathbf{H}\mathbf{H}^\dagger$ also converge almost surely (a.s.) to δ_{i-j} for any fixed pair (i, j). The empirical eigenvalue distribution, however, does not converge to a mass at 1, but rather to the Marčenko–Pastur law (Marčenko and Pastur, 1967). Integration over this asymptotic distribution yields the corresponding asymptotic capacity (normalized by the number of antennas) in closed form (Verdú and Shamai, 1999; Rapajic and Popescu, 2000).

For most channels, the mutual information normalized by the number of antennas converges a.s. to its expectation asymptotically.[†] In contrast, the unnormalized mutual information, which determines the outage capacity, still suffers from nonvanishing random fluctuations. Numerous authors have observed (through simulation) that the distribution of the unnormalized mutual information resembles a Gaussian law once n_T and n_R become large. Asymptotic normality has been rigorously established (Tulino and Verdú, 2005) for arbitrary SNR in the presence of correlation at either transmitter or receiver as a direct application of recent random matrix results (Bai and Silverstein, 2004). For other multiantenna channels, asymptotic normality has been conjectured (Moustakas *et al.*, 2000; Kang and Alouini, 2003).

5.9 Intercell and multiuser issues

A popular way to deal with multiuser interference in general, and out-of-cell interference in particular, is to model it as white Gaussian noise. It is well known that, for single-antenna systems, such a model leads to severe capacity underestimations. This conclusion holds when the interference emanates from multiantenna arrays and the receiver is also equipped with an array.

Myth 12 *Out-of-cell interference can be modeled as additional white Gaussian noise.*

It is possible to effectively exploit the structure of the out-of-cell interference even if the codebooks used in other cells are unknown. To that end, consider the scenario where the MIMO channel from the strongest out-of-cell base station array to an in-cell mobile array is known (e.g., through pilot monitoring). The following example illustrates the kind of improvements achievable in the low-SNR regime by taking this knowledge into account.

[†]Exceptions are channels with finite-rank correlations and the keyhole channel.

Example 5.9 *Let E_b and E_b^{awgn} be the minimum energy per bit achievable in the presence of one out-of-cell interferer (and no thermal noise), and in the presence of white Gaussian noise of identical power, respectively. Both the desired transmitter and the interferer have n_T antennas. From Lozano et al. (2003), if $n_T > n_R$, then*

$$\frac{E_b}{E_b^{\text{awgn}}} = 1 - \frac{n_R}{n_T}. \tag{5.22}$$

Thus, for $n_T = 2\,n_R$, $(E_b/N_0)_{\min}$ is 3 dB lower than if the interference was white Gaussian noise with the same power. For $n_T \leq n_R$, the $(E_b/N_0)_{\min}$ is the same as if the interference did not exist.

In the high-SNR regime, with a large ratio between in-cell and out-of-cell powers, the following evidences how the power offset can be considerably reduced by exploiting the structure of the interference.

Example 5.10 *Let $n_T = n_R = n$, and denote by $\mathcal{L}_\infty^{\text{awgn}}$ the high-SNR power offset with white Gaussian noise. In the presence of one interferer equipped with n_T antennas (and no thermal noise),*

$$\mathcal{L}_\infty = \mathcal{L}_\infty^{\text{awgn}} + \log_2 n + \left(\gamma - \sum_{\ell=2}^{n} \frac{1}{\ell} \right) \log_2 e$$

in 3 dB units. For $n = 4$, for instance, the difference in power required to achieve a certain capacity approaches 3.8 dB as the SNR grows.

Another important concept in multiuser problems is the principle of multiuser diversity. One of its earliest embodiments (for a flat-fading single-antenna setting) is that, to maximize the uplink sum capacity with fading known to the scheduler, only the strongest user should be allowed to transmit (Knopp and Humblet, 1995). Likewise for the downlink, the sum capacity is maximized by transmitting only to the strongest user (Tse, 1999), a result that was a central theme in the design of CDMA2000® 1xEV-DO and UMTS-HSDPA third generation data-only wireless systems. Time sharing on the strongest user ceases to be optimal in a multiantenna setting. A less coarse misconception, but perhaps more generalized, is

---

Myth 13 *In either uplink or downlink, it is optimal to schedule as many simultaneous transmissions as the number of antennas at the base station.*

This belief may originate partially from Myth 7 and from the fact that, with a linear decorrelating receiver, each additional antenna enables to maintain the same performance, but with one additional interferer (e.g., Verdú, 1998, Section 5.7.2).

For both uplink and downlink with an n_T-antenna base station, Yu and Rhee (2004) show that it is strictly suboptimal to transmit from or to only n_T mobiles if $n_T > 1$. A (generally loose) bound in Yu and Rhee (2004) indicates that maximizing the sum rate may entail scheduling up to n_T^2 users at once.

Despite the fact that scheduling up to n_T users is strictly suboptimal, it generally incurs a small loss in sum capacity. For example, consider an uplink with four statistically equivalent Rayleigh-fading mobiles and a base station with two independently faded antennas. To achieve a sum capacity of 7.5 bits/s/Hz, the optimum policy schedules more than two simultaneous users in 55% of the realizations. Using the suboptimal policy that allows only the two strongest users to transmit, achieving that same sum capacity requires an additional 0.25 dB power expenditure per user, in addition to incurring higher latency and reduced fairness.

5.10 Transmit-receive architectures

One of the earliest multiantenna transmit-receive techniques is the widely popular Alamouti scheme designed for $n_T = 2$ (Alamouti, 1998).

Myth 14 *In conjunction with a scalar code, the Alamouti scheme achieves capacity if $n_T = 2$ and $n_R = 1$.*

A signal constructed according to the Alamouti scheme is spatially isotropic. Accordingly, the statement in Myth 14 fails to hold for those channels where the identity input covariance is suboptimal. For those channels, a linear interface should be inserted to properly color the signal.

Another technique that helped propel the interest in multiantenna communication is the BLAST layered architecture, which, in both its diagonal and vertical versions (D-BLAST and V-BLAST), was designed to operate with scalar codes and without transmit side information (Foschini, 1996; Foschini *et al.*, 1999).

Myth 15 *The vertical layered architecture (V-BLAST) cannot achieve the capacity of the canonical channel.*

If the rates at which the various transmit antennas operate are not constrained to be identical, and these rates can be communicated to the transmitter, then those rates can be chosen such that V-BLAST with MMSE filtering achieves capacity (Ariyavisitakul, 2000; Chung *et al.*, 2004); see also Varanasi and Guess (1998). By feeding the independently encoded signals onto appropriate signaling eigenvectors, the optimality of V-BLAST can be extended to noncanonical channels. In contrast, a vertically layered architecture cannot attain the diversity-multiplexing trade-off of Zheng and Tse (2003). D-BLAST, on the other hand, can achieve both capacity and the diversity-multiplexing trade-off.

5.11 Noncoherent communication

When the sequence of channel matrices $\mathbf{H}_k$ is not known exactly at the receiver, the study of the capacity is much more challenging, and neither a counterpart to the log-determinant formula nor the capacity-achieving signaling are yet known, even for $n_T = n_R = 1$.

At high SNR, the analysis of the capacity of both scalar and multiantenna noncoherent channels has received considerable attention, under various models of the fading dynamics. For scalar channels with memoryless fading, Taricco and Elia (1997) showed that the capacity grows as

$$C(\text{SNR}) = \log_2 \log \text{SNR} + O(1), \tag{5.23}$$

a behavior that has been shown to hold for a broad class of scalar and vector channels where the fading coefficients cannot be perfectly predicted from their past (Lapidoth and Moser, 2003). According to (5.23), transmit power is essentially wasted in the noncoherent high-SNR regime, suggesting that the system perhaps ought to be redesigned to operate in a wider-bandwidth, lower-SNR regime. In contrast, when the fading coefficients are perfectly predictable from their past, the capacity grows—as in the coherent regime—logarithmically with SNR (Lapidoth, 2003, 2005). One such noncoherent model, which has attracted much attention in the literature, is the nonstationary block fading model, where the channel coefficients remain constant for blocks of T symbols (Marzetta and Hochwald, 1999). In that case,

Zheng and Tse (2002) showed

$$S_\infty = m \left(1 - \frac{m}{T}\right),$$

with $m = \min(n_R, n_T, \lfloor T/2 \rfloor)$. Thus, although slight modeling variations usually lead to small disparities in capacity at any given SNR, for SNR $\to \infty$ the qualitative behavior of the capacity is quite sensitive to the model.

One could espouse the view that every wireless channel is noncoherent and that any measure (such as pilot symbols) taken to provide the receiver with an estimate of the channel should be modeled as part of the input. This view is particularly acute at low SNR, where the channel coefficients are harder to estimate (Rao and Hassibi, 2004).

Myth 16 *At low SNR, coherent communication is not feasible.*

Although, as SNR $\to 0$, the channel can indeed be learned with diminishing precision, coherent operation also requires diminishing accuracy (Lapidoth and Shamai, 2002). The classical analysis of capacity in the low-power regime predicts that, regardless of whether coherence is feasible or not, it is useless, since it does not decrease the minimum energy per bit. In contrast, the view propounded by Verdú (2002b) is that in the low-power regime coherence plays a key role in reducing the bandwidth required for reliable communication. In terms of the power-bandwidth trade-off, a quasi-coherent channel (known with little error at the receiver) can indeed be judiciously analyzed as a coherent channel using the tools developed by Verdú (2002b).

Acknowledgment

The authors gratefully acknowledge valuable comments from Giuseppe Caire, Jerry Foschini, Andrea Goldsmith, Babak Hassibi, Shlomo Shamai, David Tse, and Wei Yu.

References

Alamouti, S. M. (1998). A simple transmit diversity technique for wireless communications. *IEEE J. Sel. Areas Commun.*, **16** (8), 1451–1458.

Alfano, G., Lozano, A., Tulino, A. M., and Verdú, S. (2004). Mutual information and eigenvalue distribution of MIMO Ricean channels. In *Proc. Int. Symp. Inf. Theory and its Applications*, 1040–1045.

Ariyavisitakul, S. L. (2000). Turbo space-time processing to improve wireless channel capacity. *IEEE Trans. Commun.*, **48** (8), 1347–1358.

Bai, Z. D. and Silverstein, J. W. (2004). CLT of linear spectral statistics of large dimensional sample covariance matrices. *Ann. Probab*, **32** (1), 553–605.

Biglieri, E., Caire, G., and Taricco, G. (2001). Limiting performance of block-fading channels with multiple antennas. *IEEE Trans. Inf. Theory*, **47** (4), 1273–1289.

Brandenburg, L. H. and Wyner, A. D. (1974). Capacity of the Gaussian channel with memory: The multivariate case. *Bell Syst. Tech. J.*, **53** (5), 745–778.

Bronk, B. V. (1965). Exponential ensembles for random matrices. *J. Math. Physics*, **6** (2), 228–237.

Chizhik, D., Foschini, G. J., and Valenzuela, R. A. (2000). Capacities of multi-element transmit and receive antennas: Correlations and keyholes. *IEE Electr. Lett.*, **36** (13), 1099–1100.

Chung, S. T., Lozano, A., Huang, H. C., Sutivong, A., and Cioffi, J. M. (2004). Approaching the MIMO capacity with a low-rate feedback channel in V-BLAST. *EURASIP J. Appl. Signal Process.*, **2004** (5), 762–771.

Cover, T. M. and Thomas, J. A. (1990). *Elements of Information Theory* (New York, Wiley).

Foschini, G. J. (1996). Layered space-time architecture for wireless communications in a fading environment when using multi-element antennas. *Bell Labs Tech. J.*, **1** (2), 41–59.

Foschini, G. J., Golden, G. D., Valenzuela, R. A., and Wolnianski, P. W. (1999). Simplified processing for high spectral efficiency wireless communication employing multi-element arrays. *IEEE J. Sel. Areas Commun.*, **17** (11), 1841–1852.

Goldsmith, A. and Varaiya, P. (1997). Capacity of fading channels with channel side information. *IEEE Trans. Inf. Theory*, **43** (6), 1986–1997.

Hoesli, D. and Lapidoth, A. (2004a). The capacity of a MIMO Ricean channel is monotonic in the singular values of the mean. In *5th Int. ITG Conf. Source and Channel Coding, Erlangen, Germany*.

—— (2004b). How good is an isotropic input on a MIMO Ricean channel? In *Proc. IEEE Int. Symp. Inf. Theory*, 291.

Ivrlac, M. T., Utschick, W., and Nossek, J. A. (2003). Fading correlations in wireless MIMO communication systems. *IEEE J. Sel. Areas Commun.*, **21** (5), 819–828.

Jafar, S. A. and Goldsmith, A. J. (2003). Multiple-antenna capacity in correlated Rayleigh fading with channel covariance information. In *Proc. IEEE Int. Symp. Inf. Theory*, 470.

Jafar, S. A., Vishwanath, S., and Goldsmith, A. J. (2004). Transmitter optimization and optimality of beamforming for multiple antenna systems. *IEEE Trans. Wireless Commun.*, **3** (4), 1165–1175.

Jorswieck, E. and Boche, H. (2004a). Channel capacity and capacity-range of beamforming in MIMO wireless systems under correlated fading with covariance feedback. *IEEE Trans. Wireless Commun.*, **3** (5), 1543–1553.

—— (2004b). Optimal transmission strategies and impact of correlation in multiantenna systems with differeny types of channel state information. *IEEE Trans. Signal Process.*, **52** (12), 3440–3453.

Kang, M. and Alouini, M. S. (2002). On the capacity of MIMO Rician channels. In *Proc. 40th Allerton Conf. Commun., Contr., Comput.*, 936–945.

—— (2003). Impact of correlation on the capacity of MIMO channels. In *Proc. IEEE Int. Conf. Commun.*, vol. 4, 2623–2627.

Kiessling, M. and Speidel, J. (2004). Exact ergodic capacity of MIMO channels in correlated Rayleigh fading environments. In *Proc. Int. Zurich Seminar Commun.*, 128–131.

Knopp, R. and Humblet, P. A. (1995). Information capacity and power control in single-cell multi-user communications. In *Proc. IEEE Int. Conf. Commun.*, 331–335.

Lapidoth, A. (2003). On the high SNR capacity of stationary Gaussian fading channels. In *Proc. Allerton Conf. Commun., Contr., Comput.*, 410–419.

—— (2005). On the asymptotic capacity of stationary Gaussian fading channels. *IEEE Trans. Inf. Theory*, **51** (2), 437–446.

Lapidoth, A. and Moser, S. M. (2003). Capacity bounds via duality with applications to multi-antenna systems on flat-fading channels. *IEEE Trans. Inf. Theory*, **49** (10), 2426–2467.

Lapidoth, A. and Shamai, S. (2002). Fading channels: How perfect need "perfect side information" be? *IEEE Trans. Inf. Theory*, **48** (5), 1118–1134.

Lebrun, G., Faulkner, M., Shafi, M., and Smith, P. J. (2004). MIMO Ricean channel capacity. In *Proc. IEEE Int. Conf. Commun.*, 2939–2943.

Lozano, A., Tulino, A. M., and Verdú, S. (2003). Multiple-antenna capacity in the low-power regime. *IEEE Trans. Inf. Theory*, **49** (10), 2527–2544.

—— (2005a). High-SNR power offset in multi-antenna Ricean channels. In *Proc. IEEE Int. Conf. Commun.*, vol. 1, 683–687.

—— (2005b). Mercury/waterfilling: Optimum power allocation with arbitrary input constellations. In *Proc. IEEE Int. Symp. Inf. Theory*, 1773–1777.

Marčenko, V. A. and Pastur, L. A. (1967). Distributions of eigenvalues for some sets of random matrices. *Math USSR-Sbornik*, **1**, 457–483.

Marzetta, T. L. and Hochwald, B. H. (1999). Capacity of a mobile multiple-antenna communication link in Rayleigh flat fading. *IEEE Trans. Inf. Theory*, **45** (1), 139–157.

Mestre, X., Fonollosa, J. R., and Pagès-Zamora, A. (2003). Capacity of MIMO channels: asymptotic evaluation under correlated fading. *IEEE J. Sel. Areas Commun.*, **21** (5), 829–838.

Moustakas, A., Baranger, H., Balents, L., Sengupta, A., and Simon, S. (2000). Communication through a diffusive medium: Coherence and capacity. *Science*, **287** (5451), 287–290.

Negi, R. and Cioffi, J. M. (2002). Delay-constrained capacity with causal feedback. *IEEE Trans. Inf. Theory*, **48** (9), 2478–2494.

Pinsker, M. S. (1964). *Information and Information Stability of Random Variables and Processes* (Holden-Day, San Francisco, CA).

Raleigh, G. and Cioffi, J. M. (1998). Spatio-temporal coding for wireless communications. *IEEE Trans. Commun.*, **46** (3), 353–366.

Rao, C. and Hassibi, B. (2004). Analysis of multiple-antenna wireless links at low SNR. *IEEE Trans. Inf. Theory*, **50** (9), 2123–2130.

Rapajic, P. and Popescu, D. (2000). Information capacity of a random signature multiple-input multiple-output channel. *IEEE Trans. Commun.*, **48** (8), 1245–1248.

Root, W. L. and Varaiya, P. P. (1968). Capacity of classes of Gaussian channels. *SIAM J. Appl. Math.*, **16** (6), 1350–1393.

Shamai, S. and Verdú, S. (2001). The impact of frequency-flat fading on the spectral efficiency of CDMA. *IEEE Trans. Inf. Theory*, **47** (4), 1302–1327.

Shannon, C. E. (1949). Communication in the presence of noise. *Proc. IRE*, **37** (1), 10–21.

Shin, H. and Lee, J. H. (2003). Capacity of multiple-antenna fading channels: Spatial fading correlation, double scattering and keyhole. *IEEE Trans. Inf. Theory*, **49** (10), 2636–2647.

Simeone, O. and Spagnolini, U. (2003). Combined linear pre-equalization and BLAST equalization with channel correlation feedback. *IEEE Commun. Lett.*,

7 (10), 487–489.

Simon, S. H. and Moustakas, A. L. (2003). Optimizing MIMO antenna systems with channel covariance feedback. *IEEE J. Sel. Areas Commun.*, **21** (3), 406–417.

Taricco, G. and Elia, M. (1997). Capacity of fading channels with no side information. *IEE Electr. Lett.*, **33** (16), 1368–1370.

Telatar, I. E. (1999). Capacity of multi-antenna Gaussian channels. *Eur. Trans. Telecommun.*, **10**, 585–595.

Tse, D. (1999). Forward-link multiuser diversity through rate adaptation and scheduling.

Tse, D. and Viswanath, P. (2005). *Fundamentals of Wireless communications* (Cambridge Univ. Press).

Tsybakov, B. S. (1965). The capacity of a memoryless Gaussian vector channel. *Probl. Inf. Transmission*, **1**, 18–29.

Tulino, A., Lozano, A., and Verdú, S. (2004). High-SNR power offset in multiantenna communication. In *Proc. IEEE Int. Symp. Inf. Theory*, 288.

—— (2006). Capacity-achieving input covariance for single-user multi-antenna channels. *IEEE Trans. Wireless Commun.* In press.

Tulino, A. and Verdú, S. (2004). Random matrix theory and wireless communications. *Foundations & Trends in Commun. and Inf. Theory*, **1** (1), 1–182.

—— (2005). Asymptotic outage capacity of multiantenna channels. In *Proc. IEEE Int. Conf. Acoust., Speech, Signal Process.*, 825–828.

Tulino, A., Verdú, S., and Lozano, A. (2003). Capacity of antenna arrays with space, polarization and pattern diversity. In *Proc. IEEE Inf. Theory Workshop*.

Varanasi, M. K. and Guess, T. (1998). Optimum decision-feedback multiuser equalization with successive decoding achieves the total capacity of the Gaussian multiple-access channel. In *Proc. Asilomar Conf. Signals, Syst., Comput.*, 1405–1409.

Venkatesan, S., Simon, S. H., and Valenzuela, R. A. (2003). Capacity of a Gaussian MIMO channel with nonzero mean. In *Proc. IEEE Vehic. Tech. Conf.*, vol. 3, 1767–1771.

Verdú, S. (1986). Capacity region of Gaussian CDMA channels: The symbol synchronous case. In *Proc. Allerton Conf. Commun., Contr., Comput.*, 1025–1034.

—— (1998). *Multiuser Detection* (Cambridge Univ. Press).

—— (2002a). Saving bandwidth in the wideband regime. Plenary talk, Intl. Symp. Inf. Theory.

—— (2002b). Spectral efficiency in the wideband regime. *IEEE Trans. Inf. Theory*, **48** (6), 1319–1343.

Verdú, S. and Shamai, S. (1999). Spectral efficiency of CDMA with random spreading. *IEEE Trans. Inf. Theory*, **45** (3), 622–640.

Yu, W. and Rhee, W. (2004). Degrees of freedom in multi-user spatial multiplex systems with multiple antennas. *IEEE Trans. Commun.* Submitted.

Zheng, L. and Tse, D. (2003). Diversity and multiplexing: A fundamental tradeoff in multiple antenna channels. *IEEE Trans. Inf. Theory*, **49** (5), 1073–1096.

Zheng, L. and Tse, D. N. C. (2002). Communication on the Grassmann manifold: A geometric approach to the noncoherent multiple-antenna channel. *IEEE Trans. Inf. Theory*, **48** (2), 359–383.

Zhou, S. and Giannakis, G. B. (2003). Optimal transmitter eigen-beamforming and space-time block coding based on channel correlations. *IEEE Trans. Inf. Theory*, **49** (7), 1673–1690.

6

The role of feedback, CSI, and coherence in MIMO systems

Gwen Barriac

Qualcomm, Inc.

Noah Jacobsen and Upamanyu Madhow

University of California, Santa Barbara

6.1 Introduction

The growth in wireless communication over the past decade has been fueled by the demand for high-speed wireless data, in addition to the basic cellular telephony service that is now an indispensable part of our lives. Cellular operators are upgrading their networks to support higher data rates, and the imminent completion of the 802.16 and 802.20 standards is precipitating the move toward ubiquitous broadband wireless access. Increasing the capacity of current wireless links is perhaps the most essential step in realizing the vision of high-speed wireless data on demand, and adding multiple antennas at both the transmitter and the receiver is known to dramatically increase capacity. In this chapter, we explore the role of channel knowledge at the transmitter in multiple-input multiple-output (MIMO) systems. While feedback produces marginal gains in single-antenna communication, even partial channel knowledge at the transmitter is known to produce large performance gains in MIMO systems. We also consider the benefits of partial channel knowledge at the receiver in noncoherent systems.

For indoor wireless local area network (WLAN) systems with MIMO capabilities, such as the BLAST prototype, and emerging 802.11n standards efforts, the system bandwidth is typically smaller than the channel coherence bandwidth, which is large because of small indoor delay spreads. On the other hand, emerging high-speed outdoor wireless metropolitan area network (WMAN) communication systems, such as 802.16 and 802.20, can easily span a band that is several times the channel coherence bandwidth, which is smaller due to larger delay spreads in outdoor channels. Moreover, the angular spread in paths from transmitter to receiver in outdoor channels is often much smaller than for indoor channels because of the typically high elevation of the base station. Thus, outdoor spatial channels have a relatively small number of dominant spatial modes, and can therefore benefit more from channel knowledge at the transmitter regarding these modes. Our emphasis in this chapter, therefore, is on outdoor communication systems, where implicit or explicit feedback regarding the channel is expected to be most effective. In particular, we show that certain types of channel state information (CSI) can be obtained robustly and without training overhead in wideband systems. Such "implicit" feedback is particularly useful for outdoor channels, where it leads to both large performance gains *and* simpler transceivers.

An important design consideration is the exploitation of the asymmetry inherent in outdoor cellular or fixed wireless applications, where the base station is significantly more capable than the subscriber unit. Specifically, the base station can potentially have a large number of antennas, whereas the subscriber unit may have no more than one or two, and the base station is capable of more complex signal processing. Thus, subscriber units with a single antenna lead to the important special cases of multiple-input single-output (MISO) models for downlink communication, and single-input multiple-output (SIMO) models for uplink communication. For concreteness, we focus on orthogonal frequency-division multiplexing (OFDM), which has been designated as the physical layer in emerging outdoor WMAN standards such as 802.16a and 802.20, as well as indoor WLAN standards such as 802.11n. Our approach is to characterize information-theoretic limits, with the understanding that rapid advances in turbo-like coded modulation have brought such limits within reach using relatively standard architectures.

6.1.1 Feedback in narrowband systems

When both transmitter and receiver have perfect knowledge of the channel, the seminal work by Telatar (1995) shows that the capacity-achieving transmit strategy is to send independent Gaussian symbols along the channel eigenvectors, with the power on each symbol being determined by the classical "waterfilling" solution. While the resulting capacity gains over single-antenna channels are impressive (Telatar, 1995; Biglieri *et al.*, 2001), in practice, perfect channel knowledge is difficult to obtain.

Receiver CSI In many scenarios, the transmitter can send sufficiently many training, or pilot, symbols such that the receiver can accurately estimate the channel. Hence, the approximation that the receiver has perfect CSI is often reasonable, especially for downlink communication, where a common pilot can be employed for channel estimation by a large number of subscribers. We revisit this standard assumption of *coherent* reception in

Section 6.4, which focuses on uplink communication, in which pilots cannot be shared and are, therefore, more expensive.

Transmitter CSI The role of CSI at the transmitter is the main focus of this chapter. Two standard mechanisms for obtaining CSI at the transmitter are as follows:

(a) *Implicit feedback using reciprocity.* If the same frequency band is employed for both uplink and downlink, as in a time division duplex (TDD) system, then the instantaneous channels for uplink and downlink are identical. Thus, estimates of the channel on the uplink can be employed for downlink transmission, and vice versa. Inaccuracy in such implicit feedback occurs because the channel may change between the time that the channel estimate is obtained and the time when the resulting implicit feedback is employed.

(b) *Explicit feedback:* If the uplink and downlink employ different frequency bands, as in a frequency division duplex (FDD) system, or if the implicit feedback due to reciprocity in a TDD system is unreliable due temporal variation of the channel, then channel information can be sent back to the transmitter using explicit feedback. The challenge with this approach is the design of economical and robust explicit feedback mechanisms.

For a time-varying wireless channel, it is unrealistic to expect either of these two mechanisms to yield perfect CSI at the transmitter, so that it is important to design systems and evaluate their performance under the assumption of *partial* CSI. While CSI yields marginal performance gains for single-antenna communication, for MIMO systems, even partial CSI is known to yield large potential performance gains (Narula *et al.*, 1998; Visotsky and Madhow, 2001; Medles *et al.*, 2003).

Research on improving the capacity of narrowband MIMO channels via a feedback channel to the transmitter includes the work by Jongren *et al.* (2002), Lau *et al.* (2003), Mukkavilli *et al.* (2003b). Lau *et al.* (2003) consider optimizing the CSI sent over a feedback channel, imposing a constraint on the maximum number of bits sent per fading block. It is shown that the optimal feedback scheme is equivalent to the design of a vector quantizer with a modified distortion measure. Jongren *et al.* (2002) assume that quantized channel information is available at the transmitter, and use it to guide the design of space-time block codes preceded by CSI-dependent precoding matrices. Given the cost of sending back quantized channel values, many researchers have looked at scenarios where the receiver sends back information that designates the transmission strategy to be used. Mukkavilli *et al.* (2003b) investigate outage in MISO channels using beamforming, where the beamforming vector is determined by a finite capacity feedback channel carrying the index of the desired beamformer. The construction of near-optimal beamformer codebooks for this purpose is also considered by Mukkavilli *et al.* (2003a).

Beamforming to maximize the received signal-to-noise ratio (SNR), using CSI obtained by *reciprocity*, is studied by Cavers (2000). It is shown that for outdoor urban models, the time between the uplink and downlink should be limited to 10 ms in order for the uplink measurements to be useful for capacity enhancement on the downlink.

For wireless mobile channels, second-order channel statistics vary much more slowly than the channel realization itself. Thus, an important model for robust channel CSI at the transmitter is that of spatial *covariance feedback*. Information-theoretic computations show that such covariance feedback greatly improves capacity when the spatial channel is strongly colored (Visotsky and Madhow, 2001; Jafar and Goldsmith, 2001; Jafar *et al.*, 2001). Visotsky and Madhow (2001) show that the optimal strategy for MISO systems with covariance feedback is a form of waterfilling, subject to a sum-power constraint, along the eigenvectors of the covariance matrix. This result is extended to systems with multiple receive antennas by Jafar *et al.* (2001), modeling the channel responses for different receive antenna elements as uncorrelated. The efficient computation of the waterfilling solution is considered by Simon and Moustakas (2002) and Boche and Jorswieck (2003a,b). The ergodic and outage capacities for narrowband MIMO channels with covariance feedback is considered by Kang and Alouini (2003). The waterfilling strategy with covariance feedback can be interpreted as a linear precoding matrix that directs energy along the eigenvectors, followed by a space-time or space-frequency code for diversity or multiplexing, with the eigendirections forming "virtual" antenna elements. Examples of constructive space-time coding schemes preceded by linear precoding based on covariance feedback include BLAST-like spatial multiplexing (Simeone and Spagnolini, 2003), and diversity using space-time block codes (Zhou and Giannakis, 2003). The results confirm the performance improvements from covariance feedback predicted by information-theoretic computations.

6.1.2 Feedback in wideband systems

OFDM provides a convenient method to decompose a wideband channel into a collection of parallel narrowband channels, or subcarriers. In principle, narrowband space-time communication techniques can be applied on a per-subcarrier basis in OFDM systems. Obtaining and using channel feedback per subcarrier, however, can be computationally complex, expensive in terms of training overhead, and sensitive to channel estimation errors. Recent work on MIMO-OFDM systems with per-subcarrier channel feedback includes the one by Vook *et al.* (2003) and Xia *et al.* (2004). Xia *et al.* (2004) consider a feedback model where the transmitter knows each subcarrier's channel up to some uncertainty. A precoder based on the available CSI is used with a space-time code on a per-subcarrier basis, and an adaptive power and bit loading scheme across subcarriers is also used. Vook *et al.* (2003) consider the performance of MIMO OFDM under various assumptions on the information available at the transmitter. This work reports on simulation-based results for specific code constructions, and indicates a sensitivity to errors in the transmitter's CSI.

The performance of channel estimation and feedback can be improved by exploiting the high degree of correlation between channels on neighboring subcarriers. In particular, it can be shown (Barriac and Madhow, 2004a) that the channel spatial covariance is invariant across frequency. Thus, implicit feedback regarding the downlink covariance matrix can be obtained by suitably averaging uplink measurements, for both TDD and FDD systems (Barriac and Madhow, 2004b). This concept, which we term *statistical reciprocity,* is therefore more general than the deterministic reciprocity employed to obtain implicit feedback in TDD systems. The covariance feedback obtained from statistical reciprocity is robust, since it is invariant across frequency and varies very slowly with time. This is in contrast to the fragility of both implicit and explicit feedback regarding the channel realization per subcarrier, which varies relatively rapidly across both frequency and time. Implicit covariance feedback is particularly effective for outdoor channels for several reasons. First, accurate estimation of the covariance by averaging across frequency is possible because of the smaller coherence bandwidth. Second, covariance feedback is especially useful when the spatial covariance is strongly colored, as is typically the case for the narrow power-angle profiles seen in outdoor environments. Third, since the covariance is invariant across subcarriers, so is the space-time communication strategy based on covariance feedback, which significantly reduces transceiver complexity. For the remainder of this chapter, therefore, we focus on system designs centered around the "free" availability of the spatial covariance matrix at the base station in a cellular OFDM system.

The notion of using covariance information on the uplink to optimize downlink transmission has previously been applied in the context of FDD systems using TDMA or DS-CDMA (Raleigh and Jones, 1997; Hochwald and Marzetta, 2001; Liang and Chin, 2001; Morgan, 2003). The covariance is obtained either by averaging uplink channel responses across *time*, or by estimating the directions of arrival (DOA) of the incoming paths and directly determining the covariance from these measurements. DOA estimation is known to be computationally intensive (Liang and Chin, 2001), and may be infeasible if there are too many multipath components or an insufficient number of antenna elements. On the other hand, time averaging has the disadvantage that the amount of time necessary to construct an accurate estimate of the covariance matrix may exceed the allotted uplink transmit time.

Statistical reciprocity applies directly to both TDD systems and FDD systems in which the uplink and downlink bands are close enough that the array response for a given DOA is approximately the same on both uplink and downlink. For FDD systems in which the uplink and downlink bands are widely separated, this assumption may break down. However, it is possible to transform the uplink covariance matrix to obtain the downlink covariance matrix using a frequency calibration matrix (Liang and Chin, 2001), or by use of a clever antenna configuration (Hochwald and Marzetta, 2001) that attains identical beampatterns at both uplink/downlink wavelengths. These techniques, originally developed for single-carrier systems, are directly applicable to OFDM systems as well.

6.1.3 Chapter organization

The remainder of this chapter is organized as follows. Section 6.2 presents a MIMO-OFDM model, infers statistical reciprocity, and describes estimation of the spatial covariance without any training overhead. Downlink optimization, including choice of antenna spacing, using this implicit covariance feedback is described in Section 6.3. Section 6.4 describes uplink optimization based on the spatial covariance estimate, using a novel noncoherent technique that provides beamforming gain without explicit estimation of the channel realization for each subcarrier. Section 6.5 provides our conclusions.

6.2 Modeling

We start with industry-standard statistical models for simulating outdoor space-time channels (Saleh and Valenzuela, 1987; Pedersen *et al.*, 1999). Such measurement-based models specify the power delay profile (PDP) and the power angle profile (PAP), as well as the distribution of the delays and the angles of arrival and departure of various multipath components. The PDP specifies the power distribution versus time, while the PAP specifies the power distribution as a function of the angle of arrival. A valid transceiver design must exhibit good performance at the nominal signal-to-noise ratio (SNR) for "most" random channel realizations consistent with such a statistical model. For our information-theoretic investigation, we show that simulation-based statistical models can be replaced by bandwidth-dependent tapped delay line (TDL) models that are more amenable to analytical insight.

6.2.1 Vector tap delay line channel model

We consider outdoor channels in which the base station (BS) is located high enough and far enough away from the mobiles, so that signals reaching a particular mobile leave the BS in a narrow spatial cone. As in the classic Saleh-Valenzuela model (Saleh and Valenzuela, 1987), the channel response is decomposed into clusters. Experimental measurements of outdoor channels (Pedersen *et al.*, 1997, 1999; Martin, 2002) indicate that the number of clusters is small, usually one or two. The PDP within each cluster is well modeled as exponential, and the PAP for each cluster as Laplacian. Thus, a "single-cluster channel" would have an exponential PDP and a Laplacian PAP, while a "two-cluster channel" would have a PDP comprised of the sum of two exponential profiles (each with a different start time, rate of decay, and total power), and a PAP comprised of the sum of two Laplacian profiles.

For a system bandwidth of W, the taps in a TDL channel model are spaced apart by $1/W$. Assuming a large enough number of paths, each such tap is composed of a number of unresolvable taps. The phases of the unresolvable taps are well modeled as independent and identically distributed (i.i.d.) uniformly over $[0, 2\pi]$. This is because small changes in delay produce large changes in carrier phase, under the standard assumption that the carrier frequency is much larger than the signal bandwidth. Application of the central limit theorem now leads to the classical Rayleigh fading model, in which the resolvable taps are modeled as zero mean, circular Gaussian. The variance of these resolvable taps is the sum of strengths of the unresolvable constituent taps, and therefore depends on the power-delay profile, $P_\tau(\cdot)$. For a multiantenna system, the channels seen by different antenna elements are modeled as correlated and jointly complex Gaussian, again applying the central limit theorem.

As an example, consider a MISO channel modeling a typical downlink. Letting $P_\tau(\cdot)$ and $P_\Omega(\cdot)$ denote the channel PDP and PAP, respectively, we obtain the following vector TDL model (ignoring the effect of channel time variations):

$$\mathbf{h}_W(\tau) = \sum_{l=0}^{\infty} A_l \mathbf{v}_l \delta\left(\tau - \frac{l}{W}\right), \tag{6.1}$$

where we set

$$A_l \propto \sqrt{P_\tau\left(\frac{l}{W}\right)}, \quad l = 0, \dots, \infty \tag{6.2}$$

to capture the dependence of tap strength on the PDP, and where the i.i.d. complex Gaussian vectors $\mathbf{v}_l \sim \mathcal{CN}(\mathbf{0}, \mathbf{C})$, with the spatial covariance matrix $\mathbf{C}$ determined by the array manifold and the channel PAP. Specifically, the spatial covariance is given by

$$\mathbf{C} = E[\mathbf{a}(\Omega)\mathbf{a}(\Omega)^H] = \int_{-\pi}^{\pi} \mathbf{a}(\Omega')\mathbf{a}(\Omega')^H P_\Omega(\Omega') d\Omega', \tag{6.3}$$

where $\mathbf{a}(\Omega)$ is the base station array response as a function of the angle of departure Ω. As a running example, we consider a linear array, for which

$$\mathbf{a} = [a_1 \cdots a_{N_T}]^T, \quad a_l(\Omega) = e^{j(l-1)2\pi\frac{d}{\lambda}\sin(\Omega)}, \quad l = 1, \dots, N_T,$$

where d is the antenna array spacing, and λ the carrier wavelength. N_T is the number of transmit antennas. This corresponds to a one-dimensional, equally-spaced antenna array with spacing d.

If the mobile has multiple antennas, then, assuming that there is sufficiently rich scattering around the mobile, the channels from the BS to the different mobile antennas are well modeled as i.i.d. realizations of the preceding MISO model.

6.2.2 Spatial covariance estimation from uplink measurements

We first observe that uplink measurements can be employed to estimate the spatial covariance matrix $\mathbf{C}$. For simplicity of notation, we assume in the following that the mobile has one antenna and communicates with a BS with an N element antenna array. Since the responses from the base station array to different antenna elements at the mobile are modeled as i.i.d., more mobile antennas would provide even more averaging when estimating the covariance matrix on the uplink.

The mobile employs K subcarriers (which may not be contiguous), and the received signal vector on the kth subcarrier is given by

$$\mathbf{s}_k = \mathbf{h}_k x_k + \mathbf{n}_k, \tag{6.4}$$

where $\mathbf{h}_k$ is the $N \times 1$ channel frequency response, $\mathbf{n}_k$ is AWGN with $E[\mathbf{n}_j \mathbf{n}_k^H] = 2\sigma^2 \delta_{jk} \mathbf{I}_N$, and $\mathbf{I}_N$ denotes the $N \times N$ identity matrix. We know from Section 6.2.1 that the $\mathbf{h}_k$ are identically distributed, with $\mathbf{h}_k \sim \mathcal{CN}(\mathbf{0}, \mathbf{C})$.

A spectral decomposition of the channel covariance yields

$$\mathbf{C} = \mathbf{U} \mathbf{\Lambda} \mathbf{U}^H, \tag{6.5}$$

where the eigenvector matrix $\mathbf{U} = [\mathbf{u}_1 \cdots \mathbf{u}_N]$ is unitary, and $\mathbf{\Lambda}$ is diagonal with eigenvalues $\{\lambda_l\}$ arranged in decreasing order. The eigenvalue λ_l represents the strength of the channel on its lth eigenmode $\mathbf{u}_l$.

For the large delay spreads typical of outdoor environments, the coherence bandwidth is small, and the correlation between the channel responses at different frequencies dies out quickly with their separation. Thus, the base station can accurately estimate $\mathbf{C}$ by measuring the channel over a rich enough set of frequencies on the uplink (Barriac and Madhow, 2003, 2004b). Averaging over frequency bins, the base station forms the empirical autocorrelation matrix $\mathbf{R}$:

$$\mathbf{R} = \frac{1}{K} \sum_{k=1}^{K} \mathbf{s}_k \mathbf{s}_k^H. \tag{6.6}$$

With $E[|x_k|^2] = 1$, it is easy to show that $\mathbf{R}$ is an estimate of $\mathbf{C} + 2\sigma^2 \mathbf{I}_N$, where σ^2 is the noise variance per dimension. Thus, if $\{\lambda_l\}$ are the eigenvalues of $\mathbf{C}$, the eigenvalues of $\mathbf{R}$ are $\{\lambda_l + 2\sigma^2\}$. The eigenvectors of the two matrices are the same. An eigendecomposition of $\mathbf{R}$ therefore yields the dominant channel eigenmodes. Typically, the number of dominant eigenmodes is small for an outdoor channel because of the narrow PAP corresponding to signals received from a given mobile.

In the succeeding sections, we see how the preceding covariance estimate can be employed for both downlink and uplink optimization.

6.3 Downlink optimization with implicit covariance feedback

The empirical correlation matrix of the uplink signal from a given mobile, as computed in Section 6.2.2, provides implicit feedback regarding the downlink covariance matrix from the BS to that mobile.

6.3.1 Shannon-theoretic performance evaluation

Once the spatial covariance matrix $\mathbf{C}$ is known with sufficient accuracy, downlink transmission for a system with N_T BS antennas and N_R mobile antennas ($N_T \gg N_R$) is optimized by sending i.i.d. Gaussian inputs for each subcarrier, so that the ergodic capacity is that of a narrowband system with covariance feedback, as considered by Visotsky and Madhow (2001), Jafar *et al.* (2001). The optimal policy for each subcarrier is to send independent Gaussian inputs along the eigenvectors of $\mathbf{C}$, with the power allocated to each eigenmode determined by a waterfilling strategy. In practice, for an outdoor channel, the transmitted power can be concentrated along a small number K_T ($K_T \ll N_T$) of dominant eigenmodes using a linear precoding matrix along with a space-time code designed for a virtual $K_T \times N_R$ MIMO system in which the number K_T of virtual transmit antenna elements equals the number of eigenmodes with nonzero transmitted power. If λ_i is the eigenvalue for the ith eigenmode (in decreasing order) that is employed for transmission, and p_i is the power allocated to the ith eigenmode, then the mutual information along a given subcarrier is a random variable, which can be written as

$$I(\mathbf{p}) = \log \left| \mathbf{I}_{N_R} + \frac{1}{\sigma_n^2} \sum_{i=1}^{K_T} \mathbf{z}_i \mathbf{z}_i^H p_i \lambda_i \right|, \qquad (6.7)$$

where the $\mathbf{z}_i$ are independent $N_R \times 1$ vectors whose entries are i.i.d. $\mathcal{CN}(0,1)$, and σ_n^2 is the noise variance per dimension. We normalize the channel eigenvalues $\{\lambda_i\}$ such that $\sum_{i=1}^{N_T} \lambda_i = N_T$, and the powers such that $\sum_{i=1}^{K_T} p_i = P$.

We now discuss the Shannon-theoretic performance attained by the preceding strategy. For a MISO system employing multiple subcarriers on the downlink, spanning a bandwidth W, the spectral efficiency I_W, or the mutual information averaged across subcarriers, is given by averaging (6.7) across the subcarriers employed. Under mild conditions on the PDP, the channels seen by different subcarriers decorrelate rapidly enough that such averages obey a central limit theorem. The spectral efficiency is therefore well modeled as Gaussian, with mean given by the expectation of (6.7). This is simply the ergodic capacity of a single subcarrier, given by

$$E[I_W] = E[I(\mathbf{p})]. \qquad (6.8)$$

For a single-cluster channel, the variance of the spectral efficiency can be estimated as follows (Barriac and Madhow, 2004a):

$$\text{var}[I_W] \approx \frac{N_R \sum_{i=1}^{K_T}(\lambda_i p_i)^2}{(1 + \sum_{i=1}^{K_T} \lambda_i p_i)^2} \frac{1}{W} \int P^2(\tau)\, d\tau, \qquad (6.9)$$

where the PDP is normalized as $\int P(\tau) d\tau = 1$. Note that the variance is inversely proportional to the system bandwidth.

Knowing the mean and variance of the spectral efficiency, we can now provide simple analytical estimates of the outage rate based on the Gaussian approximation. Define the rate $R(\epsilon)$ as the largest transmission rate R (normalized by the system bandwidth W) so that the following condition holds:

$$P[I_W \leq R] \leq \epsilon.$$

Modeling I_W as Gaussian, we obtain the following approximation for $R(\epsilon)$,

$$\hat{R}(\epsilon) \approx E[I_W] - \sqrt{\mathrm{var}[I_W]}Q^{-1}(\epsilon), \qquad (6.10)$$

where $Q(x)$ is the complementary cumulative distribution function of a standard Gaussian random variable. The absolute outage rate, of course, is given by $R(\epsilon)W$.

From (6.9), we note that the variance of I_W decreases with increasing W, regardless of the power allocation across eigenmodes. Thus, for large system bandwidths, the outage rate is approximately maximized by maximizing the ergodic capacity $E[I_W]$. We will refer to this observation later, when we discuss downlink optimization by varying antenna spacing.

6.3.2 Accuracy of implicit covariance feedback

There are two key issues affecting the accuracy of estimating the covariance matrix $\mathbf{C}$ as in the previous section:

(i) The covariance must vary slowly enough, so that, when the base station transmits to user k, the estimate $\hat{\mathbf{C}}_k$ is still valid.

(ii) A mobile must employ enough subcarriers and a wide enough separation among the subcarriers, so that the empirical average (6.6) provides an accurate covariance estimate.

Both of these conditions are met for a wide variety of resource-sharing models, including FDD systems where the uplink and downlink are contiguous (Barriac and Madhow, 2004b). As an example, we consider a TDD system with TDMA on the uplink and TDM on the downlink, which implies that there is a significant delay between acquiring covariance feedback based on uplink measurements and employing it on the downlink.

We assume an OFDM system with 1024 subcarriers spaced 25 kHz apart. The PAP is initially $L(0°, 5°)$, where $L(M, \alpha)$ denotes a Laplacian distribution with mean M and variance $2\alpha^2$, and the PDP is exponential with an rms value of 0.5 μs. SNR is set to 10 dB. The BS has six antennas, with a typical antenna spacing of $d/\lambda = 0.5$. At this spacing, beamforming is the optimal transmit strategy for the given PAP.

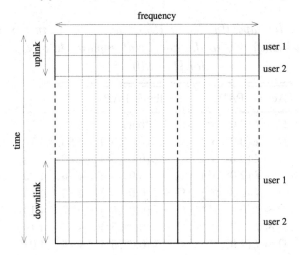

Fig. 6.1. A TDD system with TDMA on the uplink and TDM on the downlink.

The TDD system under consideration is shown in Fig. 6.1. Each user transmits to the base station using the entire frequency band for a certain amount of time, and subsequently the base station takes turns transmitting to the mobiles over the whole band. For such a system, K in (6.6) equals the entire set of frequency bins for all users. If the bandwidth is large, $\hat{\mathbf{C}}_k$ is clearly a good approximation for $\mathbf{C}_k$, but the question remains as to whether this covariance will remain valid until the BS is ready to reply to that mobile on the downlink. The longest a user will have to wait until it hears back from the BS is approximately the number of users in the system multiplied by the time the BS transmits to each user. For a rate of 20 Mbps and 10 packet payloads of 10 000 bits each, the time the BS sends to each mobile is approximately 5 ms. If there are ten users, this means the total delay is around 50 ms. However, even if the channel is fast fading, the covariance need not change much in this length of time, since it depends only on the PAP, which is slowly varying in general. It is shown by Nicoli *et al.* (2002) that for a mobile 500 m from the base station traveling less than 1000 km/h, and a BS station with eight antennas spaced half a wavelength apart, the channel statistics can be considered stationary for around 100 ms. Thus, the PAP, and hence the covariance, would also be stationary for that time length.

We now consider how variations (we can assume they are small) in the PAP would affect system performance. For a mobile moving away from the BS at 100 km/h, the angle between the BS and the mobile will change approximately 0.08° in 50 ms. If the center angle of the PAP changes by a

Table 6.1. *Ergodic capacity C and 1% outage rate R_o in b/s/Hz when the BS station beamforms to the dominant eigenmode of $\hat{\mathbf{C}}_k$, computed for $\Omega \sim L(0°, 5°)$*

Actual PAP	Feedback PAP	C	R_o
$\Omega \sim L(0.0°, 5°)$	no feedback	3.12	2.70
$\Omega \sim L(0.0°, 5°)$	$\Omega \sim L(0.0°, 5°)$	4.83	4.13
$\Omega \sim L(0.6°, 5°)$	$\Omega \sim L(0.0°, 5°)$	4.83	4.14
$\Omega \sim L(2.9°, 5°)$	$\Omega \sim L(0.0°, 5°)$	4.76	4.09
$\Omega \sim L(0.0°, 9°)$	$\Omega \sim L(0.0°, 5°)$	4.83	4.13
$\Omega \sim L(0.0°, 1°)$	$\Omega \sim L(0.0°, 5°)$	4.82	4.13
$\Omega \sim L(2.9°, 9°)$	$\Omega \sim L(0.0°, 5°)$	4.77	4.10
$\Omega \sim L(2.9°, 1°)$	$\Omega \sim L(0.0°, 5°)$	4.76	4.08

corresponding amount, we would like to know how this impacts performance results. Table 6.1 gives the 1% outage rate and ergodic capacity of a wideband system when the actual PAP differs from the PAP used to estimate the covariance. The outage rates are computed using the transmit strategy that maximizes ergodic capacity. It is assumed that the PAP remains Laplacian, and that only the mean and angular spread change with time. The first row shows the capacity and outage rate when there is no feedback and the transmitter employs a full-blown space-time code (the optimal transmit strategy when no feedback is available). The second row shows the capacity and outage rate when the BS has perfect covariance feedback information and beamforms in the direction of the covariance's dominant eigenmode (beamforming is the optimal strategy in this scenario for the given parameters). The following rows display the resulting capacity when the BS beamforms using imperfect covariance information. It can be seen that even if the base station uses covariance information obtained from a Laplacian whose mean has since shifted 2.9° and whose variance has doubled, deleterious effects on performance are minimal. Even in this case, where the changes in the PAP are much larger than one might expect, both the ergodic capacity and outage rate are much higher than the corresponding quantities when no feedback is available.

6.3.3 Optimal antenna spacing

We have seen that the spatial covariance depends on the PAP and the array manifold, with the latter determined by the array geometry. Now that we know that covariance feedback is readily available, a natural question to ask is the following: how should we choose the antenna array geometry so as to optimize performance? When there is no feedback, a reasonable strategy is to send i.i.d. Gaussian input from each transmit antenna, and the best performance is attained by spacing the antennas far enough apart that they see uncorrelated responses (Barriac and Madhow, 2004a). However, when the BS knows the channel covariance, the optimal antenna spacing is expected to be much smaller. For example, if we plan to beamform along the dominant eigenmode, then it makes sense to space the antennas such that the eigenvalue of this eigenmode is maximized.

We will focus on optimizing antenna spacing so as to maximize ergodic capacity (assuming optimal transmission with covariance feedback). As we have seen in Section 6.2.2, for large enough system bandwidth, this also approximately maximizes the outage rate. Maximization of ergodic capacity can be achieved by considering a single subcarrier, since the expected mutual information achieved by a given strategy is the same across subcarriers. From (6.7), the ergodic capacity with optimal power allocation is given by

$$
C = \max_{p_i : \sum_{i=1}^{N_T} p_i = P, p_i \geq 0} E\left[\log\left|\mathbf{I}_{N_R} + \frac{1}{\sigma_n^2}\sum_{i=1}^{N_T} \mathbf{z}_i \mathbf{z}_i^H p_i \lambda_i\right|\right], \tag{6.11}
$$

where the $\mathbf{z}_i$ are independent N_R-dimensional vectors whose entries are i.i.d. $\mathcal{CN}(0,1)$.

Ideally, we would like to maximize C by optimizing the antenna spacing, given the PAP and the SNR. This problem is difficult to solve because the eigenvalues $\{\lambda_i\}$, and thus the capacity, depend in a complex fashion on the PAP and the array geometry. We therefore consider a simplified thought experiment in which we consider a system with K eigenmodes with equal nonzero eigenvalues, and with the remaining eigenmodes corresponding to zero eigenvalues. The question then becomes: what is the optimal value $K = K_{opt}$, as a function of the number of receive elements N_R and the SNR. The results of such a thought experiment provides valuable guidance on antenna spacing, even though it may not always be possible to implement its prescriptions. For example, for $K_{opt} = 1$, we should space the antennas closely enough to create a single dominant eigenmode. But if there are two clusters with very different angles of departure from the BS, then there will be two dominant eigenmodes for any reasonable value of antenna spacing.

The results of the thought experiment can be paraphrased as follows: create a number of eigenmodes that is roughly equal to the number of receive elements N_R (except at very low SNR, where the optimal number of eigen-

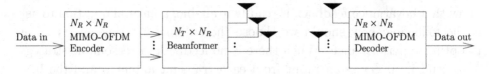

Fig. 6.2. MIMO-OFDM system with beamforming, where the number of transmit elements $N_T = N_b$, and the number of receive elements $N_R = N_s$.

modes is one).[†] Beamforming along these eigenmodes then creates an effective $N_s \times N_s$ MIMO system. While beamforming gains can be increased by increasing the number N_T of antennas at the BS transmitter, the complexity of OFDM processing at the transmitter scales as N_R, the much smaller number of antennas in the mobile receiver, since the beamforming weights are independent of the subcarriers. See Fig. 6.2. More importantly, the receiver in the subscriber unit only sees the effective $N_R \times N_R$ MIMO system, so that downlink performance can be improved by scaling up N_T, without any additional burden on the less capable receiver in the subscriber unit.

From a practical viewpoint, it is usually possible to space the transmit antennas so as to roughly follow the prescriptions of the thought experiment: in general, it is *not* possible to make the eigenvalues of the dominant eigenmodes equal, but spacing the antennas such that the number of dominant modes is close to K_{opt} still gives large capacity gains, as demonstrated in the following example. Consider a BS with six antennas transmitting to a mobile whose PAP is $L(0°, 5°)$. The SNR, P/σ_n^2, is set to 10 dB. As the antenna spacing changes, so does $\mathbf{\Lambda}$, and hence the optimal values of p_i, which can be solved for numerically. Fig. 6.3a shows how the ergodic capacity changes for different values of d/λ (the antenna spacing over the wavelength). At $d/\lambda = 8$, all channel eigenvalues are equal; hence, the capacity at this point corresponds to the maximum capacity attainable when there is no feedback. As d/λ decreases, the channel energy becomes concentrated in fewer eigenmodes, until only one eigenmode is dominant. Below $d/\lambda = 0.5$, beamforming is optimal. (Refer to the paper by Jafar and Goldsmith (2001) for the necessary and sufficient conditions for the optimality of beamforming.) We do not consider values of d/λ smaller than 0.4 because at very close spacing, the different antennas can no longer be treated as separate elements due to electro-magnetic coupling.

It is evident that beamforming with the BS antennas spaced at 0.4λ is superior to using a full-blown space time code with $d/\lambda = 8$, giving a gain of

---

[†]Refer to the paper by Barriac and Madhow (2004b) for details and caveats. Also, refer to the paper by Boche and Jorswieck (2003a) for results for $N_R = 1$.

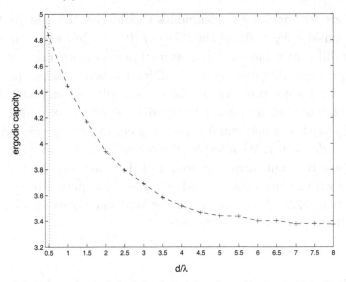

(a) $N_R = 1$: to the left of the dotted line, beamforming is optimal.

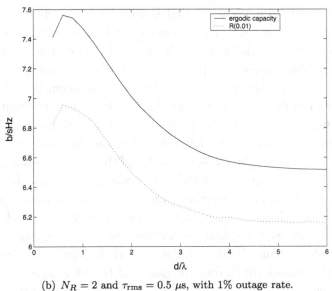

(b) $N_R = 2$ and $\tau_{\rm rms} = 0.5$ μs, with 1% outage rate.

Fig. 6.3. Ergodic capacity vs. d/λ when $N_T = 6$ and $\Omega \sim L(0°, 5°)$.

over 1.5 b/s/Hz. Not only is capacity increased by using a smaller spacing, but complexity is reduced dramatically by using beamforming instead of space-time codes.

Now, suppose that there are two receive elements. Noting that $K_{\rm opt} = 2$ when $N_R = 2$ at moderate SNR, we expect that spacing the antennas such that there are two dominant eigenmodes should give the best performance. Again, we consider $N_T = 6$ and the PAP~ $L(0°, 5°)$, but with N_R now equal to two. For different values of d, the optimal powers p_i are calculated numerically by approximating derivatives by differentials and using the projected gradient descent algorithm. Values for the ergodic capacity are plotted vs. d/λ in Fig. 6.3(b). Below $d/\lambda = 0.82$, sending along two eigenmodes is optimal. As expected, the maximum capacity occurs when two eigenmodes are dominant, at $d/\lambda = 0.7$. We also plot the outage rate in the figure, assuming that there are 1024 subcarriers spaced 25 kHz apart, and that the PDP is exponential with an rms delay spread of 0.5 μs. As expected from the discussion in Section 6.2.2, the antenna spacing that maximizes ergodic capacity also roughly maximizes the outage rate.

6.4 Uplink optimization using noncoherent eigenbeamforming

In this section, we discuss how implicit knowledge of the channel covariance at the BS can be used with noncoherent demodulation on the *uplink*, even when no other channel information is available. The complexity remains practicable even as the receiver (BS) enjoys an SNR advantage from scaling up the number of antennas. Noncoherent communication is well suited for the uplink of a cellular system in which the base station must estimate a time-varying channel to each mobile. Pilot-symbol-based channel estimation is potentially more efficient on the downlink, since the mobiles are able to share a common pilot channel. Accurate estimates of the spatial covariance matrix, available through averaging in wideband systems, allow eigenbeamforming (Jacobsen *et al.*, 2004) at the receiver along the dominant channel modes. For a typical outdoor channel, where the number of dominant modes is small, this allows the receiver to increase its SNR by scaling up the number of antennas, while limiting the demodulation and decoding complexity, which scales with the number of channel modes used by the receiver.

Once the channel covariance is obtained at the BS by averaging uplink measurements, the BS will project the received signal along the L dominant eigenvectors of the covariance matrix, obtained from its factorization (6.5). This yields L parallel uncorrelated OFDM channels $\{\langle \mathbf{s}, \mathbf{u}_l \rangle\}_{l=1}^{L}$, where $\mathbf{s}$ denotes the $N_b \times 1$ received signal, as in (6.4), and L is typically much smaller than the number of antenna elements N_b. As a rough measure of the performance gain relative to a single-antenna system, we define *beamforming gain* as the SNR if the signal power is summed over the L chosen eigenmodes, relative to the SNR for a single antenna element. This yields the following

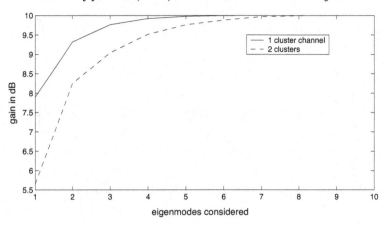

Fig. 6.4. Eigenbeamforming gain over a single antenna receiver.

formula for the beamforming gain as a function of L:

$$G(L) = 10 \log_{10} \left(\frac{N_b \sum_{l=1}^{L} \lambda_l}{\sum_{l=1}^{N_b} \lambda_l} \right), \qquad (6.12)$$

where λ_l are the channel eigenvalues.

Fig. 6.4 shows the beamforming gain as a function of the number of eigenmodes used for a ten-antenna system. The upper curve is for a single-cluster channel whose PAP is Laplacian with zero mean and angular spread $10°$ (the angular spread is defined as $\sqrt{2 \operatorname{var}(\Omega)}$, where $\operatorname{var}(\Omega)$ is the variance of Ω). The lower curve is for a two-cluster channel where the first cluster's PAP is as above, and the second cluster's PAP is also Laplacian with angular spread $10°$, but has its mean at $45°$ (both clusters have the same power). The total receive power is normalized to be the same for both plots. Note that beamforming gain as a function of L quickly saturates. Thus, beamforming along the dominant eigenmode captures most of the received energy for the one cluster channel, while using the first two eigenmodes captures most of the energy in the two cluster channel. Thus, for typical outdoor channels, estimation of the channel covariance enables the use of a small number of eigenmodes by the demodulator and decoder, limiting complexity while preserving the SNR advantage from scaling up the number of receive elements.

The signals for the L eigenmodes can be combined in a number of ways. The gain on the kth subcarrier along the lth eigenmode is given by $g_k(l) = \langle \mathbf{h}_k, \mathbf{u}_l \rangle$. One possibility is to explicitly estimate the scalar channel gains $\{g_k(l), l = 1, \ldots, L\}$ using pilots, and then perform coherent diversity combining of the L branches to obtain an estimate of x_k. The advantage that

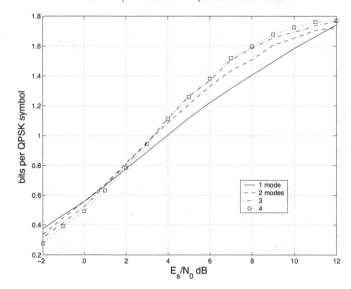

Fig. 6.5. Noncoherent block fading capacity with varying number of dominant, equal-strength eigenmodes.

may have over estimation of the original $N_b \times 1$ channel vector $\mathbf{h}_k$ is that fewer gains may need to be explicitly estimated. Another possibility is non-coherent diversity combining, which is consistent with the goal of reducing the overhead in the uplink transmission. Jacobsen *et al.* (2004) use a serial concatenation of an outer binary channel code with an inner differential modulation code to approach the noncoherent capacity on the uplink of a wideband cellular channel. A simple, yet effective combining strategy with iterative noncoherent processing is used: parallel noncoherent demodulators with extrinsic information from the channel decoder process L dominant channel modes. The soft outputs of the demodulators are then combined and sent back to the decoder as priors, setting up the next round of parallel demodulation and decoding.

In addition to beamforming gain, various levels of diversity are attained by processing the channel modes in parallel. Fig. 6.5 shows the effect of the diversity level on the noncoherent capacity when the received power is normalized to one and distributed equally amongst $L = \{1, 2, 3, 4\}$ dominant eigenmodes. To illustrate beamforming gain in this context, consider a ten-element BS array with one dominant channel mode. The SNR per antenna element in such a system is 10 dB less than that of a single-antenna system operating at the same rate; for example, we see from Fig. 6.5 that a single-antenna system requires an SNR of 2 dB for a spectral efficiency of 0.8 bits/symbol; the corresponding SNR per antenna element for a ten-element array with a single dominant mode is -8 dB.

6.5 Conclusions

Space-time communication systems that leverage uplink/downlink asymmetry in addition to statistical reciprocity inherent to wideband outdoor channels enjoy large performance gains, while at the same time reducing signal processing complexity at both the base station and mobile radio. Specifically, the techniques presented in this chapter enable an increase in capacity by increasing the number of antennas at the base station, without *any* impact on transceiver complexity at the mobile. While the performance gains from these techniques are evaluated in Shannon-theoretic terms, advances in turbo-like coded modulation imply that these information theoretic-limits are achievable at reasonable complexity.

References

Barriac, G. and Madhow, U. (2003). Wideband space-time communication with implicit channel feedback. In *Proc. IEEE Int. Symp. Signal Process. and Appl.*, 225–228.

—— (2004a). Characterizing outage rates for space-time communication over wideband channels. *IEEE Trans. Commun.*, **52** (12), 2198–2208.

—— (2004b). Space-time communication for OFDM with implicit channel feedback. *IEEE Trans. Inf. Theory*, **50** (12), 3111–3129.

Biglieri, E., Caire, G., and Taricco, G. (2001). Limiting performance of block-fading channels with multiple antennas. *IEEE Trans. Inf. Theory*, **47** (4), 1273–1289.

Boche, H. and Jorswieck, E. (2003a). Optimum power allocation, and complete characterization of the impact of correlation on the capacity of MISO systems with different CSI at the transmitter. In *Proc. IEEE Int. Symp. Inf. Theory*, 353.

—— (2003b). Optimum power allocation for MISO systems, and complete characterization of the impact of correlation on the capacity. In *Proc. IEEE Int. Conf. Acoust., Speech, Signal Process.*, vol. 4, 373–376.

Cavers, J. (2000). Single-user and multiuser adaptive maximal ratio transmission for Rayleigh channels. *IEEE Trans. Veh. Technol.*, **49** (6), 2043–2050.

Hochwald, B. and Marzetta, T. (2001). Adapting a downlink array form uplink measurements. *IEEE Trans. Signal Process.*, **49** (3), 642–653.

Jacobsen, N., Barriac, G., and Madhow, U. (2004). Noncoherent eigenbeamforming for a wideband cellular uplink. In *Proc. IEEE Int. Symp. Inf. Theory*, 280.

Jafar, S. and Goldsmith, A. (2001). On optimality of beamforming for multiple antenna systems with imperfect feedback. In *Proc. IEEE Int. Symp. Inf. Theory*, 321.

Jafar, S., Vishwanath, S., and Goldsmith, A. (2001). Channel capacity and beamforming for multiple transmit and receive antennas with covariance feedback. In *Proc. IEEE Int. Conf. Commun.*, 2266–2270.

Jongren, G., Skoglund, M., and Ottersten, B. (2002). Combining beamforming and orthogonal space-time block coding. *IEEE Trans. Inf. Theory*, **48** (3), 611–627.

Kang, M. and Alouini, M. (2003). Water-filling capacity and beamforming performance of MIMO systems with covariance feedback. In *Proc. IEEE Workshop Signal Process.*, 556–560.

Lau, K., Liu, Y., and Chen, T. (2003). Optimal partial feedback design for MIMO block fading channels with causal noiseless feedback. In *Proc. IEEE Int. Symp. Inf. Theory*, 65.

Liang, Y. and Chin, F. (2001). Downlink channel covariance matrix estimation and its applications in wireless DS-CDMA systems. *IEEE J. Sel. Areas Commun.*, **19** (2), 222–232.

Martin, U. (2002). A directional radio channel model for densely built up urban areas. In *Proc. 2nd EMPCC*, 237–244.

Medles, A., Visuri, S., and Slock, D. T. M. (2003). On MIMO capacity for various types of partial channel knowledge at the transmitter. In *Proc. IEEE Inf. Theory Workshop*, 99–102.

Morgan, D. (2003). Downlink adaptive array algorithms for cellular mobile communications. *IEEE Trans. Commun.*, **51** (3), 476–488.

Mukkavilli, K. K., Sabharwal, A., and Erkip, E. (2003a). Beamformer design with feedback rate constraints: criteria and constructions. In *Proc. IEEE Int. Symp. Inf. Theory*, 414.

Mukkavilli, K. K., Sabharwal, A., Erkip, E., and Aazhang, B. (2003b). On beamforming with finite rate feedback in multiple-antenna systems. *IEEE Trans. Inf. Theory*, **49** (10), 2562–2579.

Narula, A., Lopez, M., Trott, D., and Wornell, G. (1998). Efficient use of side information in multiple antenna data transmission over fading channels. *IEEE J. Sel. Areas Commun.*, **16** (8), 1423–1435.

Nicoli, M., Simeone, O., and Spagnolini, U. (2002). Multislot estimation of fast-varying space-time channels in TD-CDMA systems. *IEEE Commun. Lett.*, **6** (9), 376–378.

Pedersen, K. I., Mogensen, P. E., and Fleury, B. H. (1997). Power-azimuth spectrum in outdoor environments. *Electron. Lett.*, **33** (18), 1583–1584.

—— (1999). Dual-polarized model of outdoor propagation environments for adaptive antennas. In *Proc. IEEE Veh. Technol. Conf.*, vol. 2, 990–995.

Raleigh, G. and Jones, V. (1997). Adaptive antenna transmission for frequency duplex digital wireless communication. In *Proc. IEEE Int. Conf. Commun.*, vol. 6, 641–646.

Saleh, A. and Valenzuela, R. (1987). A statistical model for indoor multi-path propagation. *IEEE J. Sel. Areas Commun.*, **5** (2), 128–137.

Simeone, O. and Spagnolini, U. (2003). Combined linear pre-equalization and BLAST equalization with channel correlation feedback. *IEEE Commun. Lett.*, **7** (10), 2663–2667.

Simon, S. and Moustakas, A. (2002). Optimizing MIMO antenna systems with channel covariance feedback. In *Bell Labs Tech. Memorandum*.

Telatar, E. (1995). Capacity of multi-antenna Gaussian channels. *AT&T Bell Labs Internal Tech. Memo # BL0112170-950615-07TM*.

Visotsky, E. and Madhow, U. (2001). Space-time transmit precoding with imperfect channel feedback. *IEEE Trans. Inf. Theory*, **47** (6), 2632–2639.

Vook, F. W., Thomas, T. A., and Zhuang, X. (2003). Transmit diversity and transmit adaptive arrays for broadband mobile OFDM systems. In *Proc. IEEE*

Wireless Commun. Netw. Conf., vol. 1, 44–49.

Xia, P., Zhou, S., and Giannakis, G. (2004). Adaptive MIMO OFDM based on partial channel state information. *IEEE Trans. Signal Process.*, **52** (1), 202–213.

Zhou, S. and Giannakis, G. (2003). Optimal transmitter eigen-beamforming and space-time block coding based on channel correlations. *IEEE Trans. Inf. Theory*, **49** (7), 1673–1690.

Part II
Space-time modulation and coding

7

Introduction to space-time codes

A. Robert Calderbank

Princeton University

Ayman F. Naguib

Qualcomm, Inc.

7.1 Introduction

Information-theoretic analysis by Foschini (1996) and by Telatar (1999) shows that multiple antennas at the transmitter and receiver enable very high rate wireless communication. Space-time codes, introduced by Tarokh *et al.* (1998), improve the reliability of communication over fading channels by correlating signals across different transmit antennas. Design criteria developed for the high-SNR regime in Tarokh *et al.* (1998) and Guey *et al.* (1999) are presented in Section 7.3 from the perspective of typical error events (following the exposition by Tse and Viswanath (2005)). Techniques for multiple access and broadcast communication are described very briefly in Sections 7.9 and 7.10, where algebraic structure enables simple implementation. The emphasis throughout is on low cost, low complexity mobile receivers.

Section 7.2 provides a description of set partitioning, which was developed by Ungerboeck (1982) as the basis of code design for the additive white Gaussian noise (AWGN) channel. The importance of set partitioning to code design for the AWGN channel is that it provides a lower bound on squared Euclidean distance between signals that depends only on the binary sum of signal labels. Section 7.9 describes the importance of set partitioning to code design for wireless channels, where it provides a mechanism for translating constraints in the binary domain into lower bounds on diversity protection in the complex domain.

Section 7.4 describes space-time trellis codes, starting from simple delay diversity, and then using intuition about the product distance to realize additional coding gain. Section 7.5 uses the Alamouti code to introduce space-time block codes. Tarokh *et al.* (1999) used orthogonal designs to create analogs of the Alamouti code for more than two transmit antennas. Their

aim was maximum-likelihood decoding with only linear processing at the
receiver, and this is the function of the orthogonal structure. Section 7.6
starts with examples of orthogonal designs in dimensions 2 and 4, and then
provides a survey of a classical algebraic theory that provides fundamental
limits on transmission rates. This treatment follows the approach of Calderbank and Naguib (2001), highlights the work of Wolfe (1976) on amicable
pairs of real orthogonal designs, and is elementary in the sense that it emphasizes sets of anticommuting real matrices rather than representations of
Clifford Algebras. Others have independently made the connection to Clifford Algebras (Tirkkonnen and Hottinen, 2002).

As the number of transmit antennas increases, the data rate available
with orthogonal designs becomes unattractive. Hence, the recent focus on
nonorthogonal linear code designs such as Linear Dispersion Codes (Hassibi
and Hochwald, 2002) (and most recently the Golden Code (Belfiore et al.,
2005)) for which decoding (Damen et al., 2000) is efficient, albeit not linear
complexity. Quasi-orthogonal codes (described in Section 7.7) are a different solution, where Jafarkhani (2001) relaxes the requirement that linear
processing at the receiver be able to separate all transmitted symbols. Section 7.8 describes superorthogonal codes, which significantly improve upon
the performance of the original space-time trellis codes by concatenating an
outer trellis code with an inner space-time block code.

Naguib and Seshadri were the first to combine interference cancellation
and maximum-likelihood decoding of space-time block codes. The interference cancellation algorithm described in Section 7.10 serves as the starting point for modular design of space-time codes for larger numbers of
antennas. Integration of two Alamouti space-time users with two receive
antennas makes it possible to approach throughputs of 1 Mbps within the
GSM/EDGE cellular standard (see Al-Dhahir et al., 2002, for more details).

7.2 Set partitioning of QAM

Let $\Gamma_0, \Gamma_2, \ldots, \Gamma_L$ be an L-level partition, where the partition Γ_i is a re-
finement of the partition Γ_{i-1}. We view the L-level partition as a rooted
tree, where the root Γ_0 is the entire signal constellation, and the vertices
at level k are the subsets that constitute the partition Γ_k. We shall only
consider binary partitions here, so that the subsets of the partition Γ_k can
be labeled by binary strings $a_1 a_2 \cdots a_k$ specifying the path from the root to
the corresponding node in the tree.

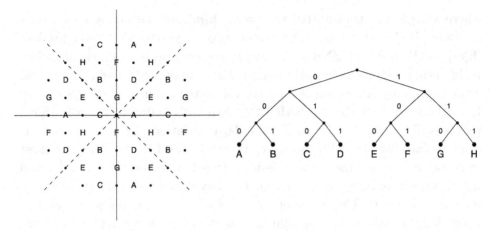

Fig. 7.1. A binary partition of a 32-point QAM constellation.

QAM constellations Signal points in QAM constellations are drawn from some realization of the integer lattice $\mathbf{Z}^2$. Let $\mathbf{i} = \sqrt{-1}$. We start with the Gaussian integers $\mathbf{Z}[\mathbf{i}]$, multiply by $1 + \mathbf{i}$ (this is just a scalar multiple of rotation through $90°$ in the complex plane) and translate by $\mathbf{i}$ to obtain the constellation shown in Fig. 7.1. The binary partition is based on the chain of sublattices $(1 + \mathbf{i})^k \mathbf{Z}[\mathbf{i}]$, where $k = 1, 2, \ldots$. If signals $\mathbf{x}$ and $\mathbf{y}$ are in the same subset at level k, then $\mathbf{x} - \mathbf{y}$ is divisible by $(1 + \mathbf{i})^{k+1}$ and the first $k + 1$ terms in the binary expansions of $\mathbf{x} - 1$ and $\mathbf{y} - 1$ are identical. The coefficients a_i in the binary expansion

$$\mathbf{x} - 1 = a_1(1 + \mathbf{i}) + a_2(1 + \mathbf{i})^2 + \cdots + a_k(1 + \mathbf{i})^k \quad (\text{mod } (1 + \mathbf{i})^{k+1})$$

specify the binary label of the path from the root to the subset at level k. Note that if $\mathbf{x}$ and $\mathbf{y}$ are in the same subset at level k, then $\|\mathbf{x} - \mathbf{y}\|^2$ is divisible by 2^{k+1}. Implicit in Fig. 7.1 is a binary partition of QPSK, where the points $1, -1, \mathbf{i}, -\mathbf{i}$ are labeled by 00, 01, 11, and 10, respectively.

Binary partitions of PSK constellations can be defined using the algebra of cyclotomic extensions of the rational numbers.

7.3 Coding gain and diversity gain

Error probability on an AWGN channel decays exponentially with SNR, and the challenge of communication over Rayleigh fading channels is that error probability decays only inversely with SNR. The received symbols $\mathbf{y}[m]$ are given by

$$\mathbf{y}[m] = \mathbf{h}[m]\mathbf{x}[m] + \mathbf{w}[m],$$

where $\mathbf{x}[m]$ is the transmitted codeword, $\mathbf{h}[m]$ are the fading coefficients, and $\mathbf{w}[m]$ is Gaussian noise. The instantaneous received SNR is the product $|\mathbf{h}[m]|^2 \, \mathrm{SNR}$. If $|\mathbf{h}[m]|^2 \, \mathrm{SNR} \gg 1$, then the separation between signal points is significantly larger than the standard deviation of the Gaussian noise, and error probability is very small since the tail of the Q function decays rapidly. If on the other hand $|\mathbf{h}[m]|^2 \, \mathrm{SNR} \ll 1$, then the separation between signal points is of the same order as the standard deviation of the noise, and the error probability is significant. Error events in the high SNR regime most often occur because the channel is in a deep fade ($|\mathbf{h}[m]|^2 \, \mathrm{SNR} < 1$), and not as a result of high additive noise. The way to improve performance is to introduce *diversity*. This is the provisioning of multiple signal paths, each of which fades independently, so that reliable communication is possible if one of the paths is strong.

A space-time codeword $\mathbf{x}$ is an $N \times K$ array where the rows are indexed by transmit antennas, the columns are indexed by time slots in a data frame, and the entries are the symbols to be transmitted. We assume a quasi-static model, where the fading coefficients are constant over a frame and change independently from one frame to the next. Information is distributed across the different transmit antennas so as to realize the benefit of independent fading across the N paths to a given receive antenna. If we normalize $\mathbf{x}$ so that the average energy per complex symbol is 1 ($\mathrm{SNR} = 1/N_0$), then the probability of error given perfect channel state information is

$$\mathbb{P}\{\mathbf{x_i} \to \mathbf{x_j}|\mathbf{h}\} = Q\left(\frac{\|h(\mathbf{x}_i - \mathbf{x}_j)\|}{2\sqrt{N_0/2}}\right),$$

and averaging over the channel statistics gives

$$\mathbb{P}\{\mathbf{x_i} \to \mathbf{x_j}\} = \mathbb{E}\left[Q\left(\frac{\sqrt{[\mathrm{SNR}\, \mathbf{h}(\mathbf{x}_i - \mathbf{x}_j)(\mathbf{x}_i - \mathbf{x}_j)^*\mathbf{h}^*]}}{\sqrt{2}}\right)\right].$$

The matrix $(\mathbf{x_i} - \mathbf{x_j})(\mathbf{x_i} - \mathbf{x_j})^*$ is Hermitian. So there is a unitary matrix $\mathbf{U}$ for which

$$\mathbf{U}(\mathbf{x_i} - \mathbf{x_j})(\mathbf{x_i} - \mathbf{x_j})^*\mathbf{U}^* = \mathrm{diag}[\lambda_1^2, \lambda_2^2, \ldots, \lambda_N^2],$$

where the entries λ_i are the singular values of $\mathbf{x_i} - \mathbf{x_j}$. The change of basis $\mathbf{h} \to \mathbf{U}\mathbf{h}$ preserves the Rayleigh distribution and if $\lambda_1, \lambda_2, \ldots, \lambda_R$ are nonzero, then

$$\mathbb{P}\{\mathbf{x_i} \to \mathbf{x_j}\} \leq (\mathrm{SNR}/2)^{-R}[1/(\lambda_1^2 \lambda_2^2 \cdots \lambda_R^2)].$$

When $R = N$, the matrix $\mathbf{x_i} - \mathbf{x_j}$ is nonsingular and maximal diversity is achieved. In general, we have the following criteria for the design of space-time codes.

Rank criterion Maximize the minimum rank of the difference $\mathbf{x_i} - \mathbf{x_j}$ over all distinct pairs of space-time codewords $\mathbf{x_i}, \mathbf{x_j}$. If the minimum rank is R, then the space-time code achieves a *diversity gain* of R.

Determinant criterion For a given diversity R, maximize the minimum product of the nonzero singular values of the difference $\mathbf{x_i} - \mathbf{x_j}$ over all distinct pairs of space-time codewords $\mathbf{x_i}, \mathbf{x_j}$. This minimum product determines the *coding gain* of the space-time code.

If there are M receive antennas, then the probability of error is given by

$$\mathbb{P}\{\mathbf{x_i} \to \mathbf{x_j}\} \leq (\text{SNR}/2)^{-RM}[1/(\lambda_1^2 \lambda_2^2 \cdots \lambda_R^2)]^M.$$

7.4 Space-time trellis codes

The simplest form of transmit diversity is the delay diversity scheme proposed by Wittneben (1991) for two transmit antennas, where a signal is transmitted from the second antenna, then delayed one time slot and transmitted from the first antenna. Fig. 7.2 provides a trellis representation of delay diversity for 8-PSK, where the eight states correspond to the eight possible prior symbols. The edge label xy means that symbol x is transmitted from the first antenna and symbol y from the second antenna. The diversity gain is two, since labels on edges entering a given state disagree in the first position and labels on edges leaving a given state disagree in the second position.

Observe that the probability of confusing two input sequences that differ in a single symbol is identical to the probability of error for a repetition code of length two when the two symbols are subject to independent fading.

Tse and Viswanath (2005, Chapter 3) provide a clear exposition of how to choose codewords $\mathbf{x_i} = (x_{i1}, \ldots, x_{iR})$ for transmission over R independent fading channels. If the path gains are $h_1, h_2, \ldots, h_R$, then the typical way for $\mathbf{x_i}$ to be confused with $\mathbf{x_j}$ is that

$$|h_1|^2 |x_{i1} - x_{j1}|^2 + \cdots + |h_R|^2 |x_{iR} - x_{jR}|^2$$

is of order 1/SNR. This event roughly holds when each term is of order 1/SNR and this happens with probability

$$\left(\prod_{\ell=1}^{R} \frac{1}{|x_{i\ell} - x_{j\ell}|^2} \right) \text{SNR}^{-R}.$$

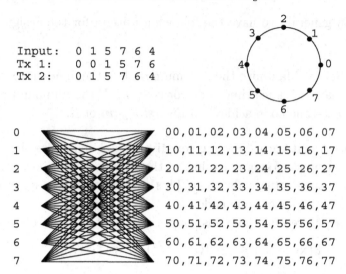

```
Input:    0 1 5 7 6 4
Tx 1:     0 0 1 5 7 6
Tx 2:     0 1 5 7 6 4
```

0	00,01,02,03,04,05,06,07
1	10,11,12,13,14,15,16,17
2	20,21,22,23,24,25,26,27
3	30,31,32,33,34,35,36,37
4	40,41,42,43,44,45,46,47
5	50,51,52,53,54,55,56,57
6	60,61,62,63,64,65,66,67
7	70,71,72,73,74,75,76,77

Fig. 7.2. Trellis representation of delay diversity for 8 PSK.

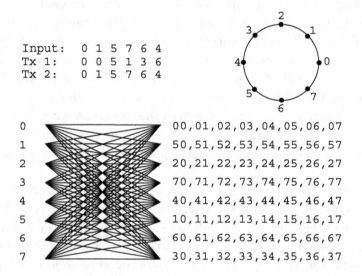

```
Input:    0 1 5 7 6 4
Tx 1:     0 0 5 1 3 6
Tx 2:     0 1 5 7 6 4
```

0	00,01,02,03,04,05,06,07
1	50,51,52,53,54,55,56,57
2	20,21,22,23,24,25,26,27
3	70,71,72,73,74,75,76,77
4	40,41,42,43,44,45,46,47
5	10,11,12,13,14,15,16,17
6	60,61,62,63,64,65,66,67
7	30,31,32,33,34,35,36,37

Fig. 7.3. A space-time code that differs from delay diversity in that for odd numbered states the label on the first antenna is negated.

The intuition is that product distance provides coding gain in addition to diversity gain, and it is consistent with the results reported by Tarokh *et al.* (1998).

For example, the code $C = \{00, 15, 22, 37, 44, 51, 66, 73\}$ maximizes the product distance among all block codes of length two for the 8-PSK constellation. We transmit the first symbol in a codeword from the second antenna, and then after a delay of one time slot we transmit the second symbol from the first antenna. The trellis is shown in Fig. 7.3 and it is apparent that we are less likely to confuse two input sequences that differ in a single symbol (we can think of the single symbol as encoded using the code C with the two outputs appearing on different channels). In fact, simulations show that there is a coding gain of 2.5 dB in addition to the diversity gain.

A space-time code is one of many components in an end to end wireless system, and concatenation with powerful outer codes [Reed–Solomon (Naguib *et al.*, 1998b), turbo codes (Liu *et al.*, 1999), or LDPC codes (Lu *et al.*, 2002)] is standard practice. Integration of inner/outer codes with channel estimation is described by Naguib *et al.* (1998b) in the context of narrowband TDMA (30 kHz channelization).

7.5 Space-time block codes

The most famous example was discovered by Alamouti (1998) and is described by a 2×2 matrix where the columns represent different time slots, the rows represent different antennas, and the entries are the symbols to be transmitted. The encoding rule is

$$(c_1, c_2) \rightarrow \begin{pmatrix} c_1 & c_2 \\ -c_2^* & c_1^* \end{pmatrix}.$$

The signals r_1, r_2 received over two consecutive time slots are given by

$$\begin{pmatrix} r_1 \\ -r_2^* \end{pmatrix} = \begin{pmatrix} h_1 & h_2 \\ -h_2^* & h_1^* \end{pmatrix} \begin{pmatrix} c_1 \\ -c_2^* \end{pmatrix} + \begin{pmatrix} w_1 \\ -w_2^* \end{pmatrix},$$

where h_1, h_2 are the path gains from the two transmit antennas to the mobile, and the noise samples w_1, w_2 are independent samples of a zero-mean complex Gaussian random variable with noise energy N_0 per complex dimension. Thus,

$$\mathbf{r} = \mathbf{H}\mathbf{c} + \mathbf{w},$$

where the matrix $\mathbf{H}$ is orthogonal. The reason for broad commercial interest in the Alamouti code is that both coherent and noncoherent detection are remarkably simple. If the path gains are known at the mobile (typically this is accomplished at some sacrifice in rate by inserting pilot tones into the data frame for channel estimation), then the receiver is able to form

$$\mathbf{H}^*\mathbf{r} = \|\mathbf{h}\|^2 \mathbf{c} + \mathbf{w}'.$$

The new noise term $\mathbf{w}'$ is still white, so that c_1, c_2 can be decoded separately rather than jointly, which is far more complex. If the path gains are not known at the mobile, then the receiver forms channel estimates using the code symbols c_1, c_2. Given

$$\mathbf{r} = \begin{pmatrix} r_1 \\ r_2 \end{pmatrix} = \begin{pmatrix} c_1 & c_2 \\ -c_2^* & c_1^* \end{pmatrix} \begin{pmatrix} h_1 \\ h_2 \end{pmatrix} + \begin{pmatrix} w_1 \\ w_2 \end{pmatrix} = \mathbf{C}\mathbf{h} + \mathbf{w},$$

the estimates h_1, h_2 are given by

$$\tilde{\mathbf{h}} = \begin{pmatrix} \tilde{h}_1 \\ \tilde{h}_2 \end{pmatrix} = \frac{1}{\|\mathbf{c}\|^2} \mathbf{C}^* \mathbf{r} = \begin{pmatrix} h_1 + \tilde{w}_1 \\ h_2 + \tilde{w}_2 \end{pmatrix},$$

where

$$\tilde{w}_1 = \frac{c_1^* w_1 - c_2 w_2}{\|\mathbf{c}\|^2} \quad \text{and} \quad \tilde{w}_2 = \frac{c_2^* w_1 + c_1 w_2}{\|\mathbf{c}\|^2}.$$

The channel estimate is used to detect the next pair of code symbols, and these symbols are in turn used to update the channel estimate. When the channel variation is slow, the receiver improves stability of the decoding algorithm by averaging old and new channel estimates. Noncoherent detection requires that transmission begin with a pair of known symbols and will perform within 3 dB of SNR as compared with coherent detection, where ideal channel state information is available at the receiver.

7.6 Real and complex orthogonal designs

Let $u_0, u_1, \ldots, u_{s-1}$ be positive integers, and let $x_0, x_1, \ldots, x_{s-1}$ be variables. A *real orthogonal design* of type $(u_0, u_1, \ldots, u_{s-1})$ and size N is an $N \times N$ matrix $\mathbf{X}$ with entries $0, \pm x_0, \pm x_1, \ldots, \pm x_{s-1}$ satisfying

$$\mathbf{X}\mathbf{X}^T = \left(\sum_{j=0}^{s-1} u_j x_j^2 \right) \mathbf{I}_N.$$

There are s variables and N time slots, so we define the rate of the orthogonal design to be s/N. There is a second notion of rate that is constellation specific and more forgiving. If we start with N antennas and a constellation of size 2^e, then the entries of the space-time array are constrained, making it easier to achieve a given level of diversity protection (see Section 7.7).

The case $N = 2$ A real orthogonal design of type (1,1) and size $N = 2$ corresponds to the representation of the complex numbers $\mathbf{C}$ as a 2×2 matrix algebra over the real numbers $\mathbf{R}$. The complex number $x_0 + ix_1$ corresponds to the matrix $\begin{pmatrix} x_0 & x_1 \\ -x_1 & x_0 \end{pmatrix}$.

The case $N = 4$ A real orthogonal design of type $(1,1,1,1)$ and size $N = 4$ corresponds to the representation of the quaternions $\mathfrak{Q}$ as a 4×4 matrix algebra over the real numbers $\mathbf{R}$. The quaternion $x_0 + \mathbf{i}x_1 + \mathbf{j}x_2 + \mathbf{k}x_3$ corresponds to the matrix

$$
\begin{bmatrix}
x_0 & x_1 & x_2 & x_3 \\
-x_1 & x_0 & -x_3 & x_2 \\
-x_2 & x_3 & x_0 & -x_1 \\
-x_3 & -x_2 & x_1 & x_0
\end{bmatrix}
= x_0 \mathbf{I}_4 + x_1
\begin{bmatrix}
& 1 & & \\
-1 & & & \\
& & & -1 \\
& & 1 &
\end{bmatrix}
$$

$$
+ x_2
\begin{bmatrix}
& & 1 & \\
& & & 1 \\
-1 & & & \\
& -1 & &
\end{bmatrix}
+ x_3
\begin{bmatrix}
& & & 1 \\
& & -1 & \\
& 1 & & \\
-1 & & &
\end{bmatrix}.
$$

A *complex orthogonal design* of size N and type $(u_0, u_1, \ldots, u_{s-1}; v_1, v_2, \ldots, v_t)$ is a matrix $\mathbf{Z} = \mathbf{X} + \mathbf{iY}$, where $\mathbf{X}$ and $\mathbf{Y}$ are real orthogonal designs of type $(u_0, u_1, \ldots, u_{s-1})$ and $(v_1, v_2, \ldots, v_t)$ respectively, and where

$$
\mathbf{ZZ}^* = \left(\left(\sum_{j=0}^{s-1} u_j x_j^2 \right) + \left(\sum_{j=1}^{t} v_j y_j^2 \right) \right) \mathbf{I}_N.
$$

Since

$$
\mathbf{ZZ}^* = (\mathbf{X} + \mathbf{iY})(\mathbf{X}^T + \mathbf{iY}^T)
$$
$$
= (\mathbf{XX}^T + \mathbf{YY}^T) + \mathbf{i}(\mathbf{YX}^T - \mathbf{XY}^T),
$$

it follows that $\mathbf{YX}^T = \mathbf{XY}^T$. A pair of real orthogonal designs that is connected in this way is called an *amicable pair* (see Geramita and Seberry, 1979, for more information). Note that if $t = s$, then the entries of $\mathbf{X} + \mathbf{iY}$ are linear combinations of the complex variables $z_k = x_k + \mathbf{i}y_k$ and their complex conjugates $z_k^* = x_k - \mathbf{i}y_k$. In fact, the definition of a complex orthogonal design found in Tarokh *et al.* (1999) is given in terms of these variables. The rate of a complex orthogonal design is defined to be $(s + t)/2N$.

A complex design of size N with $t = s + 1$ determines a real orthogonal design of size $2N$ through the substitution

$$
x_0 + \mathbf{i}x_1 \rightarrow \begin{pmatrix} x_0 & x_1 \\ -x_1 & x_0 \end{pmatrix}.
$$

The case $N = 2$ This is the Alamouti space-time block code. We may view quaternions as pairs of complex numbers, where the product of quaternions (a, b) and (c, d) is given by $(ac - bd^*, ad + bc^*)$. These are Hamilton's

Biquaternions, and if we associate the pair (a, b) with the 2×2 complex matrix $\left(\begin{smallmatrix} a & b \\ -b^* & a^* \end{smallmatrix}\right)$, then we see that the rule for multiplying biquaternions coincides with the rule for matrix multiplication. The algebraic structure of the Alamouti code (closure under addition, multiplication and taking inverses) plays an essential role in detection of two separate space-time coded transmissions at a mobile terminal that is able to employ a second receive antenna (see Section 7.10).

The case $N = 4$ The representation of the octonions or Cayley numbers as 4-tuples of complex numbers provides an example of an extremal complex design. The product $\mathbf{c} = \mathbf{ab}$ of octonions $\mathbf{a} = (a_0, a_1, a_2, a_3)$ and $\mathbf{b} = (b_0, b_1, b_2, 0)$ is given by

$$
\begin{aligned}
c_0 &= a_0 b_0 - b_1^* a_1 - b_2^* a_2 - a_3^* b_3 \\
c_1 &= b_1 a_0 + a_1 b_0^* - a_3 b_2^* + b_3 a_2^* \\
c_2 &= b_2 a_0 - a_1^* b_3 + a_2 b_0^* + b_1^* a_3 \\
c_3 &= b_3 a_0^* + a_1 b_2 - b_1 a_2 + a_3 b_0.
\end{aligned}
$$

It follows that right multiplication of an octonion $\mathbf{a}$ by octonions of the form $\mathbf{b} = (b_0, b_1, b_2, 0)$ can be represented as $\mathbf{ab} = \mathbf{a}R(b_0, b_1, b_2, 0)$, where

$$
R(b_0, b_1, b_2, 0) = \begin{pmatrix} b_0 & b_1 & b_2 & 0 \\ -b_1^* & b_0^* & 0 & b_2 \\ -b_2^* & 0 & b_0^* & -b_1 \\ 0 & -b_2^* & b_1^* & b_0 \end{pmatrix}.
$$

The columns of this matrix are orthogonal. Hence, $R(b_0, b_1, b_2, 0)$ is a rate-3/4 complex orthogonal design.

7.6.1 Fundamental limits on the rate of real orthogonal designs

We may relax the definition of a real orthogonal design $\mathbf{X}$ to allow entries that are linear combinations of variables, and to allow identities of the form

$$
\mathbf{X}\mathbf{X}^T = \mathrm{diag}[d_{i0}x_0^2 + \cdots + d_{is-1}x_{s-1}^2] = \sum_{i=0}^{s-1} x_i^2 \mathbf{D}_i,
$$

where $\mathbf{D}_i$ is a diagonal matrix with positive entries $d_{ij}, j = 0, 1, \ldots, N - 1$. We write $\mathbf{x} = \sum_{i=0}^{s-1} x_i A_i$, so that

$$
\begin{aligned}
\mathbf{A}_i \mathbf{A}_i^T &= \mathbf{D}_i, & \text{for } i = 0, 1, \ldots, s - 1 \\
\mathbf{A}_i \mathbf{A}_j^T &= -\mathbf{A}_j \mathbf{A}_i^T, & \text{for } 0 \leq i < j \leq s - 1.
\end{aligned}
$$

Let $\mathbf{D}_i^{1/2}$ be the diagonal matrix with positive entries $d_{ij}^{1/2}, j = 0, 1, \ldots, s-1$, and let $\mathbf{A}_i = \mathbf{B}_i \mathbf{D}_i^{1/2}$. Then we have

$$\mathbf{B}_i \mathbf{B}_i^T = \mathbf{I}_N, \qquad \text{for } i = 0, 1, \ldots, s-1 \tag{7.1}$$

$$\mathbf{B}_i \mathbf{B}_j^T = -\mathbf{B}_j \mathbf{B}_i^T, \qquad \text{for } 0 \le i < j \le s-1. \tag{7.2}$$

We may take $\mathbf{B}_0 = \mathbf{I}_N$ (since replacing $\mathbf{B}_i$ by $\mathbf{B}_i \mathbf{B}_0^T$ provides a representation of the same design with respect to a different orthogonal basis), so that

$$\mathbf{B}_i = -\mathbf{B}_i^T \text{ for } i = 0, 1, \ldots, s-1. \tag{7.3}$$

Equations (7.1), (7.2), and (7.3) are the defining conditions for a *Hurwitz–Radon family of matrices*. The next theorem, proved by Radon (1922), provides an upper bound on the rate of a real orthogonal design.

Note that the matrix $\mathbf{M} = \sum_{i=0}^{s-1} x_i B_i$ is a real orthogonal design of type $(1, 1, \ldots, 1)$ since

$$\mathbf{M} \mathbf{M}^T = (x_0^2 + \cdots + x_{s-1}^2) \mathbf{I}_N.$$

Hence, our relaxation of the original definition does not extend the range of values of N for which there exists a real orthogonal design with a given rate. A similar analysis applies to complex designs.

Theorem 7.1 *Given $N = 2^{4a+b} N_0$, where N_0 is odd, define $\rho(N) = 8a + 2^b$. Then*

(i) *the size of a Hurwitz–Radon Family of real $N \times N$ matrices is at most $\rho(N) - 1$, and*

(ii) *there exists a Hurwitz–Radon Family containing exactly $s = \rho(N) - 1$ integer $N \times N$ matrices.*

The optimal orthogonal designs are constructed from an extraspecial 2-group that plays a central role in quantum error correction (see Calderbank and Naguib, 2001, for more details).

7.6.2 Fundamental limits on the rate of complex orthogonal designs

We consider amicable pairs $\mathbf{X}, \mathbf{Y}$ of real orthogonal designs of size N, where $\mathbf{X}$ has type $(u_0, \ldots, u_s)$ and $\mathbf{Y}$ has type $(v_1, \ldots, v_t)$. We write $\mathbf{X} = \sum_{i=0}^{s} x_i A_i$ and $\mathbf{Y} = \sum_{i=1}^{t} y_i B_i$, so that

$$\mathbf{A}_i \mathbf{A}_i^T = u_i \mathbf{I}_N, \text{ for } i = 0, 1, \ldots, s, \qquad \mathbf{B}_i \mathbf{B}_i^T = v_i \mathbf{I}_N, \text{ for } i = 1, \ldots, t,$$

$$\mathbf{A}_i \mathbf{A}_j^T = -\mathbf{A}_j \mathbf{A}_i^T, \text{ for } 0 \le i < j \le s, \qquad \mathbf{B}_i \mathbf{B}_j^T = -\mathbf{B}_j \mathbf{B}_i^T, \text{ for } 1 \le i < j \le t,$$

and $\quad \mathbf{A}_i \mathbf{B}_j^T = \mathbf{B}_j \mathbf{A}_i^T$, for $i = 0, 1, \ldots, s$ and $j = 1, \ldots, t$.

We now view the amicable pair with respect to a different orthogonal basis by setting

$$\boldsymbol{\alpha_i} = \mathbf{A}_i \mathbf{A}_0^T / (u_0 u_i)^{\frac{1}{2}}, \text{ for } i = 0, 1, \ldots, s,$$

$$\boldsymbol{\beta_j} = \mathbf{B}_j \mathbf{A}_0^T / (u_0 v_j)^{\frac{1}{2}}, \text{ for } j = 1, \ldots, t.$$

Now $\boldsymbol{\alpha_0} = \mathbf{I}_N$,

$$2\boldsymbol{\alpha_i} = -\boldsymbol{\alpha_i^T}, \text{ for } i = 1, \ldots, s, \quad \boldsymbol{\beta_i} = \boldsymbol{\beta_i^T}, \text{ for } i = 1, \ldots, t, \quad (7.4)$$

$$\boldsymbol{\alpha_i^2} = -\mathbf{I}_N, \text{ for } i = 1, \ldots, s, \quad \boldsymbol{\beta_i^2} = \mathbf{I}_N, \text{ for } i = 1, \ldots, t, \quad (7.5)$$

$$\boldsymbol{\alpha_i \alpha_j} + \boldsymbol{\alpha_j \alpha_i} = 0, \text{ for } 1 \le i < j \le s, \quad (7.6)$$

$$\boldsymbol{\beta_k \beta_1} + \boldsymbol{\beta_1 \beta_k} = 0, \text{ for } 1 \le k < l \le t, \quad (7.7)$$

$$\text{and } \boldsymbol{\alpha_i \beta_j} = -\boldsymbol{\beta_j \alpha_i}, \text{ for } i = 1, \ldots, s \text{ and } j = 1, \ldots, t. \quad (7.8)$$

Equations (7.4)–(7.8) are a form of the relations found by Clifford (1878) in his attempt to generalize the quaternions. Formal algebras over $\mathbf{R}$ that satisfy these relations are called *Clifford Algebras*.

Given t symmetric, anticommuting orthogonal matrices of size N, let $\rho_t(N) - 1$ be the number of skew-symmetric, anticommuting orthogonal matrices of size N that anticommute with the initial set of t matrices. The next two theorems are due to Wolfe (1976).

Theorem 7.2 *There exists an amicable pair* $\mathbf{X}, \mathbf{Y}$ *of real orthogonal designs of size* N, *where* $\mathbf{X}$ *has type* $(1, \ldots, 1)$ *on variables* $x_0, x_1, \ldots, x_{s-1}$ *and* $\mathbf{Y}$ *has type* $(1, \ldots, 1)$ *on variables* $y_1, y_2, \ldots, y_t$, *if and only if* $s \le \rho_t(N) - 1$.

Theorem 7.3 *Let* $\mathbf{X}, \mathbf{Y}$ *be an amicable pair of real orthogonal designs of size* $N = 2^h N_0$, *where* N_0 *is odd. Then, the total number of real variables in* $\mathbf{X}$ *and* $\mathbf{Y}$ *is at most* $2h + 2$ *and this bound is achieved by designs* $\mathbf{X}, \mathbf{Y}$ *that each involve* $h + 1$ *variables.*

In fact, the extraspecial group mentioned at the end of Section 7.6.1 can be used to construct pairs $\mathbf{X}, \mathbf{Y}$ where the entries of $\mathbf{X}$ are $0, \pm x_0, \ldots, \pm x_s$ and the entries of $\mathbf{Y}$ are $0, \pm y_1, \ldots, \pm y_t$.

7.7 Quasi-orthogonal space-time block codes

Given the importance of high data rates and the scarcity of full-rate orthogonal designs, it is natural to relax the requirement that linear processing at the receiver be able to separate all transmitted symbols. The lack of a full rate complex design for 4 transmit antennas motivated Jafarkhani (2001) to consider the code

$$\mathbf{Z} = \left[\begin{array}{cc|cc} z_1 & z_2 & z_3 & z_4 \\ -z_2^* & z_1^* & -z_4^* & z_3^* \\ \hline -z_3^* & -z_4^* & z_1^* & z_2^* \\ z_4 & -z_3 & -z_2 & z_1 \end{array} \right],$$

where the structure of the 2×2 blocks is parallel to that of the Alamouti code. The columns $\mathbf{Z_i}, i = 1, 2, 3, 4$ of the matrix $\mathbf{Z}$ are subdivided into two groups $\{\mathbf{Z_1}, \mathbf{Z_4}\}$ and $\{\mathbf{Z_2}, \mathbf{Z_3}\}$ such that columns from different groups are orthogonal. Jafarkhani shows that the maximum-likelihood decision metric is a sum $f(z_1, z_4) + g(z_2, z_3)$ where f is independent of z_2, z_3 and g is independent of z_1, z_4. The decoder finds the pair (c_1, c_4) that minimizes $f(z_1, z_4)$ and (independently) the pair (c_2, c_3) that minimizes $g(z_2, z_3)$. For this example, the decoding complexity is quadratic in the size of the signal constellation.

Jafarkhani abstracted the notion of a *quasi-orthogonal space-time block code*, where the columns of the design are divided into groups, and columns from different groups are orthogonal. Decoding complexity is polynomial in the size of the signal constellation with the exponent equal to the size of the largest group of columns.

Many different 4×4 designs are possible and all achieve diversity two. Su and Xia (2004) (see also references therein) showed that it is often possible to achieve full diversity by selectively rotating constituent signal constellations. We illustrate this process for the quasi-orthogonal design

$$\mathbf{Z} = \left[\begin{array}{cc|cc} z_1 & z_2 & z_3 & z_4 \\ z_2 & z_1 & z_4 & z_3 \\ \hline -z_3^* & -z_4^* & z_1^* & z_2^* \\ -z_4^* & -z_3^* & z_2^* & z_1^* \end{array} \right], \tag{7.9}$$

for which

$$\det(\mathbf{Z}) = \prod_{\varepsilon = \pm 1} (|z_1 + \varepsilon z_2|^2 + |z_3 + \varepsilon z_4|^2).$$

Here, it is possible to achieve full diversity by rotating the signal constellation for z_1 and z_3 appropriately. For example, if z_2 and z_4 are taken from QPSK, then rotating z_1 and z_3 through $90°$ provides full diversity.

7.8 Superorthogonal space-time trellis codes

Improvements have been made to the original space-time trellis codes, but for one receive antenna, the performance gains are modest (Baro *et al.*, 2000;

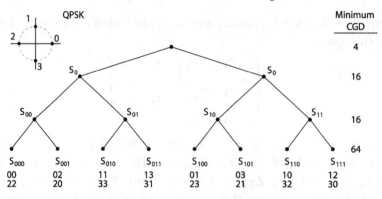

Fig. 7.4. Set partitioning for QPSK, where ab represents $\left(\begin{smallmatrix} a & b \\ -b^* & a^* \end{smallmatrix}\right)$.

Chen *et al.*, 2001)). Siwamogsatham and Fitz (2002b) showed how to achieve larger gains and greater robustness to correlated fading by concatenating an outer trellis code with an inner space-time block code. The drawback to this approach is the loss in data rate, which motivated the independent construction of superorthogonal space-time trellis codes by Jafarkhani and Seshadri (2003) and by Siwamogsatham and Fitz (2002a). (The earliest example of a superorthogonal space-time trellis code for 4-PSK is due to Ionescu *et al.* (2001).) The essential idea is to expand the universe of available orthogonal designs without expanding the size of the signal constellation. It applies to arbitrary numbers of antennas and to arbitrary signal constellations, though our treatment here is focused on QPSK transmission with two transmit antennas.

If rotation through ϕ fixes a complex signal constellation, then the translation $\mathbf{C}(a, b, \phi)$ of an Alamouti codeword $\mathbf{C}(a, b, 0)$ is available for transmission.

$$\mathbf{C}(a, b, \phi) = \begin{bmatrix} a\,e^{i\phi} & b \\ -b^*\,e^{i\phi} & a^* \end{bmatrix} = \begin{bmatrix} a & b \\ -b^* & a^* \end{bmatrix} \begin{bmatrix} e^{i\phi} & \\ & 1 \end{bmatrix}.$$

Superorthogonal codes designed for two transmit antennas employ a set partitioning of Alamouti codewords, and the partition for QPSK is shown in Fig. 7.4. The probability of confusing codewords on a fading channel motivates the following definition of *coding gain distance* (CGD):

$$\mathrm{CGD}[\mathbf{C}(a, b, 0), \mathbf{C}(c, d, 0)] = \det(\mathbf{C}(a - c, b - d, 0)\mathbf{C}(a - c, b - d, 0)^*).$$

The minimum coding gain distance between codewords in the same subset of a partition increases with depth in the partition tree.

The trellis representation of a 4-state superorthogonal code for QPSK is

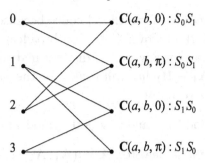

Fig. 7.5. A 4-state superorthogonal trellis code for QPSK.

shown in Fig. 7.5. Orthogonal matrices are selected from $\mathbf{C}(a, b, 0)$ when the trellis encoder is in state 0 or 2, and from $\mathbf{C}(a, b, \pi)$ when the trellis encoder is in state 1 or 3. The rate is 2 bits/s/Hz, which is the best possible for a diversity-2 space-time code employing QPSK. The probability of confusing codewords $\mathbf{C}(a_i, b_i, \phi_i)$, $i = 1, 2, \ldots, K$, and $\mathbf{C}(c_i, d_i, \eta_i)$, $i = 1, 2, \ldots, K$, is inversely proportional to

$$\det\left[\sum_{i=1}^{K}\left(\mathbf{C}(a_i, b_i, \phi_i) - \mathbf{C}(c_i, d_i, \eta_i)\right)\left(\mathbf{C}(a_i, b_i, \phi_i) - \mathbf{C}(c_i, d_i, \eta_i)\right)^*\right].$$

Since each 2×2 matrix appearing in this sum is positive semidefinite, it follows that coding gain distance determines a metric that can be tracked by the Viterbi algorithm (Siwamogsatham and Fitz, 2002a). The minimum CGD for this code is 16 (Jafarkhani and Seshadri, 2003), which is better by a factor 4 than the 4-state code for QPSK reported by Tarokh *et al.* (1998).

7.9 Multilevel space-time codes based on binary partitions of QAM and PSK constellations

Given an L-level binary partition of a QAM or PSK signal constellation, a space-time codeword is an array $\mathbf{G} = (\mathbf{G_1}, \mathbf{G_2}, \ldots, \mathbf{G_L})$ determined by a sequence $\mathbf{G_1}, \mathbf{G_2}, \ldots, \mathbf{G_L}$ of binary matrices, where the matrix $\mathbf{G_i}$ specifies the space-time array at level i. A *multilevel space-time code* is determined by constituent binary codes $\mathcal{C}_1, \mathcal{C}_2, \ldots, \mathcal{C}_L$, and for $i = 1, \ldots, L$ the binary matrix $\mathbf{G_i}$ is required to be a codeword in $\mathcal{C}_i$.

El-Gamal and Hammons (2001) showed that it is possible to design codes with a given level of diversity in the complex domain by placing constraints on the constituent binary codes $\mathcal{C}_i$, and their approach has been taken up by many authors (see the work by Lu and Kumar (2005) and the references therein). Instead of coding across all levels at once, we develop the theory from the perspective of multilevel codes because we want to make a connection with the work of Cover (1972) on broadcast channels.

For simplicity, suppose for any pair of distinct matrices $\mathbf{A_1}, \mathbf{B_1}$ in $\mathcal{C}_1$ that the column space of $\mathbf{A_1} - \mathbf{B_1}$ has full rank over the binary field. Then, by definition, for a binary vector $\mathbf{v}$,

$$\mathbf{v}(\mathbf{A_1} - \mathbf{B_1}) \equiv \mathbf{0} \pmod 2 \quad \text{and} \quad \mathbf{v} \not\equiv \mathbf{0} \pmod 2 \quad \text{implies } \mathbf{A_1} = \mathbf{B_1}.$$

Now consider space-time codewords $\mathbf{A} = (\mathbf{A_1}, \mathbf{A_2}, \ldots, \mathbf{A_L})$ and $\mathbf{B} = (\mathbf{B_1}, \mathbf{B_2}, \ldots, \mathbf{B_L})$ for QAM, and suppose there is a nonzero complex vector $\mathbf{w}$ in $\mathbf{C^N}$ that is orthogonal to every column of $\mathbf{A} - \mathbf{B}$. We may suppose that the entries of $\mathbf{w}$ are Gaussian integers (this is the analog for the field $\mathbf{Q(i)}$ of the well known fact that an integer matrix that is singular over the rational numbers has a singular vector with integer entries) and that not every entry is divisible by $1 + \mathbf{i}$. We start with the equation $\mathbf{w}(\mathbf{A} - \mathbf{B}) = \mathbf{0}$ in the Gaussian integers $\mathbf{Z[i]}$ and reduce modulo the prime $1 + \mathbf{i}$ to obtain

$$\mathbf{v}(\mathbf{A_1} - \mathbf{B_1}) \equiv \mathbf{0} \pmod 2,$$

where $\mathbf{v}$ is the reduction of $\mathbf{w}$ modulo $1 + \mathbf{i}$. The quotient of the Gaussian integers $\mathbf{Z[i]}$ by the ideal generated by $1 + \mathbf{i}$ is the binary field, and since $\mathbf{v} \not\equiv \mathbf{0} \pmod 2$, we conclude that $\mathbf{A_1} = \mathbf{B_1}$. The input bits that select the codeword from the first constituent code $\mathcal{C}_1$ are protected by full diversity in the complex domain. This argument generalizes as follows:

Theorem 7.4 *Let $\mathcal{C}$ be a multilevel space-time code for QAM transmission with N transmit antennas that is determined by constituent binary codes $\mathcal{C}_1, \mathcal{C}_2, \ldots, \mathcal{C}_L$. Let $d_1 \leq d_2 \leq \cdots \leq d_L$, and suppose for any pair of distinct matrices $\mathbf{A_i}, \mathbf{B_i}$ in $\mathcal{C}_i$ that the column space of $\mathbf{A_i} - \mathbf{B_i}$ has rank at least $N - d_i$ over the binary field. Then the input bits that select the codeword from the ith constituent code $\mathcal{C}_i$ are protected by diversity $N - d_i$ in the complex domain.*

Our initial argument used a complex vector with some entry not divisible by $1 + \mathbf{i}$, and the generalization used in the proof of Theorem 7.4 is a sublattice of $\mathbf{Z[i]}^N$, for which reduction modulo the Gaussian prime $1 + \mathbf{i}$ yields a binary vector space of dimension d_i. A similar result applies to multilevel space-time codes for PSK constellations.

A natural choice for a component code is a set $\mathbf{K_d}$ of binary $N \times N$ matrices such that for any distinct pair of matrices $\mathbf{A}, \mathbf{B} \in \mathbf{K_d}$ the rank of $\mathbf{A} - \mathbf{B}$ is at least $N - d$. The size of $\mathbf{K_d}$ is at most $2^{(d+1)N}$, since the first $d + 1$ rows of $\mathbf{A}$ and $\mathbf{B}$ must be distinct, and there is a classical example that achieves the bound. Let $f : x \to x^2$ be the Frobenius map on $GF(2^m)$ and let

$$\mathbf{K_d} = \{\alpha_0 + \alpha_1 f + \cdots + \alpha_d f^d | \alpha_i \in GF(2^N)\}.$$

Then $\mathbf{K_d}$ is closed under binary addition, and given a choice of basis for $GF(2^N)$ over $GF(2)$, each map in $\mathbf{K_d}$ may be viewed as an $N \times N$ binary matrix. There are at most 2^d elements of the null space of $\mathbf{A} - \mathbf{B}$, since each element is the root of a polynomial

$$\gamma_0 + \gamma_1 x^2 + \cdots + \gamma_d x^{2^d}$$

with degree at most 2^d. Hence, the rank of $\mathbf{A} - \mathbf{B}$ is at least $N - d$.

7.9.1 Broadcast channels

Rate and diversity impose a fundamental trade-off in space-time coding. High rate space-time codes come at the cost of lower diversity and high reliability (high diversity) implies a lower rate. Diggavi *et al.* (2004) have departed from the standard practice of designing to a single point on the rate-diversity trade-off curve by constructing high rate space-time codes that have a high diversity code embedded within them. The space-time block code

$$\left[\begin{array}{cc|cc} a_1 & a_2 & b_3 & b_4 \\ -a_2^* & a_1^* & b_4^* & -b_3^* \\ \hline b_1 & b_2 & a_1^* & -a_2 \\ -b_2^* & b_1 & a_2^* & a_1 \end{array} \right]$$

achieves diversity 3 for variables a_1 and a_2, and diversity 2 for variables b_1, b_2, b_3 and b_4. Embedded diversity can employ multilayer space-time codes, and the effect is to change spatial diversity into a fine grained resource that can provide opportunistic communication when the channel is good and reliable communication with latency guarantees when it is less benign. Embedded diversity plays the role of superposition coding in the landmark paper by Cover (1972) on broadcast channels. Here, codewords are organized into clusters with the idea that a receiver with a poor signal-to-noise ratio (SNR) will only be able to separate clusters, whereas a receiver with a good SNR will be able to separate codewords within a cluster. Diversity in the wireless world then plays the role of distance on the Gaussian broadcast channel. Different levels of diversity protection are a natural fit to many speech and image coding schemes, where some of the coded bits are very sensitive to channel errors, while others exhibit very little sensitivity.

7.10 The decorrelating detector and MMSE interference cancellation

We describe how a receiver can employ a second antenna to separate two cochannel users, each using the Alamouti space-time block code. Consider vectors $\mathbf{r}_1, \mathbf{r}_2$, where the entries of $\mathbf{r}_i$ are the signals received at antenna i over two consecutive time slots. If $\mathbf{c} = (c_1, c_2)$ and $\mathbf{s} = (s_1, s_2)$ are the codewords transmitted by the first and second users, then

$$\mathbf{r} = \begin{pmatrix} \mathbf{r}_1 \\ \mathbf{r}_2 \end{pmatrix} = \begin{pmatrix} \mathbf{H}_1 & \mathbf{G}_1 \\ \mathbf{H}_2 & \mathbf{G}_2 \end{pmatrix} \begin{pmatrix} \mathbf{c} \\ \mathbf{s} \end{pmatrix} + \begin{pmatrix} \mathbf{w}_1 \\ \mathbf{w}_2 \end{pmatrix},$$

where the vectors $\mathbf{w}_1$ and $\mathbf{w}_2$ are complex Gaussian random variables with zero mean and covariance $N_0 \mathbf{I}_2$. The matrices $\mathbf{H}_1$ and $\mathbf{H}_2$ capture the path gains from the first user to the first and second receive antennas. The matrices $\mathbf{G}_1$ and $\mathbf{G}_2$ capture the path gains from the second user to the first and second receive antennas. What is important is that all these matrices share the Alamouti structure. Define

$$\mathbf{D} = \begin{pmatrix} \mathbf{I}_2 & -\mathbf{G}_1 \mathbf{G}_2^{-1} \\ -\mathbf{H}_2 \mathbf{H}_1^{-1} & \mathbf{I}_2 \end{pmatrix}$$

and observe that

$$\mathbf{D}\mathbf{r} = \begin{pmatrix} \mathbf{H} & 0 \\ 0 & \mathbf{G} \end{pmatrix} \begin{pmatrix} \mathbf{c} \\ \mathbf{s} \end{pmatrix} + \begin{pmatrix} \tilde{\mathbf{w}}_1 \\ \tilde{\mathbf{w}}_2 \end{pmatrix},$$

where $\mathbf{H} = \mathbf{H}_1 - \mathbf{G}_1 \mathbf{G}_2^{-1} \mathbf{H}_2$ and $\mathbf{G} = \mathbf{G}_2 - \mathbf{H}_2 \mathbf{H}_1^{-1} \mathbf{G}_1$.

The matrix $\mathbf{D}$ transforms the problem of joint detection of two cochannel users into separate detection problems of two space-time users. It plays the role of the decorrelating detector in CDMA systems. Detection of the codeword $\mathbf{c}$ is performed through projection onto the orthogonal complement of $[\mathbf{G}_1^{\mathbf{T}}, \mathbf{G}_2^{\mathbf{T}}]$. The algebraic structure of the Alamouti code (closure under addition, multiplication and taking inverses) implies that the matrices $\mathbf{H}$ and $\mathbf{G}$ have the same structure as $\mathbf{H}_1, \mathbf{H}_2, \mathbf{G}_1,$ and $\mathbf{G}_2$.

Remark Given K synchronous cochannel users, each employing a space-time block code designed for N transmit antennas, there will be KN interfering signals arriving at the receiver. Winters et al. (1994) have shown that $N(K-1)+1$ antennas at the receiver are able to suppress $N(K-1)$ interfering signals and provide diversity order N to the desired user. This assumes no correlation between the interfering signals. The value of correlation in space-time communication is that only K receive antennas are needed to suppress $K-1$ space-time users, while providing diversity order N to the desired user. This means that for the purpose of interference suppression at the receiver, each transmitter can be viewed as having only a single transmit antenna.

Next, we show how the algebraic structure of the Alamouti code leads to a single receiver structure that cancels interference when it is present and delivers increased diversity gain when it is not. The covariance matrix $\mathbf{M}$ of the received signal is given by

$$\mathbf{M} = \mathbb{E}[\mathbf{r}\mathbf{r}^*] = \underbrace{\begin{pmatrix} \mathbf{H}_1 \\ \mathbf{H}_2 \end{pmatrix} (\mathbf{H}_1^*\mathbf{H}_2^*)}_{\substack{\text{orthogonal proj}^n \\ \text{on } \langle \mathbf{h}_1, \mathbf{h}_2 \rangle}} + \underbrace{\begin{pmatrix} \mathbf{G}_1 \\ \mathbf{G}_2 \end{pmatrix} (\mathbf{G}_1^*\mathbf{G}_2^*)}_{\substack{\text{orthogonal proj}^n \\ \text{on } \langle \mathbf{g}_1, \mathbf{g}_2 \rangle}} + \frac{1}{\text{SNR}}\mathbf{I}_4$$

and if

$$\begin{pmatrix} \mathbf{H}_1 \\ \mathbf{H}_2 \end{pmatrix} = (\mathbf{h}_1, \mathbf{h}_2) \quad \text{and} \quad \begin{pmatrix} \mathbf{G}_1 \\ \mathbf{G}_2 \end{pmatrix} = (\mathbf{g}_1, \mathbf{g}_2)$$

then it can be shown that if $i \neq j$, then for all integers k, we have $\mathbf{h}_i\mathbf{M}^k\mathbf{h}_j^* = \mathbf{g}_i\mathbf{M}^k\mathbf{g}_j^* = 0$ (Naguib *et al.*, 1998a; Naguib and Seshadri, 1998; Calderbank and Naguib, 2001, Section 4).

The MMSE receiver looks for a linear combination $\boldsymbol{\alpha}^*\mathbf{r}$ of received signals that is close to some linear combination $\beta_1 c_1 + \beta_2 c_2$ of the codeword $\mathbf{c}$. The solution turns out to be

$$\alpha_1 = (\mathbf{m} - \mathbf{h}_2\mathbf{h}_2^*)^{-1}\mathbf{h}_1, \qquad \beta_1 = 1, \qquad \beta_2 = \frac{\mathbf{h}_2^*\mathbf{M}^{-1}\mathbf{h}_1}{1 - \mathbf{h}_2^*\mathbf{M}^{-1}\mathbf{h}_1},$$

$$\alpha_2 = (\mathbf{M} - \mathbf{h}_1\mathbf{h}_1^*)^{-1}\mathbf{h}_2, \qquad \beta_2 = 1, \qquad \beta_1 = \frac{\mathbf{h}_1^*\mathbf{M}^{-1}\mathbf{h}_2}{1 - \mathbf{h}_1^*\mathbf{M}^{-1}\mathbf{h}_2}.$$

Either $\beta_1 = 0$ and $\beta_2 = 1$ or $\beta_2 = 0$ and $\beta_1 = 1$! The MMSE interference canceler maintains the separate detection feature of space-time block codes. Errors in decoding $\mathbf{c}_1$ do not influence the decoding of $\mathbf{c}_2$ and vice versa.

References

Al-Dhahir, N., Fragouli, C., Stamoulis, A., Younis, W., and Calderbank, A. R. (2002). Space-time processing for broadband wireless access. *IEEE Commun. Mag.*, **40** (9), 136–142.

Alamouti, S. (1998). Space-block coding: A simple transmitter diversity technique for wireless communications. *IEEE J. Sel. Areas Commun.*, **16** (8), 1451–1458.

Baro, S., Bauch, G., and Hansmann, A. (2000). Improved codes for space-time trellis-coded modulation. *IEEE Commun. Lett.*, **4** (1), 20–22.

Belfiore, J.-C., Rekaya, G., and Viterbo, E. (2005). The Golden code: A 2 × 2 full-rate space-time code with non-vanishing determinants. *IEEE Trans. Inf. Theory*, **51** (4), 1432–1436.

Calderbank, A. R. and Naguib, A. F. (2001). Orthogonal designs and third generation wireless communication. In Hirschfeld, J. W. P. (ed.), *Surveys in Combinatorics 2001*, London Math. Soc. Lecture Note Series 288, 75–107 (Cambridge Univ. Press).

Chen, Z., Yuan, J., and Vucetic, B. (2001). Improved space-time trellis coded modulation scheme on slow Rayleigh fading channels. *Electron. Lett.*, **37** (7), 440–441.

Clifford, W. K. (1878). Applications of Grassman's extensive algebra. *Am. J. Math.*, **1**, 350–358.

Cover, T. M. (1972). Broadcast channels. *IEEE Trans. Inf. Theory*, **18** (1), 2–14.

Damen, M. O., Chkeif, A., and Belfiore, J.-C. (2000). Lattice codes decoder for space-time codes. *IEEE Commun. Lett.*, **4** (5), 161–163.

Diggavi, S. N., Al-Dhahir, N., and Calderbank, A. R. (2004). Diversity embedding in multiple antenna communications. In *Advances in Network Information Theory*, vol. 66 of *DIMACS Series in Discrete Math. and Theoretical Comput. Science*, 285–301 (AMS).

El-Gamal, H. and Hammons, A. R. (2001). A new approach to layered space-time coding and signal processing. *IEEE Trans. Inf. Theory*, **47** (6), 2321–2334.

Foschini, G. J. (1996). Layered space-time architecture for wireless communication in a fading environment when using multi-element antennas. *Bell Labs Tech. J.*, **1** (2), 41–59.

Geramita, A. V. and Seberry, J. (1979). Orthogonal designs, quadratic forms and Hadamard matrices. In *Lecture Notes in Pure and Appl. Math.*, vol. 43 (Marcel Dekker, New York).

Guey, J.-C., Fitz, M. P., Bell, M. R., and Kuo, W.-Y. (1999). Signal design for transmitter diversity wireless communication systems over Rayleigh fading channels. *IEEE Trans. Commun.*, **47** (4), 527–537.

Hassibi, B. and Hochwald, B. M. (2002). High rate codes that are linear in space and time. *IEEE Trans. Inf. Theory*, **48** (7), 1804–1824.

Ionescu, M., Mukkavilli, K. K., Yan, Z., and Lilleberg, J. (2001). Improved 8- and 16-state space-time codes for 4PSK with two transmit antennas. *IEEE Commun. Lett.*, **5** (7), 301–305.

Jafarkhani, H. (2001). A quasi-orthogonal space-time block code. *IEEE Trans. Commun.*, **49** (1), 1–4.

Jafarkhani, H. and Seshadri, N. (2003). Super-orthogonal space-time trellis codes. *IEEE Trans. Inf. Theory*, **49** (4), 937–950.

Liu, Y., Fitz, M. P., and Takeshita, O. Y. (1999). Full rate space-time turbo codes. *IEEE J. Sel. Areas Commun.*, **19** (5), 969–980.

Lu, B., Wang, X., and Narayanan, K. (2002). LDPC based space-time coded OFDM systems over correlated fading channels: Performance analysis and receiver design. *IEEE Trans. Commun.*, **50** (2), 74–88.

Lu, H. F. and Kumar, P. V. (2005). A unified construction of space-time codes with optimal rate-diversity tradeoff. *IEEE Trans. Inf. Theory*, **51** (5), 1709–1730.

Naguib, A. and Seshadri, N. (1998). Combined interference cancellation and ML decoding of space-time block codes. In *Proc. IEEE Commun. Theory Mini Conf. (held in conjunction with GLOBECOM '98)*, 7–15.

Naguib, A., Seshadri, N., and Calderbank, A. R. (1998a). Applications of space-time block codes and interference suppression for high capacity and high data rate wireless systems. In *32nd Asilomar Conf. Signals, Syst., Comput.*, vol. 2, 1803–1810.

Naguib, A., Tarokh, V., Seshadri, N., and Calderbank, A. R. (1998b). A space-time coding based modem for high data rate wireless communications. *IEEE J. Sel. Areas Commun.*, **16** (2), 1459–1478.

Radon, J. (1922). Lineare Scharen orthogonaler Matrizen. *Abh. Math. Sem. Hamburg*, **1**, 1–14.

Siwamogsatham, S. and Fitz, M. P. (2002a). Improved high-rate space-time codes via orthogonality and set partitioning. In *Proc. IEEE Wireless Commun. and Netw. Conf.*, vol. 1, 264–270.

—— (2002b). Robust space-time signal processing for correlated Rayleigh fading channels. *IEEE Trans. Signal Process.*, **50** (10), 2408–2416.

Su, W. and Xia, X.-G. (2004). Signal constellations for quasi-orthogonal space-time block codes with full diversity. *IEEE Trans. Inf. Theory*, **50** (10), 2331–2347.

Tarokh, V., Jafarkhani, H., and Calderbank, A. R. (1999). Space-time block codes from orthogonal designs. *IEEE Trans. Inf. Theory*, **45** (5), 1456–1467.

Tarokh, V., Seshadri, N., and Calderbank, A. R. (1998). Space-time codes for high data rate wireless communications: Performance criteria and code construction. *IEEE Trans. Inf. Theory*, **44** (2), 744–765.

Telatar, E. (1999). Capacity of multi-antenna Gaussian channels. *Eur. Trans. Commun.*, **10** (6), 585–596.

Tirkkonnen, O. and Hottinen, A. (2002). Square-matrix embeddable space-time block codes for complex signal constellations. *IEEE Trans. Inf. Theory*, **48** (2), 384–395.

Tse, D. and Viswanath, P. (2005). *Fundamentals of Wireless Communication* (Cambridge Univ. Press).

Ungerboeck, G. (1982). Channel coding with multilevel/phase signals. *IEEE Trans. Inf. Theory*, **28** (1), 55–67.

Winters, J. H., Salz, J., and Gitlin, R. D. (1994). The impact of antenna diversity on the capacity of wireless communication systems. *IEEE Trans. Commun.*, **42** (2/3/4), 1740–1751.

Wittneben, A. (1991). Base station modulation diversity for digital SIMULCAST. In *Proc. IEEE Veh. Technol. Conf. (St. Louis, USA)*, vol. 1, 848–853.

Wolfe, W. (1976). Amicable orthogonal designs—existence. *Canadian J. Math.*, **28** (5), 1006–1020.

8

Perspectives on the diversity-multiplexing trade-off in MIMO systems

Huan Yao, Lizhong Zheng, and Gregory W. Wornell

Massachusetts Institute of Technology

8.1 Introduction

Multiple-element antenna arrays have an important role to play in both improving the robustness of wireless transmission through spatial diversity and increasing the throughput of wireless links through spatial multiplexing. Indeed, a wide variety of coding and signal processing techniques have been developed to realize one or the other of these benefits, or combinations of the two.

In recent years there has been an increasing appreciation that the two types of gains are inherently coupled. In particular, for a given antenna configuration, large robustness gains preclude the possibility of large throughput gains, and vice-versa. While the detailed nature of the fundamental trade-off between these gains is rather complicated, in the high signal-to-noise ratio regime, the scaling behavior of this diversity-multiplexing trade-off takes a particularly simple form.

To facilitate the use of this trade-off by communication engineers in system design, this chapter provides an intuitive development of the trade-off. In addition, we illustrate how a number of systems make these trade-offs efficiently in various regimes of interest. In the process, a variety of additional interpretations and perspectives on the underlying relationships are developed.

An outline of the chapter is as follows. We begin with a basic channel model and some definitions. We then proceed to develop and interpret the efficient frontier of diversity-multiplexing trade-offs. The remainder of the chapter focuses on how to achieve different portions of the frontier, what it means to operate on the frontier, and on how performance degrades when operating away from the frontier.

154

8.2 The diversity-multiplexing trade-off

8.2.1 Channel models and system definitions

We consider a multiple-input multiple-output (MIMO) wireless link with N_t transmit and N_r receive antennas in a flat-fading environment, so that the N_r-dimensional complex baseband received signal takes the form

$$\mathbf{y} = \sqrt{\frac{\mathsf{SNR}}{N_t}} \cdot \mathbf{H}\mathbf{x} + \mathbf{w} \tag{8.1}$$

with $\mathbf{x}$ denoting the N_t-dimensional transmitted signal, $\mathbf{w}$ denoting the white Gaussian noise, and $\mathbf{H}$ denoting the $N_r \times N_t$-dimensional random channel matrix. Finally, we normalize the noise $\mathbf{w}$ and the transmitted signal $\mathbf{x}$ to have unit variance entries, so that SNR is the overall signal-to-noise ratio of the channel. We also use the notation $\rho = \mathsf{SNR}/N_t$ to denote the SNR per transmit antenna for convenience.

In the scenario of interest, the channel matrix $\mathbf{H}$ varies with time. While there are a number of different models of such time variation, we adopt the block-fading model in this chapter. Specifically, $\mathbf{H}$ is constant within a block of T symbol periods, and varies independently from block to block. We further assume that the blocks are long enough that the receiver has, effectively, perfect knowledge of each realization of the channel matrix. However, we restrict our attention to the case where there is no feedback, so the transmitter only knows the channel statistics.

Multiple-antenna channels provide *spatial diversity*, which can be used to improve the reliability of the link. In Tarokh *et al.* (1998), it is shown that for the multiple antenna channel, the pairwise error probability can be made to decay as $\mathsf{SNR}^{-N_r N_t}$ at high SNR. The improvement of the reliability is thus dictated by the SNR exponent of the error probability, which is referred to as the *diversity gain*. Intuitively, the diversity gain corresponds to the number of independent fading paths that a symbol passes through; in other words, the number of independent fading coefficients that can be averaged over to detect the symbol.

In addition to the spatial diversity gain, it is also well known that using multiple antennas at both the transmitter and the receiver allows higher data rates to be supported (Telatar, 1995). To see this, first note that with an input of equal power on each antenna element, and for a fixed realization of the channel, the mutual information between input and output is, in b/s/Hz,

$$I(\mathbf{x}, \mathbf{y}|\mathbf{H}) = \log \det\left(I + \frac{\mathsf{SNR}}{N_t}\mathbf{HH}^\dagger\right) = \sum_{i=1}^{\min\{N_t, N_r\}} \log\left(1 + \frac{\mathsf{SNR}}{N_t}\lambda_i\right), \quad (8.2)$$

where the second equality follows from the eigenvalue decomposition $\mathbf{HH}^\dagger = \mathbf{U\Lambda U}^\dagger$ with $\mathbf{\Lambda} = \mathrm{diag}(\lambda_1, \lambda_2, \ldots, \lambda_{\min\{N_t, N_r\}})$ and unitary $\mathbf{U}$. From (8.2) we see immediately that the effect of having multiple antennas is to provide $\min\{N_t, N_r\}$ separate *spatial subchannels* over which to multiplex the data. The ergodic capacity, with the operational meaning of the maximum rate that can be reliably transmitted in a fast fading scenario, is given by

$$C(\mathsf{SNR}) = E[I(\mathbf{x}, \mathbf{y}|\mathbf{H})] = \sum_{i=1}^{\min\{N_t, N_r\}} E\left[\log\left(1 + \frac{\mathsf{SNR}}{N_t}\lambda_i\right)\right]. \quad (8.3)$$

Thus, at high SNR, the ergodic capacity increases linearly with $\log \mathsf{SNR}$ according to the number of antennas, i.e.,

$$\lim_{\mathsf{SNR} \to \infty} \frac{C(\mathsf{SNR})}{\log \mathsf{SNR}} = \min\{N_t, N_r\}.$$

When we restrict our attention to codes over a single block (channel realization), as will be our focus, we have to back off from this ergodic capacity (8.3) to achieve small error rates, but can still achieve significant gains. To explore this, we consider rates that are a fixed fraction of this ergodic capacity at high SNR, i.e., rates of the form $R = r \log \mathsf{SNR}$ with $0 < r < \min\{N_t, N_r\}$, and evaluate the associated error behavior. The value r can therefore be viewed as the *spatial multiplexing gain*.

Using (M, SNR) to denote a code with M codewords for a single block of a multiple antenna channel having the specified SNR, and focusing on the high-SNR regime, the following more formal definitions are rather natural.

Definition 8.1 *A spatial multiplexing gain r is said to be* achievable *if there exists a sequence of $(2^{TR(\mathsf{SNR})}, \mathsf{SNR})$ codes, with*

$$\lim_{\mathsf{SNR} \to \infty} \frac{R(\mathsf{SNR})}{\log \mathsf{SNR}} = r \quad (8.4)$$

such that the probability of error tends to 0 as $\mathsf{SNR} \to \infty$.

Definition 8.2 *A spatial multiplexing gain r and a diversity gain d are said to be* simultaneously achievable *if there exists a sequence of $(2^{TR(\mathsf{SNR})}, \mathsf{SNR})$*

codes, satisfying (8.4) and the error probability $P_e(\text{SNR})$ satisfying

$$\lim_{\text{SNR}\to\infty} \frac{\log P_e(\text{SNR})}{\log \text{SNR}} = -d. \tag{8.5}$$

For future convenience, we rewrite (8.5) as $P_e(\text{SNR}) \doteq \text{SNR}^{-d}$ with $\doteq$ denoting exponential equality.[†]

Before proceeding, note that the two gains concern inherently different aspects of the channel, and thus two channels with the same multiplexing gain may support vastly different diversity gains. As an illustration, consider a pair of parallel scalar ergodic fading channels, with channel gains h_1 and h_2 that are each equally likely to be 0 or 1. Such channels have an ergodic capacity of $\log \text{SNR}$ at high SNR, corresponding to unit multiplexing gain. However, if $h_2 = h_1$, i.e., the two channels fade in unison, the error probability is $1/2$ when coding over a single block, corresponding to zero diversity gain. By contrast, if $h_0 = 1 - h_1$, a simple repetition scheme over the single block can achieve the Gaussian channel error probability, which decays exponentially in SNR, and therefore has an infinite diversity gain, according to Definition 8.2.

Our focus in this chapter is to study the ability of the multiple antenna channels to provide combinations of the diversity and the spatial multiplexing gains and explore the fundamental trade-off between how much each type of gain one can achieve. As one extreme point, the maximum diversity gain one can achieve is $d_{\max} = N_t N_r$. This can be achieved when transmitting at a fixed data rate, corresponding to a multiplexing gain $r = 0$. On the other hand, from our earlier discussion the maximum spatial multiplexing gain is $r_{\max} = \min\{N_t, N_r\}$. Communicating at a rate so close to the ergodic capacity results in large error probability, corresponding to a diversity gain $d = 0$. In the sequel, we explore the achievable (r, d) pairs between these two extremes.

One can interpret the diversity-multiplexing trade-off in the context of traditional rate-reliability trade-offs in communication theory. In particular, while in traditional communication theory one examines achievable rate at fixed SNR in the limit of large block lengths (Cover and Thomas, 1991), in this case we are examining achievable rate at a fixed block length in the limit of high SNR. Accordingly, we refer to the sequence of codes in the above

---

[†]Formally, we write $f(\text{SNR}) \doteq \text{SNR}^b$ when

$$\lim_{\text{SNR}\to\infty} \frac{\log f(\text{SNR})}{\log \text{SNR}} = b.$$

The symbols $\dot{\geq}$ and $\dot{\leq}$ are defined analogously.

definition, indexed by SNR, with $2^{Tr \log SNR}$ codewords in each code, as a *scheme*. By studying the performance of a scheme, we focus not on the design of specific codes, but rather on the structure of MIMO codes. Examples of such schemes include random codes with the number of codewords increasing with SNR, algebraic codes over QAM symbols with the size of the QAM constellation increasing with SNR, etc.

With this interpretation, the multiplexing gain can be viewed as the number of bits reliably conveyed per dB of the SNR, where reliable means the error probability can be driven to zero with increasing $\log$ SNR rather than the block length L. Thus, while traditional communication theory examines the exponential decay of error probability with block length, here we examine the exponential decay of error probability with $\log$ SNR. Furthermore, the number of bits that can be conveyed is linear in $\log$ SNR just as it is in block length L.

In our setup, we focus on transmitting at a fixed fraction of the maximum multiplexing gain and examine the exponent in the decay of error probability with $\log$ SNR for a fixed block length. By comparison, the traditional error exponent analysis of Gallager focuses on transmitting at a fixed fraction of capacity and examines the exponent in the decay of error probability with block length L. Thus, there is a natural connection to traditional capacity and error exponent analysis in communication theory; further discussion can be found in, e.g., Zheng and Tse (2003).

Before proceeding, we should emphasize at this point that the measures of reliability and rate provided by the definitions are rather coarse in some respects. In particular, the reliability measure is invariant to changes in coding gain, which scales the SNR inside the logarithm, and other such constant factors. As such, two systems that achieve a particular diversity-multiplexing operating point may not be equally attractive when subjected to a finer scale analysis in which the effects of factors like coding gain manifest themselves.

8.2.2 System outage characteristics

When the transmitter does not know the channel matrix and codes within the coherence time of the channel, two types of errors are possible at the receiver. One type is due to noise realization, and the other due to the channel realization. As we now develop, it is the latter that dominates the error behavior in our scenario of interest.

Errors due to the channel realization occur as follows. Because the transmitter must choose a transmission rate without knowledge of the channel matrix, it may choose a rate that even with the best possible coding and long

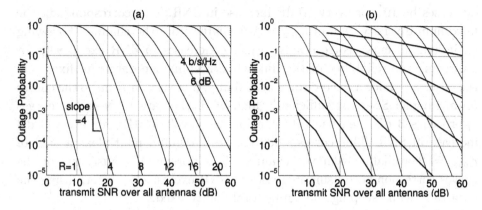

Fig. 8.1. (a) Family of outage probability curves as functions of SNR for various target rates R in the $N_t = N_r = 2$ case. (b) Trade-off between increasing the data rate and decreasing the error probability.

coherence times cannot be supported by the channel. This corresponds to the rate R being higher than the mutual information of the realized channel, and the corresponding error event, termed *outage*, is defined via

$$\{I(x; y|\mathbf{H}) < R\}. \tag{8.6}$$

Since the channel mutual information is a random variable $I = I(x; y|\mathbf{H})$, there is a probability associated with this error event, which we denote using $P_{\text{out}}(R, \text{SNR})$.

The overall error probability can be expressed in terms of the outage probability according to

$$P_e(\text{SNR}) = P(\text{error}|\text{outage})P_{\text{out}}(R, \text{SNR})$$
$$+ P(\text{error}|\text{no outage})(1 - P_{\text{out}}(R, \text{SNR})).$$

Conditioned on the outage event, the error probability is necessarily close to one and thus the outage probability is an asymptotic lower bound on the error probability, i.e., $P_e(\text{SNR}) \geq P_{\text{out}}(R, \text{SNR})$. It is for this reason that traditional space-time code design focuses on approaching the outage behavior as closely as possible.

We can visualize the relationship between SNR, rate, and outage probability by plotting P_{out} as a function of SNR for various rates R. This is depicted in Fig. 8.1(a) for the case of two transmit and two receive antennas. Each curve represents how outage probability decays with SNR for a fixed rate R.

Observe that for each curve, corresponding to a fixed rate, at sufficiently high SNR the slope of the curve approaches 4, i.e., the outage probability decreases by 10^4 for every 10 dB increase in SNR. This corresponds to the maximum diversity gain of $d_{\max} = 4$ available from the channel. Following an individual curve corresponds to using increases in SNR exclusively for improving the reliability of the link while keeping the data rate fixed. On the other hand, for a fixed outage probability, increases in R move the curves toward higher SNR. At sufficiently high SNR, the gaps between the curves approaches 3 dB, i.e., the rate can increase by 2 b/s/Hz for every 3 dB increase in SNR. This corresponds to the maximum multiplexing gain of $r_{\max} = 2$ available from the channel. This horizontal transection of the curves corresponds to using increases in SNR exclusively for increasing the data rate, while keeping the outage probability fixed.

More generally, we are interested in transecting these curves not horizontally, but with some downward slope. Indeed, a coding scheme in our parlance corresponds to the resulting sequence of points on these plots whereby as the SNR increases, the data rate increased and the error probability is decreased so as to achieve both diversity and multiplexing gains simultaneously. Fig. 8.1(b) shows these cross-cutting curves of various downward slopes. As the figure reflects, there is a trade-off between how much of each type of gain can be obtained: a higher multiplexing gain, corresponding to increase the data rate faster with SNR, will result in a lower diversity gain, and vice versa.

8.2.3 The efficient frontier

The optimal trade-off between the diversity gain and the multiplexing gain is described by a function $d^*(\cdot)$ whereby $d^*(r)$ gives the maximum achievable diversity gain at each multiplexing gain r.

The optimal trade-off in the case of independent, identically-distributed Rayleigh fading channels between each pair of antennas, and sufficiently large coherence times, i.e., $T \geq N_t + N_r - 1$, is developed in Zheng and Tse (2003). For this case, the trade-off is given by the piecewise linear curve connecting the points $(k, d^*(k))$, $k = 0, 1, \ldots, \min(N_t, N_r)$, where

$$d^*(k) = (N_t - k)(N_r - k). \tag{8.7}$$

The function $d^*(r)$ is plotted in Fig. 8.2(a).

Before proceeding, it is worth noting that it is also shown that at any r, the optimum achievable diversity gain $d^*(r)$ is the same as the SNR exponent of the outage probability, i.e., for any r,

$$P_e(\mathrm{SNR}) \doteq P_{\mathrm{out}}(R = r \log \mathrm{SNR}, \mathrm{SNR}) \doteq \mathrm{SNR}^{-d^*(r)}. \tag{8.8}$$

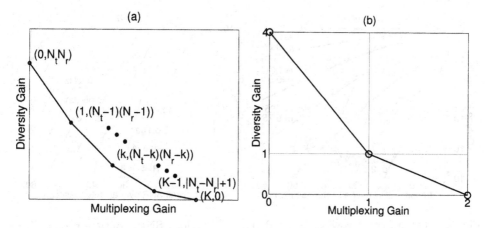

Fig. 8.2. (a) The efficient frontier $d^*(r)$ of diversity-multiplexing trade-offs for a system with N_t transmit antennas and N_r receive antennas, where $K \triangleq \min(N_t, N_r)$. (b) The efficient frontier for $N_t = N_r = 2$.

This means that the upper limit of the performance given by the outage probability can actually be achieved. While this is not surprising in the limit of infinite coherence time since one can drive noise-induced errors to zero using good long codes, it is noteworthy that it also holds for finite coherence times. In essence, it tells us that when using good signaling schemes, the probability of error is asymptotically dominated by the outage probability, and that the typical way of making an error in MIMO fading channels is from the channel being in a deep fade. In the following, we will develop some intuition for this result. For this purpose, it will be sufficient to focus on 2×2 systems, for which the optimal trade-off curve is plotted in Fig. 8.2(b).

We begin by relating the shape of the frontier to the dependence of error probability on rate and SNR for optimal systems. To this end, in Fig. 8.3(a) we plot the outage probability (8.8) as a function of SNR for different values of r. To facilitate its interpretation, the curves of Fig. 8.1(a) are overlaid as gray lines. Dashed lines with slopes $d^*(r)$ are also drawn to show the match between the asymptotic expression (8.8) and the exact outage probability.

The asymptotes corresponding to outage probability (8.8) are depicted in the Bode-style plot of Fig. 8.3(b). Such a plot reveals two distinct regions of operation. The case where the rate is high relative to the SNR corresponds to a *heavily-loaded* system, and its performance is captured by the region of the plot above and to the right of the dashed line. Here the slopes of the outage probability curves are flatter. By contrast, performance in the *lightly-*

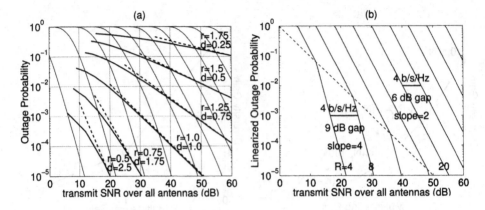

Fig. 8.3. (a) As rate grows with SNR, i.e., $R = r \log_2(\text{SNR})$, outage probability $P_{\text{out}}(R, \text{SNR})$ decays with SNR with slope $d^*(r)$. (b) Linearized approximation of Fig. 8.1(a), which emphasizes the two distinct regions of the P_{out}-SNR space with differing curve slopes and horizontal spacings.

loaded regime is captured by the region below and to the left of the dashed line. Here the slopes of the outage curves are steeper. The dashed boundary is the line with $P_{\text{out}} = \text{SNR}^{-1}$, and corresponds to the $(1,1)$ point of the frontier curve in Fig. 8.2(b). More generally, there are $\min\{N_t, N_r\}$ distinct regions of operation, one corresponding to each piecewise linear segment of the frontier $d^*(r)$.

Note that associated with each of these regions is a set of inherently different *local* trade-offs around each operating point in the region. To see this, observe from Fig. 8.3(b) that in the heavily-loaded region, the lines have slope 2 and spacing of 1.5 dB for every 1 b/s/Hz increment in rate. Thus 3 dB of additional SNR can be used to either increase rate by 2 b/s/Hz or reduce error probability by a factor of $(1/2)^2$. More generally, one can achieve any linear combination of the two, i.e., for any $\alpha \in (0,1)$, increase rate by 2α b/s/Hz while decreasing the error probability by $(1/2)^{2(1-\alpha)}$. In terms of Fig. 8.2(b), this local trade-off corresponds to a straight line connecting $(r,d) = (0,2)$ and $(2,0)$, which is an extension of the lower segment of the frontier down to $r = 0$. As we would expect, the maximum diversity gain of 4 is not achievable in this heavily-loaded region. Similarly, when we operate in the lightly-loaded region, the local trade-off corresponds to an extension of the upper segment of the frontier down to $d = 0$, i.e., the straight line connecting $(0,4)$ and $(4/3,0)$. In this region, it is the maximum multiplexing gain of 2 that is not achieved.

To understand why the efficient frontier of diversity-multiplexing trade-offs takes the particular form it does, it is convenient to rewrite the outage event (8.6) in the form

$$\left\{ \prod_{i=1}^{\min\{N_t, N_r\}} \left(1 + \frac{\text{SNR}}{N_t}\lambda_i\right) \stackrel{.}{<} \text{SNR}^r \right\}. \tag{8.9}$$

If all the eigenvalues λ_i are large relative to the noise threshold, then each corresponding spatial subchannel is operating in the high-SNR regime, contributing a factor SNR to the left-hand side of (8.9), and no outage occurs. However, if some of the eigenvalues are small relative to the noise threshold, then those subchannels are operating in the low-SNR regime and do not contribute a factor of SNR. Thus, if there are not at least r large eigenvalues, outage occurs. From this perspective, the piecewise linear nature of the diversity-multiplexing frontier reflects the fact that eigenvalues of $\mathbf{HH}^\dagger$ are generally not all equal, i.e., that the corresponding spatial subchannels are not all equally strong.

This behavior underlies the distinct operating regions in Fig. 8.3(b). In particular, in this case, when the eigenvalues are ordered according to $\lambda_1 \geq \lambda_2$, the smaller eigenvalue λ_2 takes values close to zero with a much larger probability than λ_1 does. Consequently, when the system is heavily-loaded ($r \geq 1$), the trade-off curve is flatter since an outage occurs whenever λ_2 alone is small. By contrast, when the system is lightly-loaded ($r \leq 1$), both λ_1 and λ_2 must be small to get an outage; thus, the trade-off curve is steeper. While this can be quantified by deriving the joint distribution of ordered eigenvalues (Zheng and Tse, 2003), in the sequel we make use of a different decomposition of the channel instead, which also provides useful insights into code design.

For a 2×2 system, the mutual information (8.2) of the realized channel can be rewritten as

$$I(x; y|\mathbf{H}) = \log\left(1 + \rho\|\mathbf{H}\|^2 + \rho^2|\det(\mathbf{H})|^2\right). \tag{8.10}$$

From this we see that $\det(\mathbf{H})$ being large relative to the noise is sufficient for both spatial subchannels to be in the high-SNR regime (enabling a multiplexing gain of two), and that $\|\mathbf{H}\|$ being large relative to the noise is sufficient for at least one subchannel to be in the high-SNR regime (enabling a multiplexing gain of one).

Using the factorization $\mathbf{H} = \mathbf{QR}$, where $\mathbf{Q}$ is unitary and

$$\mathbf{R} = \begin{bmatrix} r_{11} & r_{12} \\ 0 & r_{22} \end{bmatrix}, \tag{8.11}$$

we see

$$\|\mathbf{H}\|^2 = r_{11}^2 + |r_{12}|^2 + r_{22}^2 \quad \text{and} \quad |\det(\mathbf{H})|^2 = r_{11}^2 r_{22}^2.$$

Letting $\mathbf{h}_1$ and $\mathbf{h}_2$ denote the columns of $\mathbf{H}$, and $\mathbf{h}_{2\|1}$ and $\mathbf{h}_{2\perp1}$ denote the components of $\mathbf{h}_2$ in the direction along and perpendicular to $\mathbf{h}_1$, we see that

$$r_{11}^2 = \|\mathbf{h}_1\|^2 \quad \text{and} \quad |r_{12}|^2 = \|\mathbf{h}_{2\|1}\|^2 \quad \text{and} \quad r_{22}^2 = \|\mathbf{h}_{2\perp1}\|^2.$$

Thus for the case of interest where $\mathbf{H}$ has independent identically distributed circularly symmetric complex Gaussian entries, r_{11}^2, $|r_{12}|^2$, and r_{22}^2 are independent chi-squared distributed random variables of order 4, 2, and 2, respectively. Since for $\alpha < 1$ we have $\Pr[\chi < \alpha] \doteq \alpha^{k/2}$ when χ is chi-squared of order k, having a small r_{11}^2 is less likely than having either small $|r_{12}|^2$ or r_{22}^2. Therefore, all of r_{11}^2, $|r_{12}|^2$, and r_{22}^2 need to be small in order for $\|\mathbf{H}\|^2$ to be small, while $|\det(\mathbf{H})|^2$ being small is most likely due to r_{22}^2 being small.

Now we are ready to observe the difference between the typical outage events for the cases $r > 1$ and $r < 1$, the heavily loaded and the lightly loaded regimes, respectively. For $r > 1$, the outage event

$$\left\{ \log\left(1 + \rho \|\mathbf{H}\|^2 + \rho^2 r_{11}^2 r_{22}^2\right) \leq r \log \rho \right\}$$

occurs when $\rho^2 r_{11}^2 r_{22}^2 \leq \rho^r$. This happens typically when r_{11}^2 is of order 1 and r_{22}^2 is as small as $\rho^{-(2-r)}$, with a probability of the order $\rho^{-(2-r)}$. Intuitively, when $\mathbf{h}_1$ and $\mathbf{h}_2$ are closely aligned with each other, i.e., $\|\mathbf{h}_{2\perp1}\|^2 = r_{22}^2$ is abnormally small, the channel matrix $\mathbf{H}$ becomes close to singular, and less rate is supported by the weaker spatial channel. In the extreme, when $r_{22}^2 \doteq \rho^{-1}$, with a probability ρ^{-1}, the weaker spatial channel supports zero multiplexing gain, and can be viewed as completely shut off. For even lower rate $r < 1$, the outage event occurs only if both $\rho \|\mathbf{H}\|^2$ and $\rho^2 r_{11}^2 r_{22}^2$ are less than ρ^r. Typically, this happens when the weaker spatial channel is shut off, $r_{22}^2 \doteq \rho^{-1}$, and the stronger spatial channel can barely support the rate, $r_{11}^2 + |r_{12}|^2 \doteq \rho^{-(1-r)}$. Since the stronger channel is better protected, there is a steeper slope in the trade-off curve when $r < 1$.

For a general $N_r \times N_t$ channel, the trade-off result suggests that k spatial subchannels are operating in the high-SNR regime, supporting a multiplexing gain of $r = k$, with probability $\mathsf{SNR}^{-(N_r-k)(N_t-k)}$, corresponding to the typical outage event that the $\min\{N_t, N_r\} - k$ smallest eigenvalues of $\mathbf{H}\mathbf{H}^\dagger$ are all of the same order as the noise.

8.3 Achieving the diversity-multiplexing frontier

To develop additional insights into diversity-multiplexing trade-offs, we now turn our attention to developing MIMO coding schemes that achieve different portions of the efficient frontier. In the process we see how the structure of the associated schemes determines their diversity-multiplexing characteristics. While this section focuses on the two-transmit two-receive antenna case, many of the design principles apply to larger systems.

8.3.1 Achieving full diversity gain or full multiplexing gain

This section examines two well-known low-complexity coding schemes for MIMO, each achieving exactly one point of the diversity-multiplexing frontier. One achieves the full multiplexing gain, while the other achieves full diversity.

Achieving full diversity gain

The Alamouti orthogonal space-time block code (OSTBC) (Alamouti, 1998) employs a smart repetition to achieve the full diversity gain. It only applies to systems with two transmit antennas. It encodes two independent information symbols, s_1 and s_2, into the transmit signal matrix $\mathbf{X}$ as

$$\mathbf{X} = \begin{bmatrix} s_1 & -s_2^* \\ s_2 & s_1^* \end{bmatrix}, \tag{8.12}$$

where $(\cdot)^*$ indicates conjugation. The OSTBC effectively transforms a 2×2 MIMO channel to a scalar channel with channel gain $\|\mathbf{H}\|$. When OSTBC is combined with capacity-achieving error correction codes (ECC) designed for scalar channels, it is possible to communicate at rates

$$R < \log_2\left(1 + \rho\|\mathbf{H}\|^2\right). \tag{8.13}$$

The associated diversity-multiplexing trade-off can be derived using

$$
\begin{aligned}
P_e &= \Pr\left[\log_2\left(1 + \rho\|\mathbf{H}\|^2\right) < R = r\log_2\rho\right] \\
&\doteq \Pr\left[\|\mathbf{H}\|^2 < \rho^{r-1}\right] = \rho^{4(r-1)}.
\end{aligned} \tag{8.14}
$$

Therefore, the trade-off achieved is $d(r) = 4(1 - r)$, i.e., a straight line between $(0, 4)$ and $(1, 0)$, which is plotted in Fig. 8.4(a). It shows that only the full diversity point of the frontier is achieved. The lesson is that by transmitting each information symbol twice using different antennas at different times, each symbol interacts with all entries of the channel matrix $\mathbf{H}$. As a result, the dominant error event is when all entries of $\mathbf{H}$ are small. Thus, the

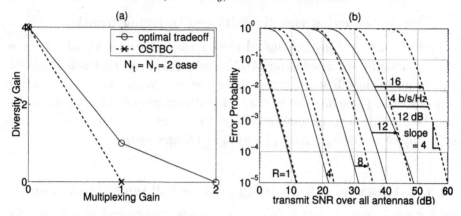

Fig. 8.4. (a) Diversity-multiplexing trade-off achieved by OSTBC. (b) Comparison of the OSTBC error probability curves (dashed) and the channel outage probability curves (thin solid) as functions of SNR for rates 1, 4, 8, 12, and 16 b/s/Hz.

full diversity gain is achieved. This is true even when there are more than two receive antennas. The drawback of OSTBC is that the maximum multiplexing gain achieved is only 1 since only one symbol is effectively transmitted at a time due to the repetition. It is only in the two-transmit one-receive antenna case that OSTBC achieves the optimal diversity-multiplexing trade-off (Zheng and Tse, 2003).

The error probability performance of OSTBC can be visualized by plotting $\Pr\left[\log_2\left(1 + \rho\|\mathbf{H}\|^2\right) < R\right]$ as functions of SNR for various R, via Monte-Carlo simulation, as shown in Fig. 8.4(b) with dashed lines. These curves approximately form a set of parallel lines with slopes of 4 and horizontal gaps of 12 dB per 4 b/s/Hz, i.e., 1 b/s/Hz for every 3 dB. The slopes and gaps are consistent with the trade-off points $(0, 4)$ and $(1, 0)$.

The channel outage probability curves are also plotted in Fig. 8.4(b) as thin solid lines for comparison. Below $R = 4$ b/s/Hz, the OSTBC curves are actually very close to the outage performance limit. This is because (8.13) differs from the channel capacity expression in (8.10) only by missing the second order term, which is negligible at low SNR. However, as rate increases, the performance gaps increase indefinitely.

Achieving full multiplexing gain

The vertical Bell LAbs LAyered Space-Time (V-BLAST) code transmits two independently encoded codewords simultaneously using the two antennas as depicted in Fig. 8.5. By transmitting two independent symbols per channel use, it achieves the full multiplexing gain. Unlike the Alamouti OSTBC, V-BLAST applies to systems with any number of transmit antennas.

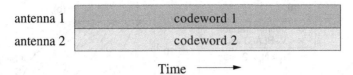

Fig. 8.5. V-BLAST, where coding is restricted to one row of the transmitted signal matrix.

The basic form of V-BLAST scheme employs a low-complexity nulling and cancellation decoding. We refer to this scheme as V-BLAST-nulling. The received vector y is multiplied with the $\mathbf{Q}^\dagger$ matrix in (8.11), making the effective channel matrix $\mathbf{Q}^\dagger\mathbf{H} = \mathbf{R}$ upper triangular. Consequently, codeword 2 experiences no interference and can be decoded first with channel gain r_{22}^2. Assuming codeword 2 is correctly decoded, its interference on codeword 1 can then be canceled, and codeword 1 can be decoded with channel gain r_{11}^2. Therefore, V-BLAST-nulling transforms a 2×2 MIMO channel to two parallel scalar channels with channel gains r_{11}^2 and r_{22}^2, respectively.

Assuming the data rate R is split evenly between the two codewords and the codes are capacity achieving, the necessary and sufficient condition for successful decoding is that both subchannels can support rate $R/2$, i.e.,

$$\frac{R}{2} < \log_2\left(1 + \rho r_{22}^2\right) \quad \text{and} \quad \frac{R}{2} < \log_2\left(1 + \rho r_{11}^2\right).$$

The associated diversity-multiplexing trade-off can be derived using

$$P_e = \Pr\left[\log_2\left(\rho r_{22}^2 + 1\right) < \frac{R}{2} \text{ or } \log_2\left(\rho r_{11}^2 + 1\right) < \frac{R}{2}\right]$$

$$\doteq \Pr\left[r_{22}^2 < \rho^{\frac{r}{2}-1} \text{ or } r_{11}^2 < \rho^{\frac{r}{2}-1}\right] \doteq \Pr\left[r_{22}^2 < \rho^{\frac{r}{2}-1}\right] = \rho^{\frac{r}{2}-1}. \quad (8.15)$$

Therefore, the trade-off achieved is $d(r) = 1 - r/2$, i.e., a straight line between $(0,1)$ and $(2,0)$, which is plotted in Fig. 8.6(a). It shows that only the full multiplexing gain point of the frontier is achieved. The drawback of V-BLAST-nulling is that the maximum diversity gain achieved is only 1, because r_{22}^2 is distributed like the energy of one entry of $\mathbf{H}$. The dominant error event is having a small r_{22}^2 and failing to decode the first codeword.

The error probability performance of V-BLAST-nulling can be visualized by plotting $\Pr\left[R > 2\log_2\left(1 + \rho r_{22}^2\right) \text{ or } R > 2\log_2\left(1 + \rho r_{11}^2\right)\right]$ as functions of SNR for various R, via Monte-Carlo simulation, as shown in Fig. 8.6(b) with dashed lines. These curves approximately form a set of parallel lines with slopes of 1 and horizontal gaps of 6 dB per 4 b/s/Hz, i.e., 2 b/s/Hz for every 3 dB. The slopes and gaps are consistent with the trade-off points $(0,1)$

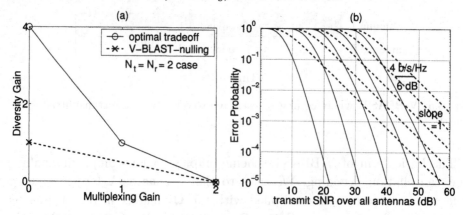

Fig. 8.6. (a) Diversity-multiplexing trade-off achieved by V-BLAST encoding with nulling and cancellation decoding. (b) Comparison of the V-BLAST-nulling error probability curves (dashed) and the channel outage probability curves (thin solid) as functions of SNR for rates 4, 8, 12, 16, and 20 b/s/Hz.

and $(2,0)$. Comparing to the channel outage probability curves plotted with thin solid lines in Fig. 8.6(b), the performance gaps in SNR increase quickly as the target error rate decreases due to the difference in slopes.

V-BLAST decoding can be improved by choosing the decoding order to optimize the effective channel gains for each codeword and employing MMSE decoding rather than nulling. However, these variations do not significantly affect the diversity-multiplexing trade-off achieved.

8.3.2 *Achieving the high-multiplexing-gain or high-diversity-gain frontiers*

This section discusses two coding schemes each achieving one segment of the diversity-multiplexing frontier in Fig. 8.2(b). We refer to the two segments as the high-diversity-gain frontier and the high-multiplexing gain frontier. The high-diversity-gain frontier corresponds to lightly-loaded systems where the multiplexing gain is less than one and rate is relatively small compared with SNR. The high-multiplexing gain frontier corresponds to heavily-loaded systems. We note that while the coding schemes discussed in this section achieve more points on the frontier than OSTBC and V-BLAST-nulling, they require higher complexity joint decoding.

Achieving the high-multiplexing-gain frontier

The V-BLAST encoding scheme presented in Section 8.3.1 when combined with joint decoding can achieve the high-multiplexing-gain segment of the diversity-multiplexing frontier by optimally handling the interference between the codewords. We refer to this scheme as V-BLAST-joint. Again assuming the data rate R is split evenly between the two separately-encoded codewords and the codes are capacity achieving, with joint decoding, the necessary and sufficient condition for successful decoding is that the total rate can be supported by the channel and that the individual codewords can be decoded in the absence of interference from the other codeword, i.e.,

$$R < \log_2\left(1 + \rho\|\mathbf{H}\|^2 + \rho^2|\det(\mathbf{H})|^2\right),$$

$$\frac{R}{2} < \log_2\left(1 + \rho\|\mathbf{h}_1\|^2\right), \text{ and } \frac{R}{2} < \log_2\left(1 + \rho\|\mathbf{h}_2\|^2\right). \quad (8.16)$$

with $\mathbf{h}_1$ and $\mathbf{h}_2$ continuing to denote the columns of $\mathbf{H}$. Equation (8.16) shows that the dominant error event for V-BLAST-joint is either $\|\mathbf{h}_1\|^2$ or $\|\mathbf{h}_2\|^2$ being small. The associated diversity-multiplexing trade-off can be derived using

$$P_e \doteq \Pr\left[\|\mathbf{h}_1\|^2 < \rho^{\frac{r}{2}-1} \text{ or } \|\mathbf{h}_1\|^2 < \rho^{\frac{r}{2}-1}\right] = \rho^{2\left(\frac{r}{2}-1\right)} = \rho^{-d(r)}. \quad (8.17)$$

Therefore, the trade-off achieved is $d(r) = 2 - r$, i.e., a straight line between $(0, 2)$ and $(2, 0)$, which is plotted in Fig. 8.7(a). It is clear that the high-multiplexing-gain segment of the frontier is achieved. The key idea is that with V-BLAST encoding, each transmitted symbol interacts with one column of the channel matrix $\mathbf{H}$. By fully taking advantage of this with joint decoding, entire columns of $\mathbf{H}$ has to be small for decoding to fail. Since each column of $\mathbf{H}$ has two entries, the maximum diversity gain achieved is 2.

The error probability performance of V-BLAST-joint can be visualized by plotting the probability that the condition in (8.16) fails as functions of SNR for various R, via Monte-Carlo simulation, as shown in Fig. 8.7(b) with dashed lines. These curves approximately form a set of parallel lines with slopes of 2 and horizontal gaps of 6 dB per 4 b/s/Hz, i.e., 2 b/s/Hz for every 3 dB. The slopes and gaps are consistent with the trade-off points $(0, 2)$ and $(2, 0)$. Comparing to the channel outage probability curves plotted with thin solid lines in Fig. 8.7(b), V-BLAST-joint performs quite well in the region above $P_e = \mathsf{SNR}^{-1}$. The reason is this region corresponds to the high-multiplexing segment of the trade-off curve.

Achieving the high-diversity-gain frontier

The high-diversity-gain segment of the diversity-multiplexing frontier can be achieved with a two-layer diagonal-BLAST (D-BLAST) encoding scheme with joint decoding. We refer to it as 2L-D-BLAST-joint. While V-BLAST

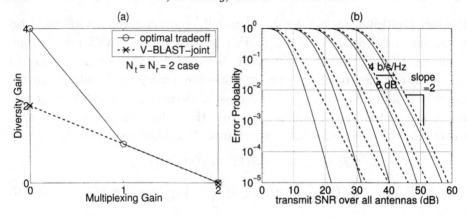

Fig. 8.7. (a) Diversity-multiplexing trade-off achieved by V-BLAST encoding with joint decoding. (b) Comparison of V-BLAST-joint error probability curves (dashed) and the channel outage probability curves (thin solid) as functions of SNR for rates 4, 8, 12, 16, and 20 b/s/Hz.

antenna 1	codeword 1 – 1st half	codeword 2 – 1st half	0 0 ... 0
antenna 2	0 0 ... 0	codeword 1 – 2nd half	codeword 2 – 2nd half

Time $\longrightarrow$

Fig. 8.8. D-BLAST encoding with two layers. Two halves of two codewords are transmitted by a certain antenna at a certain time as indicated. 0 indicates silence.

places separately encoded codewords on different antennas, D-BLAST transmits each codeword using different antennas at different times as illustrated in Fig. 8.8 for the two-layer case. Note that the antennas are sometimes silent. Similar to the V-BLAST case, the codewords are assumed to be capacity achieving and have equal rate. Although decoding can be done one codeword at a time via successive cancellation, we consider joint decoding in this section, since successive-cancellation decoding does not achieve the high-diversity-gain segment of the frontier.

The necessary and sufficient condition for successful joint decoding is that the total rate can be supported by the channel and the individual codeword can be decoded in the absence of interference from the other codeword, i.e.,

$$3R < \log_2\left(1+\rho\|\mathbf{h}_1\|^2\right) + \log_2\left(1+\rho\|\mathbf{H}\|^2+\rho^2|\det(\mathbf{H})|^2\right) + \log_2\left(1+\rho\|\mathbf{h}_2\|^2\right) \quad (8.18)$$

$$\text{and } \frac{3R}{2} < \log_2\left(1+\rho\|\mathbf{h}_1\|^2\right) + \log_2\left(1+\rho\|\mathbf{h}_2\|^2\right). \quad (8.19)$$

Since $\log_2\left(1+\rho\|\mathbf{H}\|^2+\rho^2|\det(\mathbf{H})|^2\right) < \log_2\left(1+\rho\|\mathbf{h}_1\|^2\right) + \log_2\left(1+\rho\|\mathbf{h}_2\|^2\right),$

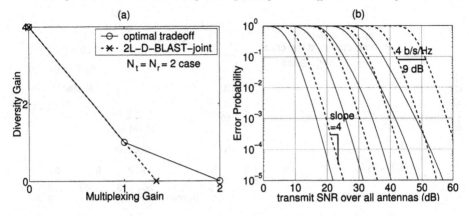

Fig. 8.9. (a) Diversity-multiplexing trade-off achieved by two-layer D-BLAST en-coding with joint decoding. (b) Comparison of 2L-D-BLAST-joint error probability curves (dashed) and the channel outage probability curves (thin solid) as functions of SNR for rates 4, 8, 12, 16, and 20 b/s/Hz.

(8.19) is always satisfied when (8.18) is. Therefore, (8.18) by itself is the necessary and sufficient condition for successful decoding.

The dominant event for the condition in (8.18) to fail is $\|\mathbf{H}\|^2 < \rho^{\frac{3}{4}r-1}$, which implies that all of $\|\mathbf{h}_1\|^2$, $\|\mathbf{h}_2\|^2$, and $|\det(\mathbf{H})|$ are just as small. The associated diversity-multiplexing trade-off can be derived using

$$P_e \doteq \Pr\left[\|\mathbf{H}\|^2 < \rho^{\frac{3}{4}r-1}\right] = \rho^{4\left(\frac{3}{4}r-1\right)} = \rho^{-d(r)}. \qquad (8.20)$$

Therefore, the trade-off achieved is $d(r) = 4 - 3r$, i.e., a straight line be-tween $(0,4)$ and $(4/3,0)$, which is plotted in Fig. 8.9(a). It is clear that the high-diversity-gain segment of the frontier is achieved. The 2L-D-BLAST-joint scheme achieves the full diversity gain because each codeword interacts with all entries of the channel matrix $\mathbf{H}$, just like in OSTBC. Furthermore, the encoding structure is such that four information symbols are transmit-ted in three channel uses, which results in the maximum multiplexing gain of $4/3$.

The error probability performance of 2L-D-BLAST-joint can be visualized by plotting the probability that the condition in (8.18) fails as functions of SNR for various R, via Monte-Carlo simulation, as shown in Fig. 8.9(b) with dashed lines. These curves approximately form a set of parallel lines with slopes of 4 and horizontal gaps of 9 dB per 4 b/s/Hz, i.e., 4/3 b/s/Hz per 3 dB. The slopes and gaps are consistent with the trade-off points $(0,4)$ and $(4/3,0)$. Compared with the channel outage probability curves plotted with thin solid lines in Fig. 8.9(b), 2L-D-BLAST-joint performs quite well in

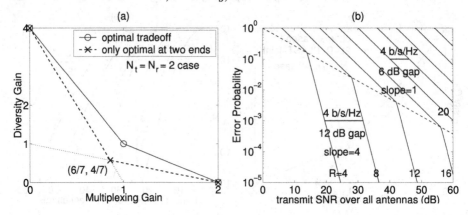

Fig. 8.10. (a) A hypothetical diversity-multiplexing trade-off that only meets the optimal trade-off at the end points, $d(r) = \max(4 - 4r, 1 - r/2)$. (b) A family of error probability curves consistent with the trade-off for rates $4, 8, \cdots 36$ b/s/Hz.

the region below $P_e = \mathrm{SNR}^{-1}$. The reason is this region corresponds to the high-diversity segment of the trade-off curve as discussed in Section 8.2.3.

8.3.3 Full-diversity full-multiplexing schemes

We next examine the implications of code constructions where both the full diversity gain and the full multiplexing gain are achieved, but not the rest of the frontier. Our focus in this section is not on the description of schemes with this property. Rather, it is to demonstrate how performance can degrade rather severely relative to schemes that achieve the entire frontier, to help code and system designers understand the underlying issues.

Let us consider the trade-off $d(r) = \max(4 - 4r, 1 - r/2)$, $0 \leq r \leq 2$, which is suboptimal everywhere except at the end points, as shown in Fig. 8.10(a). The error probability function that is consistent with this trade-off curve is $P_e(R, \mathrm{SNR}) = \mathrm{SNR}^{-d(R/\log_2(\mathrm{SNR}))}$, which is plotted in Fig. 8.10(b) for a range of R and SNR, similar to Fig. 8.3(b) for the optimal trade-off.

Corresponding to the two segments of the trade-off curve, there are two regions in the $P_e(R, \mathrm{SNR})$ plot. The boundary $P_e = \mathrm{SNR}^{-4/7}$, marked with a dashed line, corresponds to the knee in the trade-off. The slopes of and the gaps between the curves in each region are labeled in the figure. They agree with the intercepts of the two trade-off segments with the axes. Compared with Fig. 8.3(b) for the optimal trade-off, the slopes in the upper region are less and the gaps in the lower region are wider. These deficiencies lead to increasing performance gaps as SNR increases.

8.3.4 Achieving the entire frontier using rotation-based codes

In this section, we identify code properties that enable the full efficient frontier of diversity-multiplexing trade-offs to be achieved. Then, as an illustration, we construct a particular *tilted-QAM* code that meets these conditions and examine its performance characteristics more generally to develop additional insights.

Tilted-QAM codes are code designs based on applying unitary transformations to multidimensional QAM constellations. Such designs have a rich history, going back at least as far as Lang (1963). In the more recent wireless literature, such codes were used in, e.g., Boulle and Belfiore (1992), Wornell (1995), Boutros *et al.* (1996), Giraud *et al.* (1997), and Bayer-Fluckiger *et al.* (2004) for single-antenna communication over multiple fades. More recently still, they have been considered as space-time code candidates for multiple-antenna communication within a single fade. For example, in Sethuraman *et al.* (2003) and Sharma and Papadias (2004), such codes are used to obtain the maximum diversity subject to a multiplexing gain constraint of unity. As another set of examples, Damen *et al.* (2002) and Ma and Giannakis (2003) focus on using such codes to achieve the two end-points of the diversity-multiplexing frontier. Finally, Yao and Wornell (2003), Dayal and Varanasi (2003), Belfiore *et al.* (2004), and Elia *et al.* (2004) develop the role of such codes in achieving the entire trade-off frontier, as is of interest in this section, particularly in the two-transmit two-receive antenna case.

We begin with a natural sufficient condition for achieving the entire frontier. This condition is encapsulated in the following theorem (Yao, 2003):

Theorem 8.1 *For a system with two transmit and at least two receive antennas and code length $T \geq 2$, consider a family of codebooks indexed by rate R that is scaled such that the peak and average codeword energy grow with R as $E_s \doteq \max_{\mathbf{X}} \|\mathbf{X}\|^2 \doteq 2^{R/2}$. Then a sufficient condition for achieving the diversity-multiplexing frontier is*

$$\min_{\mathbf{X_1} \neq \mathbf{X_2}} |\det(\mathbf{X_1} - \mathbf{X_2})| \gtrsim 1. \tag{8.21}$$

Equation (8.21) means that either the worst-case codeword-difference determinants do not decay to zero with rate or decay at most subexponentially.

As we now develop, a tilted-QAM code can indeed achieve a constant worst-case determinant. A construction is as follows. Given a transmission rate $R = r \log_2(\text{SNR})$, a constellation of size $M^2 = 2^{R/2} = \text{SNR}^{r/2}$ is carved from $\mathbb{Z} + \mathbb{Z}j$. Four information symbols are chosen from this constellation and encoded into a codeword matrix $\mathbf{X} = \begin{bmatrix} x_{11} & x_{12} \\ x_{21} & x_{22} \end{bmatrix}$ via two rotations,

$$\begin{bmatrix} x_{11} \\ x_{22} \end{bmatrix} = \begin{bmatrix} \cos(\theta_1) & -\sin(\theta_1) \\ \sin(\theta_1) & \cos(\theta_1) \end{bmatrix} \begin{bmatrix} s_{11} \\ s_{22} \end{bmatrix},$$

$$\begin{bmatrix} x_{21} \\ x_{12} \end{bmatrix} = \begin{bmatrix} \cos(\theta_2) & -\sin(\theta_2) \\ \sin(\theta_2) & \cos(\theta_2) \end{bmatrix} \begin{bmatrix} s_{21} \\ s_{12} \end{bmatrix},$$

(8.22)

where $(\theta_1, \theta_2) = ((1/2)\arctan(1/2), (1/2)\arctan(2))$. Like OSTBC, each information symbol s_{ij} effectively appears in both rows and columns of the codeword matrix $\mathbf{X}$, which is essential for achieving the full diversity gain. Unlike OSTBC, with rotation instead of repetition, two information symbols are transmitted per channel use, so there is no sacrifice of multiplexing gain. It is shown in Yao (2003) that this tilted-QAM code satisfies Theorem 8.1. In particular, the worst-case determinant is $1/(2\sqrt{5})$ for arbitrarily large rates.

To illustrate some intuition behind why this rotation leads to a constant worst-case determinant, let us set $s_{12} = s_{21} = 0$ and limit s_{11} and s_{22} to real numbers. In this case, $\det(\mathbf{X}) = x_{11}x_{22}$. The rotation of s_{11} and s_{22} to obtain x_{11} and x_{22} is shown in Fig. 8.11. We first note that since $\sin(\theta_1)$ and $\cos(\theta_1)$ are irrational numbers, all points except the origin are kept off the x_{11} and x_{22} axes. So the determinant is always nonzero. While a formal proof of the lower bound on the worst-case determinant appears in Yao (2003), the intuition comes from focusing on a series of points circled in Fig. 8.11. We see that while the points get closer and closer to the x_{11} axis, they also move further and further away from the x_{22} axis. As a result, the product $|x_{11}x_{22}|$ actually remains constant. More generally, when all four symbols s_{ij} are complex integers, the worst-case determinant is still lower bounded.

Numerical simulations with this tilted-QAM encoding and maximum likelihood (ML) decoding, implemented using sphere decoding, are performed for rates $R = 4, 8, 12, \ldots, 32$ b/s/Hz. The resulting family of block error rate curves are plotted in Fig. 8.12 with dashed lines. The channel outage probability curves for those rates are plotted with thin solid lines for comparison. We see that the tilted-QAM block error rate curves match the channel outage probability curves closely, especially at higher rates. Since diversity-multiplexing trade-off is a high-SNR characteristic, it is possible for two systems with the same trade-off to have different low-SNR performances. The slopes of and the horizontal gaps between the tilted-QAM block error rate curves above and below the $P_{\text{out}} = \text{SNR}^{-1}$ line are labeled in Fig. 8.12. They agree with the slopes and gaps labeled in Fig. 8.3(b), which corresponds to the optimal diversity-multiplexing trade-off. This shows that the tilted-

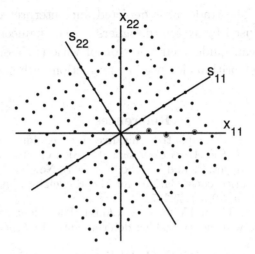

Fig. 8.11. Rotate (s_{11}, s_{22}) to obtain (x_{11}, x_{22}). All points except the origin are off the x_{11} and x_{22} axes. Points circled have constant $|x_{11}x_{22}|$ values.

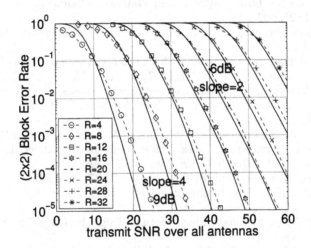

Fig. 8.12. Comparison of the tilted-QAM block error rate curves (dashed) and the channel outage probability curves (thin solid) as functions of SNR for rates $4, 8, 12, \ldots, 32$ b/s/Hz. They match quite well.

QAM encoding defined in (8.22) together with ML decoding can achieve the entire diversity-multiplexing frontier.

8.4 Summary

The focus of this chapter was on an intuitive development of the diversity-multiplexing trade-off inherent in the use of multiple-element antenna arrays for wireless links. The trade-off is depicted and interpreted in a variety of ways that can be used by system designers in the engineering of communication links, and can guide their selection of various classes of coding and signal processing algorithms for use in conjunction with such arrays.

References

Alamouti, S. M. (1998). A simple transmit diversity technique for wireless communications. *IEEE J. Sel. Areas Commun.*, **16** (8), 1451–1458.

Bayer-Fluckiger, E., Oggier, F., and Viterbo, E. (2004). New algebraic constructions of rotated $\mathbf{Z}^n$-lattice constellations for the Rayleigh fading channel. *IEEE Trans. Inf. Theory*, **50** (4), 702–714.

Belfiore, J.-C., Rekaya, G., and Viterbo, E. (2004). The golden code: a 2x2 full rate space-time code with non-vanishing determinants. In *Proc. IEEE Int. Symp. Inf. Theory*, 308.

Boulle, K. and Belfiore, J.-C. (1992). Modulation schemes designed for the Rayleigh fading channel. In *Proc. Conf. Inf. Science and Syst.*, 288–293.

Boutros, J., Viterbo, E., Rastello, C., and Belfiore, J.-C. (1996). Good lattice constellations for both Rayleigh and Gaussian channels. *IEEE Trans. Inf. Theory*, **42** (2), 502–518.

Cover, T. M. and Thomas, J. A. (1991). *Elements of Information Theory* (Wiley, New York).

Damen, M. O., Tewfik, A., and Belfiore, J.-C. (2002). A construction of a space-time code based on number theory. *IEEE Trans. Inf. Theory*, **48** (3), 753–760.

Dayal, P. and Varanasi, M. (2003). An optimal two transmit antenna space-time code and its stacked extension. In *Proc. Asilomar Conf. Signals, Syst., Comput.*, vol. 1, 987–991.

Elia, P., Kumar, K. R., Pawar, S. A., Kumar, P. V., and Lu, H. F. (2004). Explicit construction of space-time block codes: Achieving the diversity-multiplexing gain tradeoff. *Submitted to IEEE Trans. Inf. Theory*.

Giraud, X., Boutillon, E., and Belfiore, J. C. (1997). Algebraic tools to build modulation schemes for fading channels. *IEEE Trans. Inf. Theory*, **43** (3), 938–952.

Lang, G. R. (1963). Rotational transformation of signals. *IEEE Trans. Inf. Theory*, **9** (3), 191–198.

Ma, X. and Giannakis, G. B. (2003). Full-diversity full-rate complex-field space-time coding. *IEEE Trans. Signal Process.*, **51** (11), 2917–2930.

Sethuraman, B. A., Rajan, B. S., and Shashidhar, V. (2003). Full-diversity, high-rate space-time block codes from division algebras. *IEEE Trans. Inf. Theory*, **49** (10), 2596–2616.

Sharma, N. and Papadias, C. B. (2004). Full-rate full-diversity linear quasi-orthogonal space-time codes for any number of transmit antennas. *EURASIP J. Appl. Signal Process. (Special Issue on Advances in Smart Antennas)*, **2004** (9), 1246–1256.

Tarokh, V., Seshadri, N., and Calderbank, A. R. (1998). Space-time codes for high data rate wireless communication: performance criterion and code construction. *IEEE Trans. Inf. Theory*, **44** (2), 744–765.

Telatar, E. (1995). Capacity of multi-antenna Gaussian channels. AT&T Bell Labs Internal Tech. Memo.

Wornell, G. W. (1995). Spread-signature CDMA: Efficient multiuser communication in the presence of fading. *IEEE Trans. Inf. Theory*, **41** (5), 1418–1438.

Yao, H. (2003). *Efficient Signal, Code, and Receiver Designs for MIMO Communication Systems*. Ph.D. thesis, Massachusetts Institute of Technology.

Yao, H. and Wornell, G. W. (2003). Achieving the full MIMO diversity-multiplexing frontier with rotation based space-time codes. In *Proc. Allerton Conf. Commun., Contr., Comput.*, 400–409.

Zheng, L. and Tse, D. N. C. (2003). Diversity and multiplexing: a fundamental tradeoff in multiple antenna channels. *IEEE Trans. Inf. Theory*, **49** (5), 1073–1096.

9

Linear precoding for MIMO channels

Anna Scaglione

Cornell University

Petre Stoica

Uppsala University

9.1 Precoding versus coding

A pivotal function of the physical transmission layer is that of making available a linear space of functions that meet transmission constraints. Each dimension of the coordinate system used in this space is called *one degree of freedom*. Except when orthogonal waveforms are used, the number of modulation waveforms is greater than the number of degrees of freedom. The transmission of a signature waveform colinear with one dimension is what is commonly referred to as one *channel use*. The constraints in designing modulation signals for MIMO systems are the classic ones: the transmit power is limited and the signals used should have their energy concentrated in a predefined window in time and frequency. The latter pair of constraints are coupled and the number of orthogonal signals with finite bandwidth-time product (WT) that can be constructed is $\leq WT$. The formal statement of this fact is generally attributed to Gabor (1946).

Ideally, the selection of the signal space should be in continuous time and space. In practice, the digital to analog converters and antenna arrays and the RF propagation channel transduce the digital stream into signals with undesired properties. The transceivers' frontends can transform digitally the coordinate into a more suitable system of coordinates compared with that determined by the physical transducers. The linear mapping that establishes the new set of coordinates is referred to as linear precoding and its design can be carried out in discrete time (see Section 9.2 and Scaglione *et al.*, 1999).

The question that may arise is why we select a signal space and not simply signal codes. Veeravalli and Mantravadi (1999) tried to formalize the difference between coding and designing modulating waveforms. The different roles of precoding and coding are illustrated in Fig. 9.1, where it is schemat-

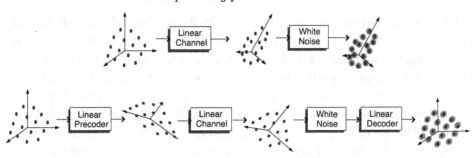

Fig. 9.1. Illustration of the roles of coding and linear precoding.

ically shown what is the impact of optimizing the coordinate system. Coding can be thought as setting up a lattice of points in the signal space while decoding consists in defining decision regions in it. Precoding is a change of coordinate system; in general it entails more than a simple rotation since the code lattices must lie in a limited region of in the precoder coordinate system due to the energy and power constraints. Therefore, the precoding operation will be a linear mapping F that in general can be partitioned as $F = \Phi V$ where V is a rotation and Φ is necessary to distribute the power so as to satisfy the power constraint. The position of the new precoded coordinate simple 3-D scheme in Fig. 9.1 is not chosen arbitrarily: one can note that the direction of the coordinate vectors is not changed by the passage through the channel. A number of different criteria studied throughout this chapter lead to the structure $F = \Phi V$ where V are singular vectors of the channel matrix and the discriminating element of the design is the matrix Φ, which is diagonal and loads power optimally in each direction independently.

In MIMO transmission, the code design is preferably described as the optimization of a matrix code whose transmission requires several channel uses and it is therefore called space-time code (e.g., Alamouti, 1998; Naguib *et al.*, 1998, to cite some of the earliest designs). The fact that the codes are matrices and the precoders are matrix transformations does not make their roles interchangeable. The matrix structure of space-time codes is mathematically convenient: in fact, code-design guidelines are easier to specify when the codes are viewed as matrices rather then vectors or strings. In principle, there is no benefit in searching for a coordinate system but only a need to find good lattices, which perform optimally for a specific channel. Optimal precoding does not add *performance benefits* that cannot be obtained otherwise—its inclusion has purely *architectural benefits*. Precoding can adapt preexisting coding strategies designed for ideal AWGN channel to the real channel characteristics (Forney and Eyuboğlu, 1991) via a simple linear transformation, reducing the complexity of the coding task and increasing the modularity of the design. In MIMO systems, each *channel use* can mix spatial and temporal dimensions and, in this context, the framework of redundant linear precoding is a powerful way of optimizing the coordinate system. This is why an increasing number of emerging communication standards are including precoding modules. As explained in detail by Scaglione *et al.* (1999), precoding generalizes a number of methods used in practical cellular and wireless local area networks, such as multicarrier methods and spread-spectrum techniques.

The subject of this chapter is the optimization of linear precoders and decoders for the case where the channel state is fully known at the transmitter side. Channel state information (CSI) can be acquired at the transmitter either if a feedback channel is present or when the transmitter and receiver operate in time-division duplex (TDD), so that the time-invariant MIMO channel transfer function is the same in both ways. In some applications (e.g., wireless local-area networks), the channel coherence time is large enough that the CSI is valid for long periods and can be used for optimization purposes. The design goal is to select an optimal pair of linear transformations $\boldsymbol{F}$ (precoder) and $\boldsymbol{G}$ (decoder) of blocks of the transmit symbols and receive samples, respectively, that *operate jointly* and *linearly* on the time and space dimensions. In all designs, the paradigm of linear precoding/decoding exploits the channel eigendecomposition in constructing the optimal $\boldsymbol{F}, \boldsymbol{G}$. The distinct solutions are characterized by how the power is loaded on each channel eigenfunction. The precoder designs capitalize on the available knowledge about the channel by investing the available power wisely in each dimension. The capacity of MIMO flat fading channels has been studied originally by Foschini (1996), and for frequency selective channels by Raleigh and Cioffi (1998) and the discussions in these papers are applicable to a linearly precoded system. Optimal linear designs developed in the past, which were based on specific multiple-input multiple-output (MIMO) models, such as proposed by Brandenburg and Wyner (1974), Salz (1985), Lechleider (1990), Kasturia *et al.* (1990), Yang and Roy (1994b,a) are related to the results that this chapter exposes in a systematic way.

Notation All boldface letters indicate vectors (lowercase) or matrices (uppercase). The $\operatorname{tr}(\boldsymbol{A}), |\boldsymbol{A}|, \lambda(\boldsymbol{A})$ are the trace, determinant and eigenvalues of matrix $\boldsymbol{A}$, $\boldsymbol{a} = \operatorname{vec}(\boldsymbol{A})$ is the column vector formed by stacking the columns of $\boldsymbol{A}$ on top of each other. Continuous-time multidimensional signals are written as $\boldsymbol{a}(t)$, discrete-time vector sequences as $\boldsymbol{a}[n]$, while sequences of vectors obtained by stacking consecutive blocks are characterized by a subscript $(\ )_i$, for example $\boldsymbol{a}_i = \operatorname{vec}([\boldsymbol{a}[iM], \ldots, \boldsymbol{a}[iM + M - 1]])$.

9.2 System model

We consider a MIMO system with K transmit and R receive antennas; the transmitted complex baseband signal vector $\boldsymbol{x}(t) := (x_1(t), \ldots, x_K(t))^T$ is:

$$\boldsymbol{x}(t) = \sum_{n=-\infty}^{+\infty} \boldsymbol{x}[n] g_T(t - nT),$$

where $g_T(t)$ is the transmit pulse and $1/T$ is the rate with which the data $\boldsymbol{x}[n]$ are transmitted. Correspondingly, $\boldsymbol{z}(t) = \boldsymbol{y}(t) + \boldsymbol{n}(t)$ is the received $R \times 1$ vector, which contains the channel output $\boldsymbol{y}(t)$ and additive noise $\boldsymbol{n}(t)$. For a linear (generally time varying) channel, the vector of received samples $\boldsymbol{y}[k] := \boldsymbol{y}(kT)$ can be written as

$$\boldsymbol{y}[k] = \sum_{n=-\infty}^{\infty} \mathbf{H}[k, k-n] \boldsymbol{x}[n]. \tag{9.1}$$

If the discrete-time, time-varying impulse response $\mathbf{H}[k, n]$ is causal and has finite memory L, we can write the I/O relationship (9.1) in block finite impulse response (FIR) form. Specifically, stacking $P = M + L$ transmit snapshots in a $PK \times 1$ vector $\boldsymbol{x}_i := \text{vec}([\boldsymbol{x}[iP], \ldots, \boldsymbol{x}[iP + P - 1]])$ and M received snapshots in an $MR \times 1$ vector $\boldsymbol{y}_i := \text{vec}([\boldsymbol{y}[iP + L], \ldots, \boldsymbol{y}[iP + P - 1]])$, where we eliminated the first L vectors to cancel the interblock interference (IBI), we have

$$\boldsymbol{y}_i = \boldsymbol{H} \boldsymbol{x}_i, \tag{9.2}$$

where $\boldsymbol{H}$ is an $RM \times KP$ block-banded matrix and M is the number of receive samples per block per antenna. Alternatively, defining $\boldsymbol{x}_i := \text{vec}([\boldsymbol{x}[iP], \ldots, \boldsymbol{x}[iP + M - 1]])$ and $\boldsymbol{y}_i := \text{vec}([\boldsymbol{y}[iP], \ldots, \boldsymbol{y}[iP + P - 1]])$ and padding with L zero samples the tail of every block $\boldsymbol{x}_i$ we could have written an equation analogous to (9.2) but with $\boldsymbol{H}$ of dimension $RP \times KM$. For simplicity, we will assume that IBI is removed at the receiver and thus $\boldsymbol{H}$ is $RM \times KP$, but most of the derivations in the following are valid in both cases if one replaces M by P and vice versa.

If the channel is also time invariant (then called linear, time-invariant, or LTI), i.e., $\mathbf{H}[n, l] \equiv \mathbf{H}[l]$, where $\{\mathbf{H}[l]\}_{r,k}$ is the lth sample of the impulse response characterizing the channel between the kth transmit element and the rth receive element, then $\boldsymbol{H}$ in (9.2) becomes a block Toeplitz matrix. The designs we are going to present use full knowledge of the matrix $\boldsymbol{H}$.

9.3 Constrained linear precoding/decoding designs

The transmitter side For all $i \in \mathbb{Z}$ the vector $\boldsymbol{x}_i$ contains the coordinates of the transmitted signal with respect to the set of functions $\mathbf{E}_{i,jK+k}(t) = \mathbf{e}_k g_T(t - (iP + j)T)$ $(j = 0, \ldots, P - 1, \ k = 0, \ldots, K - 1)$, where $\mathbf{e}_k$ is a canonical vector in $\mathbb{R}^K$. Note that the square-root raised cosine functions $g_T(t - nT)$ are orthogonal and such that $WT \geq 1$. In the linear space spanned by these basis functions, precoding corresponds to linearly mapping $N \times 1$ vectors of data symbols $\boldsymbol{s}_i$ onto $\boldsymbol{x}_i$, that is:

$$\boldsymbol{x}_i = \boldsymbol{F}\boldsymbol{s}_i. \tag{9.3}$$

Because the transmission of $\boldsymbol{x}_i$ takes P periods, each equal to T, the symbol rate is $N/(PT)$. The dimension of $\boldsymbol{s}_i$ is chosen to satisfy $N \leq \min(PK, RM)$ to guarantee that the lattice points of the code are mapped in a space of equal or greater number of dimensions.

The time-bandwidth constraint is met by the basis $\mathbf{E}_{i,jK+k}(t)$. A transmit power constraint must be imposed on $\boldsymbol{F}$ otherwise any optimization aimed at improving the receiver performance will lead to increasing the norm of $\boldsymbol{F}$ to infinity. We consider two power constraints: one limiting the average transmit power per block and the other limiting the peak power. The average constrained power (CP) is met by bounding the expected norm of the transmit vector $E\{\|\boldsymbol{x}_i\|^2\} = \text{tr}(\boldsymbol{F}\boldsymbol{R}_{ss}\boldsymbol{F}^H)$, where $\boldsymbol{R}_{ss} = E\{\boldsymbol{s}_i\boldsymbol{s}_i^H\}$, thus:

$$(\text{CP}): \quad \text{tr}(\boldsymbol{F}\boldsymbol{R}_{ss}\boldsymbol{F}^H) = \mathcal{P}_0. \tag{9.4}$$

An alternative is to constrain the maximum eigenvalue $\boldsymbol{F}\boldsymbol{F}^H$,

$$(\text{C}\lambda_{\max}): \quad \lambda_{\max}(\boldsymbol{F}\boldsymbol{F}^H) = \mathcal{L}_0, \tag{9.5}$$

which also limits the power since $\text{tr}(\boldsymbol{F}\boldsymbol{R}_{ss}\boldsymbol{F}^H) \leq \lambda_{\max}(\boldsymbol{F}\boldsymbol{F}^H) \lambda_{\max}(\boldsymbol{R}_{ss})N$. Note that the peak of the transmitted signal corresponds to the largest (in absolute value) entry of $\boldsymbol{x}_i = \boldsymbol{F}\boldsymbol{s}_i$. The constraint (9.5) limits the peak power because the following inequalities for $\max_{i,k}(|\{\boldsymbol{x}_i\}_k|^2)$ apply

$$\max_{i,k}(|\{\boldsymbol{F}\boldsymbol{s}_i\}_k|^2) \leq \max_i(\|\boldsymbol{F}\boldsymbol{s}_i\|^2) \leq \lambda_{\max}(\boldsymbol{F}\boldsymbol{F}^H) \max_i(\|\boldsymbol{s}_i\|^2),$$

where $\|\boldsymbol{s}_i\|^2$ is bounded since the symbols are all bounded in amplitude. Geometrically, the constraint (9.4) is equivalent to requiring that the average output be constrained in an ellipsoid (see Fig. 9.1), while (9.5) forces the average signal to be in a hypersphere.

The receiver side The received vector signal $z(t)$ is projected by the receiver RF frontend onto the $M = P - L$ basis functions $E'_{i,jK+k}(t) = e_r g_R(t - (iP+j)T)$ $(j = L, \ldots, P-1, r = 0, \ldots, R-1)$, where $g_R(t)$ is the receive filter, matched to the transmit pulse $g_T(t)$. The index j runs from L to $P-1$, and L basis functions are left unused to avoid IBI and retain a block memoryless channel. As M increases, the loss of LR dimensions per block becomes negligible. The coefficients of this orthogonal projections are the elements of the received vector $z_i = y_i + n_i$ where n_i is additive Gaussian noise (AGN) with covariance R_{nn}. The linear receiver performs the mapping G on the vector[†] $z_i = \text{vec}[z[iP+L], \ldots, z[iP+P-1]]$, estimating the symbols as:

$$\hat{s}_i = G z_i = G H F s_i + G n_i. \tag{9.6}$$

Contrary to F, G does not have to satisfy any constraint.

9.4 Optimal design criteria

Assumptions[‡]

a0) *The size N of the block s_i of encoded symbols satisfies $N \leq \text{rank}(H)$.[§]*
a1) *The transmit symbols are white, i.e., $R_{ss} = \sigma_{ss}^2 I$. The noise n_i is Gaussian with covariance R_{nn} that is positive definite, and n_i and s_i are uncorrelated.*

Let us introduce the following eigenvalue decomposition (EVD):

$$H^H R_{nn}^{-1} H = \bar{V} \bar{\Lambda} \bar{V}^H, \tag{9.7}$$

where $\bar{V}$ may be tall if $H^H R_{nn}^{-1} H$ is rank deficient and $\bar{\Lambda}$ is a $Q \times Q$ diagonal matrix, where $Q := \text{rank}(H^H R_{nn}^{-1} H) = \text{rank}(H)$. We assume that the elements $\{\lambda_{qq}\}_{q=1}^{Q}$ in the diagonal of matrix $\bar{\Lambda}$, which are the nonzero eigenvalues of $H^H R_{nn}^{-1} H$, are arranged in decreasing order. Note that a0) requires that $N \leq Q$. For convenience, we will denote by Λ the $N \times N$ diagonal matrix with diagonal entries $\{\lambda_{qq}\}_{q=1}^{N}$, and by V the first N columns of $\bar{V}$, which are the eigenvectors corresponding to the N largest eigenvalues $\{\lambda_{qq}\}_{q=1}^{N}$ of $H^H R_{nn}^{-1} H$.

[†]The definition of z_i would change consistently with the definition of y_i, and H if zero guards are used at the transmitter side.

[‡]Though less restrictive assumptions are possible for some designs, proofs and notation become cumbersome and we avoid them for the sake of simplicity.

[§]Note that if the channel is LTI then if $\text{rank}(H) < \min(KP, RM)$, the channels between the K transmitters and the R receivers have common zeros (Kailath, 1980, p.142).

Design criteria Our design criteria are based on two matrices. The first is called the mean squared error (MSE) matrix:

$$E\{(\hat{\boldsymbol{s}}_i - \boldsymbol{s}_i)(\hat{\boldsymbol{s}}_i - \boldsymbol{s}_i)^H\} = \text{MSE}(\boldsymbol{F}, \boldsymbol{G}) \tag{9.8}$$

$$\text{MSE}(\boldsymbol{F}, \boldsymbol{G}) := \sigma_{ss}^2(\boldsymbol{GHF} - \boldsymbol{I})(\boldsymbol{GHF} - \boldsymbol{I})^H + \boldsymbol{GR}_{nn}\boldsymbol{G}^H. \tag{9.9}$$

The second is called the signal-to-noise ratio (SNR) matrix:

$$\text{SNR}(\boldsymbol{F}, \boldsymbol{G}) := \boldsymbol{F}^H\boldsymbol{H}^H\boldsymbol{G}^H(\boldsymbol{GR}_{nn}\boldsymbol{G}^H)^{-1}\boldsymbol{GHF}\sigma_{ss}^2. \tag{9.10}$$

In the following sections it will become apparent that our optimization criteria will minimize/maximize functions of $\text{MSE}(\boldsymbol{F}, \boldsymbol{G})$ or $\text{SNR}(\boldsymbol{F}, \boldsymbol{G})$ that depend exclusively on their eigenvalues. With this fact in mind, we can find a matrix $\boldsymbol{G}_{\text{opt}}$ that applies to all $\text{MSE}(\boldsymbol{F}, \boldsymbol{G})$ or $\text{SNR}(\boldsymbol{F}, \boldsymbol{G})$ designs, in fact:

$$\overline{\text{MSE}}(\boldsymbol{F}) \triangleq \text{MSE}(\boldsymbol{F}, \boldsymbol{G}_{\text{opt}}) \leq \text{MSE}(\boldsymbol{F}, \boldsymbol{G})$$

$$\overline{\text{SNR}}(\boldsymbol{F}) \triangleq \text{SNR}(\boldsymbol{F}, \boldsymbol{G}_{\text{opt}}) \geq \text{SNR}(\boldsymbol{F}, \boldsymbol{G}) \tag{9.11}$$

and the inequalities indicate that $\forall \, \boldsymbol{G} \neq \boldsymbol{G}_{\text{opt}}$, $[\text{MSE}(\boldsymbol{F}, \boldsymbol{G}) - \overline{\text{MSE}}(\boldsymbol{F})]$ and $[\overline{\text{SNR}}(\boldsymbol{F}) - \text{SNR}(\boldsymbol{F}, \boldsymbol{G})]$ are positive semidefinite matrices.

The $\boldsymbol{G}_{\text{opt}}$ that minimizes the matrix $\text{MSE}(\boldsymbol{F}, \boldsymbol{G})$ can be shown to be the same as the minimum mean squared error (MMSE) receiver, i.e., the Wiener receiver (Scaglione *et al.*, 2002):

$$\boldsymbol{G}_{\text{opt}} = \sigma_{ss}^2\boldsymbol{F}^H\boldsymbol{H}^H(\sigma_{ss}^2\boldsymbol{HFF}^H\boldsymbol{H}^H + \boldsymbol{R}_{nn})^{-1}. \tag{9.12}$$

$$\overline{\text{MSE}}(\boldsymbol{F}) = \sigma_{ss}^2(\boldsymbol{I} + \sigma_{ss}^2\boldsymbol{F}^H\boldsymbol{H}^H\boldsymbol{R}_{nn}^{-1}\boldsymbol{HF})^{-1}. \tag{9.13}$$

For the SNR matrix we have that:

$$\text{SNR}(\boldsymbol{F}, \boldsymbol{G}) = \sigma_{ss}^2\boldsymbol{F}^H\boldsymbol{H}^H\boldsymbol{R}_{nn}^{-1/2}\boldsymbol{\Pi}\boldsymbol{R}_{nn}^{-1/2}\boldsymbol{HF} \leq \sigma_{ss}^2\boldsymbol{F}^H\boldsymbol{H}^H\boldsymbol{R}_{nn}^{-1}\boldsymbol{HF}, \tag{9.14}$$

where $\boldsymbol{\Pi} = \boldsymbol{R}_{nn}^{1/2}\boldsymbol{G}^H(\boldsymbol{GR}_{nn}\boldsymbol{G}^H)^{-1}\boldsymbol{GR}_{nn}^{1/2} \leq \boldsymbol{I}$ is the orthogonal projector onto the range space of $\boldsymbol{R}_{nn}^{1/2}\boldsymbol{G}^H$. The upper-bound in (9.14) is reached if and only if

$$\boldsymbol{G}_{\text{opt}} = \tilde{\boldsymbol{\Gamma}}\boldsymbol{F}^H\boldsymbol{H}^H\boldsymbol{R}_{nn}^{-1}, \tag{9.15}$$

where $\tilde{\boldsymbol{\Gamma}}$ is an $N \times N$ invertible arbitrary matrix. Thus:

$$\overline{\text{SNR}}(\boldsymbol{F}) = \sigma_{ss}^2\boldsymbol{F}^H\boldsymbol{H}^H\boldsymbol{R}_{nn}^{-1}\boldsymbol{HF}. \tag{9.16}$$

Hence, in the following, we will determine $\boldsymbol{F}_{\text{opt}}$ for the MSE based criteria using directly $\overline{\text{MSE}}(\boldsymbol{F})$ in (9.13) while for the SNR criteria we will use directly $\overline{\text{SNR}}(F)$ in (9.16). All the proofs of the lemmas and corollaries in the following sections are derived by Scaglione *et al.* (2002).

9.4.1 The MSE criterion

The MMSE design minimizes $\text{tr}(\text{MSE}(F, G))$ jointly with respect to G and F under the transmit-power constraint. The corresponding joint transmit and receive design can be obtained by minimizing $\text{tr}(\overline{\text{MSE}}(F))$ with respect to F. The solution for F_{opt} is given in the following lemma, and G_{opt} can be obtained by replacing F with F_{opt} in (9.12).

Lemma 9.1 *The solution to the optimization problem:*

$$F_{\text{opt}} = \arg\min_{F} \ \text{tr}(\overline{\text{MSE}}(F)) \qquad subject\ to \qquad \text{tr}(F_{\text{opt}} F_{\text{opt}}^{H}) \sigma_{ss}^{2} = \mathcal{P}_0$$

is given by $F_{\text{opt}} = V\Phi$, *where* Φ *is an* $N \times N$ *diagonal matrix with the following* (i, i) *entry:*[†]

$$|\phi_{ii}|^2 = \left(\frac{\mathcal{P}_0 + \sum_{n=1}^{\bar{N}} \lambda_{nn}^{-1}}{\sigma_{ss}^2 \sum_{n=1}^{\bar{N}} \lambda_{nn}^{-1/2}} \lambda_{ii}^{-1/2} - \frac{1}{\lambda_{ii} \sigma_{ss}^2} \right)^{+} \tag{9.17}$$

where $(x)^{+} := \max(x, 0)$ *and* $\bar{N} \leq N$ *is such that* $|\phi_{nn}|^2 > 0$ *for* $n \in [1, \bar{N}]$ *and* $|\phi_{nn}|^2 = 0$ *for all other* n.

Note that $\bar{N}$ is a function of the eigenvalues as well: for given Λ and N, $\bar{N}$ can be found by calculating (9.17) iteratively starting with $\bar{N} = N$ and decreasing $\bar{N}$ by one if not all the values inside the $(\ )^{+}$ in (9.17) are positive, as explained by Scaglione *et al.* (2002). Interestingly, in the following lemma (and its corollary) we show that the minimization of the determinant, in lieu of the trace, of the $\overline{\text{MSE}}(F)$ matrix with respect to F is equivalent to maximizing the information rate. The capacity of a MIMO channel was first derived by Brandenburg and Wyner (1974); the general model introduced in this chapter includes several other cases (Foschini, 1996; Marzetta and Hochwald, 1999; Raleigh and Cioffi, 1998).

Lemma 9.2 *The solution to the optimization problem*

$$F_{\text{opt}} = \arg\min_{F} |\overline{\text{MSE}}(F)| \qquad subject\ to \qquad \text{tr}(F_{\text{opt}} F_{\text{opt}}^{H}) \, \sigma_{ss}^{2} = \mathcal{P}_0$$

[†]Note that only the amplitude of ϕ_{ii} is fixed while the phase is arbitrary; thus, ϕ_{ii} can be a real number.

is $F_{\mathrm{opt}} = V\Phi$, where Φ is an $N \times N$ diagonal matrix with the (i,i) entry

$$|\phi_{ii}|^2 = \left(\frac{\mathcal{P}_0 + \sum_{k=1}^{\bar{N}} \lambda_{kk}^{-1}}{\bar{N}\sigma_{ss}^2} - \frac{1}{\lambda_{ii}\sigma_{ss}^2} \right)^+, \qquad (9.18)$$

and $\bar{N} \leq N$ is the number of strictly positive $|\phi_{ii}|^2$.

The power loading on the eigenvectors V of $H^H R_{nn}^{-1} H$ in Lemma 9.2 is identical to the so called "waterfilling" power loading, obtained from the maximization of the mutual information on parallel Gaussian channels (e.g., Gallager, 1968; Cover and Thomas, 1991; Foschini, 1996). More specifically the waterfilling solution is obtained in the following optimization (Scaglione *et al.*, 2002):

Corollary 9.1 *For a Gaussian input s_i, if G has the following structure (see (9.15)):*[†]

$$G = \tilde{\Gamma} F^H H^H R_{nn}^{-1},$$

where $\tilde{\Gamma}$ is an arbitrary $N \times N$ matrix, the mutual information $I(\hat{s}, x)$ per block is given by:

$$I(\hat{s}, x) = \log_2 |\sigma_{ss}^2 H F F^H H^H R_{nn}^{-1} + I|. \qquad (9.19)$$

The F_{opt} in (9.18) and G_{opt} in (9.12) also maximize the mutual information between transmit and receive data.

The two criteria lead to the same power loading because

$$I(\hat{s}, x) = \log_2 |\overline{\mathrm{MSE}}(F)| - \log_2(\sigma_{ss}^2)$$

(see (9.13)) and, since logarithm is a monotonic function, the maximum points of $I(\hat{s}, x)$ and $|\overline{\mathrm{MSE}}(F)|$ coincide. Using the eigenvalue constraint in lieu of the average power constraint leads to the following:

Lemma 9.3 *The solution to the optimization problem*

$$F_{\mathrm{opt}} = \arg\min_{F} \mathrm{tr}(\overline{\mathrm{MSE}}(F)) \qquad \text{subject to} \quad \lambda_{\max}(F F^H)\sigma_{ss}^2 = \mathcal{L}_0$$

is given by $F_{\mathrm{opt}} = \sqrt{\mathcal{L}_0/\sigma_{ss}^2}\, V$.

Lemma 9.4 *The solution to the optimization problem*

$$F_{\mathrm{opt}} = \arg\min_{F} |\overline{\mathrm{MSE}}(F)| \qquad \text{subject to} \quad \lambda_{\max}(F F^H)\sigma_{ss}^2 = \mathcal{L}_0$$

[†] Note that the receiver selection is not completely defined by some of the optimal design criteria discussed in the following.

is given by $\boldsymbol{F}_{\mathrm{opt}} = \sqrt{\mathcal{L}_0/\sigma_{ss}^2}\,\boldsymbol{V}$.

As with Lemma 9.2, it is worth noting that, because of (9.19), the solution in Lemma 9.4 also provides the maximum information rate under (9.5).

9.4.2 The SNR criteria

Designs minimizing the probability of error are rarely solvable in closed form. Here, we propose design criteria that can come close to the desired goal of minimizing the bit error rate (BER) although their optimization is *alphabet independent*. Based on (9.6), the optimal decision rule is the maximum-likelihood (ML) detector, provided that the noise is Gaussian and that the symbols are i.i.d.[†] Specifically, if we let $\boldsymbol{s}_i(\mathcal{H}_k)$ denote the symbol vector corresponding to hypothesis $\mathcal{H}_k$, and let $\mathcal{D}_i$ denote the decision on the ith symbol block, then the ML decision rule is:

$$\mathcal{D}_i = \min_{\mathcal{H}_k}\ [\hat{\boldsymbol{s}}_i - \boldsymbol{GHF}\boldsymbol{s}_i(\mathcal{H}_k)]^H (\boldsymbol{GR}_{nn}\boldsymbol{G}^H)^{-1}[\hat{\boldsymbol{s}}_i - \boldsymbol{GHF}\boldsymbol{s}_i(\mathcal{H}_k)]. \quad (9.20)$$

The detection rule (9.20) in AWGN leads to a pairwise probability of error that decreases monotonically with the minimum distance between the hypotheses (Proakis, 2001). Thus, an indirect way of reducing the probability of error is to maximize the minimum distance between hypotheses, and this is done through the appropriate selection of the code vectors $\boldsymbol{s}_i$. Here, instead, we want to search for the optimal $\boldsymbol{F}$ and $\boldsymbol{G}$ without changing $\boldsymbol{s}_i$, to retain the modularity of the system design that only focuses on the choice of $\boldsymbol{F}$ and $\boldsymbol{G}$. Note that the distance between hypotheses is not the Frobenius norm; the distance is defined as $[\boldsymbol{s}_i(\mathcal{H}_h) - \boldsymbol{s}_i(\mathcal{H}_k)]^H \boldsymbol{F}^H \boldsymbol{H}^H \boldsymbol{G}^H (\boldsymbol{GR}_{nn}\boldsymbol{G}^H)^{-1} \boldsymbol{GHF}[\boldsymbol{s}_i(\mathcal{H}_h) - \boldsymbol{s}_i(\mathcal{H}_k)]$, and the norm matrix is represented by $\mathrm{SNR}(\boldsymbol{F},\boldsymbol{G})$ matrix defined in (9.10). Hence, the design criterion that targets the minimum distance without changing $\boldsymbol{s}_i$ can be written as:

$$\max_{\boldsymbol{F},\boldsymbol{G}} \left\{ \min_{h,k;\ h\neq k} [\boldsymbol{s}_i(\mathcal{H}_h) - \boldsymbol{s}_i(\mathcal{H}_k)]^H \mathrm{SNR}(\boldsymbol{F},\boldsymbol{G})[\boldsymbol{s}_i(\mathcal{H}_h) - \boldsymbol{s}_i(\mathcal{H}_k)] \right\}, \quad (9.21)$$

under some constraint on $\boldsymbol{F}$. The solution to (9.21) changes depending on the symbol alphabet. Instead of solving (9.21), a cost function based only on $\mathrm{SNR}(\boldsymbol{F},\boldsymbol{G})$ may provide suboptimal but more general design solutions that are not tied to a certain symbol alphabet. The criterion that we will adopt

---

[†]If this assumption is not satisfied, one has to use the maximum a posteriori probability (MAP) detector.

is based on the observation that the minimum eigenvalue $\lambda_{\min}(\text{SNR}(\boldsymbol{F}, \boldsymbol{G}))$ provides a lower bound for the minimum distance:

$$\min_{h,k;\ h \neq k} [\boldsymbol{s}_i(\mathcal{H}_h) - \boldsymbol{s}_i(\mathcal{H}_k)]^H \text{SNR}(\boldsymbol{F}, \boldsymbol{G})[\boldsymbol{s}_i(\mathcal{H}_h) - \boldsymbol{s}_i(\mathcal{H}_k)] \qquad (9.22)$$

$$\geq \lambda_{\min}(\text{SNR}(\boldsymbol{F}, \boldsymbol{G})) \min_{h,k;\ h \neq k} \|\boldsymbol{s}_i(\mathcal{H}_h) - \boldsymbol{s}_i(\mathcal{H}_k)\|^2.$$

Maximizing the lower bound in (9.22) will possibly force (9.22) to higher values. The corresponding solutions are given in the following two lemmas.

Lemma 9.5 *The solution to the optimization problem*

$$(\boldsymbol{F}_{\text{opt}}, \boldsymbol{G}_{\text{opt}}) = \underset{F,G}{\text{argmax}} \ \lambda_{\min}(\text{SNR}(\boldsymbol{F}, \boldsymbol{G}))$$

$$\textit{subject to} \quad \text{tr}(\boldsymbol{F}_{\text{opt}} \boldsymbol{F}_{\text{opt}}^H)\sigma_{ss}^2 = \mathcal{P}_0$$

is given by $\boldsymbol{F}_{\text{opt}} = \boldsymbol{V}\boldsymbol{\Phi}$, *with* $\boldsymbol{\Phi}$ *diagonal* $N \times N$ *having diagonal entries*

$$|\phi_{ii}|^2 = \frac{\mathcal{P}_0}{\sigma_{ss}^2 \sum_{k=1}^{N} \lambda_{kk}^{-1}} \lambda_{ii}^{-1},$$

and $\boldsymbol{G}_{\text{opt}} = \tilde{\boldsymbol{\Gamma}} \boldsymbol{V}^H \boldsymbol{H}^H \boldsymbol{R}_{nn}^{-1}$ *with the* $N \times N$ *matrix* $\tilde{\boldsymbol{\Gamma}}$ *being invertible.*

Note that the solution leads to

$$\text{SNR}(\boldsymbol{F}_{\text{opt}}, \boldsymbol{G}_{\text{opt}}) = \sigma_{ss}^2 \boldsymbol{\Phi}^H \boldsymbol{\Lambda} \boldsymbol{\Phi} = \frac{\mathcal{P}_0}{\sum_{k=1}^{N} \lambda_{kk}^{-1}} \boldsymbol{I},$$

which ensures that the lower bound in (9.22) is met with equality.

Lemma 9.6 *The solution to the optimization problem*

$$(\boldsymbol{F}_{\text{opt}}, \boldsymbol{G}_{\text{opt}}) = \underset{F,G}{\text{argmax}} \ \lambda_{\min}(\text{SNR}(\boldsymbol{F}, \boldsymbol{G}))$$

$$\textit{subject to} \quad \lambda_{\max}(\boldsymbol{F}\boldsymbol{F}^H)\sigma_{ss}^2 = \mathcal{L}_0$$

is given by $\boldsymbol{F}_{\text{opt}} = \boldsymbol{V}\boldsymbol{\Phi}$, *with* $\boldsymbol{\Phi}$ *diagonal* $N \times N$ *such that*

$$|\phi_{ii}|^2 = \frac{\mathcal{L}_0 \lambda_{NN}}{\sigma_{ss}^2} \lambda_{ii}^{-1},$$

and $\boldsymbol{G}_{\text{opt}} = \tilde{\boldsymbol{\Gamma}} \boldsymbol{F}^H \boldsymbol{H}^H \boldsymbol{R}_{nn}^{-1}$ *with the* $N \times N$ *arbitrary matrix* $\tilde{\boldsymbol{\Gamma}}$ *being invertible.*

Note that, similar to Lemma 9.5, $\text{SNR}(\boldsymbol{F}_{\text{opt}}, \boldsymbol{G}_{\text{opt}}) = \mathcal{L}_0 \boldsymbol{I}$. Interestingly, the solution of Lemma 9.5 coincides with the MMSE solution under the zero forcing (ZF) constraint (Scaglione *et al.*, 1999, Th.3). As discussed in the following section, the selection of $\boldsymbol{F} = \boldsymbol{V}\boldsymbol{\Phi}$ and $\boldsymbol{G} = \tilde{\boldsymbol{\Gamma}} \boldsymbol{F}^H \boldsymbol{H}^H \boldsymbol{R}_{nn}^{-1}$ with

parallel subchannels

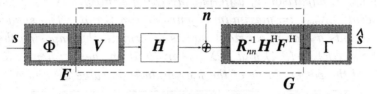

Fig. 9.2. Optimal transceivers: matrix model

$(\mathbf{\Phi}, \tilde{\mathbf{\Gamma}})$ diagonal matrices, leads to diagonalizing the overall channel $\boldsymbol{GHF}$ and the noise covariance $\boldsymbol{GR}_{nn}\boldsymbol{G}^H$.

It is interesting to observe that for arbitrary $\boldsymbol{F}$ and $\boldsymbol{G}$ we can extend to the Gaussian MIMO case the capacity formula of the SISO AWGN case:

$$I(\hat{\boldsymbol{s}}, \boldsymbol{x}) = \log |\boldsymbol{I} + \mathrm{SNR}(\boldsymbol{F}, \boldsymbol{G})|,$$

which is the matrix version of the well-known formula for the scalar AWGN channel (Cover and Thomas, 1991).

9.5 Performance of the optimal designs

In this section, we will derive expressions for performance measures such as the mutual information, the probability of error and the mean square error achievable with the optimal precoding/decoding schemes presented so far. As mentioned before, all optimal designs lead invariably to loading the power across the eigenvectors of $\boldsymbol{H}^H \boldsymbol{R}_{nn}^{-1} \boldsymbol{H}$.

Lemma 9.7 *All optimal designs we described so far are such that:*

$$\boldsymbol{F}_{\mathrm{opt}} = \boldsymbol{V}\boldsymbol{\Phi}, \qquad \boldsymbol{G}_{\mathrm{opt}} = \boldsymbol{\Gamma}\boldsymbol{\Lambda}^{-1}\boldsymbol{V}^H \boldsymbol{H}^H \boldsymbol{R}_{nn}^{-1}, \qquad (9.23)$$

where $\boldsymbol{\Phi}$ and $\boldsymbol{\Gamma}$ are diagonal matrices.

The matrices $\boldsymbol{F}_{\mathrm{opt}}$ and $\boldsymbol{G}_{\mathrm{opt}}$ in (9.23) cascaded with the channel matrix $\boldsymbol{H}$ in between are depicted in Fig. 9.2. The equivalent matrix of the cascade inside the box of Fig. 9.2 is

$$\boldsymbol{\Lambda}^{-1}\boldsymbol{V}^H \boldsymbol{H}^H \boldsymbol{R}_{nn}^{-1} \boldsymbol{H}\boldsymbol{V} = \boldsymbol{\Lambda}^{-1}\boldsymbol{V}^H \bar{\boldsymbol{V}} \bar{\boldsymbol{\Lambda}} \bar{\boldsymbol{V}}^H \boldsymbol{V} = \boldsymbol{I} \,,$$

and the noise correlation at the output of the box is:

$$\boldsymbol{\Lambda}^{-1}\boldsymbol{V}^H \boldsymbol{H}^H \boldsymbol{R}_{nn}^{-1} \boldsymbol{R}_{nn} \boldsymbol{R}_{nn}^{-1} \boldsymbol{H}\boldsymbol{V}\boldsymbol{\Lambda}^{-1} = \boldsymbol{\Lambda}^{-1} \,. \qquad (9.24)$$

Table 9.1. *Comparison of different precoding strategies (CP := power constraint; $C\lambda_{\max} := $ maximum eigenvalue constraint). The symbol $\mathcal{K}$ represents a scaling constant that is determined by imposing the constraint (CP) in (9.4) and ($C\lambda_{\max}$) in (9.5) on the ϕ_{ii}; hence, $\mathcal{K}$ is a function of the power $\mathcal{P}_0$ or of the maximum eigenvalue $\mathcal{L}_0$ and of all λ_{ii}.*

	Criterion	Constraint	$\text{SNR}_i = \|\phi_{ii}\|^2 \sigma_{ss}^2 \lambda_{ii}$
1	$\min(\text{tr}(\text{MSE}))$	CP	$\left(\mathcal{K}_{\text{tr}(\text{MSE})} \; \lambda_{ii}^{1/2} - 1\right)^+$
2	$\min(\|\text{MSE}\|)$	CP	$\left(\mathcal{K}_{\|\text{MSE}\|} \; \lambda_{ii} - 1\right)^+$
3	$\min(\text{tr}(\text{MSE}))$	$C\lambda_m$	$\mathcal{L}_0 \, \lambda_{ii}$
4	$\min(\|\text{MSE}\|)$	$C\lambda_m$	$\mathcal{L}_0 \, \lambda_{ii}$
5	$\max(\lambda_m \; (\text{SNR}))$	CP	$\mathcal{K}_{\text{SNR}}$
6	$\max(\lambda_m \; (\text{SNR}))$	$C\lambda_m$	$\mathcal{K}_{\text{SNR}} \quad \lambda_{\max}$

Thus, the matrix (or block) channel has a diagonal transfer matrix equal to $\boldsymbol{\Gamma}\boldsymbol{\Phi}$ and an additive noise with correlation matrix $\boldsymbol{\Gamma}^H \boldsymbol{\Gamma} \boldsymbol{\Lambda}^{-1}$. Hence, the N subchannels are decoupled and Fig. 9.2 becomes equivalent a set of parallel flat-fading subchannels corresponds to the diagonal elements of $\boldsymbol{\Gamma}\boldsymbol{\Phi}$. The noise components $\{\boldsymbol{\beta}_i\}_k$, $k = 1, \ldots, N$, are uncorrelated and have variance equal to λ_{kk}^{-1}.

Corollary 9.2 *The optimal transceivers in (9.23) convert the MIMO linear AGN channel with memory into an equivalent set of N parallel independent ISI-free subchannels, each with flat-fading gain equal to $\phi_{kk}\gamma_{kk}$, an AGN $\{\boldsymbol{\beta}_i\}_k$ with variance $1/\lambda_{kk}$, and with $\{\boldsymbol{\beta}_i\}_k$, $\{\boldsymbol{\beta}_i\}_j$ uncorrelated for $k \neq j$; i.e.,*

$$\{\hat{\boldsymbol{s}}_i\}_k = \phi_{kk} \, \gamma_{kk} \, \{\boldsymbol{s}_i\}_k + \gamma_{kk} \, \{\boldsymbol{\beta}_i\}_k, \qquad k = 1, \ldots, N.$$

The SNR at the output of the kth subchannel is:

$$\text{SNR}_k = \sigma_{ss}^2 \|\phi_{kk}\|^2 \lambda_{kk}.$$

Table 9.1 gives the SNR_i characterizing the optimal designs of Section 9.4. Comparing the SNR_i expressions in this table, we observe the similarity between the MMSE and waterfilling solutions of Lemmas 9.1 and 9.2: both solutions tend to exclude the most noisy subchannels (corresponding to small λ_{ii}'s) and the SNR_i grows as $\lambda_{ii}^{1/2}$ and λ_{ii} respectively. The other designs do not include mechanisms to *adapt* the number of subchannels used to the average signal-to-noise ratio and, thus, they waste power over the most noisy subchannels. Consequently, their associated performance may degrade for N close to its maximum value of $Q = \text{rank}(\boldsymbol{H}^H \boldsymbol{R}_{nn} \boldsymbol{H})$ when the smallest eigenvalues are close to zero. Finally, we note that the independence of the parallel subchannels implies the following:

Corollary 9.3 *For the linear transceivers of (9.23)*

$$I(\hat{\boldsymbol{s}}; \boldsymbol{x}) = \sum_{i=1}^{N} \log_2(1 + \text{SNR}_i).$$

Corollary 9.4 *The MSE for the $\boldsymbol{F}$ and $\boldsymbol{G}$ in (9.23) is:*

$$\text{MSE}(\boldsymbol{\Gamma}, \boldsymbol{\Phi}) = \sum_{i=1}^{N} \left[\frac{|\gamma_{ii}|^2}{\lambda_{ii}} + |\gamma_{ii}\phi_{ii} - 1|^2 \sigma_{ss}^2 \right].$$

For the MMSE receiver, we have $\boldsymbol{\Gamma} = \sigma_{ss}^2 \boldsymbol{\Phi}^H (\boldsymbol{\Lambda}^{-1} + \sigma_{ss}^2 \boldsymbol{\Phi}\boldsymbol{\Phi}^H)^{-1}$, and:

$$\overline{\text{MSE}}(\boldsymbol{\Phi}) = \sum_{i=1}^{N} \frac{\sigma_{ss}^2}{1 + |\phi_{ii}|^2 \sigma_{ss}^2 \lambda_{ii}} = \sum_{i=1}^{N} \frac{\sigma_{ss}^2}{1 + \text{SNR}_i}.$$

For the ZF receiver, we have $\boldsymbol{\Gamma} = \boldsymbol{\Phi}^\dagger$, and the corresponding MSE is:

$$\widetilde{\text{MSE}}(\boldsymbol{\Phi}) = \sum_{i=1}^{N} \frac{1}{|\phi_{ii}|^2 \lambda_{ii}} = \sum_{i=1}^{N} \frac{\sigma_{ss}^2}{\text{SNR}_i}.$$

9.6 Performance analysis

In random fading, performance is a function of the random eigenvalues of the matrix $\boldsymbol{H}^H \boldsymbol{R}_{nn}^{-1} \boldsymbol{H}$. This is technically a tractable problem. In fact, for any $n \times n$ matrix $\boldsymbol{A}$ in the form $\boldsymbol{B}^H \boldsymbol{B}$, where $\boldsymbol{B}$ is $m \times n$ with $m > n$ the probability density function of the eigenvalues is as follows:

$$p_{\boldsymbol{\Lambda}}(\boldsymbol{\Lambda}) = 2^{-n} \prod_{1 \le i < k \le n} (\lambda_k - \lambda_i)^2 \left(\prod_{i=1}^{n} \lambda_i \right)^{m-n} \Psi(\lambda_1, \ldots, \lambda_n) \quad (9.25)$$

$$\Psi(\lambda_1, \ldots, \lambda_n) \triangleq \int p_{\boldsymbol{B}}(\boldsymbol{Q}\sqrt{\boldsymbol{\Lambda}}\boldsymbol{U}^H)(\bar{\boldsymbol{Q}}^H d\boldsymbol{Q})(\boldsymbol{U}^H d\boldsymbol{U}),$$

where $\boldsymbol{Q}$ is an $m \times m$ unitary matrix and $\bar{\boldsymbol{Q}}$ is semiunitary and contains the first m columns of $\boldsymbol{Q}$, $\boldsymbol{U}$ is $n \times n$ unitary, $(\bar{\boldsymbol{Q}}^H d\boldsymbol{Q})$ and $(\boldsymbol{U}^H d\boldsymbol{U})$ are the elements of volume of the so-called Stiefel manifold of the unitary matrix

group. $(\bar{Q}^H dQ)$ and $(U^H dU)$ are each functions of mn and n^2 independent parameters respectively. Replacing B by $HR_{nn}^{-1/2}$, the calculation of
(9.25) is simpler, if the distribution of $B = HR_{nn}^{-1/2}$ is invariant under left
and right multiplication by a unitary matrix, i.e., $p_B(Q\sqrt{\Lambda}U^H) = p_B(\Lambda)$;
the remaining integral is the product of the volumes of the Stiefel manifold, whose expression is known. The case of frequency-selective channels
is further complicated by the fact that matrix the $B = HR_{nn}^{-1/2}$ has a
block-Toeplitz structure; although is tractable under some approximations
(e.g., Scaglione, 2002), the expression remains quite complex and therefore
we illustrate the performance hereafter through numerical simulations done
under the following channel model.

Channel model The FIR channel taps $h_{k,r}(l)$ are uncorrelated complex
Gaussian random variables (Rayleigh fading). The variance of the taps
$\sigma_{k,r}^2(l) = \sigma^2(l)$ follows the channel power-delay profile named "Channel A"
operating at 5.2 GHz, with $B = 100$ MHz ($T = 10$ ns) in Commitee (1998).
The channel order is $L \approx 19$ since the impulse response energies beyond
the 19th (i.e., after 190 ns) are statistically very small. The results are averaged over hundred random channels with white complex stationary additive
Gaussian noise having the same variance N_0 for each antenna. The horizontal axis in the plots that follow is the average block SNR (in dB) defined as
SNR $:= \mathrm{tr}(FF^H)\sigma_{ss}^2/N_0$, which does not include in its definition possible
small scale fading gain/attenuation.

Note that, especially if the multipath is dominated by a few strong reflec-
tors in the far field (Raleigh and Cioffi, 1998), the assumption of uncorrelated
scattering is optimistic. Nevertheless, the model is sufficiently accurate to
provide some insight.

Example 9.1 The first set of curves in Figs. 9.3 showcase the performance
of the designs developed in this chapter in terms of MSE, BER, and mutual
information. Note that only five curves are visible on each figure because
two criteria in Lemmas 9.5 and 9.6 (see Table 9.1, lines 5 and 6) lead to
identical designs. We set (9.5) and (9.4) so that $\mathcal{L}_0 = \mathcal{P}_0/N$. Because the
designs in Lemmas 9.3 and 9.4 (see Table 9.1, lines 3 and 4) are such that
$\Phi = \mathcal{L}_0 I$, we have that $\mathrm{tr}(F_{\mathrm{opt}}F_{\mathrm{opt}}^H) = \mathcal{P}_0$ for every channel. For the design
in Lemma 9.6, instead, if $\lambda_{\max}(F_{\mathrm{opt}}F_{\mathrm{opt}}^H) = \mathcal{P}_0/N$ then $\mathrm{tr}(F_{\mathrm{opt}}F_{\mathrm{opt}}^H) \neq$
$\mathcal{P}_0$. The curves corresponding to Lemma 9.6 are given as a function of the
corresponding $\mathrm{tr}(F_{\mathrm{opt}}F_{\mathrm{opt}}^H)/\sigma_n^2$ averaged over the channels. It appears that
the design in Lemma 9.1, which minimizes the MSE subject to the average

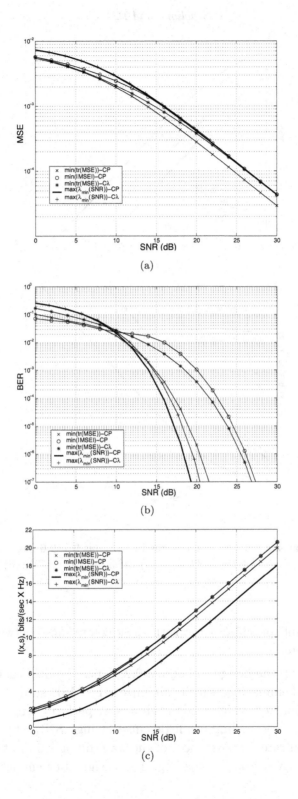

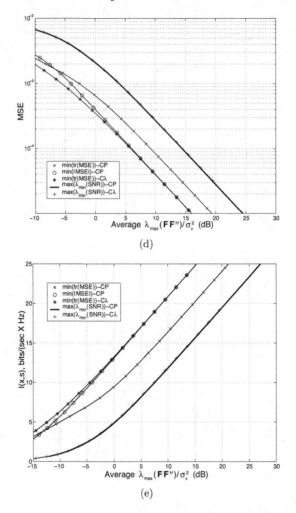

Fig. 9.3. Comparison between the designs in Lemmas 9.1–9.6 (see Table 9.1) $M = 32, K = R = 4, N = 112, \mathcal{P}_0 = 32$: (a), (d) MSE=tr($\overline{\text{MSE}}(\boldsymbol{F})$); (b) average BER; (c), (e) $I(\boldsymbol{x}, \hat{\boldsymbol{s}})$.

power constraint, provides perhaps the best compromise between BER and information rate.

In Figs. 9.3(c), the design in Lemma 9.2, which requires that $\boldsymbol{\Phi} \neq \boldsymbol{I}$, and the design in Lemma 9.3, which minimizes the MSE subject to the peak power constraint and which leads to $\boldsymbol{\Phi} = \mathcal{K}\boldsymbol{I}$, perform very similarly; this confirms that power loading has little effect at high SNR when maximizing the information rate, because most of the waterfilling gain in terms of information rate derives from bit loading. The designs in Lemmas 9.5 and 9.6,

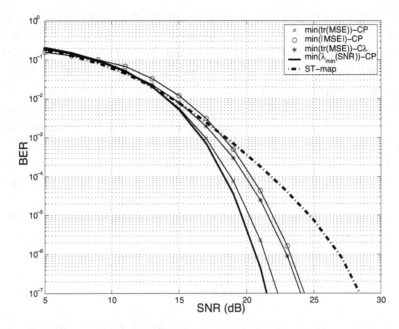

Fig. 9.4. BER obtained with the designs in Lemmas 9.1–9.6 ($K = R = 2, M = 32, N = 32$) and the design of Alamouti (1998) (ST-map).

which maximize a lower bound on the minimum distance under the power and maximum eigenvalue constraints, are ZF designs and tend to be superior to other criteria only in terms of BER. The reason why the design in Lemma 9.6 with eigenvalue constraint performs slightly worse than the design in Lemma 9.5 (with the average power constraint) is that we fixed the same $\mathcal{L}_0 = \mathcal{P}_0/N$ for the design in Lemma 9.6 as used for Lemma 9.3, instead of fixing the average power. Had we fixed the average power to be the same, the two algorithms in Lemmas 9.5 and 9.6 would have performed identically. In Figs. 9.3(d) and (e), we show the MSE and information rate curves, versus the ratio of the maximum eigenvalue (which is an upper bound for the peak power) over the noise variance. The curves are obtained by normalizing the value in the abscissa $\lambda_{\max}(\boldsymbol{F}\boldsymbol{F}^H)/\sigma_n^2$ to the same constant $\mathcal{L}_0$ for all criteria in all iterations and averaging over hundred random channels. The plots clearly show the gain obtained by the designs that use the eigenvalue constraint over the corresponding criteria using the average power constraint (see Table 9.1). The designs in Lemmas 9.5 and 9.6 (see Table 9.1 lines 5 and 6), perform identically because of the normalization.

Example 9.2 Compared with schemes that do not make use of the CSI, the flexibility of the optimal linear designs can justify their extra complexity.

Designs of space-time codes that do not make use or take advantage of the CSI, such as the Alamouti codes (Alamouti, 1998) cannot be generalized to an arbitrary number of antennas if available. Methods such as proposed by Alamouti (1998), Naguib *et al.* (1998) or their extension to frequency-selective fading, operate as diversity schemes, which exploit multiple antennas to increase the symbol SNR but not the symbol rate. Therefore, the number of transmitted symbols does not increase with the number of antennas and remains equal to the size of the block in time, i.e., $N = M = 32$. Setting the parameters to match the values required by the design of Alamouti (1998), for example $K = 2, R = 2, N = M$ symbols can be transmitted over M orthogonal subcarriers as in MIMO-OFDM. In this case, it can be shown that the SNR_k of the symbols received in each frequency bin for the scheme of Alamouti (1998) is:

$$\text{SNR}_k = \sum_{k,r=1}^{2} |H_{k,r}(i)|^2 \mathcal{P}_0/(2M\sigma_{nn}^2),$$

where $H_{k,r}(i) = \text{FFT}[h_{k,r}(l)]$ is the k,r channel transfer function at frequency bin i/M. We compare the BER of our scheme with that of Alamouti (1998) in Fig. 9.4.

References

Alamouti, S. M. (1998). A simple transmit diversity technique for wireless communications. *IEEE J. Sel. Areas Commun.*, **16** (8), 1451–1458.

Brandenburg, L. H. and Wyner, A. D. (1974). Capacity of the Gaussian channel with memory: The multivariate case. *Bell Syst. Tech. J.*, **53**, 745–778.

Commitee, E. N. (1998). *Norme ETSI, doc. 3ERI085B*. ETSI, Sophia-Antipolis, Valbonne, France.

Cover, T. M. and Thomas, J. A. (1991). *Elements of Information Theory* (Wiley, New York).

Forney, G. D., Jr. and Eyuboğlu, M. V. (1991). Combined equalization and coding using precoding. *IEEE Commun. Mag.*, **29** (12), 25–34.

Foschini, G. J. (1996). Layered space-time architecture for wireless communication in a fading environment when using multielement antennas. *Bell Labs. Tech. J.*, **1** (2), 41–59.

Gabor, D. (1946). Theory of communication. *J. Inst. Elec. Eng.*, **93**, 429–457.

Gallager, R. G. (1968). *Information Theory and Reliable Communication* (Wiley, New York).

Kailath, T. (1980). *Linear Systems* (Prentice-Hall, Englewood Cliffs, NJ).

Kasturia, S., Aslanis, J. T., and Cioffi, J. M. (1990). Vector coding for partial response channels. *IEEE Trans. Inf. Theory*, **36** (4), 741–761.

Lechleider, J. W. (1990). The optimum combination of block codes and receivers for arbitrary channels. *IEEE Trans. Commun.*, **38** (5), 615–621.

Marzetta, T. L. and Hochwald, B. H. (1999). Capacity of a mobile multiple-antenna communication link in Rayleigh flat fading. *IEEE Trans. Inf. Theory*, **45** (1), 139–157.

Naguib, A., Tarokh, V., Seshadri, N., and Calderbank, A. R. (1998). A space-time coding modem for high-data-rate wireless communication. *IEEE J. Sel. Areas Commun.*, **16** (8), 1459–1478.

Proakis, J. G. (2001). *Digital Communications* (McGraw-Hill, New York), sixth edn.

Raleigh, G. G. and Cioffi, J. M. (1998). Spatio-temporal coding for wireless communications. *IEEE Trans. Commun.*, **46** (3), 357–366.

Salz, J. (1985). Digital transmission over cross-coupled linear channels. *AT&T Tech. J.*, **64**, 1147–1159.

Scaglione, A. (2002). Statistical analysis of the capacity of MIMO frequency selective Rayleigh fading channels with arbitrary number of inputs and outputs. In *Proc. IEEE Int. Symp. Inf. Theory*, 278.

Scaglione, A., Giannakis, G. B., and Barbarossa, S. (1999). Redundant filterbank precoders and equalizers part I: unification and optimal designs. *IEEE Trans. Signal Process.*, **47** (7), 1988–2006.

Scaglione, A., Stoica, P., Barbarossa, S., Giannakis, G. B., and Sampath, H. (2002). Optimal designs for space-time linear precoders and decoders. *IEEE Trans. Signal Process.*, **50** (5), 1051–1064.

Veeravalli, V. V. and Mantravadi, A. (1999). The coding-spreading tradeoff in CDMA systems. *IEEE J. Sel. Areas Commun.*, **20** (2), 396–408.

Yang, J. and Roy, S. (1994a). Joint transmitter–receiver optimization for multi-input-multi-output systems with decision feedback. *IEEE Trans. Inf. Theory*, **40** (5), 1334–1347.

—— (1994b). On joint transmitter and receiver optimization for multiple-input–multiple-output (MIMO) transmission systems. *IEEE Trans. Commun.*, **42** (12), 3221–3231.

10

Space-time coding for noncoherent channels

Jean-Claude Belfiore and Antonio Maria Cipriano

École Nationale Supérieure des Télécommunications

This chapter presents some constructions of noncoherent space-time codes, that is, codes for MIMO systems when the channel is known neither at the transmitter nor at the receiver. Based on the generalized likelihood ratio test (GLRT) detector, we introduce some optimization criteria and describe the obtained codes.

10.1 Introduction

To achieve high spectral efficiency on wireless channels, we need multiple antennas at both the transmitter and the receiver. Information-theoretic results promise considerable capacity gains for wireless communication systems that use multiple transmit and receive antennas for coherent and noncoherent reception. Coherent reception means that the receiver knows the channel response, but the transmitter does not. Noncoherent reception means that neither the transmitter nor the receiver knows the channel response. In this chapter, we propose to show a nonexhaustive presentation of the space-time codes we can use in the noncoherent case, when the communication system uses M transmit antennas and N receive antennas.

General assumptions as well as notations are the following: we assume a Rayleigh flat-fading channel to separate space-time processing and multipath problems. Moreover, this channel is also assumed quasi-static. That means that channel coefficients do not vary during the transmission of a codeword with temporal length T. In that case, the received signal can be expressed as

$$\mathbf{Y}_{T \times N} = \alpha \mathbf{X}_{T \times M} \mathbf{H}_{M \times N} + \mathbf{W}_{T \times N}, \qquad (10.1)$$

where $\mathbf{X}$ is the transmitted codeword, $\mathbf{H}$ is the channel response, $\mathbf{W}$ is the i.i.d. Gaussian noise, and α is a normalizing factor. Subscripts indicate the respective dimensions of the complex-valued matrices. In the following, we will assume symmetric communication ($M = N$), and a code length $T \geq 2M$.

10.2 Results from information theory

10.2.1 General results

First investigations into the power-constrained ergodic capacity in the MIMO case when the channel is not known by the receiver are due to Marzetta and Hochwald (1999), Hochwald and Marzetta (2000), and Hassibi and Marzetta (2002). We report the following fundamental results:

(i) **Capacity Dependence on T.** For any block length T, any number of receive antennas N, and any SNR ρ, the capacity obtained with $M > T$ and $M = T$ are equal.

(ii) **Optimal signal structure.** For each value ρ of the SNR, the capacity-achieving signal matrix can be written as

$$\mathbf{X} = \mathbf{\Phi}\mathbf{\Lambda},$$

where $\mathbf{\Phi} \in \mathcal{U}_{T,M}$, and $\mathcal{U}_{T,M}$ denotes the set of all $T \times N$ unitary matrices, i.e., $\mathbf{\Phi}^\dagger\mathbf{\Phi} = \mathbf{I}_M$, and

$$\mathbf{\Lambda} = \begin{bmatrix} \lambda_1 & 0 & \cdots & 0 \\ 0 & \lambda_2 & \cdots & 0 \\ \vdots & & \ddots & \vdots \\ 0 & 0 & \cdots & \lambda_M \end{bmatrix}.$$

The matrix $\mathbf{\Phi}$ is a $T \times M$ isotropically distributed unitary matrix,[†] and $\mathbf{\Lambda}$ is an independent $M \times M$ real, nonnegative, diagonal matrix (Throughout the chapter, $(\cdot)^\dagger$ stands for the Hermitian transpose). In other words, the optimal signal consists of M *orthogonal vectors whose norms are* $\lambda_m \geq 0$, $m = 1, \ldots, M$. The general analytical form of the joint density of $[\lambda_1, \ldots, \lambda_M]$ is still unknown.

(iii) **Asymptotic capacity for $T \to \infty$.** When $T \to \infty$, the capacity tends to the capacity of the coherent case with perfect channel state information (CSI) at the receiver. Moreover, all the densities $p(\lambda_m)$ converge to a Dirac delta, centered at $\sqrt{T}$, showing that *the information is carried by the direction of the vectors* and not by their norms (see also the paper by Warrier and Madhow (2002)).

[†]An isotropically distributed unitary matrix has a probability density function that is invariant under left-multiplication by deterministic unitary matrices.

10.2.2 Asymptotic results at high SNR

Zheng and Tse (2002) investigated the ergodic capacity in the high SNR regime. They proved the following results.

(i) **Optimal signal structure.** For high SNR ρ, the random variables λ_m converge to the deterministic constant $\sqrt{T}$, as in the case $T \gg 1$. So, once again, the optimal signal structure is a matrix whose columns are isotropically distributed orthogonal vectors,

$$\mathbf{X} = [\mathbf{x}_1 \cdots \mathbf{x}_M] = \sqrt{T}\,\boldsymbol{\Phi}, \qquad \boldsymbol{\Phi} \in \mathcal{U}_{T,M}. \qquad (10.2)$$

(ii) **The information is carried by subspaces.** Equation (10.2) states that a good codebook should encode information in the directions of the vectors $\mathbf{x}_m$, $m = 1, \ldots, M$. However, Zheng and Tse (2002) show that the best strategy is to encode information in the whole *subspace*

$$\Omega_\mathbf{X} = \mathrm{span}(\mathbf{x}_1, \ldots, \mathbf{x}_M) = \mathrm{span}(\mathbf{X}). \qquad (10.3)$$

In other words, at high SNR and for $\mathbf{H}$ and $\mathbf{W}$ with i.i.d. circularly symmetric Gaussian entries, they show that the mutual information is $I(\mathbf{X}; \mathbf{Y}) = I(\Omega_\mathbf{X}; \mathbf{Y})$. An intuitive explanation is that, at high SNR, we can write

$$\mathbf{Y} = \alpha\mathbf{X}\mathbf{H} + \mathbf{W} \simeq \alpha\mathbf{X}\mathbf{H} \qquad (10.4)$$

to see that the channel stretches and rotates the basis $\mathbf{X}$ of $\Omega_\mathbf{X}$. Since the channel is unknown, the receiver cannot recover the particular basis $\mathbf{X}$, but the subspace $\Omega_\mathbf{X}$ is unchanged.

(iii) **Capacity and degrees of freedom.** The asymptotic capacity (in bit *per channel use*) is

$$C = K_{nc} \log_2 \rho + c_{M,N} + o(1), \qquad (10.5)$$

where $c_{M,N}$ is a constant depending on M, N, and T, while

$$K_{nc} = M^*(1 - M^*/T), \quad \text{with} \quad M^* = \min(M, N, \lfloor T/2 \rfloor), \qquad (10.6)$$

is called the number of *degrees of freedom*, or *multiplexing gain* of the noncoherent ergodic MIMO channel (Zheng, 2002). Since $\log_2(\rho)$ is the high-SNR behavior of a classical AWGN SISO channel, K_{nc} can be interpreted as the number of parallel spatial channels that can be used at the same time (Zheng and Tse, 2003). The optimal number of transmit antennas that should be used to communicate is M^*. Using more that M^* transmit antennas does not yield any benefit in terms of capacity.

From the previous observations, we can draw some useful conclusions. Since K_{nc} is the leading coefficient, and a 3 dB SNR increase implies a capacity increase of K_{nc} bit/s/Hz, the number of degrees of freedom should be maximized. This means that M^* should be $T/2$ or close to this value, if possible. Moreover, T should satisfy

$$T \geq \min(2M, 2N). \tag{10.7}$$

Otherwise, the additional antennas will not be useful (from a capacity perspective).

10.2.3 Asymptotic results for low SNR

In the low-SNR regime, results are quite different. In fact, if $\rho \to 0$, the coherent and noncoherent capacities are asymptotically equal in general, and their limit is (Zheng and Tse, 2002)

$$\lim_{\rho \to 0} C(\rho)/\rho = N \log_2 e, \qquad \text{bit/s/Hz.} \tag{10.8}$$

So, the relationship between the capacity and the number of degrees of freedom vanishes. Only an increase in the number of receive antennas can increase the capacity in order to collect the small power of the information signal. Moreover, even if T and M are larger than one, the optimal strategy consists in allocating all the transmit power to only one antenna during one symbol period.

10.3 Introduction to subspace representations

As it has been seen in Section 10.2.2, for systems with unknown channel at high SNR, the information is mainly carried by subspaces. We will recall here some basic definitions about subspaces.

10.3.1 Basics about subspaces

Let $\Omega_{\mathbf{X}}$ be an M-dimensional (vector) subspace of $\mathbb{C}^T$, with $T > M$. Given one of its bases $\mathbf{X}$, we recall (10.3)

$$\Omega_{\mathbf{X}} = \text{span}(\mathbf{X}) \quad \text{with} \quad \text{rank}(\mathbf{X}) = M. \tag{10.9}$$

The basis $\mathbf{X}$ is not unique. In fact, for any nonsingular $M \times M$ complex matrix $\mathbf{A}$, another valid basis of the same subspace is $\mathbf{X}\mathbf{A}$. The thin singular value decomposition[†] (TSVD) (Golub and Loan, 1996, p. 72) gives

---

[†]When $T = M$, the thin singular value decomposition becomes the common singular value decomposition, where all matrices in (10.10) are square.

interesting insights in the structure of a generic basis

$$\mathbf{X} = \mathbf{V} \boldsymbol{\Lambda} \mathbf{U}^\dagger, \qquad \mathbf{V} \in \mathcal{U}_{T,M}, \ \mathbf{U} \in \mathcal{U}_M, \ \boldsymbol{\Lambda} = \mathrm{diag}([\lambda_1 \ \cdots \ \lambda_M]), \qquad (10.10)$$

where $\mathcal{U}_M$ is the set of unitary $M \times M$ matrices, and $\boldsymbol{\Lambda}$ is a diagonal matrix whose entries are positive and arranged in decreasing order. A brief summary is

$$\mathbf{V}^\dagger \mathbf{V} = \mathbf{I}_M, \qquad \mathbf{U}^\dagger \mathbf{U} = \mathbf{I}_M, \qquad \lambda_1 \geq \lambda_2 \geq \ldots \geq \lambda_M. \qquad (10.11)$$

10.3.2 Basics on the Grassmann manifold

We give the following basic definition

Definition 10.1 *The set of all the M-dimensional complex (real) vector subspaces $\Omega_\mathbf{X}$ of $\mathbb{C}^T$ ($\mathbb{R}^T$) with $T > M$ is called the Grassmann manifold, or Grassmannian. It is denoted by $G_{T,M}$.*

The concepts of biorthonormal bases and principal angles are very important to characterize a couple of subspaces (Golub and Loan, 1996, p. 603).

Definition 10.2 *Given two subspaces $\Omega_\mathbf{X}, \Omega_\mathbf{Y} \in G_{T,M}$, two corresponding bases $\mathbf{X}$ and $\mathbf{Y}$ are said to be biorthonormal if they are orthonormal and*

$$\mathbf{x}_m^\dagger \mathbf{y}_{m'} = 0 \quad \text{for all } m \neq m', \qquad \mathbf{x}_m^\dagger \mathbf{y}_m = c_m \text{ with } 0 < c_m \leq 1, \qquad (10.12)$$

i.e., for all $m = 1, \ldots, M$, the mth vector of the first basis is orthogonal to all the vectors of the other basis, except the mth one.

A pair of biorthonormal bases can be obtained by means of the SVD of any two *orthonormal* bases. Let $\mathbf{X}, \mathbf{Y}$ be two orthonormal, but not necessarily biorthonormal, bases of $\Omega_\mathbf{X}$ and $\Omega_\mathbf{Y}$, and let

$$\mathbf{X}^\dagger \mathbf{Y} = \mathbf{U}_X \mathbf{C} \mathbf{U}_Y^\dagger \quad \mathbf{U}_X, \mathbf{U}_Y \in \mathcal{U}_M, \quad \mathbf{C} = \mathrm{diag}([c_1 \ \cdots \ c_M]), \qquad (10.13)$$

where $0 < c_m \leq 1$ for all m. Then, two biorthonormal bases are $\mathbf{X} \mathbf{U}_X$ and $\mathbf{Y} \mathbf{U}_Y$.

Definition 10.3 *Given two biorthonormal bases of the subspaces $\Omega_\mathbf{X}, \Omega_\mathbf{Y} \in G_{T,M}$, the real positive inner products c_m (10.12) are uniquely written in the following form:*

$$c_m = \cos\theta_m, \ \theta_m \in [0, \pi/2), \qquad m = 1, \ldots, M. \qquad (10.14)$$

The angles $\theta_1, \ldots, \theta_M$ are called the principal angles between subspaces $\Omega_\mathbf{X}$ and $\Omega_\mathbf{Y}$.

Principal angles are unique, while the biorthonormal bases are not (for example, a common permutation of two biorthonormal bases gives again two valid biorthonormal bases).

Definition 10.4 *Two subspaces $\Omega_{\mathbf{X}}, \Omega_{\mathbf{Y}} \in G_{T,M}$ are called intersecting subspaces if the dimension of their intersection is nonzero, i.e., $\dim(\Omega_{\mathbf{X}} \cap \Omega_{\mathbf{Y}}) > 0$. Otherwise, they are called nonintersecting.*

When two M-dimensional subspaces are intersecting, the dimension of their intersection equals the number of principal angles that are zero. In this case, some vectors of the two biorthonormal bases coincide: they span the intersection (Golub and Loan, 1996, p. 604).

Several distances between subspaces can be defined over the Grassmannian (Edelman *et al.*, 1998; Barg and Nogin, 2002). Here, we recall the most often used ones.

Definition 10.5 *Let $\boldsymbol{\theta} = (\theta_1, \ldots, \theta_M)$. Let $\mathbf{X}$ and $\mathbf{Y}$ two orthonormal bases of two different subspaces, and let $\mathbf{X}^\dagger \mathbf{Y} = \mathbf{U}_X \mathbf{C} \mathbf{U}_Y^\dagger$, the SVD as in (10.13) with $\mathbf{C} = \cos \boldsymbol{\theta}$. Then, we can define the following distances:*

(i) *The geodesic distance*

$$d_{\mathrm{arc}}(\Omega_{\mathbf{X}}, \Omega_{\mathbf{Y}}) = \|\boldsymbol{\theta}\| = \left(\sum_{m=1}^{M} \theta_m^2 \right)^{1/2}. \qquad (10.15)$$

(ii) *The chordal distance*

$$d_c(\Omega_{\mathbf{X}}, \Omega_{\mathbf{Y}}) = \|\sin \boldsymbol{\theta}\| = \left(\sum_{m=1}^{M} \sin^2 \theta_m \right)^{1/2}. \qquad (10.16)$$

A pseudodistance, called product distance, is often used in the literature in the case of the MIMO Rayleigh fading channel. It is

$$d_p(\Omega_{\mathbf{X}}, \Omega_{\mathbf{Y}}) = \left(\prod_{m=1}^{M} \sin \theta_m \right)^{1/M}. \qquad (10.17)$$

In the following, we will sometimes use the notation $d(\mathbf{X}, \mathbf{Y})$, meaning $d(\Omega_{\mathbf{X}}, \Omega_{\mathbf{Y}})$.

10.4 Detection criteria

10.4.1 The noncoherent ML criterion

When the statistics of the noise and fading channel, but not the realization of the channel itself, are known at the receiver, the maximum-likelihood (ML) criterion can be used for noncoherent detection (Proakis, 2000).

With these assumptions, the ML detector is a quadratic receiver. It can be stated as

$$\hat{\mathbf{X}}_{\text{ML}} = \arg \min_{i=1,\dots,L} [-\mathbf{Y}^\dagger \mathbf{F}_i \mathbf{Y} + c_i], \qquad (10.18)$$

where

$$\mathbf{F}_i = \frac{1}{\sigma^2} \mathbf{X}_i \left(\frac{\sigma^2}{\alpha^2} \mathbf{I}_{NM} + \mathbf{X}_i^\dagger \mathbf{X}_i \right)^{-1} \mathbf{X}_i^\dagger, \quad c_i = \ln \left| \frac{\sigma^2}{\alpha^2} \mathbf{I}_{NM} + \mathbf{X}_i^\dagger \mathbf{X}_i \right|. \quad (10.19)$$

L is the code size, and α is the normalization factor defined in (10.1).

10.4.2 The GLRT

The GLRT requires neither the knowledge of the fading and noise statistics, nor the knowledge of their realizations (Warrier and Madhow, 2002; Lapidoth and Narayan, 1998). It is defined as

$$\hat{\mathbf{X}}_{\text{GLRT}} = \arg \max_{i=1,\dots,L} \sup_{\mathbf{H}} p(\mathbf{Y}|\mathbf{X}_i, \mathbf{H}). \qquad (10.20)$$

The criterion simplifies to

$$\hat{\mathbf{X}}_{\text{GLRT}} = \arg \max_{i=1,\dots,L} \text{tr}\left[\mathbf{Y}^\dagger \mathbf{X}_i (\mathbf{X}_i^\dagger \mathbf{X}_i)^{-1} \mathbf{X}_i^\dagger \mathbf{Y}\right]. \qquad (10.21)$$

From (10.21) we see that the GLRT projects the received signal $\mathbf{Y}$ on the different subspaces $\Omega_{\mathbf{X}_i}$ and then calculates the energies of these projections and chooses the projection that maximizes this energy (see Fig. 10.1).

From the perspective of the average supersymbol error probability, also called frame error rate (FER), minimization of the GLRT in general gives a suboptimal result with respect to the ML criterion. However, the GLRT's independence on any kind of fading information makes it an excellent detection-rule candidate when the receiver cannot estimate channel correlations, or when the channel has nonstationary statistics.

In the case of i.i.d. fading and unitary codebooks, the ML and the GLRT criteria are equivalent [this results from (10.19)]. In this case, the decision is made according to

$$\hat{\mathbf{X}} = \max_{i=1,\dots,L} \text{tr}(\mathbf{Y}^\dagger \mathbf{X}_i \mathbf{X}_i^\dagger \mathbf{Y}) = \max_{i=1,\dots,L} \|\mathbf{Y}^\dagger \mathbf{X}_i\|_F^2, \qquad (10.22)$$

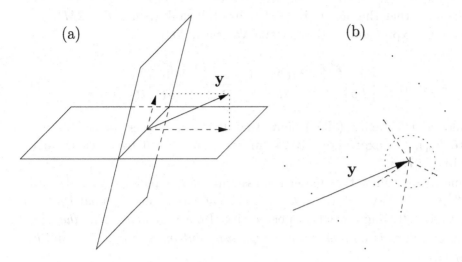

Fig. 10.1. (a) The GLRT chooses the subspace with the highest projection energy. (b) The coherent ML chooses the closest point.

for which the remarks of (10.22) hold as well.

10.5 Error probability bounds

Let P_{ij} be the pairwise error probability (PEP) between the two codewords $\mathbf{X}_i$ and $\mathbf{X}_j$, $i \neq j$. The expression for P_{ij} gives useful indication on how to design the code.

10.5.1 Unitary codebooks

In the case of unitary codebooks, and when the ML criterion and GLRT are equivalent, Hochwald and Marzetta (2000) report an exact closed-form analytical expression of the PEP P_{ij}, as well as a Chernoff bound that only depends on the principal angles between $\Omega_{\mathbf{X}_i}$ and $\Omega_{\mathbf{X}_j}$.

10.5.2 Nonunitary codebooks

In the case of nonunitary codebooks or correlated fading, the noncoherent ML criterion and the GLRT do not coincide. We give here the asymptotic expression of the PEP for the GLRT criterion. Let

$$\begin{bmatrix} \mathbf{X}_i^\dagger \\ \mathbf{X}_j^\dagger \end{bmatrix} \begin{bmatrix} \mathbf{X}_i & \mathbf{X}_j \end{bmatrix} = \begin{bmatrix} \mathbf{R}_{ii} & \mathbf{R}_{ij} \\ \mathbf{R}_{ji} & \mathbf{R}_{jj} \end{bmatrix}, \tag{10.23}$$

and assume that the matrix in (10.23) has full rank (hence $T \geq 2M$). The asymptotic expression is (Brehler and Varanasi, 2001)

$$P_{ij,\text{GLRT}}^{\infty} = \left[\frac{M}{T\rho}\right]^{NM} \binom{2MN-1}{MN} \frac{\left(1 + \frac{|\mathbf{R}_{ii}|}{|\mathbf{R}_{jj}|}\right)}{|\mathbf{R}_{ii} - \mathbf{R}_{ij}\mathbf{R}_{jj}^{-1}\mathbf{R}_{ji}|^N}. \qquad (10.24)$$

Brehler and Varanasi (2001) show that if the fading is correlated, $\mathbf{h} \sim \mathcal{CN}(\mathbf{0}, \mathbf{K_h})$, the expression (10.24) must be multiplied by a scalar factor equal to $1/|\mathbf{K_h}|$.

Finally, they show that under the assumption of equal-energy codewords $(\text{tr}(\mathbf{X}_i^\dagger \mathbf{X}_i) = P, \ \forall i = 1, \ldots, L)$, unitary codebooks are optimal from an asymptotic PEP-minimization perspective. Hence, *at high SNR, the same signal structure is optimal from a capacity point of view and from a PEP point of view.*

10.6 Diversity for the noncoherent case

In the literature, we can find three different definitions of diversity for non-coherent MIMO block fading systems.

10.6.1 PEP-based diversity

The most widespread definition of diversity, which can be used for coherent system, too (Tarokh *et al.*, 1998), is based on the asymptotic PEP or on its Chernoff bound

Definition 10.6 (PEP-based diversity) *Let $P_{ij}(\rho)$ be the pairwise error probability, and let $P_{ij}^{CB}(\rho)$ be the Chernoff bound on the PEP between the codewords $\mathbf{X}_i$ and $\mathbf{X}_j$, as a function of the SNR ρ. Let $\mathbf{X}_i$ and $\mathbf{X}_j$ belong to a codebook $\mathcal{C}$ of size L. The codebook $\mathcal{C}$ is said to achieve the diversity gain d (or briefly, to have diversity d) if and only if (iff)*

- *for the PEP:*

$$\min_{i,j:i\neq j} \lim_{\rho\to\infty} \frac{\ln P_{ij}(\rho)}{\ln \rho} = -d;$$

- *for the Chernoff bound:*

$$\min_{i,j:i\neq j} \lim_{\rho\to\infty} \frac{\ln P_{ij}^{CB}(\rho)}{\ln \rho} = -d. \qquad (10.25)$$

Let $m_{d,ij}$ be the number of nonzero principal angles between subspaces $\Omega_{\mathbf{X}_i}$ and $\Omega_{\mathbf{X}_j}$, the exponent of the SNR is $Nm_{d,ij}$. Thus, we can say that

$$N \le d = N m_d \le MN, \qquad m_d = \min_{i,j:i\ne j} m_{d,ij}. \tag{10.26}$$

When the subspaces are all nonintersecting, then d reaches its maximum, NM, and the code is called a *full-diversity* code. A necessary condition to have a full diversity code is $T \ge 2M$. It can be shown that the minimization of the PEP is equivalent to the maximization of the minimum product distance (10.17).

Brehler and Varanasi (2001) prove the following proposition, which holds for every kind of codebook:

Proposition 10.1 *If for all tuples of codewords $\mathbf{X}_i$ and $\mathbf{X}_j$, $i \ne j$, belonging to $\mathcal{C}$, the matrices*

$$\begin{bmatrix} \mathbf{X}_i^\dagger \\ \mathbf{X}_j^\dagger \end{bmatrix} \begin{bmatrix} \mathbf{X}_i & \mathbf{X}_j \end{bmatrix} = \begin{bmatrix} \mathbf{R}_{ii} & \mathbf{R}_{ij} \\ \mathbf{R}_{ji} & \mathbf{R}_{jj} \end{bmatrix} \tag{10.27}$$

have full rank, then the codebook $\mathcal{C}$ achieves full PEP-based diversity. However, it is necessary that $T \ge 2M$.

10.6.2 Error probability-based diversity

A definition, which relates the diversity to the concept of multiplexing gain (see Section 10.2.2) is given by Zheng and Tse (2003), Zheng (2002). This definition is based on the average supersymbol error probability, and not on the PEP; moreover, its is defined for families of codes whose rate scales logarithmically with the SNR. This kind of definition is useful to study the trade-off between diversity and multiplexing gain.

10.6.3 Algebraic diversity

A definition based on algebraic properties of the codebook is introduced by El Gamal *et al.* (2003). We state it in the case of no coding between different fading blocks, and we give a slightly different, but equivalent, definition. Let us assume that the receiver is equipped with a single antenna ($N = 1$), and additive noise is absent. Let us define the subspace of channel realizations $\mathcal{H}_{nc}(i,j)$ that makes the GLRT unable to distinguish between two possible transmitted symbols $\mathbf{X}_i$ and $\mathbf{X}_j$ as

$$\mathcal{H}_{nc}(i,j) = \{\mathbf{h} \in \mathbb{C}^M : \exists \, \mathbf{h}_1 \in \mathbb{C}^M, \, \mathbf{X}_i\mathbf{h} = \mathbf{X}_j\mathbf{h}_1\} \tag{10.28}$$
$$= \{\mathbf{h} \in \mathbb{C}^M : \exists \, \tilde{\mathbf{h}} \in \Omega_{\mathbf{X}_i} \cap \Omega_{\mathbf{X}_j}, \text{ where } \tilde{\mathbf{h}} = \mathbf{X}_i\mathbf{h}\} \tag{10.29}$$

Definition 10.7 (Algebraic Diversity Gain)   *The codebook $\mathcal{C}$ is said to achieve the algebraic diversity d if*

$$d = N[M - \max_{i,j:i\neq j} \dim \mathcal{H}_{nc}(i,j)] = N \min_{i,j:i\neq j} [\dim(\Omega_{\mathbf{X}_i} + \Omega_{\mathbf{X}_j}) - \dim \Omega_{\mathbf{X}_j}].$$
$$\tag{10.30}$$

There does not exist a formal proof of the equivalence between the classical PEP-based diversity and the algebraic diversity for generic codebooks. However, El Gamal *et al.* (2005) proved the following

Proposition 10.2   *If the codebook $\mathcal{C}$ is unitary, the algebraic diversity and the PEP-based diversity are equivalent.*

In the general case, it is also clear that when the codebook has full algebraic diversity, it has also full PEP-based diversity, because with this assumption, matrices in (10.23) have full rank, and Proposition 10.1 can be applied.

10.7 Code design criteria and propositions

Both information-theoretic criteria [see (10.2)] and error probability criteria show that, at high SNR, the optimal signals are matrices with orthonormal columns. Most research concentrated on codes of this type. However, many propositions in the literature present both unitary and nonunitary codebooks. These propositions differ in code design methods and criteria, and hence in decoding methods. The main ones are:

- Codebooks designed by numerical minimization of some cost function related to the distances of Definition 10.5 (Hochwald and Marzetta, 2000; Agrawal *et al.*, 2001; Gohary and Davidson, 2004), by numerical minimization of the union bound on the supersymbol or bit error probability, also called bit error rate (BER) (McCloud *et al.*, 2002; Brehler and Varanasi, 2003), or on the Kullback–Leibler distance (Borran *et al.*, 2003), an information-theoretic criterion (Cover and Thomas, 1991).
- Codebooks obtained by some parameterization of unitary matrices (Hochwald *et al.*, 2000; Jing and Hassibi, 2003; Wang *et al.*, 2005), or of the Grassmann manifold (Kammoun and Belfiore, 2003; Kammoun, 2004).

- Codebooks obtained by algebraic construction for some particular cases (Tarokh and Kim, 2002; Zhao *et al.*, 2004; Oggier *et al.*, 2005).
- Codebooks that follow the so-called training-based format, i.e., that estimate the channel in the first part of the supersymbol and use the second part to send information by means of a space-time code designed for coherent detection (Brehler and Varanasi, 2003; El Gamal and Damen, 2003; El Gamal *et al.*, 2003; Dayal *et al.*, 2004; El Gamal *et al.*, 2005).

In the following, we will briefly report the various propositions with advantages and drawbacks. Some characteristics of these codes are summarized in Table 10.1.

10.7.1 Numerically optimized designs

These propositions differ in their cost functions and in their (often suboptimal) minimization methods. They suffer from common shortcomings:

(i) only low-size constellations can be constructed because of the increasing complexity in the design process;
(ii) the codebook of size $L = 2^{RT}$ has to be stored in the transmitter and the receiver; hence, the required memory is exponential in RT (R is the transmit rate expressed in bits per symbol period);
(iii) in general, no simplified decoding algorithm is available, so that the GLRT or ML rule must be evaluated for all codewords. Hence, the decoding complexity is in general exponential in RT.

Small ($L \leq 64$) unitary space-time constellations were designed by Hochwald and Marzetta (2000). Agrawal *et al.* (2001) used the minimum chordal distance as a cost function,

$$d_{c,min}^2 = \min_{1 \leq l < l' \leq L} d_c^2(\Omega_{\mathbf{X}_l}, \Omega_{\mathbf{X}_{l'}}) = \min_{1 \leq l < l' \leq L} \sum_{m=1}^{M} \sin^2 \theta_{m,ll'}, \qquad (10.31)$$

where $\theta_{m,ll'}$ are the M principal angles between the two subspaces generated by $\mathbf{X}_l$ and $\mathbf{X}_{l'}$ [see (10.16)]. The minimum chordal distance can be related to the worst case upper bound of the Chernoff bound on the PEP, a quite loose approximation of the PEP. Moreover, $d_{c,min}^2$ is a good approximation of the product distance (10.17) only when all subspaces are quasiorthogonal (Hochwald *et al.*, 2000), an assumption that is not true even for codebooks of small size when T is comparable to M. The optimization technique used by Agrawal *et al.* (2001) is called relaxation method, and it generalizes Conway *et al.* (1996), who used it for real Grassmannians.

Table 10.1. *Summary of the most quoted propositions in the literature.*

Ref.	Nc. code[a]	Design method	Co. code[b]	Diversity	Dec.[c]	Dec. compl.
Agrawal *et al.* (2001)	U	num. min. $d^2_{c,min}$	no	no control	no	$O(2^{RT})$
Gohary and Davidson (2004)	U	num. min. $d^2_{cF,min}$	no	no control	S	local GLRT
McCloud *et al.* (2002), Brehler and Varanasi (2003)	U	num. min. $P_e/P_{e,bit}$	no	full	no	$O(2^{RT})$
Borran *et al.* (2003)	N/U	num. min.	no	no control	no[d]	$O(2^{RT})$
Hochwald *et al.* (2000)	U	successive rotations	no	no control	no	$O(2^{RT})$
Jing and Hassibi (2003), Wang *et al.* (2005)	U	Cayley Transf.	no	no control	S	sphere decoder
Kammoun and Belfiore (2003)	U	exponential transf.	yes	full con-jectured	S	local GLRT
Tarokh and Kim (2002)	U	algebraic/ training	yes[e]	full	O	$O(MN)/O(M^2N)$
Zhao *et al.* (2004)	U	algebraic/ training	yes	full	O	$O(2^{TR/2})$
Oggier *et al.* (2005)	N/U	algebraic	no	full	no	$O(2^{TR})$
Brehler and Varanasi (2003), Dayal *et al.* (2004), El Gamal and Damen (2003), El Gamal *et al.* (2003)	N/U	training	yes	full	S	sphere decoder

[a] Noncoherent code: U = unitary codebooks, N/U = all kinds of codebooks.
[b] Coherent code: is the noncoherent code built from a coherent code?
[c] There exists a simplified decoding? S = yes, but is it suboptimal, O = yes, and it is optimal (with respect to the noncoherent ML or GLRT).
[d] Unitary constellations with simplified decoding can be used; however, it is not the general case.
[e] A proposition not based on coherent codes is also presented.

Gohary and Davidson (2004) use another metric, the so-called chordal Frobenius distance (Edelman *et al.*, 1998),

$$d^2_{cF}(\Omega(\mathbf{X}_l), \Omega_{\mathbf{X}_{l'}}) = 4 \sum_{m=1}^{M} \sin^2(\theta_{m,ll'}/2) < d^2_c(\Omega_{\mathbf{X}_l}, \Omega_{\mathbf{X}_{l'}}),$$

which is looser than the chordal distance. A simplified decoding method is proposed, which, from some reference point on the Grassmannian, locates a list of candidate points over which the GLRT is calculated (local GLRT). This procedure makes it possible to calculate the GLRT metric for a subset of the codewords only. However, tables with coordinates of all codewords as well as the codebook, which has no algebraic structure, must be saved in memory.

McCloud *et al.* (2002) and Brehler and Varanasi (2003) obtained constellations through numerical search minimizing the asymptotic union bound on the supersymbol error rate and the asymptotic union bound on the bit error rate. While the common drawbacks persist, the advantage of this method is that, being based on the PEP, it guarantees that the constructed constellation has full diversity when $T \geq 2M$.

Motivated by the fact that unitary codebooks are not optimal at low SNR or for T comparable to M, Borran *et al.* (2003) design nonunitary codebooks with the Kullback–Leibler distance[†] criterion. This method has been proposed because of the intractability of the PEP-based design criterion when the unitarity assumption for the codewords is no longer true.

10.7.2 Parameterization designs

A more structured approach was used by Hochwald *et al.* (2000), where an initial matrix generates the whole constellation by successive rotations.[‡] The parameters of the codebook are chosen by a random search to maximize the minimal chordal distance $d_{c,min}$ as in (10.31). However, no simplified decoding algorithm is reported.

Jing and Hassibi (2003) describe a method that constructs unitary codebooks. The codeword $\mathbf{X}$ is obtained via the Cayley transform

$$\mathbf{X} = (\mathbf{I}_T + j\mathbf{A})^{-1}(\mathbf{I}_T - j\mathbf{A}) \begin{bmatrix} \mathbf{I}_M \\ \mathbf{0} \end{bmatrix}, \qquad (10.32)$$

where the Hermitian matrix $\mathbf{A}$ is calculated from a set of fixed Hermitian matrices $\mathbf{A}_q$ as $\mathbf{A} = \sum_{q=1}^{Q} \alpha_q \mathbf{A}_q$. To ensure the unitarity of $\mathbf{X}$, coefficients α_q are real scalars that belong to a discrete set $\mathcal{A}$ whose cardinality is r. Even if the transform (10.32) is invertible (for all Hermitian matrices $\mathbf{A}$ with no eigenvalue equal to -1), the Cayley Transform is nonlinear. The authors

---

[†] The Kullback–Leibler distance between two distributions $p_1(x)$ and $p_2(x)$ is defined as (Cover and Thomas, 1991)

$$D(p_1\|p_2) = \int_x p_1(x)(\ln p_1(x) - \ln p_2(x))\mathrm{d}x.$$

[‡] Marzetta *et al.* (2002) construct constellations with better statistical properties and spectral efficiencies. However, the absence of a simplified decoder prevented the authors from verifying the performances of these codes.

constrain some entries of the set of matrices $\{\mathbf{A}_q\}$, constrain the scalar Q (related to the rate of the system $R = Q\log_2(r)/T$), and make an approximation on the ML rule, so that they get a simplified suboptimal decoding problem that can be solved via the sphere decoding algorithm (Viterbo and Boutros, 1999). Many variables involved in the design of the code are found by numerical optimization; the set of matrices $\{\mathbf{A}_q\}$ is chosen to maximize a criterion that guarantees the diversity for differentially encoded unitary codebooks (see Hassibi and Hochwald (2002)) but not for noncoherent systems. Subsequently, Wang *et al.* (2005) proposed other optimization methods to enhance performance.

Another method based on a parameterization is the one proposed by Kammoun and Belfiore (2003). The unitary codewords are obtained via the *exponential parameterization*, or map, from a subset of skew-Hermitian matrices $(\mathbf{A} = -\mathbf{A}^{\dagger})$:

$$\mathbf{X} = \exp(\mathbf{A})\mathbf{I}_{T,M} = \exp\left\{\begin{bmatrix} \mathbf{0}_M & -\mathbf{B}^{\dagger} \\ \mathbf{B} & \mathbf{0}_{T-M} \end{bmatrix}\right\}\begin{bmatrix} \mathbf{I}_M \\ \mathbf{0} \end{bmatrix}, \qquad (10.33)$$

where the matrices $\mathbf{B}$ must satisfy some conditions. In fact, the matrices $\mathbf{B}$ are codewords from coherent space-time codes (from El Gamal and Damen (2003), for example), linearly scaled by a positive real factor α_o, also called *homothetic factor*, and which is the only parameter to optimize for a fixed codebook. The diversity of the unitary codebook cannot easily be linked to the diversity of the coherent code; a conjecture on this topic is made by Kammoun (2004). Codes built in this way can have high spectral efficiencies, just by choosing the appropriate coherent code. Only the homothetic factor must be optimized instead of a large number of parameters, as in the codes by Jing and Hassibi (2003) or McCloud *et al.* (2002).

10.7.3 Algebraic designs

These codes are quite different from each other, but they share the same property: it is possible to control the code parameters because of a quite constraining and powerful algebraic structure.

Tarokh and Kim (2002) presented two constructions. The first one, called *generalized PSK constellations*, can be used for $T = 2M$; it can be written as

$$\mathbf{X}_l = \begin{bmatrix} \cos(\phi_l)\mathbf{I}_M \\ \sin(\phi_l)\mathbf{I}_M \end{bmatrix}, \qquad \phi_l = l\pi/L, \quad l = 0, 1, \ldots, L-1, \qquad (10.34)$$

where L is the size of the code. An ML decoder exists whose complexity is only $O(MN)$, independent of the duration T of the frame, and the rate

$R = \log_2(L)/T$. However, this is achieved by imposing a strong structure (10.34), which only exploits M real degrees of freedom among the possible $2M(T - M)$ real degrees of freedom of the system. The principal angles between two given codewords $\mathbf{X}_k$ and $\mathbf{X}_l$ are equal to $\pi(k-l)/2^{TR}$, and are all the same. Even if the constellation has full diversity, the minimum product distance (10.17) is $d_{p,min} = \sin(\pi/L) = \sin(\pi/2^{TR})$ and decreases exponentially with the rate or the duration, so that this method is only efficient for low spectral efficiencies. A generalization, named *complex Givens codes*, is proposed (Dayal *et al.*, 2004), which exploits $2M$ among the $2M(T - M)$ degrees of freedom, and only doubles the decoding complexity of the generalized PSK codes. These codes can also be obtained by the exponential parameterization. The second proposition by Tarokh and Kim (2002) is based on the coherent space-time orthogonal designs (Tarokh *et al.*, 1999). It can be described in the framework of the training-based codes (Dayal *et al.*, 2004).

Zhao *et al.* (2004) present some unitary codes that derive from the orthogonal designs, and can be seen as training-based codes. They also present a simplified decoder to perform the ML detection with complexity $O(2^{RT/2})$, instead of $O(2^{RT})$.

Finally, Oggier *et al.* (2005) present an investigation about the maximum number of nonintersecting subspaces. Here, the condition is that the codewords' entries come from fixed small constellations. The problem is solved when these constellations coincide with the Galois field $GF(q)$, where q is a power of some prime integer. In this case, the maximum number of non-intersecting subspaces is $(q^T - 1)/(q^M - 1)$. An upper and a lower bound are given in the case of PSK constellations where the number of transmit antennas is $M = 2$. An encoder as well as a simplified decoder have not yet been proposed.

10.7.4 Training-based schemes

The codes

Recently, the research community has carried a growing interest in the so-called training-based schemes (Brehler and Varanasi, 2003; El Gamal and Damen, 2003; El Gamal *et al.*, 2003; Dayal *et al.*, 2004). In this approach, each block is divided into two parts of T_t and T_d channel uses ($T_t + T_d = T$). In the first T_t channel uses, a pilot signal, known to the transmitter and the receiver, is sent to get a rough estimate of the channel (training phase). The remaining T_d channel uses are used to send information, usually encoded via some coherent space-time code $\mathcal{B}$. So, typically, codewords are

$$\mathbf{X} = \begin{bmatrix} \sqrt{\tau}\mathbf{T} \\ \sqrt{1-\tau}\mathbf{B} \end{bmatrix}, \tag{10.35}$$

where $\mathbf{T}$ is the pilot $T_t \times M$ matrix, $\mathbf{B} \in \mathcal{B}$ is $T_d \times M$, and $\tau \in (0,1)$ is a scalar that assigns different transmit power ratios to the training part. Naturally, the codewords can extend to several blocks of the fading channel (El Gamal *et al.*, 2003).

The channel estimation performed in the training phase is rather unusual, compared with common estimation practice. In fact, for these noncoherent systems, channel estimation is performed without having different estimates of the channel coefficients, but just one. This is due to the statistical independence of channel coefficients from one block to the other ones. Moreover, the block length is so short that repeating the estimation process within the same block would cause an unacceptable decrease of the spectral efficiency. In the literature, training for these systems has been ignored for some years (as remarked by Dayal *et al.*, 2004), probably for the previous two reasons.

However, since their introduction, training-based codes seem to be one of the best competitors. The advantages of this approach are:

- theory end experience in the design of space-time codes for the coherent channel (also called coherent space-time codes) can be reused;
- it follows that simplified decoding techniques of coherent space time codes can be used to decode training-based codes;
- training-based codes, as in (10.35), achieve the diversity (in the PEP sense) of the underlying coherent space-time code $\mathcal{B}$, when $\mathbf{B}$ is full-rank and the fading coefficients are i.i.d. Gaussian random variables (Dayal *et al.*, 2004).

Simplified decoding

Let the received signal be

$$\mathbf{Y} = \begin{bmatrix} \mathbf{Y}_t \\ \mathbf{Y}_d \end{bmatrix} = \begin{bmatrix} \sqrt{\tau}\mathbf{T} \\ \sqrt{1-\tau}\mathbf{B} \end{bmatrix} \mathbf{H} + \begin{bmatrix} \mathbf{W}_t \\ \mathbf{W}_d \end{bmatrix}, \tag{10.36}$$

where the channel coefficients are i.i.d. $\mathcal{CN}(0,1)$ and each complex component of the additive noise is $\mathcal{CN}(0,\sigma^2)$.

The simplified receiver for training-based symbols performs two operations:

(i) The receiver estimates the channel coefficients via a minimum mean square error (MMSE) estimator (Hassibi and Hochwald, 2003) from the signal received during the first M channel uses:

$$\widehat{\mathbf{H}} = \sqrt{\tau}(\sigma^2 \mathbf{I}_M + \tau \mathbf{T}^\dagger \mathbf{T})^{-1} \mathbf{T}^\dagger \mathbf{Y}_t. \tag{10.37}$$

(ii) The receiver treats the channel estimation as if it was perfect and decodes—with the coherent ML rule—the signal received in the remaining $T - M$ symbol periods:

$$\widehat{\mathbf{B}} = \arg \min_{l=1,\ldots,L} \|\mathbf{Y}_d - \mathbf{B}_l \widehat{\mathbf{H}}\|_F^2 \tag{10.38}$$

The rule (10.38) corresponds to finding the closest point to a given point $\text{vec}(\mathbf{Y}_d)$ of $\mathbb{C}^{M(T-M)}$. This problem can be efficiently solved via the so-called sphere decoding algorithm (Viterbo and Boutros, 1999). There exist different search strategies (see Agrell *et al.* (2002) and references therein). We recall the Pohst strategy, which scans the points inside a hypersphere of fixed radius, and, when it find a point, decrease the radius, and the Schnorr–Euchner strategy, which also searches in a hypersphere, but scans the points in a different order (i.e., it first searches for points in the nearest hyperplanes inside the sphere).

References

Agrawal, D., Richardson, T. J., and Urbanke, R. L. (2001). Multiple-antenna signal constellations for fading channels. *IEEE Trans. Inf. Theory*, **47** (6), 2618–2626.

Agrell, E., Eriksson, T., Vardy, A., and Zeger, K. (2002). Closest point search in lattices. *IEEE Trans. Inf. Theory*, **48** (8), 2201–2214.

Barg, A. and Nogin, D. Y. (2002). Bounds on packings of spheres in the Grassmann manifold. *IEEE Trans. Inf. Theory*, **48** (9), 2450–2454.

Borran, M. J., Sabharwal, A., and Aazhang, B. (2003). On design criteria and construction of noncoherent space-time constellations. *IEEE Trans. Inf. Theory*, **49** (10), 2332–2351.

Brehler, M. and Varanasi, M. K. (2001). Asymptotic error probability analysis of quadratic receivers in Rayleigh-fading channels with applications to a unified analysis of coherent and noncoherent space-time receivers. *IEEE Trans. Inf. Theory*, **47** (6), 2383–2399.

—— (2003). Training-codes for the noncoherent multi-antenna block-Rayleigh-fading channel. In *Proc. Conf. Inf. Sciences and Syst.*

Conway, J. H., Hardin, R. H., and Sloane, N. J. A. (1996). Packing lines, planes, etc.: Packings in Grassmannian spaces. *Experimental Math.*, **5** (2), 139–159.

Cover, T. M. and Thomas, J. A. (1991). *Elements of Information Theory* (Wiley, New York).

Dayal, P., Brehler, M., and Varanasi, M. K. (2004). Leveraging coherent space-time codes for noncoherent communication via training. *IEEE Trans. Inf. Theory*, **50** (9), 2058–2080.

Edelman, A., Arias, T. A., and Smith, S. T. (1998). The geometry of algorithms with orthogonality constraints. *SIAM J. Matrix Anal. Appl.*, **20** (2), 303–353.

El Gamal, H., Aktas, D., and Damen, M. O. (2003). Coherent space-time codes for noncoherent channels. In *Proc. IEEE Global Telecommun. Conf.*, 1915–1918.

—— (2005). Noncoherent space-time coding: an algebraic perspective. *IEEE Trans. Inf. Theory*, **51** (7), 2380–2390.

El Gamal, H. and Damen, M. O. (2003). Universal space-time coding. *IEEE Trans. Inf. Theory*, **49** (5), 1097–1119.

Gohary, R. H. and Davidson, T. N. (2004). Non-coherent MIMO communication: Grassmannian constellation and efficient detection. In *Proc. IEEE Int. Symp. Inf. Theory*, 65.

Golub, G. H. and Loan, C. F. V. (1996). *Matrix Computations* (Johns Hopkins Univ. Press), third edn.

Hassibi, B. and Hochwald, B. M. (2002). Cayley differential unitary space-time codes. *IEEE Trans. Inf. Theory*, **48** (6), 1485–1503.

—— (2003). How much training is needed in multiple-antenna wireless links? *IEEE Trans. Inf. Theory*, **49** (4), 951–963.

Hassibi, B. and Marzetta, T. L. (2002). Multiple-antennas and isotropically random unitary inputs: the received signal density in closed form. *IEEE Trans. Inf. Theory*, **48** (6), 1473–1484.

Hochwald, B. M. and Marzetta, T. L. (2000). Unitary space-time modulation for multiple-antenna communications in Rayleigh flat fading. *IEEE Trans. Inf. Theory*, **46** (2), 543–564.

Hochwald, B. M., Marzetta, T. L., Richardson, T. J., Sweldens, W., and Urbanke, R. (2000). Systematic design of unitary space-time constellations. *IEEE Trans. Inf. Theory*, **46** (6), 1962–1973.

Jing, Y. and Hassibi, B. (2003). Unitary space-time modulation via Cayley transform. *IEEE Trans. Signal Process.*, **51** (11), 2891–2904.

Kammoun, I. (2004). *Codage Spatio-temporel sans connaissance à priori du canal.* Ph.D. thesis, Ecole Nationale Supérieure des Télécommunications, Paris.

Kammoun, I. and Belfiore, J.-C. (2003). A new family of Grassmannian space-time codes for non-coherent MIMO systems. *IEEE Commun. Lett.*, **7** (11), 528–530.

Lapidoth, A. and Narayan, P. (1998). Reliable communications under channel uncertainty. *IEEE Trans. Inf. Theory*, **44** (6), 2148–2177.

Marzetta, T. L., Hassibi, B., and Hochwald, B. M. (2002). Structured unitary space-time autocoding constellations. *IEEE Trans. Inf. Theory*, **48** (4), 942–950.

Marzetta, T. L. and Hochwald, B. M. (1999). Capacity of a mobile multiple-antenna communication link in Rayleigh flat fading. *IEEE Trans. Inf. Theory*, **45** (1), 139–157.

McCloud, M. L., Brehler, M., and Varanasi, M. K. (2002). Signal design and convolutional coding for noncoherent space-time communication on the block-Rayleigh-fading channel. *IEEE Trans. Inf. Theory*, **48** (5), 1186–1194.

Oggier, F. E., Sloane, N. J. A., Calderbank, A. R., and Diggavi, S. N. (2005). Nonintersecting subspaces based on finite alphabets. *IEEE Trans. Inf. Theory*, **51** (12), 4320–4325.

Proakis, J. G. (2000). *Digital Communications* (McGraw-Hill), fourth edn.

Tarokh, V., Jafarkhani, H., and Calderbank, A. R. (1999). Space-time block codes from orthogonal designs. *IEEE Trans. Inf. Theory*, **45** (5), 744–765.

Tarokh, V. and Kim, I.-M. (2002). Existence and construction of noncoherent unitary space-time codes. *IEEE Trans. Inf. Theory*, **48** (12), 3112–3117.

Tarokh, V., Seshadri, N., and Calderbank, A. R. (1998). Space-time codes for high data rate wireless communications: Performance criterion and code construction. *IEEE Trans. Inf. Theory*, **44** (2), 1456–1467.

Viterbo, E. and Boutros, J. (1999). A universal lattice decoder for fading channels. *IEEE Trans. Inf. Theory*, **45** (5), 1639–1642.

Wang, J., Wang, X., and Madihian, M. (2005). Design of minimum error-rate Cayley differential unitary space-time codes. *IEEE J. Sel. Areas Commun.*, **23** (9), 1779–1787.

Warrier, D. and Madhow, U. (2002). Spectrally efficient noncoherent communication. *IEEE Trans. Inf. Theory*, **48** (3), 651–668.

Zhao, W., Leus, G., and Giannakis, G. (2004). Orthogonal design of unitary constellations for uncoded and trellis-coded noncoherent space-time systems. *IEEE Trans. Inf. Theory*, **50** (6), 1319–1327.

Zheng, L. (2002). *Diversity-Multiplexing Tradeoff: A Comprehensive View of Multiple Antenna Systems*. Ph.D. thesis, University of California, Berkeley, CA, U.S.A. URL http://web.mit.edu/lizhong/www/.

Zheng, L. and Tse, D. N. C. (2002). Communication on the Grassmann manifold: A geometric approach to the noncoherent multiple-antenna channel. *IEEE Trans. Inf. Theory*, **48** (2), 359–383.

—— (2003). Diversity and multiplexing: a fundamental tradeoff in multiple-antenna channels. *IEEE Trans. Inf. Theory*, **49** (5), 1073–1096.

11

Space-time coding for time- and frequency-selective MIMO channels

Xiaoli Ma

Georgia Institute of Technology

Georgios B. Giannakis

University of Minnesota

Focusing on challenging propagation channels, this chapter discusses space-time coding techniques utilizing multiple antennas to facilitate high transmission rates, enhanced capacity, and robust system performance in mobile and fading environments.

11.1 Multipath and Doppler diversity

High rates come with broadband frequency-selective multipath propagation, while high mobility gives rise to Doppler-induced time-selective fading effects. The combined time-frequency selectivity of the underlying channel induces *multipath-Doppler fading*, which affects critically communication performance. Capturing multidimensional fading effects (over time, frequency, and space) requires many parameters, making it necessary for the resultant models to cope with the "curse of dimensionality." Our motivation in this chapter is, at a high-level, to turn this "curse" into a "blessing" by designing multiantenna systems capable of collecting the embedded joint multipath-Doppler-spatial diversity gains. But before tackling this design goal, it is important to understand and quantify these diversity gains emerging from time- and frequency-selective propagation.

Frequency-selective channels and multipath diversity In their wireless propagation, the emitted signal waveforms may be reflected or diffracted, before reaching the receiver through different paths—a manifestation of what is known as multipath propagation. Due to the finite speed of light, the multiple paths conveying the information content travel variable distances and arrive at the receiver at different times. This causes time dispersion of the transmitted waveforms, which is known as delay spread. As a result, each symbol spills over adjacent symbols and gives rise to so-called intersymbol interference (ISI). In this case, we say that the channel exhibits frequency selectivity.

Taking into account transmit and receive filters, the baseband equivalent frequency-selective channel is typically modeled as

$$h(\tau) = \sum_{l=0}^{L} h_l \, \delta(\tau - lT_s), \tag{11.1}$$

where T_s is the sampling period, L denotes the channel order, and $\{h_l\}_{l=0}^{L}$ is the set of impulse response coefficients (a.k.a. channel taps).

If $x(n)$ (respectively $y(n)$) denotes the transmitted (received) symbol at time n, the input-output (I/O) relationship of the channel is

$$y(n) = \sum_{l=0}^{L} h_l \, x(n-l) + w(n), \tag{11.2}$$

where $w(n)$ is the additive white Gaussian noise (AWGN) with variance N_0. Notice that for each transmitted symbol $x(n)$, $L+1$ received symbols $\{y(n+l)\}_{l=0}^{L}$ contain copies of $x(n)$ scaled with the corresponding channel taps $\{h_l\}_{l=0}^{L}$. As these copies can be averaged to enhance the received signal-to-noise ratio (SNR), this explains intuitively how multipath diversity is provided by frequency-selective channels and why this can be beneficial to improve error performance at the receiver side. The diversity gain is defined, in general, as

$$G_d := \lim_{\text{SNR} \to \infty} -\frac{\log P_e(\text{SNR})}{\log \text{SNR}}.$$

Based on the latter, multipath diversity in particular can be formally quantified as follows:

Proposition 11.1 (Wang and Giannakis, 2003)   *For an Lth-order Rayleigh fading channel, the maximum achievable diversity is given by $G_d = \text{rank}(\boldsymbol{R}_h) \le L+1 = G_{d,\max}$, where $\boldsymbol{R}_h$ denotes the correlation matrix $\boldsymbol{R}_h := E[\boldsymbol{h}\boldsymbol{h}^{\mathcal{H}}]$ of the discrete-time baseband equivalent vector of channel taps $\boldsymbol{h} := [h_0 \cdots h_L]^T$.*

This way of quantifying multipath diversity allows for random taps with arbitrary correlation profiles and can be extended to cover all distributions encountered with practical wireless multipath channels.

Time-selective channels and Doppler diversity   Frequency-selective multipath channels with I/O relationships described by (11.2) entail time-invariant channel taps. However, mobility and mismatch between transmit-receive oscillators give rise to time variations. Let us consider a channel with coherence time approximately equal to a symbol period. The discrete-time baseband equivalent model is then

$$y(n) = h(n)x(n) + w(n),$$ (11.3)

where $h(n)$ is the channel coefficient at the nth time-slot. In addition to rapidly varying channels, the I/O relationship (11.3) encompasses also quasi-static (a.k.a. block-fading) channel models, provided that $h(n)$ remains invariant over several symbol periods (or one block) and changes independently from block to block.

One widely adopted time-varying channel model is Jakes' model (Jakes, 1974). However, the huge number of parameters present in Jakes' model renders channel estimation prohibitively complex. This justifies well the finite parameter basis-expansion model (BEM) that has been introduced to approximate Jakes' model parsimoniously using a small number of parameters (Giannakis and Tepedelenlioglu, 1998). Later on, we will also show that the BEM plays for time-selective channels a role dual to the tapped-delay line model for frequency-selective channels [see (11.1)]. For a time-varying frequency-flat channel, the BEM is described by the expansion

$$h(n) := \sum_{q=0}^{Q} h_q \, e^{j\omega_q n},$$ (11.4)

where the coefficients $\{h_q\}_{q=0}^{Q}$ remain invariant over a certain time period, and the Fourier bases with frequencies $\omega_q := 2\pi(q - Q/2)/N$ capture the time variation.

Based on the BEM and similar to frequency-selective channels, one can quantify the diversity order provided by time-selective channels as follows:

Proposition 11.2 (Ma and Giannakis, 2003b) *If the correlation matrix R_h of the BEM channel coefficients $\{h_q\}_{q=0}^{Q}$ in (11.4) has rank r_h, then the Doppler diversity order of the time-selective BEM is $G_d \leq r_h$. When R_h has full rank $r_h = Q + 1$, the maximum diversity order is $G_{d,\max} = Q + 1$.*

The implication of this proposition is that the number of bases in the BEM affects the error performance in the sense of determining the maximum diversity order provided by the underlying time-selective channel.

Time-frequency duality Collecting N symbols $\{y(n)\}_{n=0}^{N-1}$, we can write the matrix-vector counterpart of (11.3) as:

$$y = D_h x + w, \tag{11.5}$$

where $y := [y(0) \cdots y(N-1)]^T$, $x := [x(0) \cdots x(N-1)]^T$, $w := [w(0) \cdots w(N-1)]^T$, and $D_h := \mathrm{diag}[h(0) \cdots h(N-1)]$, with ^{T} denoting transposition and $\mathrm{diag}[\cdot]$ representing a diagonal matrix.

It is well known that circulant matrices can be diagonalized by (I)FFT matrices (Peled and Ruiz, 1980) and (Golub and van Loan, 1996, p. 202). Using this property, and recalling that the BEM in (11.4) has its bases on the FFT grid, we can rewrite D_h as

$$D_h = \sum_{q=0}^{Q} h_q D_q = F_N H F_N^{\mathcal{H}}, \tag{11.6}$$

where $D_q := \mathrm{diag}[1 \; e^{j\omega_q} \cdots e^{j\omega_q(N-1)}]$, H is an $N \times N$ circulant matrix with first column given by $[h_{Q/2} \cdots h_0 \; 0 \cdots 0 \; h_Q \cdots h_{Q/2+1}]^T$, and F_N denotes the N-point FFT matrix with (m,n)th entry given by $[F_N]_{m,n} = (1/\sqrt{N})e^{-j2\pi(m-1)(n-1)/N}$. Performing the FFT at the transmitter and the IFFT at the receiver, we arrive at

$$y = F^{\mathcal{H}} D_h F x + w = H x + w. \tag{11.7}$$

Let us pause for a while and recall that for frequency-selective channels with I/O obeying (11.2), one can insert a cyclic prefix (CP) at the transmitter and remove it at the receiver, to render the Toeplitz matrix corresponding to a convolutional channel equivalent to a circulant matrix; see, e.g., Liu *et al.* (2002), Zhou and Giannakis (2003). Similarly, for time-selective channels, the resultant circulant matrix can be diagonalized by FFT and IFFT operations. Equations (11.6)–(11.7) suggest the converse direction: thanks to the BEM, it is possible to convert the diagonal time-selective channel D_h into circulant matrix after IFFT and FFT operations. From this point of view, the $Q+1$ BEM coefficients are dual to the $L+1$ channel taps of a frequency-selective channel. Let us recapitulate this time-frequency duality in the following property:

Property 11.1 *Based on the BEM in (11.4) for time-selective channels, a block transmission over time-selective channels can be equivalently viewed as a transmission over frequency-selective channels (11.1) after FFT processing at the transmitter and IFFT processing at the receiver. An equivalent diagonal (or circulant) channel matrix can be generated either from a* *time-selective or from a frequency-selective channel with appropriate (I)FFT operations.*

This property, establishing duality between time- and frequency-selective channels, does not imply that the overall wireless systems entailing time- or frequency-selective channels have identical properties. For example, to implement (11.5), the design for frequency-selective channels incurs longer decoding delays than its dual for time-selective channels. Having quantified the multipath diversity provided by frequency-selective channels, the Doppler diversity arising from time-varying channels, and the duality present between the two, we are ready to consider the most challenging class of wireless channels, which exhibit joint time- and frequency-selectivity.

Multidimensional diversity   For time- and frequency-selective (a.k.a. doubly-selective) channels, the BEM is given by:

$$h_l(n) := \sum_{q=0}^{Q} h_{l,q} e^{j\omega_q n}, \ l = 0, 1, \ldots, L. \tag{11.8}$$

Upon defining the channel coefficient vector as

$$\boldsymbol{h} := [h_{0,0} \ \cdots \ h_{L,0} \ \cdots \ h_{L,Q}]^T \in \mathbb{C}^{(L+1)(Q+1)\times 1}, \tag{11.9}$$

we can assess the diversity of doubly-selective channels as follows:

Proposition 11.3 (Ma and Giannakis, 2003b) *If the correlation matrix* $\boldsymbol{R}_h = E[\boldsymbol{h}\boldsymbol{h}^{\mathcal{H}}]$ *of the BEM channel vector in (11.9) has rank* r_h, *then the maximum diversity order (a.k.a. multipath-Doppler diversity order) of the doubly-selective channel in (11.8) is* $G_d = r_h$. *When* $\boldsymbol{R}_h$ *has full rank* $r_h = (L+1)(Q+1)$, *the maximum diversity gain is* $G_{d,\max} = (L+1)(Q+1)$.

Up until now, we have dealt with single-antenna wireless links. But the theme of this book and the chapters so far have motivated well the benefits brought by multiantenna transmissions over flat fading MIMO channels. However, to deliver the high-rates and resilience to high-mobility required by future generation wireless modems, MIMO transmissions should be designed capable of enabling the multiple flavors of diversity provided by the resultant doubly-selective MIMO propagation in all dimensions. At least at the intuitive level, it should be easy to appreciate that MIMO frequency- and/or time-selective channels provide joint space (multiantenna), multipath ($L+1$ taps), and Doppler ($Q+1$ bases) diversity. The aggregate diversity effect is multiplicative in the degrees of freedom available in every domain. However,

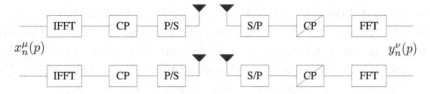

Fig. 11.1. MIMO-OFDM block diagram.

to collect this multidimensional form of diversity, one needs coding over multiple dimensions, which may accordingly increase complexity of the resultant encoders and decoders. In the remaining sections of this chapter, we show how to design space-time codes (STCs) for MIMO frequency- and/or time-selective channels in order to collect the multidimensional diversity and at the same time enable pertinent trade-offs among error performance, complexity, and transmission rate.

11.2 STCs for frequency-selective channels

Most STCs for MIMO frequency-selective channels aim at collecting the joint space-multipath diversity, while achieving desirable compromises between transmission rate and decoding complexity. A number of STCs have been designed for MIMO frequency-selective channels. In the following, we outline these designs by classifying them into two categories: single-carrier and multicarrier ones implemented with MIMO orthogonal frequency-division multiplexing (OFDM).

MIMO-OFDM The main advantage of OFDM is that it converts a frequency-selective channel into a set of parallel flat fading subchannels, thus reducing the equalization and demodulation complexity at the receiver. This is also the main reason behind OFDM's popularity in present and future generation wireless standards. In this section, we lay out a general framework for MIMO-OFDM.

Consider a multiantenna wireless communication system with N_t transmit and N_r receive antennas. The fading channel between the μth transmit antenna and the νth receive antenna is represented by the vector of impulse response coefficients $\boldsymbol{h}^{(\nu,\mu)} := [h_0^{(\nu,\mu)} \; \cdots \; h_L^{(\nu,\mu)}]^T \in \mathbb{C}^{(L+1)\times 1}$, with L denoting the channel order [see (11.1)]. The block diagram of a MIMO-OFDM system is depicted in Fig. 11.1.

Let $x_n^\mu(p)$ be the symbol transmitted on the pth subcarrier (frequency bin) from the μth transmit antenna during the nth OFDM symbol interval.

The symbols $\{x_n^\mu(p), \ \mu \in [1, N_t], \ p \in [1, N_c]\}$ are transmitted in parallel through the N_c subcarriers by N_t transmit antennas. Notice that the three variables μ, n, and p index, respectively, the transmit antenna (space), time, and frequency dimensions over which each symbol $x_n^\mu(p)$ is transmitted. (We should underscore here that n is used to index OFDM blocks.)

At the receiver, each antenna receives a noisy superposition of the multiantenna transmissions through the fading channels. The received signal $y_n^\nu(p)$ (after CP removal and FFT processing) at the νth receive antenna can be expressed as

$$y_n^\nu(p) = \sum_{\mu=1}^{N_t} H^{(\nu,\mu)}(p) \, x_n^\mu(p) + w_n^\nu(p), \qquad \forall \nu \in [1, N_r], p \in [1, N_c], \quad (11.10)$$

where $H^{(\nu,\mu)}(p)$ is the subchannel transfer function from the μth transmit antenna to the νth receive antenna evaluated at the frequency corresponding to the pth subcarrier: $H^{(\nu,\mu)}(p) := \sum_{l=0}^{L} h_l^{(\nu,\mu)} e^{-j2\pi l(p-1)/N_c}$, and $w_n^\nu(p)$ is AWGN, which is assumed to be statistically independent across time, space, and subcarriers. There are two often-used matrix representations of MIMO-OFDM systems. One can be formed by collecting N_c frequency-domain symbols (one OFDM block) as:

$$\boldsymbol{y}_n^\nu = \sum_{\mu=1}^{N_t} \boldsymbol{D}^{(\nu,\mu)} \, \boldsymbol{x}_n^\mu + \boldsymbol{w}_n^\nu, \qquad (11.11)$$

where $\boldsymbol{D}^{(\nu,\mu)} := \mathrm{diag}[H^{(\nu,\mu)}(1) \ \cdots \ H^{(\nu,\mu)}(N_c)]$. According to (11.11), received MIMO-OFDM blocks are expressed as the superposition of single-antenna OFDM blocks. The second representation results by stacking N_r received symbols on the pth subcarrier in the nth block as

$$\boldsymbol{y}_n(p) = \boldsymbol{H}(p)\boldsymbol{x}_n(p) + \boldsymbol{w}_n(p). \qquad (11.12)$$

Model (11.12) shows that on each subcarrier the channel is MIMO flat-fading. Using these representations, we will show three approaches to designing space-time-frequency (STF) codes.

Space-time OFDM designs As per (11.12), the equivalent MIMO channel on each OFDM subcarrier is flat fading. Thus, conventional space-time (ST) codes designed for flat-fading channels can be directly applied across time (index n) and space. This approach yields space-time OFDM (ST-OFDM) codes. One such ST-OFDM design adopts orthogonal ST block coding per subcarrier, as detailed in Li *et al.* (1999) and Liu *et al.* (2001). With $N_t = 2$ for instance, one can implement Alamouti's code design (Alamouti, 1998) on a per subcarrier basis; i.e., use on each subcarrier p the ST code matrix

$$X(p) = [x_1(p), x_2(p)] = \begin{bmatrix} s_1(p) & -s_2^*(p) \\ s_2(p) & s_1^*(p) \end{bmatrix}, \quad p = 1, \ldots, N_c. \quad (11.13)$$

Likewise, other ST codes for flat-fading channels can be applied here. At the receiver end, the corresponding decoders developed for flat-fading channels are implemented again on a per subcarrier basis.

The main advantage of the ST-OFDM transceiver with block-orthogonal ST codes per subcarrier is its low complexity, which is inherited by OFDM and the linear optimal demodulation, which orthogonal ST codes can afford even for channels with large order L. However, without coding across subcarriers, the potentially high multipath diversity is not enabled.

Space-frequency OFDM designs Another approach to designing codes for multiantenna OFDM transmissions is by coding across space and frequency. Notice that no coding is performed here across time. This approach yields space-frequency OFDM (SF-OFDM) codes, for which one space-frequency (SF) codeword is only one OFDM symbol long, whereas ST-OFDM codewords may span two or more OFDM symbols. This difference implies that SF-OFDM may incur smaller decoding delays and allow for faster channel variations than ST-OFDM. The first SF-OFDM code was investigated by Agrawal *et al.* (1998), while the term "SF-OFDM" was introduced by Bölcskei and Paulraj (2000). Lu and Wang (2000) confirmed that the diversity order of MIMO frequency-selective channels in Bölcskei and Paulraj (2000) is $N_t N_r (L + 1)$ and proposed some heuristic steps to construct ST trellis codes (STTCs) with the SF-OFDM structure. In the following, we introduce an SF-OFDM design that achieves joint space-multipath diversity and enjoys full transmission rate (Ma and Giannakis, 2003a).

For the sake of clarity, let us set the number of subcarriers to $N_c = N_t(L + 1)$. When the number of subcarriers $N_c > N_t(L + 1)$, we need a subcarrier grouping approach, whereby a large set of subcarriers is split into smaller ones, each with $N_t(L + 1)$ subcarriers.

Similar to the flat-fading case in Ma and Giannakis (2003a), the full-diversity full-rate (FDFR) encoder design for frequency-selective channels entails two stages: linear complex-field (LCF) encoding followed by ST multiplexing. Let us consider an $N_t^2(L + 1) \times 1$ symbol block s and demultiplex it into N_t subblocks, $\{s_k\}_{k=1}^{N_t}$, each with length $N_t(L + 1)$. Multiplying with the subblock specific LCF encoder Θ_k, we obtain the kth layer as $u_k = \Theta_k s_k$, $\forall k \in [1, N_t]$, where $\Theta_k = \beta^{k-1}\Theta$, and the $N_t(L+1) \times N_t(L+1)$ matrix Θ and the scalar β are designed as in Rainish (1996), Xin *et al.*

(2003), and Ma and Giannakis (2003a). After LCF encoding, we map the LCF encoded symbols into a space-frequency coded matrix as follows:

$$
\begin{aligned}
\boldsymbol{X} &:= \begin{bmatrix} \boldsymbol{x}(1) & \boldsymbol{x}(2) & \cdots & \boldsymbol{x}(N_c) \end{bmatrix} \\
&= \begin{bmatrix}
u_1(1) & \cdots & u_2(N_t) & u_1(N_t+1) & \cdots & u_2(2N_t) & \cdots & u_2(N_c) \\
u_2(1) & \cdots & u_3(N_t) & u_2(N_t+1) & \cdots & u_3(2N_t) & \cdots & u_3(N_c) \\
\vdots & \cdots & \vdots & \vdots & \cdots & \vdots & & \\
u_{N_t}(1) & \cdots & u_1(N_t) & u_{N_t}(N_t+1) & \cdots & u_1(2N_t) & \cdots & u_1(N_c)
\end{bmatrix}, \quad (11.14)
\end{aligned}
$$

where $u_k(p)$ is the pth symbol of the kth layer, and $\boldsymbol{x}(p)$ is defined as $\boldsymbol{x}_n(p)$ in (11.12) after dropping the index n. Observing (11.14), we notice that indeed every layer is transmitted over N_t transmit antennas, and, through each antenna, each layer is spread over at least $L + 1$ frequency bins. Intuitively, this enables joint exploitation of the space and multipath diversity dimensions. At the same time, since we transmit $N_t^2(L + 1)$ information symbols in $N_t(L + 1)$ time slots, the transmission rate is N_t symbols per channel use (pcu), without taking into account the CP insertion.

With regards to diversity and rate of this design, we have the following:

Proposition 11.4 (Ma and Giannakis, 2003a) *If the entries of $\boldsymbol{s}$ are integers drawn from a finite alphabet, there exists at least one pair $(\boldsymbol{\Theta}, \beta)$ for which the ST encoder in (11.14) enables full diversity $N_t N_r(L + 1)$ at a transmission rate $N_t N_c/(N_c + L)$ symbols pcu.*

Note that the rate loss $L/(N_c + L)$ in Proposition 11.4 is due to the CP insertion in OFDM. When $N_c \gg L$, the transmission rate is approximately N_t symbols per channel use.

Space-time-frequency (STF)-OFDM design As a natural generalization of the two-dimensional schemes (ST-OFDM and SF-OFDM), let us now consider designs where multiantenna coding is performed across space, time, and frequency. Here, we use $N_t = 2$ as an example to demonstrate the resulting STF code design.

Suppose we have drawn two information blocks $\boldsymbol{s}_1$ and $\boldsymbol{s}_2$ with length $N_c = N_{\text{sub}} N_g$. After splitting $\boldsymbol{s}_n$ equally into N_g groups, let us denote the gth group of symbols in $\boldsymbol{s}_n$ as

$$
\boldsymbol{s}_n(g) := [s_n(1 + (g - 1)N_g) \ \ s_n(2 + (g - 1)N_g) \ \cdots \ s_n(N_{\text{sub}} + (g - 1)N_g)]^T.
$$

For each group, LCF coding through an $N_{\text{sub}} \times N_{\text{sub}}$ precoder $\boldsymbol{\Theta}$, designed as in Xin *et al.* (2003), is performed to obtain

$$\boldsymbol{u}_n(g) = \boldsymbol{\Theta}\boldsymbol{s}_n(g), \ g \in [1, N_g].$$ (11.15)

Multiplexing the N_g precoded groups into a single block by equally placing the precoded symbols, we can write $\boldsymbol{u}_n = \boldsymbol{\Phi}\boldsymbol{s}_n$, $n = 1, 2$, where $\boldsymbol{\Phi}$ is defined based on the rows of $\boldsymbol{\Theta}$, denoted by $\boldsymbol{\theta}_n^T$, and the placement.

The remaining modules at the transmitter are identical to those involved in ST-OFDM. The transmitted ST matrix on the pth subcarrier is the same as the one in (11.13) (replacing s with u).

Thanks to orthogonality, the STF-OFDM code enjoys a low-complexity two-step decoding process:

Step 1: Optimal linear decoding on each subcarrier to obtain $\hat{\boldsymbol{u}}_n$;

Step 2: Exact or near-maximum likelihood (ML) demodulation using the sphere decoding (SD) algorithm to recover $\boldsymbol{s}_n(g)$ from the decoded $\hat{\boldsymbol{u}}_n$, based on (11.15).

Observe that the LCF precoder $\boldsymbol{\Theta}$ in the frequency domain enables multipath diversity, while the ST orthogonal coder enables spatial diversity. When $N_{\text{sub}} \geq L + 1$, STF-OFDM can collect the joint space-multipath diversity with a decoding complexity that depends on the group size N_{sub}. The group size N_{sub} is in fact a tuning parameter controlling the trade-off between error performance and complexity. The extension of STF-OFDM to more than two antennas is straightforward. However, similar to the orthogonal ST codes in Tarokh *et al.* (1999), the grouped STF-OFDM design incurs a rate loss (transmission rate is less than one) when $N_t > 2$.

Example 11.1 (Comparisons of MIMO-OFDM systems) Consider a frequency-selective MIMO channel with parameters $(N_t, N_r, L) = (2, 2, 1)$. The channel taps are independent, and for each channel, the power of the taps satisfies an exponentially decaying profile. We compare three schemes: VBLAST-OFDM (Piechocki *et al.*, 2001), FDFR-OFDM (Ma and Giannakis, 2003a), and Grouped (G) STF-OFDM (Liu *et al.*, 2002). The block size N_c is chosen as $N_c = N_t(L + 1) = 4$. To fix the transmission rate at $R = 8/3$ bits pcu, we use QPSK for the FDFR-OFDM and VBLAST-OFDM schemes, and 16-QAM for the GSTF scheme. At the receiver, we use SD for all three schemes. Fig. 11.3 depicts the simulated performance. Considering the slopes of the BER curves, we infer that FDFR-OFDM and GSTF achieve full diversity, while VBLAST-OFDM does not (only receive diversity N_r can be achieved). The big gap between FDFR-OFDM and STF-OFDM is due to the smaller constellation size used for FDFR. Note that the decoding complexity order of FDFR is $\mathcal{O}((N_t^2(L+1))^3)$, while VBLAST-OFDM has complexity $\mathcal{O}(N_r^3)$ and STF-OFDM $\mathcal{O}((L+1)^3)$.

Single-carrier transmissions To deal with MIMO frequency-selective channels, single-carrier ST codes have also been designed to yield full diversity and at the same time enjoy advantages not present in multicarrier systems (e.g., low peak-to-average power ratio). One such class of codes we outline next can be thought of as the single-carrier version of the STF-OFDM design. For $N_t = 2$, this class offers the vector counterpart of Alamouti's code (Alamouti, 1998), showing how the latter can be made suitable for frequency-selective MIMO links, too. Due to the dispersive nature of FIR channels, ST block coding is applied on symbol blocks rather than on individual information symbols. Specifically, consider two symbol blocks s_1 and s_2, each of length J, and construct the following $2 \times 2J$ matrix

$$X = \begin{bmatrix} (s_1)^T & -(Ps_2^*)^T \\ (s_2)^T & (Ps_1^*)^T \end{bmatrix}, \tag{11.16}$$

where P is a permutation matrix drawn from a set of permutation matrices $\{P_J^{(n)}\}_{n=0}^{J-1}$. Each $P_J^{(n)}$ performs a reverse cyclic shift (that depends on n), when applied to a $J \times 1$ vector $a := [a(0)\, a(1)\, \cdots\, a(J-1)]^T$. Specifically, the $(p+1)$st entry of $P_J^{(n)} a$ is:

$$\left[P_J^{(n)} a \right]_p = a\big((J - p + n - 1) \bmod J \big). \tag{11.17}$$

From (11.17), we observe that $P_J^{(0)} a$ performs a *time reversal* of a, while

$$P_J^{(1)} a = F_J F_J a = F_J^{\mathcal{H}} F_J^{\mathcal{H}} a, \tag{11.18}$$

corresponds to taking the J-point FFT or IFFT twice on the vector a. The unifying view in Zhou and Giannakis (2003) allows for any P from the set $\{P_J^{(n)}\}_{n=0}^{J-1}$ and subsumes the designs in Lindskog and Paulraj (2000), Al-Dhahir (2001), Barbarossa and Cerquetti (2001), and Vook and Thomas (2000).

After inserting the CP in every block, the block symbols s_1 and s_2 are transmitted. The CP is discarded at the receiver to avoid interblock interference (IBI). Let y_1 denote the received block in the first block period and y_2 the one at the second block period after CP removal. After left multiplying y_2 by P and using the orthogonality of the two blocks, decoding each individual block reduces to a block equalization problem, for which many options are available, including exhaustive ML search, SD, block decision feedback, and linear equalization.

As shown in Zhou and Giannakis (2003), instead of CP, one can also insert L zeros at the end of s_n to remove IBI (this is referred to as zero-padding (ZP)-only scheme). Interestingly, Viterbi's algorithm is applicable to ZP-only and likewise to CP-only designs, when the last L symbols in s_n are known (see Zhou and Giannakis (2003) for details). Recall that for channels of order L, the complexity of Viterbi's algorithm is $\mathcal{O}(|\mathcal{A}|^{L+1})$ per symbol, where $|\mathcal{A}|$ is the cardinality of the constellation $\mathcal{A}$. However, the complexity of ML or SD search depends on the block length J. For this reason, ML decoding with exact application of Viterbi's algorithm is particularly attractive for large block transmissions with small constellation sizes over relatively short channels.

To further improve the system performance, error control coding can be applied to either multicarrier or single-carrier systems. For example, outer convolutional codes can be used in the ZP-only design, as described in Zhou and Giannakis (2003). Other options, including trellis coded modulation (TCM) and turbo codes, are applicable here as well.

11.3 STCs for time-selective channels

As we mentioned in Section 11.1, MIMO time-selective channels provide joint space-Doppler diversity. However, in lieu of a parsimonious channel model, ST codes for time-selective channels have not been explored as well as the ones for frequency-selective channels. In this section, we rely on the time-frequency duality we saw in Section 11.1 to design so-called ST-Doppler (STDO) codes based on the ST-multipath codes we derived in Section 11.2.

Each duality-based STDO coder (respectively decoder) entails two stages. The first stage is identical to the one we saw for multicarrier STF schemes in Section 11.2 and for the single-carrier ones in Section 11.2. The second stage is an inner codec performing an IFFT (FFT) over a number of blocks from the first stage at the transmitter (receiver). The duality in (11.6) renders the time-selective channel equivalent to a frequency-selective channel after the (I)FFT operations. Since the latter enables the available space-multipath diversity order, it is not difficult to show that such duality-based STDO designs can achieve joint space-Doppler diversity (Ma and Giannakis, 2003b). But similar to the STF-OFDM schemes in Liu *et al.* (2002) and the single-carrier ones in Zhou and Giannakis (2003), STDO designs based on orthogonal ST codes yield transmission rates less than one symbol per channel use. However, the FDFR scheme we saw in Section 11.2 for frequency-selective channels can be mapped via duality to obtain its FDFR counterpart for time-selective

channels. The latter is capable of capturing space-Doppler diversity at full
transmission rate.

11.4 Phase sweeping and delay diversity—MIMO to SIMO

So far, we have presented sophisticated ST codes for time- and frequency-
selective channels, initiating their design directly from the MIMO model.
However, a number of well-tested code designs are also available for single-
input single-output (SISO) and single-input multiple-output (SIMO) channels. To migrate SISO and SIMO designs to the MIMO setup, it is important
to have means of converting multiple channels to an equivalent SISO channel. To this end, we introduce two approaches: delay diversity and phase
sweeping.

Analog phase sweeping and delay diversity The analog phase-sweeping (a.k.a. intentional frequency offset) idea was introduced in Hiroike *et al.*
(1992) and was later combined with channel coding in Kuo and Fitz (1997).
The two-transmit-antenna analog implementation, modulates the signal of
one antenna with a sweeping frequency f_s in addition to the carrier frequency
f_c that is present in both antennas (Hiroike *et al.*, 1992; Kuo and Fitz, 1997).

Suppose the two channels are $h_1(t)$ and $h_2(t)$, and the two transmitted
signals are $x_1(t) = s(t)e^{j2\pi f_c t}$ and $x_2(t) = s(t)e^{j2\pi(f_c+f_s)t}$, where $s(t)$ denotes
the information bearing waveform. The received signal is $y(t) = h_1(t)x_1(t) +
h_2(t)x_2(t) + w(t) = h(t)s(t)e^{j2\pi f_c t} + w(t)$, where $w(t)$ is the AWGN, and the
equivalent channel is given by $h(t) = h_1(t) + h_2(t)e^{j2\pi f_s t}$. The equivalent
channel has faster time variation than $h_1(t)$ and $h_2(t)$. As we have seen
in Section 11.1, the faster a channel varies, the higher Doppler diversity it
potentially provides. Phase sweeping offers a means of converting antenna
diversity to Doppler diversity.

Dual to phase-sweeping schemes, which apply to time-varying channels,
are delay-diversity schemes, which apply to flat-fading or frequency-selective
channels. They were originally developed for flat-fading channels in Seshadri
and Winters (1994), but they have been recently extended to frequency-
selective channels by transmitting one symbol over multiple antennas in
different time slots (Gore *et al.*, 2001; Naguib, 2001). If the delay in transmitting from one antenna to the next is only one symbol period (Seshadri
and Winters, 1994), multiantenna flat-fading channels can be turned into an
equivalent FIR channel with N_t taps, thus guaranteeing full space diversity.
However, if the channels are frequency-selective, this simple delay-diversity
transmission will end up with an equivalent FIR channel of $N_t + L$ taps

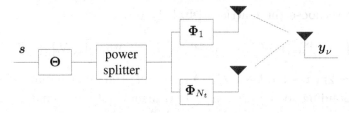

Fig. 11.2. Illustration of multiantenna transmissions with DPS.

(Naguib, 2001). As a result, the multipath diversity is not enabled. If, on the other hand, the delay per antenna is $L+1$ symbols, the equivalent channel length becomes $N_t(L+1)$ and full diversity is enabled with longer delay and lower rate.

In general, analog phase sweeping and delay diversity cost extra channel bandwidth, because of the analog delay or frequency shift. They can only yield joint space-Doppler/multipath diversity at the expense of rate loss. In the following, we present digital phase-sweeping alternatives and their block-circular delay-diversity dual schemes, which enable full diversity without compromising on spectral efficiency.

DPS/CDD for time- or frequency-selective channels The block diagram of a digital phase-sweeping (DPS) system is depicted in Fig. 11.2. The matrix Θ denotes an $N \times N$ LCF precoder, and the Φ_μ's are phase-sweeping or circular delay matrices, which will be defined. We first use time-selective channels as an example to illustrate DPS and then show the equivalence between DPS and block circular delay diversity (BCDD) using frequency-selective channels.

With reference to Fig. 11.2, the received block $\boldsymbol{y}_\nu$ and the transmitted block $\boldsymbol{s}$ can be related via

$$\boldsymbol{y}_\nu = \frac{1}{\sqrt{N_t}} \sum_{\mu=1}^{N_t} \boldsymbol{D}_H^{(\nu,\mu)} \boldsymbol{\Phi}_\mu \boldsymbol{\Theta} \boldsymbol{s} + \boldsymbol{\zeta}_\nu, \qquad \nu \in [1, N_r]. \tag{11.19}$$

Recalling (11.4), we deduce that different channels share the same exponential bases, although they have different channel coefficients. Suppose that we shift the $Q+1$ bases of each channel corresponding to one of the N_t transmit antennas so that all the bases are consecutive on the FFT grid of complex exponentials. Then, we can view the N_t channels to each receive antenna as one equivalent time-selective channel with $N_t(Q+1)$ bases. To realize this intuition, we choose the matrices $\{\boldsymbol{\Phi}_\mu\}_{\mu=1}^{N_t}$ to be

$$\boldsymbol{\Phi}_\mu = \text{diag}[1 \;\; e^{j\phi_\mu} \;\; \cdots \;\; e^{j\phi_\mu(N-1)}], \qquad \forall \mu \in [1, N_t], \qquad (11.20)$$

where $\phi_\mu = 2\pi(\mu - 1)(Q + 1)/N$. As $\boldsymbol{\Phi}_1 = \boldsymbol{I}$, the exponentials of the channel corresponding to the first ($\mu = 1$) transmit antenna remain unchanged. But those corresponding to the second channel ($\mu = 2$) are shifted from their original location in $\{\boldsymbol{D}_q\}_{q=0}^{Q}$ to $\{\boldsymbol{D}_q\}_{q=Q+1}^{2Q+1}$ after multiplication with the DPS matrix $\boldsymbol{\Phi}_2$, which applied to the second transmit-antenna, i.e., $\{\boldsymbol{D}_q\boldsymbol{\Phi}_2\}_{q=0}^{Q} = \{\boldsymbol{D}_q\}_{q=Q+1}^{2Q+1}$. Proceeding likewise with all N_t DPS matrices, it follows that (11.19) reduces to

$$\boldsymbol{y}_\nu = \frac{1}{\sqrt{N_t}} \sum_{q=0}^{N_t(Q+1)-1} h_q^{(\nu)} \boldsymbol{D}_q \boldsymbol{\Theta} s + \boldsymbol{\zeta}_\nu, \qquad \forall \nu \in [1, N_r], \qquad (11.21)$$

where $h_q^{(\nu)} := h_{q \bmod (Q+1)}^{(\nu, \lfloor q/(Q+1) \rfloor + 1)}$. Comparing (11.19) with (11.21), we obtain

Property 11.2 (Ma et al., 2005)   *DPS converts an N_t transmit-antenna system, where each channel can be expressed via $Q + 1$ exponential bases, to a single transmit-antenna system, where the equivalent channel is expressed by $N_t(Q + 1)$ exponential bases.*

Notice that since $\boldsymbol{\Phi}_\mu$ operates in the digital domain, the sweeping wraps the phases around $[-\pi, \pi)$, which explains why DPS does not incur bandwidth expansion.

Remark 11.1   To avoid overlapping of the shifted bases, we should make sure that $N > N_t(Q+1)$. From the definition of $Q := 2\lceil f_{\max} N T_s \rceil$, for fixed $f_{\max}$ and N, we can adjust the sampling period T_s to satisfy this condition. Since for each receive antenna we have $N_t(Q+1)$ unknown BEM coefficients corresponding to N_t channels every N symbols, this condition guarantees that the number of unknowns is less than the number of equations. Thus, even from a channel estimation viewpoint, this condition is reasonable.

Since the received blocks $\boldsymbol{y}_\nu$ from all N_r receive-antennas contain the information block s, we need to combine the information from all received blocks to decode s. To this end, we employ the maximum-ratio combiner (MRC), which retains decoding optimality. With the equivalence established by Property 11.2, diversity can be enabled by selecting as any single-input design an outer STDO encoder that achieves the maximum diversity gain for the SISO time-selective channels corresponding to each transmit-receive antenna pair.

To collect the maximum diversity gains at the receiver, we need ML demodulation (Ma and Giannakis, 2003b, Proposition 2), which can be efficiently implemented using SD. We summarize the diversity and coding gain results for DPS in the following proposition:

Proposition 11.5 (Ma *et al.*, 2005) *The maximum achievable STDO diversity order* $G_d = r_h$ *is guaranteed by the DPS design when the group size is selected as* $N_{\text{sub}} \geq N_t(Q+1)$. *A transmission rate of* 1 *symbol pcu is achieved by the DPS design.*

Remark 11.2 Comparing the duality based block STDO designs (STF-based, SC-based, and FDFR) with DPS, we note that: i) all schemes guarantee the maximum diversity gain; ii) FDFR design achieves the highest transmission rate (N_t symbols per channel use); iii) DPS guarantees one symbol per channel use, while designs based on orthogonal ST codes (Tarokh *et al.*, 1999) suffer from rate loss, when $N_t > 2$ antennas are signaling with complex constellations; and iv) to guarantee full diversity, in general, STF based codes (relying on grouping and orthogonal designs) come with the lowest decoding complexity.

If we view (11.19) as a MIMO-OFDM model, it is easy to verify that DPS also works for frequency-selective channels. Analogous to the duality on analog phase sweeping and delay, DPS can be shown to be equivalent to BCDD (Ma and Giannakis, 2005).

Example 11.2 In this example, we compare the simple DPS scheme with the smart greedy codes for $(N_t, N_r) = (2, 1)$. To maintain the same rate, we select BPSK for the DPS scheme and use the code in Tarokh *et al.* (1998, Example 3.9.2). Each channel has $Q + 1 = 3$ bases, and the channel coefficients are i.i.d. with mean zero and variance $1/(Q+1)$. First, we consider the uncoded setup. The information block length is $K = P = 30$. The number of groups for DPS is $N_g = 5$, so that these two schemes have comparable decoding complexity. Fig. 11.4 depicts the BER versus SNR comparison for the smart greedy code, and the DPS one (solid lines). It is evident that DPS outperforms smart greedy coding, because the former guarantees the full space-Doppler diversity.

Furthermore, we consider the coded case for both schemes. We select a $(7, 3)$ Reed–Solomon (RS) coder with block interleaving. The number of information bits is 90. Therefore, the length of each block of coded bits is 210. We select the depth of the block interleaver as 42. For the DPS design, we split the coded bits into 5 blocks. Each block is divided into 7 groups. The

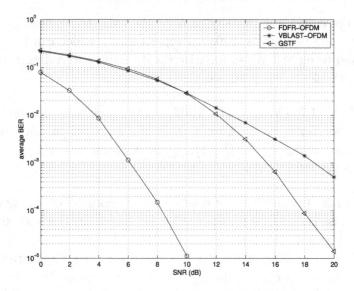

Fig. 11.3. MIMO-OFDM comparisons.

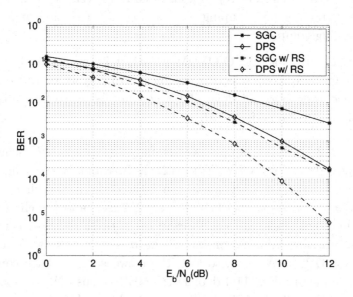

Fig. 11.4. DPS versus SGC.

simulation results are shown in Fig. 11.4 (dashed lines). Note that the DPS scheme still outperforms the smart greedy code (SGC) markedly.

The DPS/CDD schemes are useful not only for coherent designs, but also for differential and noncoherent designs. For example, the block differential scheme developed in Liu and Giannakis (2003) for single antenna OFDM is capable of collecting full multipath diversity. In conjunction with DPS, one can easily construct a differential scheme for MIMO-OFDM, with full diversity $N_t N_r (L+1)$. The price to be paid is increased complexity.

DPS for doubly-selective channels DPS can also be applied to render multiple doubly-selective channels equivalent to a single doubly-selective channel. In fact, the DPS based scheme we will outline here, combined with linear precoding, is the only multiantenna code known to date capable of enabling the full multiplicative diversity jointly in all three dimensions: space, multipath, and Doppler.

Suppose that the transmitter has been designed to remove the IBI through zero padding. The matrix-vector representation of the I/O relationship for doubly-selective MIMO channels can then be written as:

$$y_\nu = \sum_{\mu=1}^{N_t} H^{(\nu,\mu)} x_\mu + w_\nu, \qquad \forall \nu \in [1, N_r], \tag{11.22}$$

where from (11.8), the $N \times N$ channel matrix is given by

$$H^{(\nu,\mu)} := \sum_{q=0}^{Q} D_q H_q^{(\nu,\mu)}, \tag{11.23}$$

with D_q defined in (11.6) and $H_q^{(\nu,\mu)}$ denoting a lower triangular Toeplitz matrix with first column given by $[h_q^{(\nu,\mu)}(0) \ \cdots \ h_q^{(\nu,\mu)}(L) \ 0 \ \cdots \ 0]^T$. Each block y_ν is demodulated using a decoder $\mathcal{G}(\cdot)$ to obtain an estimate of s as: $\hat{s} := \mathcal{G}(\{y_\nu\}_{\nu=1}^{N_r})$.

As far as diversity of doubly-selective channels is concerned, the intuitively expected result can be asserted as a natural combination of corresponding results on frequency-selective and time-selective channels:

Proposition 11.6 (Ma and Giannakis, 2002) *Consider an $N_r \times N_t$ doubly-selective MIMO channel adhering to a BEM with $Q+1$ bases and $L+1$ taps. If the correlation matrix of the $N_t N_r (L+1)(Q+1)$ channel coefficients has rank r_h, then the maximum diversity order of transmissions in (11.22) is $G_d = r_h \le N_t N_r (L+1)(Q+1) = G_{d,\max}$.*

To enable the maximum product diversity assessed by the proposition, we will combine DPS with single-antenna linear precoding. By equally allocating power across transmit antennas, (11.22) can be written as

$$y_\nu = \frac{1}{\sqrt{N_t}} \sum_{\mu=1}^{N_t} H^{(\nu,\mu)} \Phi_\mu \Theta s + w_\nu, \qquad \forall \nu \in [1, N_r]. \tag{11.24}$$

As with time-selective channels, different channels share the same exponential bases, but have different channel coefficients. The idea again is to shift the $Q+1$ bases of each channel corresponding to one of the N_t transmit antennas so that all the bases become consecutive on the FFT grid of complex exponentials. This renders the N_t channels to each receive-antenna equivalent to a single doubly-selective channel with $N_t(Q+1)$ bases. Recalling the structure of $H_q^{(\nu,\mu)}$ in (11.23), we have $H_q^{(\nu,\mu)} \Phi_\mu = \Phi_\mu \bar{H}_q^{(\nu,\mu)}$, where $\bar{H}_q^{(\nu,\mu)}$ has the same structure as $H_q^{(\nu,\mu)}$, but its first column is $[h_q(0)\ h_q(1)e^{-j\phi_\mu}$ $\cdots\ h_q(L)e^{-j\phi_\mu L}\ 0\ \cdots\ 0]^T$. Considering the designs in (11.20), (11.24) can be rewritten as

$$y_\nu = \frac{1}{\sqrt{N_t}} \sum_{q=0}^{N_t(Q+1)-1} D_q H_q^{(\nu)} \Theta s + w_\nu, \qquad \forall \nu \in [1, N_r], \tag{11.25}$$

where

$$H_q^{(\nu)} := \bar{H}_{\mathrm{mod}(q,Q+1)}^{(\nu,\lfloor q/(Q+1)\rfloor+1)}.$$

Comparing (11.25) with the single-antenna model of Ma and Giannakis (2003b) for time-selective, frequency-flat channels, we observe that this DPS design transforms the N_t transmit-antenna scenario, where each channel can be expressed via $Q+1$ exponential bases, into a single-transmit-antenna setup, where the equivalent channel can be expressed by $N_t(Q+1)$ bases. Hence, relying on Θ and $\mathcal{G}(\cdot)$, we can now apply any SISO or SIMO method for doubly-selective channels to ensure the maximum diversity gain.

In summary, DPS is a useful tool when we want to map MIMO models to SIMO ones to simplify ST code design. Another important advantage of DPS is that it is quite robust to the time-varying model considered. It has been demonstrated in Ma *et al.* (2005) that DPS works well even in the presence of model mismatch.

11.5 Summary

In this chapter, we quantified systematically the diversity benefits of multipath fading and Doppler provided by frequency- and/or time-selective channels. The maximum diversity achievable turned out to be the product of the number of transmit antennas, the number of receive antennas, the number of taps per channel and the number of bases in the Doppler domain. We also presented space-time code designs tailored to frequency-selective and time-selective channels. Finally, we saw how a simple scheme based on digital phase sweeping can be beneficial in mapping multiple channels to an equivalent channel so that well established code designs for single-antenna systems can be utilized to design multiantenna codes as well.

References

Agrawal, D., Tarokh, V., Naguib, A., and Seshadri, N. (1998). Space-time coded OFDM for high data-rate wireless communication over wideband channels. In *Proc. IEEE Veh. Technol. Conf.*, vol. 3, 2232–2236.

Al-Dhahir, N. (2001). Single-carrier frequency-domain equalization for space-time block-coded transmissions over frequency-selective fading channels. *IEEE Commun. Lett.*, **5** (7), 304–306.

Alamouti, S. M. (1998). A simple transmit diversity technique for wireless communications. *IEEE J. Sel. Areas Commun.*, **16** (8), 1451–1458.

Barbarossa, S. and Cerquetti, F. (2001). Simple space-time coded SS-CDMA systems capable of perfect MUI/ISI elimination. *IEEE Commun. Lett.*, **5** (12), 471–473.

Bölcskei, H. and Paulraj, A. J. (2000). Space-frequency coded broadband OFDM systems. In *Wireless Commun. and Netw. Conf.*, vol. 1, 1–6.

Giannakis, G. B. and Tepedelenlioglu, C. (1998). Basis expansion models and diversity techniques for blind equalization of time-varying channels. *Proc. IEEE*, **86** (10), 1969–1986.

Golub, G. H. and van Loan, C. F. (1996). *Matrix Computations* (Johns Hopkins Univ. Press), third edn.

Gore, D., Sandhu, S., and Paulraj, A. (2001). Delay diversity code for frequency selective channels. *Electron. Lett.*, **37** (20), 1230–1231.

Hiroike, A., Adachi, F., and Nakajima, N. (1992). Combined effects of phase sweeping transmitter diversity and channel coding. *IEEE Trans. Veh. Technol.*, **41** (2), 170–176.

Jakes, W. C. (1974). *Microwave Mobile Communications* (Wiley, New York).

Kuo, W.-Y. and Fitz, M. P. (1997). Design and analysis of transmitter diversity using intentional frequency offset for wireless communications. *IEEE Trans. Veh. Technol.*, **46** (4), 871–881.

Li, Y., Chung, J. C., and Sollenberger, N. R. (1999). Transmitter diversity for OFDM systems and its impact on high-rate data wireless networks. *IEEE J. Sel. Areas Commun.*, **17** (7), 1233–1243.

Lindskog, E. and Paulraj, A. (2000). A transmit diversity scheme for channels with intersymbol interference. In *Proc. IEEE Int. Conf. Commun.*, vol. 1, 307–311.

Liu, Z. and Giannakis, G. B. (2003). Block differentially encoded OFDM with maximum multipath diversity. *IEEE Trans. Wireless Commun.*, **2** (3), 420–423.

Liu, Z., Giannakis, G. B., Muquet, B., and Zhou, S. (2001). Space-time coding for broadband wireless communications. *Wireless Commun. and Mobile Comput.*, **1** (1), 33–53.

Liu, Z., Xin, Y., and Giannakis, G. B. (2002). Space-time-frequency coded OFDM over frequency-selective fading channels. *IEEE Trans. Signal Process.*, **50** (10), 2465–2476.

Lu, B. and Wang, X. (2000). Space-time code design in OFDM systems. In *Proc. IEEE Global Telecommun. Conf.*, vol. 2, 1000–1004.

Ma, X. and Giannakis, G. B. (2002). Space-time-multipath coding using digital phase sweeping. In *Proc. IEEE Global Telecommun. Conf.*, vol. 1, 384–388.

—— (2003a). Full-diversity full-rate complex-field space-time coding. *IEEE Trans. Signal Process.*, **51** (11), 2917–2930.

—— (2003b). Maximum-diversity transmissions over doubly-selective wireless channels. *IEEE Trans. Inf. Theory*, **49** (7), 1832–1840.

—— (2005). Space-time-multipath coding using digital phase sweeping or block circular delay diversity. *IEEE Trans. Signal Process.*, **53** (3), 1121–1131.

Ma, X., Leus, G., and Giannakis, G. B. (2005). Space-time-Doppler coding for correlated time-selective fading channels. *IEEE Trans. Signal Process.*, **53** (6), 2167–2181.

Naguib, A. F. (2001). On the matched filter bound of transmit diversity techniques. In *Proc. IEEE Int. Conf. Commun.*, 596–603.

Peled, A. and Ruiz, A. (1980). Frequency domain data transmission using reduced computational complexity algorithms. In *Proc. IEEE Int. Conf. Acoust., Speech, Signal Process.*, 964–967.

Piechocki, R. J., Fletcher, P. N., Nix, A. R., Canagarajah, C. N., and McGeehan, J. P. (2001). Performance evaluation of BLAST-OFDM enhanced HIPERLAN/2 using simulated and measured channel data. *Electron. Lett.*, **37** (18), 1137–1139.

Rainish, D. (1996). Diversity transform for fading channels. *IEEE Trans. Commun.*, **44** (12), 1653–1661.

Seshadri, N. and Winters, J. H. (1994). Two signaling schemes for improving the error performance of frequency division duplex (FDD) transmission systems using transmitter antenna diversity. *Int. J. Wireless Inf. Netw.*, **1** (1), 49–60.

Tarokh, V., Jafarkhani, H., and Calderbank, A. R. (1999). Space-time block coding for wireless communications: performance results. *IEEE J. Sel. Areas Commun.*, **17** (3), 451–460.

Tarokh, V., Seshadri, N., and Calderbank, A. R. (1998). Space-time codes for high data rate wireless communication: performance criterion and code construction. *IEEE Trans. Inf. Theory*, **44** (2), 744–765.

Vook, F. W. and Thomas, T. A. (2000). Transmit diversity schemes for broadband mobile communication systems. In *Proc. IEEE Veh. Technol. Conf.*, vol. 6, 2523–2529.

Wang, Z. and Giannakis, G. B. (2003). Complex-field coding for OFDM over fading wireless channels. *IEEE Trans. Inf. Theory*, **49** (3), 707–720.

Xin, Y., Wang, Z., and Giannakis, G. B. (2003). Space-time diversity systems based on linear constellation precoding. *IEEE Trans. Wireless Commun.*, **2** (2), 294–309.

Zhou, S. and Giannakis, G. B. (2003). Single-carrier space-time block coded transmissions over frequency-selective fading channels. *IEEE Trans. Inf. Theory*, **49** (1), 164–179.

Part III
Receiver algorithms and parameter estimation

Part III

Recovery algorithms and parameter estimation

12

Array signal processing

Alex B. Gershman

Darmstadt University of Technology

In this chapter, an overview of the fundamentals of array signal processing is provided with a particular emphasis on recent advances in direction-of-arrival (DOA) estimation and adaptive beamforming. It is then highlighted that the signal models used in array processing and space-time block coded MIMO communications are rather similar. At the same time, the MIMO model enjoys some interesting properties that do not emerge in classical applications of array processing. Using the established links between the aforementioned models, several important applications of array processing philosophy and concepts to point-to-point and multiuser MIMO communications are discussed.

12.1 Introduction

Sensor array processing has a long and rich history of theoretical research and practical applications to radar (Brennan *et al.*, 1976; Haykin, 1980; Haykin *et al.*, 1992; Swindlehurst and Stoica, 1998); sonar (Cox, 1973; Morgan and Smith, 1990; Böhme, 1995; Gershman *et al.*, 1995; Krolik, 1996); wireless communications (Winters, 1982; Godara, 1997; Paulraj and Papadias, 1998; Paulraj and Gesbert, 1998); navigation (Evans *et al.*, 1982; Amin *et al.*, 2004); seismology (Capon *et al.*, 1967, 1969; Böhme, 1995; Sidorovitch and Gershman, 1998); biomedicine (Hochwald and Nehorai, 1987; Mosher and Leahy, 1998; Li *et al.*, 1998); radio astronomy (Raza *et al.*, 2002; Boonstra and van der Veen, 2003); and other fields (Spielman *et al.*, 1990; Zoubir and Böhme, 1995; Nehorai *et al.*, 1995; Gershman and Turchin, 1995; Liu *et al.*, 1996; Nehorai and Jeremic, 2000).

Direction finding and *adaptive beamforming* are the two most important areas of sensor array processing that will be considered in the remainder of this chapter. The main objective of direction finding is to obtain estimates of the number of emitting sources and their DOAs, whereas the objective of adaptive beamforming is to detect and estimate the signal-of-interest (SOI) in the presence of strong interfering sources using data-adaptive spatial filtering and interference cancellation.

Although the theory of direction finding and adaptive beamforming has been well developed over the last three decades (Monzingo and Miller, 1980; Johnson and Dugeon, 1993; Krim and Viberg, 1996; Van Trees, 2002), there is currently a renewed interest in this field mainly driven by emerging applications of array processing to wireless communications involving space division multiple access (SDMA) and MIMO technologies.

In this chapter, we provide an overview of the fundamentals and recent advances in the areas of DOA estimation and adaptive beamforming, and establish some interesting links between these areas and MIMO communications. On the one hand, it is shown that the signal models used in array processing and space-time block coded MIMO systems are rather similar. On the other hand, the MIMO model is shown to enjoy interesting properties that do not emerge in array processing. Using the established links between these two models, several important applications of array processing to single- and multiuser MIMO receivers are discussed.

The remainder of this chapter is organized as follows. Sections 12.2 and 12.3 are devoted to DOA estimation and adaptive beamforming, respectively. In Section 12.4, the aforementioned links between array processing and MIMO communications are discussed with application to MIMO receivers. Conclusions are given in Section 12.5.

12.2 DOA estimation

The geometry of one-dimensional DOA estimation problem is shown in Fig. 12.1. An antenna array of M sensors is assumed to receive the signals impinging from P far-field point plane-wavefront sources with the DOAs $\theta_1, \ldots, \theta_P$. In the narrowband source case, the $M \times 1$ array data snapshot vector can be expressed using the following familiar model (Schmidt, 1979; Johnson and Dugeon, 1993; Van Trees, 2002)

$$\mathbf{y}(t) = \mathbf{A}(\boldsymbol{\theta})\mathbf{s}(t) + \mathbf{n}(t), \tag{12.1}$$

where $\mathbf{A}(\boldsymbol{\theta}) \triangleq [\mathbf{a}(\theta_1), \ldots, \mathbf{a}(\theta_P)]$ is the $M \times P$ source direction matrix, $\mathbf{a}(\theta)$ is the $M \times 1$ array response (steering) vector, $\mathbf{s}(t)$ is the $P \times 1$ vector of source waveforms, $\mathbf{n}(t)$ is the $M \times 1$ vector of sensor noise, $\boldsymbol{\theta} = [\theta_1, \ldots, \theta_P]^T$ is the $P \times 1$ vector of source DOAs, t is the snapshot time index, and $(\cdot)^T$ denotes the transpose.

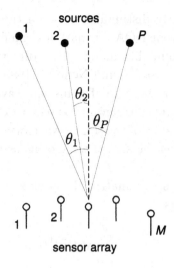

Fig. 12.1. DOA estimation problem.

The steering vector $\mathbf{a}(\theta)$ describes the nominal array response to the source impinging on the array from the direction θ. The form of this vector is entirely determined by the specific array configuration and propagation environment. For example, in the case of free space and a uniform linear array (ULA) with the interelement spacing d, the vector $\mathbf{a}(\theta)$ is given by

$$\mathbf{a}(\theta) = \left[1, e^{j(2\pi/\lambda)d\sin(\theta)}, \ldots, e^{j(2\pi/\lambda)(M-1)d\sin(\theta)}\right]^T, \tag{12.2}$$

where λ is the wavelength and $j = \sqrt{-1}$.

There are two types of snapshot models used in array processing (Stoica and Nehorai, 1990). The deterministic (or conditional) model assumes that the source waveform vectors are nonrandom, whereas the sensor noise is a random zero-mean spatially and temporally white Gaussian process with the correlation matrix $\mathrm{E}\{\mathbf{n}(t)\mathbf{n}^H(t)\} = \sigma^2\mathbf{I}$ where σ^2 is the noise variance, $\mathbf{I}$ is the identity matrix, and $(\cdot)^H$ denotes the Hermitian transpose. The stochastic (or unconditional) model assumes that both the noise and source waveforms are zero-mean Gaussian. Under these two models, $\mathbf{y}(t) \sim \mathcal{N}(\mathbf{A}(\theta)\mathbf{s}(t), \sigma^2\mathbf{I})$ and $\mathbf{y}(t) \sim \mathcal{N}(\mathbf{0}, \mathbf{R})$, respectively, where $\mathcal{N}(\cdot, \cdot)$ denotes the complex Gaussian distribution,

$$\mathbf{R} \triangleq \mathrm{E}\{\mathbf{y}(t)\mathbf{y}^H(t)\} = \mathbf{A}\mathbf{S}\mathbf{A}^H + \sigma^2\mathbf{I} \tag{12.3}$$

is the $M \times M$ array correlation matrix, $\mathbf{S} \triangleq \mathrm{E}\{\mathbf{s}(t)\mathbf{s}^H(t)\}$ is the $P \times P$ correlation matrix of the source waveforms, and $\mathbf{0}$ is the vector of zeros.

It is important to clearly distinguish between these two models because they result into two different DOA estimation Cramér–Rao bounds (CRBs), and two different maximum likelihood (ML) direction finding techniques (Böhme, 1985; Wax, 1985; Stoica and Nehorai, 1989, 1990). The stochastic model is more suitable for the case of Gaussian waveforms and large number of snapshots, but becomes irrelevant for scenarios with unknown (potentially non-Gaussian) statistics of the source waveforms and small sample size, where it is natural to treat the source waveforms as unknown deterministic sequences.

In practical applications, the matrix $\mathbf{R}$ is unknown but can be estimated from the snapshot data as

$$\hat{\mathbf{R}} = \frac{1}{J} \sum_{t=1}^{J} \mathbf{y}(t)\mathbf{y}^H(t), \tag{12.4}$$

where J is the number of snapshots available.

The simplest nonparametric DOA estimation method is usually referred to as *conventional* (or *Bartlett*) *beamformer* and is based on array beam scanning and computing the output power for each beam scan angle (Krim and Viberg, 1996; Van Trees, 2002). The output power for the angle θ is given by $E\{|\mathbf{a}^H(\theta)\mathbf{y}(t)|^2\} = \mathbf{a}^H(\theta)\mathbf{R}\mathbf{a}(\theta)$. In the finite sample case, the conventional beamformer can be expressed as

$$f_{\mathrm{CB}}(\theta) = \frac{1}{J} \sum_{t=1}^{J} |\mathbf{a}^H(\theta)\mathbf{y}(t)|^2 = \mathbf{a}^H(\theta)\hat{\mathbf{R}}\mathbf{a}(\theta). \tag{12.5}$$

The main drawback of the conventional beamformer is its low resolution, whereas the advantages are implementational simplicity and robustness.

To overcome the low-resolution limitation of conventional beamformer, Capon (1969) proposed to estimate the spatial spectrum of multiple sources by means of a spatial filter that maintains a distortionless response to the signal impinging from the direction θ while minimizing the total output array power. In the finite sample case, the resulting nonparametric Capon estimator can be written as $f_{\mathrm{CAPON}}(\theta) = \mathbf{w}^H(\theta)\hat{\mathbf{R}}\mathbf{w}(\theta)$ where the weight vector $\mathbf{w} = \mathbf{w}(\theta)$ of the spatial filter is obtained by solving the following problem:

$$\min_{\mathbf{w}} \mathbf{w}^H \hat{\mathbf{R}} \mathbf{w} \quad \text{subject to} \quad \mathbf{w}^H \mathbf{a}(\theta) = 1. \tag{12.6}$$

The solution to (12.6) is given by

$$\mathbf{w}(\theta) = \frac{1}{\mathbf{a}^H(\theta)\hat{\mathbf{R}}^{-1}\mathbf{a}(\theta)} \hat{\mathbf{R}}^{-1}\mathbf{a}(\theta) \tag{12.7}$$

and, therefore, the Capon spatial spectrum estimate can be expressed in the following familiar form (Capon, 1969; Krim and Viberg, 1996; Van Trees, 2002)

$$f_{\text{CAPON}}(\theta) = \frac{1}{\mathbf{a}^H(\theta)\hat{\mathbf{R}}^{-1}\mathbf{a}(\theta)}. \tag{12.8}$$

The Capon estimator is known to achieve better resolution than conventional beamformer and, therefore, it belongs to the class of high-resolution[†] techniques (Johnson and Dugeon, 1993; Krim and Viberg, 1996; Van Trees, 2002). At the same time, its resolution is essentially limited in the sense that it remains finite even if the number of snapshots is increased without limit. As a result, the performance of the Capon DOA estimator in scenarios with multiple closely spaced sources is usually far away from the corresponding deterministic and stochastic CRBs (Krim and Viberg, 1996). The latter fact prompted the development of improved versions of the Capon estimator (Borgiotti and Kaplan, 1979; Lagunas *et al.*, 1986), as well as parametric (model-based) DOA estimation methods (Pisarenko, 1973; Schmidt, 1979; Reddi, 1979; Wax and Kailath, 1983; Böhme, 1985; Wax, 1985; Paulraj *et al.*, 1986; Bresler and Macovski, 1986) that enjoy an improved DOA estimation accuracy as compared with the Capon technique.

The parametric direction finding techniques are entirely based on the signal parameterization according to model (12.1). Perhaps the most well known and statistically motivated approach among these techniques is the maximum likelihood method, whose essence is to maximize the *likelihood function*[‡] with respect to the unknown parameters. As the form of the likelihood function depends on whether the stochastic or deterministic model is used, there are two different ML DOA estimation techniques called the deterministic and stochastic ML estimators (Stoica and Nehorai, 1990; Krim and Viberg, 1996). Among these two techniques, the deterministic ML estimator has the most intuitively clear form (Böhme, 1985; Wax, 1985)

$$\hat{\theta}_{\text{ML}} = \arg\max_{\theta} \text{trace}\{\mathbf{P_A}(\theta)\hat{\mathbf{R}}\}, \tag{12.9}$$

where $\mathbf{P_A}(\theta) \triangleq \mathbf{A}(\theta)(\mathbf{A}^H(\theta)\mathbf{A}(\theta))^{-1}\mathbf{A}^H(\theta)$ is the orthogonal projection operator onto the subspace spanned by the columns of $\mathbf{A}(\theta)$, and trace$\{\cdot\}$ denotes the trace of a matrix. From (12.9), it follows that the deterministic ML technique obtains the source DOAs by finding, in essence, the best fit between the hypothetical source steering vectors and the array data.

[†]A DOA estimation technique is called *high-resolution* if it can resolve two or more sources within the array beamwidth.

[‡]The likelihood function can be defined as the probability density function with the abstract random variables replaced by the observed data.

In the single source case, the ML estimator in (12.9) has an appealing interpretation. In this case, the projection matrix $\mathbf{P_A}(\boldsymbol{\theta})$ simplifies to $\left(\mathbf{a}(\theta)\mathbf{a}^H(\theta)\right)/\left(\mathbf{a}^H(\theta)\mathbf{a}(\theta)\right)$ with a denominator that is independent of θ and, therefore, can be omitted. Hence, (12.9) can be rewritten as

$$\hat{\theta}_{\mathrm{ML}} = \arg\max_{\theta} \operatorname{trace}\{\mathbf{a}(\theta)\mathbf{a}^H(\theta)\hat{\mathbf{R}}\} = \arg\max_{\theta} f_{\mathrm{CB}}(\theta). \qquad (12.10)$$

According to (12.10), the deterministic ML DOA estimate in the single source case coincides with the location of the global maximum of the conventional beamformer function.

Both the deterministic and stochastic ML estimators have excellent threshold and asymptotic DOA estimation performances that are comparable to the stochastic CRB. The main shortcoming of the ML techniques is that they are based on a nonlinear optimization in a high-dimensional parameter space, and, therefore, their computational cost may be prohibitively high. Several attempts to reduce the computational costs of the global search based ML techniques have been made using iterative quadratic optimization (Bresler and Macovski, 1986), expectation-maximization (EM) algorithms (Feder and Weinstein, 1988), alternating projection techniques (Ziskind and Wax, 1988), simulated annealing and genetic algorithms (Sharman, 1988; Sharman and McClurkin, 1989), and data-supported optimization (Stoica and Gershman, 1999; Gershman and Stoica, 1999). However, all these approaches still suffer from a rather high computational cost and may not guarantee the convergence to the global extremum.

A somewhat different approach to reduced-cost ML DOA estimation is to obtain a certain (e.g., large-sample) approximation of the ML estimator. Techniques of such type have been proposed in Stoica and Sharman (1990) and Viberg and Ottersten (1991). In the ULA case, these techniques enjoy computationally efficient implementations based on eigendecomposition and polynomial rooting (Krim and Viberg, 1996; Stoica and Sharman, 1990) that achieve the same asymptotic performance as their original ML counterparts. However, their threshold performance is usually substantially worse than that of the original ML techniques (Gershman and Stoica, 1999).

A fundamental parametric approach to DOA estimation developed in the late 70s to early 80s is the multiple signal classification (MUSIC) method (Schmidt, 1979; Bienvenu and Kopp, 1980). Along with the earlier Pisarenko method (Pisarenko, 1973), it started the era of subspace techniques in array processing.

MUSIC makes an explicit use of the specific structure and properties of the array correlation matrix (12.3). The eigendecomposition of this matrix can be written as (Schmidt, 1979; Van Trees, 2002)

$$\mathbf{R} = \mathbf{U}_s \mathbf{\Lambda}_s \mathbf{U}_s^H + \mathbf{U}_n \mathbf{\Lambda}_n \mathbf{U}_n^H \qquad (12.11)$$

where the $M \times P$ matrix $\mathbf{U}_s$ contains the P signal subspace eigenvectors of $\mathbf{R}$, and the $P \times P$ diagonal matrix $\mathbf{\Lambda}_s$ contains the corresponding eigenvalues. Similarly, the $M \times (M - P)$ matrix $\mathbf{U}_n$ contains the $M - P$ noise-subspace eigenvectors of $\mathbf{R}$, while the $(M - P) \times (M - P)$ diagonal matrix $\mathbf{\Lambda}_n$ is built from the corresponding eigenvalues. Then, it can be easily shown that the noise subspace and the column space of $\mathbf{A}$ are orthogonal, that is, $\mathbf{U}_n^H \mathbf{A} = 0$ (Schmidt, 1979; Bienvenu and Kopp, 1980; Van Trees, 2002). This property enables to find the signal DOAs from the equation

$$\mathbf{a}^H(\theta) \mathbf{U}_n \mathbf{U}_n^H \mathbf{a}(\theta) = 0 \qquad (12.12)$$

or, equivalently, from the locations of the P highest peaks of the "MUSIC spectrum" function

$$f_{\text{MUSIC}}(\theta) = \frac{1}{\mathbf{a}^H(\theta) \mathbf{U}_n \mathbf{U}_n^H \mathbf{a}(\theta)}. \qquad (12.13)$$

In the finite sample case, one can use the eigendecomposition of the sample correlation matrix (Van Trees, 2002)

$$\hat{\mathbf{R}} = \hat{\mathbf{U}}_s \hat{\mathbf{\Lambda}}_s \hat{\mathbf{U}}_s^H + \hat{\mathbf{U}}_n \hat{\mathbf{\Lambda}}_n \hat{\mathbf{U}}_n^H \qquad (12.14)$$

and the MUSIC estimator can be obtained from (12.13) by replacing $\mathbf{U}_n$ with $\hat{\mathbf{U}}_n$ as

$$f_{\text{MUSIC}}(\theta) = \frac{1}{\mathbf{a}^H(\theta) \hat{\mathbf{U}}_n \hat{\mathbf{U}}_n^H \mathbf{a}(\theta)}. \qquad (12.15)$$

The MUSIC DOA estimator (12.15) is known to provide an excellent trade-off between the threshold/asymptotic performances and computational cost (Kaveh and Barabell, 1986; Krim and Viberg, 1996). In particular, in the uncorrelated source case MUSIC can be interpreted as a large-sample realization of the deterministic ML estimator (Stoica and Nehorai, 1989).

As a result, MUSIC has been extensively used in the literature as a benchmark method prompting researchers to seek for subspace techniques with further performance improvements and/or reduced computational complexity (Johnson and DeGraaf, 1982; Barabell, 1983; Kumaresan and Tufts, 1983; Paulraj *et al.*, 1986; Tufts and Melissinos, 1986; Brandwood, 1987; Oh and Un, 1989; Kaveh and Bassias, 1990; Buckley and Xu, 1990; Zoltowski *et al.*, 1993; Ermolaev and Gershman, 1994; Haardt and Nossek, 1995; Stoica *et al.*, 1995; Gershman, 1998; Pesavento *et al.*, 2000).

In the particular ULA case, a computationally efficient modification of MUSIC was proposed in Barabell (1983), where it is suggested to reformulate the MUSIC "null-spectrum" $g_{\text{MUSIC}}(\theta) \triangleq 1/f_{\text{MUSIC}}(\theta)$ in terms of $z = e^{j(2\pi/\lambda)d\sin(\theta)}$. Then, the array response vector (12.2) can be written as $\mathbf{a}(z) = [1, z, \ldots, z^{M-1}]^T$ and, hence, the MUSIC null-spectrum can be rewritten as the following polynomial (Barabell, 1983)

$$g_{\text{MUSIC}}(z) = \mathbf{a}^T(1/z)\hat{\mathbf{U}}_{\text{n}}\hat{\mathbf{U}}_{\text{n}}^H\mathbf{a}(z). \tag{12.16}$$

The signal DOA estimates can be found from the roots of (12.16) in a search-free way. Advantages of such root-MUSIC technique over the conventional MUSIC algorithm is a substantially reduced computational cost and improved threshold performance (Rao and Hari, 1989). Further improvements of the root-MUSIC technique and its extensions to more general classes of array geometry have been proposed in Zoltowski *et al.* (1993), Friedlander (1993), Pesavento *et al.* (2000), and Swindlehurst *et al.* (2001).

An extension of the MUSIC algorithm referred to as the rank reduction estimator (RARE) has been recently developed in Pesavento *et al.* (2002) and See and Gershman (2004) for the case when the array consists of L subarrays with some unknown intersubarray parameters. These parameters may include unknown (or imprecisely known) displacements between subarrays, subarray time synchronization errors, unknown perturbations of the propagation channel between subarrays, or any combination of them.

In this case, the actual array response vector can be modeled as (See and Gershman, 2004)

$$\mathbf{a}(\theta, \boldsymbol{\alpha}) = \mathbf{V}(\theta)\mathbf{h}(\theta, \boldsymbol{\alpha}), \tag{12.17}$$

where $\boldsymbol{\alpha}$ is the vector of unknown intersubarray parameters, $\mathbf{V}(\theta)$ is a sparse matrix involving all the subarray steering vectors [see See and Gershman (2004) for more details], and $\mathbf{h}(\theta, \boldsymbol{\alpha})$ is the vector that captures all unknown intersubarray perturbations of the nominal array response vector.

Inserting (12.17) into (12.12) yields $\mathbf{h}^H(\theta, \boldsymbol{\alpha})\mathbf{V}^H(\theta)\mathbf{U}_{\text{n}}\mathbf{U}_{\text{n}}^H\mathbf{V}(\theta)\mathbf{h}(\theta, \boldsymbol{\alpha}) = 0$ and, therefore, equation (12.12) can only be satisfied when the matrix $\mathbf{C}(\theta) \triangleq \mathbf{V}^H(\theta)\mathbf{U}_{\text{n}}\mathbf{U}_{\text{n}}^H\mathbf{V}(\theta)$ drops rank. This fact was used in Pesavento *et al.* (2002) and See and Gershman (2004) to estimate the source DOAs without any knowledge of the intersubarray parameter vector $\boldsymbol{\alpha}$. In the finite sample case, the DOAs can be found from the P highest peaks of the following two alternative RARE estimators:

$$f_{\text{RARE}1}(\theta) = (\det\{\hat{\mathbf{C}}(\theta)\})^{-1}, \qquad f_{\text{RARE}2}(\theta) = (\mathcal{L}\{\hat{\mathbf{C}}(\theta)\})^{-1}, \tag{12.18}$$

where $\hat{\mathbf{C}}(\theta) = \mathbf{V}^H(\theta)\hat{\mathbf{U}}_{\text{n}}\hat{\mathbf{U}}_{\text{n}}^H\mathbf{V}(\theta)$ is the sample estimate of $\mathbf{C}(\theta)$, $\det\{\cdot\}$

denotes the determinant, and $\mathcal{L}\{\cdot\}$ is the operator that yields the smallest eigenvalue.

The polynomial rooting-based modification of RARE was derived in Pesavento *et al.* (2002) assuming that the array consists of identically oriented linear subarrays whose interelement spacings are integer multiples of the known shortest baseline. Interestingly, in the case of a single subarray ($L = 1$) the spectral and root-RARE estimators reduce to the standard spectral and root-MUSIC estimators, respectively.

Another popular and computationally attractive subspace DOA estimation method is the ESPRIT (estimation of signal parameters via rotational invariance techniques) algorithm proposed in Paulraj *et al.* (1986). This approach is applicable to any array with *translational invariance*. An interesting interpretation of arrays with translational invariance is that they consist of multiple identical and identically oriented two-element subarrays with some common interelement spacing (Paulraj *et al.*, 1986; Roy and Kailath, 1989). Note that the knowledge of displacements between these two-element subarrays is not required in ESPRIT and, therefore, this technique is applicable to such arrays with unknown intersubarray displacements. As ESPRIT is usually applied to fully calibrated arrays (Van Trees, 2002), this is a noteworthy fact that also shows a certain relationship between ESPRIT and RARE.

There have been several attempts to find improved versions of the ESPRIT algorithm (Roy and Kailath, 1989; Zoltowski and Stavrinides, 1989; Haardt and Nossek, 1995), and to extend ESPRIT to more general classes of array geometries (Ramos *et al.*, 1999; Gao and Gershman, 2004). An interesting link between ESPRIT-type techniques and parallel factor (PARAFAC) analysis has been discovered in Sidiropoulos *et al.* (2000).

An important problem in DOA estimation is how to extend high-resolution methods to the case when the sources have a strong mutual correlation caused by multipath propagation effects. Several elegant solutions to this problem exist, the most popular being the so-called *spatial smoothing* technique (Evans *et al.*, 1982). Although the latter technique was originally developed for the ULA case, its extensions to other types of array geometry are also known.

Further important research trends in the field include source number detection; wideband and multidimensional DOA estimation; analysis of direction estimation CRBs; and direction finding in nonideal environments such as those with array imperfections, manifold mismatches, non-point sources, and unknown colored or spatially inhomogeneous noise fields. All interested readers are referred to Van Trees (2002), where a detailed discussion of these topics can be found.

12.3 Adaptive beamforming

The output signal of an adaptive beamformer can be written as

$$x(t) = \mathbf{w}^H \mathbf{y}(t),$$

where $\mathbf{w}$ is the $M \times 1$ complex vector of beamformer weights and, as before, $\mathbf{y}(t)$ is the $M \times 1$ complex snapshot vector of array observations. In beamforming applications, this vector can be modeled as (Monzingo and Miller, 1980; Van Trees, 2002)

$$\mathbf{y}(t) = \mathbf{y}_s(t) + \mathbf{y}_i(t) + \mathbf{n}(t), \tag{12.19}$$

where $\mathbf{y}_s(t)$, $\mathbf{y}_i(t)$, and $\mathbf{n}(t)$ are the statistically independent components of the desired signal, interference, and sensor noise, respectively. If the desired signal is a point (rank-one) source with a time-invariant wavefront, we obtain that $\mathbf{y}_s(t) = s(t)\mathbf{a}_s$ where $s(t)$ is the complex signal waveform and $\mathbf{a}_s$ is its $M \times 1$ steering vector. Note that in contrast to Section 12.2, we do not parameterize the signal steering vector as a function of DOA because in the general case, the signal source may have a nonplane wavefront.

The optimal weight vector can be obtained by means of maximizing the signal-to-interference-plus-noise ratio (SINR) (Monzingo and Miller, 1980; Van Trees, 2002)

$$\text{SINR} = \frac{\mathbf{w}^H \mathbf{R}_s \mathbf{w}}{\mathbf{w}^H \mathbf{R}_{i+n} \mathbf{w}}, \tag{12.20}$$

where $\mathbf{R}_s \triangleq \mathrm{E}\left\{\mathbf{y}_s(t)\mathbf{y}_s^H(t)\right\}$ and $\mathbf{R}_{i+n} \triangleq \mathrm{E}\left\{(\mathbf{y}_i(t) + \mathbf{n}(t))(\mathbf{y}_i(t) + \mathbf{n}(t))^H\right\}$ are the $M \times M$ signal and interference-plus-noise correlation matrices, respectively.

Assuming the rank-one signal source case, we have $\mathbf{R}_s = \sigma_s^2 \mathbf{a}_s \mathbf{a}_s^H$ where σ_s^2 is the power of the desired signal. In this case, (12.20) reduces to (Monzingo and Miller, 1980)

$$\text{SINR} = \frac{\sigma_s^2 |\mathbf{w}^H \mathbf{a}_s|^2}{\mathbf{w}^H \mathbf{R}_{i+n} \mathbf{w}}. \tag{12.21}$$

To find the optimal solution for the weight vector, the SINR in (12.21) should be maximized. This is equivalent to maintaining a distortionless response to the desired signal while minimizing the output interference-plus-noise power as

$$\min_{\mathbf{w}} \mathbf{w}^H \mathbf{R}_{i+n} \mathbf{w} \qquad \text{subject to} \qquad \mathbf{w}^H \mathbf{a}_s = 1. \tag{12.22}$$

Note that (12.22) corresponds to the Capon spatial filtering problem (Capon, 1969) in (12.6). In application to adaptive beamforming, the Capon approach is commonly referred to as the minimum-variance distortionless response (MVDR) beamformer (Monzingo and Miller, 1980; Van Trees, 2002).

The solution to (12.22) can be expressed in the following familiar form (Monzingo and Miller, 1980; Van Trees, 2002):

$$\mathbf{w}_{\mathrm{opt}} = \beta \mathbf{R}_{\mathrm{i+n}}^{-1} \mathbf{a}_{\mathrm{s}}, \tag{12.23}$$

where the scalar $\beta = (\mathbf{a}_{\mathrm{s}}^H \mathbf{R}_{\mathrm{i+n}}^{-1} \mathbf{a}_{\mathrm{s}})^{-1}$ normalizes the resulting weight vector to satisfy the distortionless response constraint in (12.22). However, such a normalization is immaterial from the SINR viewpoint because the multiplication of the weight vector by any nonzero constant does not affect (12.21). Therefore, the factor β will be omitted in the sequel.

In practice, the matrix $\mathbf{R}_{\mathrm{i+n}}$ is unavailable but can be estimated from the received data. Therefore, the sample correlation matrix (12.4) is usually used in (12.22) instead of $\mathbf{R}_{\mathrm{i+n}}$. This yields the following sample matrix inverse (SMI) beamformer (Reed *et al.*, 1974; Monzingo and Miller, 1980)

$$\mathbf{w}_{\mathrm{SMI}} = \hat{\mathbf{R}}^{-1} \mathbf{a}_{\mathrm{s}}. \tag{12.24}$$

The use of the sample correlation matrix $\hat{\mathbf{R}}$ instead of the true interference-plus-noise correlation matrix $\mathbf{R}_{\mathrm{i+n}}$ in (12.24) is known to affect the performance dramatically if the desired signal component is present in the data snapshots that are used to compute $\hat{\mathbf{R}}$ (Feldman and Griffiths, 1994; Gershman, 1999). This performance degradation becomes especially strong if $\mathbf{a}_{\mathrm{s}}$ is known imprecisely because of look direction errors, unknown propagation channel, imperfect array calibration, and so on. This effect is commonly termed as signal *self-nulling*.

A popular approach to improve the robustness of the SMI techniques and to prevent signal self-nulling is the *diagonal loading* (DL) technique (Abramovich, 1981; Cox *et al.*, 1987; Carlson, 1988). Its essence is to *regularize* the solution by adding the quadratic penalty term $\gamma \mathbf{w}^H \mathbf{w}$ to the objective function of the finite sample MVDR problem, where γ is the so-called DL factor. The resulting DL-MVDR problem takes the form

$$\min_{\mathbf{w}} \mathbf{w}^H \hat{\mathbf{R}} \mathbf{w} + \gamma \mathbf{w}^H \mathbf{w} \qquad \text{subject to} \qquad \mathbf{w}^H \mathbf{a}_{\mathrm{s}} = 1. \tag{12.25}$$

The solution to (12.25) is commonly referred to as the loaded SMI (LSMI) beamformer:

$$\mathbf{w}_{\mathrm{LSMI}} = (\hat{\mathbf{R}} + \gamma \mathbf{I})^{-1} \mathbf{a}_{\mathrm{s}}. \tag{12.26}$$

It can be seen that (12.26) differs from (12.24) only by the weighted identity matrix added to $\hat{\mathbf{R}}$ (that is where the name "diagonal loading" comes from).

This operation can be interpreted as injecting artificial white noise into the main diagonal of $\hat{\mathbf{R}}$ (Abramovich, 1981; Carlson, 1988).

The main shortcoming of the standard DL method is that there is no easy and reliable way of choosing the parameter γ. Furthermore, the fixed choice of γ can be only suboptimal because the optimal value of γ is well known to be scenario-dependent.

To avoid the aforementioned drawbacks of the standard diagonal loading technique, a theoretically rigorous *robust MVDR beamforming* approach was recently proposed in Vorobyov *et al.* (2003). It chooses the DL factor in an adaptive way, by optimally matching it to the known amount of uncertainty in the signal steering vector.

The authors of Vorobyov *et al.* (2003) consider the steering vector uncertainty $\boldsymbol{\delta} \triangleq \tilde{\mathbf{a}}_s - \mathbf{a}_s$ where $\tilde{\mathbf{a}}_s$ and $\mathbf{a}_s$ are the actual and presumed signal steering vectors, respectively, and assume that the norm of $\boldsymbol{\delta}$ is bounded from above by the known constant ε (this corresponds to the case of *spherical* uncertainty). The key idea of Vorobyov *et al.* (2003) is to add robustness to the standard MVDR beamforming problem by means of imposing the worst-case distortionless response constraint that should be satisfied for all mismatched signal steering vectors in the spherical uncertainty set. With such a constraint, the robust MVDR beamformer can be obtained by solving the following optimization problem

$$\min_{\mathbf{w}} \mathbf{w}^H \hat{\mathbf{R}} \mathbf{w} \quad \text{subject to} \quad |\mathbf{w}^H(\mathbf{a}_s + \boldsymbol{\delta})| \geq 1 \quad \text{for all} \quad \|\boldsymbol{\delta}\| \leq \varepsilon, \quad (12.27)$$

where $\| \cdot \|$ denotes the Euclidean norm of a vector or the Frobenius norm of a matrix. It was shown in Vorobyov *et al.* (2003) that (12.27) can be converted to

$$\min_{\mathbf{w}} \mathbf{w}^H \hat{\mathbf{R}} \mathbf{w} \quad \text{subject to} \quad \mathbf{w}^H \mathbf{a}_s \geq \varepsilon \|\mathbf{w}\| + 1 \quad (12.28)$$

and that the constraint in (12.28) is satisfied with equality. The problem (12.28) belongs to the class of convex second-order cone programming (SOCP) problems that can be easily solved with complexity $O(M^3)$ using standard and highly efficient interior point method software. It is also proven in Vorobyov *et al.* (2003) that (12.28) is equivalent to the MVDR problem with *adaptive* diagonal loading where the DL factor is optimally matched to the known amount ε of uncertainty in the steering vector.

Several further extensions of the robust MVDR beamformer of Vorobyov *et al.* (2003) have been recently developed by different authors. In Lorenz and Boyd (2003), this beamformer has been extended to the case of ellipsoidal (anisotropic) uncertainty. In Li *et al.* (2003), a covariance fitting interpretation of the robust MVDR problems of Vorobyov *et al.* (2003) and Lorenz and Boyd (2003) has been presented. Alternative Newton-type algorithms to solve the problem (12.28) and its generalized versions have been proposed in Lorenz and Boyd (2003), Li *et al.* (2003), and Zarifi *et al.* (2005), all with the complexity $O(M^3)$. Extensions of the worst-case approach of Vorobyov *et al.* (2003) to scenarios with nonstationary interferers and to the general-rank signal case have been considered in Vorobyov *et al.* (2004) and Shahbazpanahi *et al.* (2003), respectively.

Current research focus in the area of robust adaptive beamforming includes extending the robust MVDR beamformers of Vorobyov *et al.* (2003) and Lorenz and Boyd (2003) to the wideband case and developing computationally efficient on-line algorithms to update their weight vectors.

12.4 Links between array processing and MIMO

In this section, an array processing perspective of MIMO communications is discussed. To establish a relationship between these two areas, we will borrow the conventional MIMO model (that is addressed in much more detail in other chapters of this book) and apply this model to the case when orthogonal space-time codes (OSTBCs) (Alamouti, 1998; Tarokh *et al.*, 1999) are used as the underlying space-time encoding scheme.[†]

More specifically, let us consider a point-to-point flat block fading MIMO system with N transmit and M receive antennas. Assuming that the channel is used at the times $1, 2, \ldots, T$ with the block length T, the input-output relationship of such system can be written as (Tarokh *et al.*, 1999)

$$\mathbf{Y} = \mathbf{X}\mathbf{H} + \mathbf{N}, \tag{12.29}$$

where $\mathbf{H}$ is the $N \times M$ complex channel matrix, $\mathbf{X}$ is the $T \times N$ complex matrix of the transmitted signals, $\mathbf{Y}$ is the $T \times M$ complex matrix of the received signals, and $\mathbf{N}$ is the $T \times M$ matrix of noise.

Let us denote complex information symbols prior to space-time encoding as $\xi_1, \ldots, \xi_K$ and define the $K \times 1$ vector $\boldsymbol{\xi} \triangleq [\xi_1, \ldots, \xi_K]^T$. The $T \times N$ matrix $\mathbf{X} = \mathbf{X}(\boldsymbol{\xi})$ is called an OSTBC if (Tarokh *et al.*, 1999)

- all elements of $\mathbf{X}(\boldsymbol{\xi})$ are linear functions of the K complex variables $\xi_1, \ldots, \xi_K$ and their complex conjugates;
- for any $\boldsymbol{\xi}$, it satisfies $\mathbf{X}^H(\boldsymbol{\xi})\mathbf{X}(\boldsymbol{\xi}) = \|\boldsymbol{\xi}\|^2 \mathbf{I}$.

[†]Note that OSTBCs represent quite an attractive class of space-time coding techniques because they enjoy low decoding complexity and full diversity.

The matrix $\mathbf{X}(\boldsymbol{\xi})$ can be written as (Hassibi and Hochwald, 2002; Shahbaz-panahi *et al.*, 2004a)

$$\mathbf{X}(\boldsymbol{\xi}) = \sum_{k=1}^{K} \left(\mathbf{C}_k \text{Re}\{\xi_k\} + \mathbf{D}_k \text{Im}\{\xi_k\} \right),$$

where $\mathbf{C}_k \triangleq \mathbf{X}(\mathbf{e}_k)$, $\mathbf{D}_k \triangleq \mathbf{X}(j\mathbf{e}_k)$, and $\mathbf{e}_k$ is the $K \times 1$ vector having one in the kth position and zeros elsewhere. Using this representation, the MIMO model (12.4) can be rewritten as (Shahbazpanahi *et al.*, 2004a; Gharavi-Alkhansari and Gershman, 2005)

$$\underline{\mathbf{Y}} = \mathbb{A}(\mathbf{H})\underline{\boldsymbol{\xi}} + \underline{\mathbf{N}}, \tag{12.30}$$

where the "underline" operator for any matrix $\mathbf{P}$ is defined as

$$\underline{\mathbf{P}} \triangleq \begin{bmatrix} \text{vec}\{\text{Re}(\mathbf{P})\} \\ \text{vec}\{\text{Im}(\mathbf{P})\} \end{bmatrix} \tag{12.31}$$

and vec$\{\cdot\}$ is the vectorization operator stacking all columns of a matrix on top of each other. Here, the $2MT \times 2K$ real matrix $\mathbb{A}(\mathbf{H})$ is given by (Shahbazpanahi *et al.*, 2004a; Gharavi-Alkhansari and Gershman, 2005)

$$\mathbb{A}(\mathbf{H}) = \left[\underline{\mathbf{C}_1 \mathbf{H}}, \dots, \underline{\mathbf{C}_K \mathbf{H}}, \underline{\mathbf{D}_1 \mathbf{H}}, \dots, \underline{\mathbf{D}_K \mathbf{H}} \right]. \tag{12.32}$$

It can be seen from (12.32) that the matrix $\mathbb{A}(\mathbf{H})$ in (12.30) captures both the effects of the OSTBC and the channel, whereas the vector $\underline{\boldsymbol{\xi}}$ in (12.30) depends on the information symbols only. Clearly, there is a strong similarity between the vectorized MIMO model in (12.30) and the array processing model in (12.1), in the sense that these models are identical up to the replacements $\mathbb{A}(\mathbf{H}) \to \mathbf{A}(\boldsymbol{\theta})$, $\underline{\boldsymbol{\xi}} \to \mathbf{s}$, and $\underline{\mathbf{N}} \to \mathbf{n}$. However, the model in (12.30) is real-valued, while that in (12.1) is complex-valued. Moreover, an important property of the matrix $\mathbb{A}(\mathbf{H})$ is that its columns have the same norms and are orthogonal to each other, that is,

$$\mathbb{A}^T(\mathbf{H})\mathbb{A}(\mathbf{H}) = \|\mathbf{H}\|^2 \mathbf{I}. \tag{12.33}$$

Equation (12.33) means that the columns of $\mathbb{A}(\mathbf{H})$ are the eigenvectors of the $2MT \times 2MT$ real-valued correlation matrix $\mathbb{R} \triangleq \text{E}\{\underline{\mathbf{Y}}\,\underline{\mathbf{Y}}^T\} = \mathbb{A}\mathbb{S}\mathbb{A}^H + \sigma^2\mathbf{I}$ where $\mathbb{S} \triangleq \text{E}\{\underline{\boldsymbol{\xi}}\,\underline{\boldsymbol{\xi}}^T\}$. It is worth noting that the property equivalent to (12.33) is not satisfied in array processing, where the columns of $\mathbf{A}(\boldsymbol{\theta})$ are orthogonal with probability zero.

The established similarity between the models (12.30) and (12.1) opens an avenue for extending array processing techniques to MIMO communications. For example, the similarity between these two models has been used in Shahbazpanahi *et al.* (2004a) to develop MVDR-type receivers for the multiple-access MIMO case. These receivers are able to simultaneously decode the symbols of the user-of-interest while rejecting multiple access interference (MAI) coming from the other users. In Shahbazpanahi *et al.* (2004b), the aforementioned model similarity was used along with the orthogonality property of (12.33) to develop *blind* space-time decoding algorithms for point-to-point MIMO systems that do not require any knowledge of the channel matrix **H** at the receiver. The work of Shahbazpanahi *et al.* (2004b) has been extended to the multiple access MIMO case in Shahbazpanahi *et al.* (2005), where *semiblind* algorithms have been developed to estimate the channel matrices of multiple users. The latter algorithms use the concepts of the Capon and MUSIC array processing techniques.

The study to discover similarities between array processing and MIMO communications is currently in the very beginning and much work still remains to be done. It might become a useful research trend that can enable and enforce efficient applications of the existing array processing techniques to the MIMO area.

12.5 Conclusions

We have presented an overview of the fundamentals and recent developments in array signal processing with the main emphasis on DOA estimation and adaptive beamforming. Array processing links to MIMO communications have been briefly discussed.

References

Abramovich, Y. I. (1981). Controlled method for adaptive optimization of filters using the criterion of maximum SNR. *Radio Eng. Electron. Phys.*, **26**, 87–95.

Alamouti, S. M. (1998). A simple transmit diversity technique for wireless communications. *IEEE J. Sel. Areas Commun.*, **16** (8), 1451–1458.

Amin, M. G., Zhao, L., and Lindsey, A. R. (2004). Subspace array processing for the suppression of FM jamming in GPS receivers. *IEEE Trans. Aerosp. Electron. Syst.*, **40** (1), 80–92.

Barabell, A. J. (1983). Improving the resolution performance of eigenstructure-based direction-finding algorithms. In *Proc. IEEE Int. Conf. Acoust., Speech, Signal Process.*, 336–339.

Bienvenu, G. and Kopp, L. (1980). Adaptivity to background noise spatial coherence for high resolution passive methods. In *Proc. IEEE Int. Conf. Acoust., Speech, Signal Process.*, 307–310.

Böhme, J. F. (1985). Source parameter estimation by approximate maximum likelihood and nonlinear regression. *IEEE J. Ocean. Eng.*, **10** (3), 206–212.

—— (1995). Statistical array signal processing of measured sonar and seismic data. In *Proc. SPIE Conf. Advanced Signal Process. Algorithms*, vol. 2563, 2–20.

Boonstra, A.-J. and van der Veen, A.-J. (2003). Gain calibration methods for radio telescope arrays. *IEEE Trans. Signal Process.*, **51** (1), 25–38.

Borgiotti, G. and Kaplan, L. (1979). Superresolution of uncorrelated interference sources by using adaptive array techniques. *IEEE Trans. Antennas Propagat.*, **27** (6), 482–485.

Brandwood, D. H. (1987). Noise-space projection: MUSIC without eigenvectors. *IEE Proc. H. Microwaves, Antennas, Propagat.*, **134** (3), 303–309.

Brennan, L. E., Mallett, J. D., and Reed, I. S. (1976). Adaptive arrays in airborne MTI radar. *IEEE Trans. Antennas Propagat.*, **24** (9), 607–615.

Bresler, Y. and Macovski, A. (1986). Exact maximum likelihood parameter estimation of superimposed exponential signals in noise. *IEEE Trans. Acoust., Speech, Signal Process.*, **34** (5), 1081–1089.

Buckley, K. M. and Xu, X.-L. (1990). Spatial-spectrum estimation in a location sector. *IEEE Trans. Acoust., Speech, Signal Process.*, **38** (11), 1842–1852.

Capon, J. (1969). High-resolution frequency-wavenumber spectrum analysis. *Proc. IEEE*, **57** (8), 2408–2418.

Capon, J., Greenfield, R. J., and Kolker, R. J. (1967). Multidimensional maximum-likelihood processing for a large aperture seismic array. *Proc. IEEE*, **55** (2), 192–211.

Capon, J., Greenfield, R. J., and Lacoss, R. T. (1969). Long-period signal processing results for the large aperture seismic array. *Geophysics*, **34** (3), 305–329.

Carlson, B. D. (1988). Covariance matrix estimation errors and diagonal loading in adaptive arrays. *IEEE Trans. Aerosp. Electron. Syst.*, **24** (3), 397–401.

Cox, H. (1973). Line array performance when the signal coherence is spatially dependent. *J. Acoust. Soc. Amer.*, **54** (6), 1743–1746.

Cox, H., Zeskind, R. M., and Owen, M. H. (1987). Robust adaptive beamforming. *IEEE Trans. Acoust., Speech, Signal Process.*, **35** (10), 1365–1376.

Ermolaev, V. T. and Gershman, A. B. (1994). Fast algorithm for minimum-norm direction of arrival estimation. *IEEE Trans. Signal Process.*, **42** (9), 2389–2394.

Evans, J. E., Johnson, J. R., and Sun, D. F. (1982). Application of advanced signal processing techniques to angle of arrival estimation in ATC navigation and surveillance systems. Tech. rep., MIT Lincoln Laboratory, Lexington, MA.

Feder, M. and Weinstein, E. (1988). Parameter estimation of superimposed signals using the EM algorithm. *IEEE Trans. Acoust., Speech, Signal Process.*, **36** (4), 477–489.

Feldman, D. D. and Griffiths, L. J. (1994). A projection approach to robust adaptive beamforming. *IEEE Trans. Signal Process.*, **42** (4), 867–876.

Friedlander, B. (1993). The root-MUSIC algorithm for direction finding with interpolated arrays. *Signal Process.*, **30** (1), 15–29.

Gao, F. and Gershman, A. B. (2004). Generalized ESPRIT method for arbitrary array structure. In *Proc. IEEE Sensor Array and Multichannel Signal Process. Workshop*.

Gershman, A. B. (1998). Pseudo-randomly generated estimator banks: A new tool for improving the threshold performance of direction finding. *IEEE Trans. Signal Process.*, **46** (5), 1351–1364.

—— (1999). Robust adaptive beamforming in sensor arrays. *AEÜ – Int. J. Electron. and Commun.*, **53** (6), 305–314.

Gershman, A. B. and Stoica, P. (1999). New MODE-based techniques for direction finding with an improved threshold performance. *Signal Process.*, **76** (3), 221–235.

Gershman, A. B. and Turchin, V. I. (1995). Nonwave field processing using sensor array approach. *Signal Process.*, **44** (2), 197–210.

Gershman, A. B., Turchin, V. I., and Zverev, V. A. (1995). Experimental results of localization of moving underwater signal by adaptive beamforming. *IEEE Trans. Signal Process.*, **43** (10), 2249–2257.

Gharavi-Alkhansari, M. and Gershman, A. B. (2005). Constellation space invariance of orthogonal space-time block codes. *IEEE Trans. Inf. Theory*, **51** (1), 331–334.

Godara, L. C. (1997). Application of antenna arrays to mobile communications. II. Beam-forming and direction-of-arrival considerations. *Proc. IEEE*, **85** (8), 1195–1245.

Haardt, M. and Nossek, J. A. (1995). Unitary ESPRIT: How to obtain increased estimation accuracy with a reduced computational burden. *IEEE Trans. Signal Process.*, **43** (5), 1232–1242.

Hassibi, B. and Hochwald, B. M. (2002). High-rate codes that are linear in space and time. *IEEE Trans. Inf. Theory*, **48** (7), 1804–1824.

Haykin, S. (ed.) (1980). *Array Processing: Applications to Radar* (Dowden, Hutchinson & Ross).

Haykin, S., Litva, J., and Shepard, T. (eds.) (1992). *Radar Array Processing* (Springer-Verlag).

Hochwald, B. and Nehorai, A. (1987). Magnetoencephalography processing with diversely-oriented and multi-component sensors. *IEEE Trans. Biomed. Eng.*, **44** (1), 40–50.

Johnson, D. H. and DeGraaf, S. R. (1982). Improving the resolution of bearing in passive sonar arrays by eigenvalue analysis. *IEEE Trans. Acoust., Speech, Signal Process.*, **30** (4), 638–647.

Johnson, D. H. and Dugeon, D. E. (1993). *Array Signal Processing: Concepts and Techniques* (Pearson Int.).

Kaveh, M. and Barabell, A. (1986). The statistical performance of the MUSIC and the minimum-norm algorithms in resolving plane waves in noise. *IEEE Trans. Acoust., Speech, Signal Process.*, **34** (2), 331–341.

Kaveh, M. and Bassias, A. (1990). Threshold extension based on a new paradigm for MUSIC-type estimation. In *Proc. IEEE Int. Conf. Acoust., Speech, Signal Process.*, 2535–2538.

Krim, H. and Viberg, M. (1996). Two decades of array signal processing research: The parametric approach. *IEEE Signal Process. Mag.*, **13** (4), 67–94.

Krolik, J. L. (1996). The performance of matched-field beamformers with Mediterranean vertical array data. *IEEE Trans. Signal Process.*, **44** (10), 2605–2611.

Kumaresan, R. and Tufts, D. W. (1983). Estimating the angles of arrival of multiple plane waves. *IEEE Trans. Aerosp. Electron. Syst.*, **19** (1), 134–139.

Lagunas, M. A., Santamaria, M. E., Gasull, A., and Moreno, A. (1986). Maximum likelihood filters in spectral estimation problems. *Signal Process.*, **10** (1), 7–18.

Li, J., Stoica, P., and Wang, Z. (2003). On robust Capon beamforming and diagonal loading. *IEEE Trans. Signal Process.*, **51** (7), 1702–1715.

Li, Y., Razavilar, J., and Liu, K. J. R. (1998). A high-resolution technique for multidimensional NMR spectroscopy. *IEEE Trans. Biomed. Eng.*, **45** (1), 78–86.

Liu, Q. G., Champagne, B., and Kabal, P. (1996). A microphone array processing technique for speech enhancement in a reverberant space. *Speech Commun.*, **18** (4), 317–334.

Lorenz, R. G. and Boyd, S. P. (2003). Robust minimum variance beamforming. In *Proc. 37th Asilomar Conf. Signals, Syst., Comput.*, 1345–1352.

Monzingo, R. A. and Miller, T. W. (1980). *Introduction to Adaptive Arrays* (Wiley).

Morgan, D. R. and Smith, T. M. (1990). Coherence effects on the detection performance of quadratic array processors with application to large-array matched-field beamforming. *J. Acoust. Soc. Amer.*, **87** (2), 737–747.

Mosher, J. and Leahy, R. (1998). Recursive MUSIC: A framework for EEG and MEG source localization. *IEEE Trans. Biomed. Eng.*, **45** (11), 1342–1354.

Nehorai, A. and Jeremic, A. (2000). Landmine detection and localization using chemical sensor array processing. *IEEE Trans. Signal Process.*, **48** (5), 1295–1305.

Nehorai, A., Porat, B., and Paldi, E. (1995). Detection and localization of vapor-emitting sources. *IEEE Trans. Signal Process.*, **43** (1), 243–253.

Oh, S. K. and Un, C. K. (1989). An improved MUSIC algorithm for high-resolution array processing. *Electron. Lett.*, **25** (22), 1523–1525.

Paulraj, A. and Gesbert, D. (1998). Smart antennas for mobile communications. In Webster, J. (ed.), *Encyclopedia on Electrical and Electronics Engineering* (Wiley).

Paulraj, A. and Papadias, C. B. (1998). Array processing for mobile communications. In Madisetti, V. K. and Williams, D. B. (eds.), *Digital Signal Processing Handbook* (CRC/IEEE Press).

Paulraj, A., Roy, R., and Kailath, T. (1986). A subspace rotation approach to signal parameter estimation. *Proc. IEEE*, **74** (7), 1044–1046.

Pesavento, M., Gershman, A. B., and Haardt, M. (2000). Unitary root-MUSIC with a real-valued eigendecomposition: A theoretical and experimental performance study. *IEEE Trans. Signal Process.*, **48** (5), 1306–1314.

Pesavento, M., Gershman, A. B., and Wong, K. M. (2002). Direction finding using partly calibrated sensor arrays composed of multiple subarrays. *IEEE Trans. Signal Process.*, **50** (9), 2103–2115.

Pisarenko, V. F. (1973). The retrieval of harmonics from a covariance function. *Geophys. J. Roy. Astronom. Soc.*, **33**, 347–366.

Ramos, J., Mathews, C. P., and Zoltowski, M. D. (1999). FCA-ESPRIT: A closed-form 2-D angle estimation algorithm for filled circular arrays with arbitrary sampling lattices. *IEEE Trans. Signal Process.*, **47** (1), 213–217.

Rao, B. D. and Hari, K. V. S. (1989). Performance analysis of root-MUSIC. *IEEE Trans. Acoust., Speech, Signal Process.*, **37** (12), 1939–1949.

Raza, J., Boonstra, A.-J., and van der Veen, A.-J. (2002). Spatial filtering of RF interference in radio astronomy. *IEEE Signal Process. Lett.*, **9** (2), 64–67.

Reddi, S. (1979). Multiple source location—A digital approach. *IEEE Trans. Aerosp. Electron. Syst.*, **15** (1), 95–105.

Reed, I. S., Mallett, J. D., and Brennan, L. E. (1974). Rapid convergence rate in adaptive arrays. *IEEE Trans. Aerosp. Electron. Syst.*, **10** (6), 853–863.

Roy, R. and Kailath, T. (1989). ESPRIT—Estimation of signal parameters via rotational invariance techniques. *IEEE Trans. Acoust., Speech, Signal Process.*, **37** (7), 984–995.

Schmidt, R. O. (1979). Multiple emitter location and signal parameter estimation. In *Proc. RADC Spectral Estimation Workshop*, 234–258.

See, C. M. S. and Gershman, A. B. (2004). Direction-of-arrival estimation in partly calibrated subarray-based sensor arrays. *IEEE Trans. Signal Process.*, **52** (2), 329–338.

Shahbazpanahi, S., Beheshti, M., Gershman, A. B., Gharavi-Alkhansari, M., and Wong, K. M. (2004a). Minimum variance linear receivers for multiaccess MIMO wireless systems with space-time block coding. *IEEE Trans. Signal Process.*, **52** (12), 3306–3313.

Shahbazpanahi, S., Gershman, A. B., and Giannakis, G. B. (2005). Semi-blind multi-user MIMO channel estimation based on Capon and MUSIC techniques. In *Proc. IEEE Int. Conf. Acoust., Speech, Signal Process.*, vol. 4, 773–776.

Shahbazpanahi, S., Gershman, A. B., Luo, Z.-Q., and Wong, K. M. (2003). Robust adaptive beamforming for general-rank signal models. *IEEE Trans. Signal Process.*, **51** (9), 2257–2269.

Shahbazpanahi, S., Gershman, A. B., and Manton, J. (2004b). Closed-form channel estimation for blind decoding of orthogonal space-time block codes. In *Proc. IEEE Int. Conf. Commun.*, vol. 1, 603–607.

Sharman, K. C. (1988). Maximum likelihood parameter estimation by simulated annealing. In *Proc. IEEE Int. Conf. Acoust., Speech, Signal Process.*, 2741–2744.

Sharman, K. C. and McClurkin, G. D. (1989). Genetic algorithms for maximum likelihood parameter estimation. In *Proc. IEEE Int. Conf. Acoust., Speech, Signal Process.*, 2716–2719.

Sidiropoulos, N. D., Bro, R., and Giannakis, G. B. (2000). Parallel factor analysis in sensor array processing. *IEEE Trans. Signal Process.*, **48** (8), 2377–2388.

Sidorovitch, D. V. and Gershman, A. B. (1998). 2-D wideband interpolated root-MUSIC applied to measured seismic data. *IEEE Trans. Signal Process.*, **46** (8), 2263–2267.

Spielman, D., Paulraj, A., and Kailath, T. (1990). Eigenstructure approach to directions-of-arrival estimation in IR detector arrays. *Appl. Optics*, **26** (2), 199–202.

Stoica, P. and Gershman, A. B. (1999). Maximum likelihood DOA estimation by data-supported grid search. *IEEE Signal Process. Lett.*, **6** (10), 273–275.

Stoica, P., Handel, P., and Nehorai, A. (1995). Improved sequential MUSIC. *IEEE Trans. Aerosp. Electron. Syst.*, **31** (4), 1230–1239.

Stoica, P. and Nehorai, A. (1989). MUSIC, maximum likelihood and Cramér-Rao bound. *IEEE Trans. Signal Process.*, **37** (5), 720–741.

—— (1990). Performance study of conditional and unconditional direction-of-arrival estimation. *IEEE Trans. Signal Process.*, **38** (10), 1783–1795.

Stoica, P. and Sharman, K. C. (1990). Novel eigenanalysis method for direction estimation. *IEE Proc. F*, **137** (1), 19–26.

Swindlehurst, A. L. and Stoica, P. (1998). Maximum likelihood methods in radar array signal processing. *Proc. IEEE*, **86** (2), 421–441.

Swindlehurst, A. L., Stoica, P., and Jansson, M. (2001). Exploiting arrays with multiple invariances using MUSIC and MODE. *IEEE Trans. Signal Process.*, **49** (11), 2511–2521.

Tarokh, V., Jafarkhani, H., and Calderbank, A. R. (1999). Space-time block codes from orthogonal designs. *IEEE Trans. Inf. Theory*, **45** (5), 1456–1467.

Tufts, D. W. and Melissinos, C. D. (1986). Simple, effective computation of principal eigenvectors and their eigenvalues and applications to high-resolution estimation of frequences. *IEEE Trans. Acoust., Speech, Signal Process.*, **34** (5), 1046–1052.

Van Trees, H. L. (2002). *Optimum Array Processing* (Wiley).

Viberg, M. and Ottersten, B. (1991). Sensor array processing based on subspace fitting. *IEEE Trans. Signal Process.*, **39** (5), 1110–1121.

Vorobyov, S., Gershman, A. B., and Luo, Z.-Q. (2003). Robust adaptive beam-forming using worst-case performance optimization: A solution to the signal mismatch problem. *IEEE Trans. Signal Process.*, **51** (2), 313–324.

Vorobyov, S., Gershman, A. B., Luo, Z.-Q., and Ma, N. (2004). Adaptive beam-forming with joint robustness against mismatched signal steering vector and interference nonstationarity. *IEEE Signal Process. Lett.*, **11** (2), 108–111.

Wax, M. (1985). *Detection and Estimation of Superimposed Signals.* Ph.D. thesis, Stanford University, Stanford, CA.

Wax, M. and Kailath, T. (1983). Optimum localization of multiple sources by passive arrays. *IEEE Trans. Acoust., Speech, Signal Process.*, **31** (5), 1210–1217.

Winters, J. (1982). Spread spectrum in a four-phase communication system em-ploying adaptive antennas. *IEEE Trans. Commun.*, **30** (5), 929–936.

Zarifi, K., Shahbazpanahi, S., Gershman, A. B., and Luo, Z.-Q. (2005). Robust blind multiuser detection based on the worst-case performance optimization of the MMSE receiver. *IEEE Trans. Signal Process.*, **53** (1), 295–305.

Ziskind, I. and Wax, M. (1988). Maximum likelihood localization of multiple sources by alternating projection. *IEEE Trans. Acoust., Speech, Signal Process.*, **36** (10), 1553–1560.

Zoltowski, M. D., Kautz, G. M., and Silverstein, S. D. (1993). Beamspace root-MUSIC. *IEEE Trans. Signal Process.*, **41** (1), 344–364.

Zoltowski, M. D. and Stavrinides, D. (1989). Sensor array signal processing via a Procrustes rotations based eigenanalysis of the ESPRIT data pencil. *IEEE Trans. Acoust., Speech, Signal Process.*, **37** (6), 832–861.

Zoubir, A. M. and Böhme, J. F. (1995). Bootstrap multiple tests applied to sensor location. *IEEE Trans. Signal Process.*, **43** (6), 1386–1396.

13

Optimal subspace techniques for DOA estimation in the presence of noise and model errors

Magnus Jansson and Björn Ottersten

KTH—Royal Institute of Technology

Mats Viberg

Chalmers University of Technology

A. Lee Swindlehurst

Brigham Young University

Signal parameter estimation, and specifically direction of arrival (DOA) estimation for sensor array data is encountered in a number of applications ranging from electronic surveillance to wireless communications. Subspace-based methods have shown to provide computationally as well as statistically efficient algorithms for DOA estimation. Estimator performance is ultimately limited by model disturbances like measurement noise and model errors. Herein, we review a recently proposed framework that allows the derivation of optimal subspace methods, taking both finite sample effects (noise) and model errors into account. We show how this generic estimator reduces to well-known techniques for cases when one disturbance completely dominates the other.

13.1 Introduction

Subspace-based techniques have been shown to be powerful tools in many signal processing applications where the observed data consist of low-rank signals in noise. Some examples include sensor array signal processing, harmonic retrieval, factor analysis, timing estimation, frequency offset estimation, image processing, system identification, and blind channel identification. By appropriate use of the underlying low-rank data model and the associated signal and noise characteristics, subspace estimation techniques can often be made computationally and/or statistically efficient.

This chapter focuses on subspace techniques for direction of arrival (DOA) estimation from data collected by a sensor array. This is quite a mature field of research by now; many tutorial papers and books have been presented, some detailing specific aspects and others giving broader views (e.g., Krim and Viberg (1996), Van Trees (2002), and the references therein). We have no ambition whatsoever to give a comprehensive account of the development of the field of DOA estimation in this chapter. Rather, we will pursue a specific view of the problem, and consider the sensitivity of subspace DOA estimation methods to noise and errors in the array model. Although this may seem to be a very narrow view of the field, we will argue that many existing DOA estimation techniques can be seen as special cases of the estimator presented later in this chapter.

An important aspect of high-resolution DOA estimation is algorithm sensitivity to noise and model errors. Performance of DOA detection and estimation algorithms is ultimately limited by noise in the array measurements and by errors in the array model. In many cases, the array response is not exactly known, and the deviations of the true response from that of the model may severely influence the performance. Indeed, many authors have studied the quantitative effects of model errors on both DOA and signal waveform estimation (Ramsdale and Howerton, 1980; Compton, 1982; Quazi, 1982; Zhu and Wang, 1988; Wong *et al.*, 1988; Friedlander, 1990a,b; Swindlehurst and Kailath, 1990, 1992, 1993; Li and Vaccaro, 1992; Soon and Huang, 1992; Friedlander and Weiss, 1994; Kangas *et al.*, 1994; Viberg and Swindlehurst, 1994a; Yang and Swindlehurst, 1995; Kangas *et al.*, 1996).

If the array response is known to depend on some unknown factors, a natural approach is to parameterize the array model not only by the DOA parameters but also by some additional calibration-dependent parameters. These parameters can include, for example, sensor element positions, gain and phase offsets, mutual coupling, element directivity, and others. Given such a model, it is natural to attempt to estimate the unknown (nuisance) model parameters simultaneously with the signal parameters. This approach is often referred to as autocalibration (Paulraj and Kailath, 1985; Rockah and Schultheiss, 1987a,b; Weiss and Friedlander, 1989; Wahlberg *et al.*, 1991; Wylie *et al.*, 1994; Viberg and Swindlehurst, 1994b; Flieller *et al.*, 1995; Gustafsson *et al.*, 1996; Swindlehurst, 1996). Using autocalibration techniques may, however, be problematic in certain cases. One obvious reason is that the number of unknown parameters that need to be estimated from the data can be quite large, which may lead to difficulties in numerically calculating the optimal solution. An even more critical issue to consider when autocalibration techniques are employed is whether or not both the DOA and the model error parameters are simultaneously identifiable. For example, it is clearly not possible to estimate both DOAs and sensor positions simultaneously without the use of additional information

(Lo and Marple, 1987; Weiss and Friedlander, 1989; Ng and Nehorai, 1993; Koerber and Fuhrmann, 1993; McArthur and Reilly, 1994; Yip and Zhou, 1995).

An alternative to autocalibration approaches is to use techniques that assume the model error parameters to be realizations of some underlying random vector with a known a priori distribution. With this modification of the problem formulation using a Bayesian framework, difficulties with parameter identifiability may be alleviated. It also allows a systematic approach to estimator design such as, for example, maximum a posteriori approaches (Wahlberg *et al.*, 1991; Viberg and Swindlehurst, 1994b). Another way to exploit the assumption of random model error parameters is to consider, at least implicitly, the random model error as an additional noise term in the data model. To reduce the sensitivity of the DOA estimator to the model errors, statistically optimal weighting matrices can then be derived. Methods that follow this latter approach were presented by Swindlehurst and Kailath (1992, 1993), who studied MUSIC and subspace fitting methods under the assumption that the estimation errors due to model imperfections dominate the effects of additive noise.

The best performance is achieved by appropriately taking into account both noise and model error effects. Optimally weighted methods treating the combined effects of model errors and noise were presented by Viberg and Swindlehurst (1994a), Jansson *et al.* (1998). This may be viewed as a pragmatic approach even when the model errors cannot be considered random (which very well may be the case), and we will pursue this approach herein. The perturbation parameter covariance matrix can be seen as a design variable reflecting the expected level of parameter variability.

The remainder of this chapter is organized as follows: in the next section, the data model and some fundamental facts are introduced. Section 13.3 presents the general array error model that is used herein, along with three explicit examples of typical model errors. In Section 13.4, DOA estimation using subspace techniques is formulated in a generalized signal subspace fitting framework. This framework is then utilized in Section 13.5 to present an optimally weighted DOA estimation method that accounts for both finite sample effects because of the noise and small array model errors. Section 13.5 continues with a discussion of the performance of this general estimator, and how it is related to many existing optimal and suboptimal DOA estimators in different special cases. Finally, the presented results are illustrated in a numerical example in Section 13.6. The chapter ends with some concluding remarks in Section 13.7.

13.2 Data model

Assume that the output $\mathbf{x}(t)$ of an array of m sensors is given by the model

$$\mathbf{x}(t) = \mathbf{A}(\boldsymbol{\theta}, \boldsymbol{\rho})\mathbf{s}(t) + \mathbf{n}(t),$$

where $\mathbf{s}(t) \in \mathbb{C}^{d \times 1}$ contains the emitted signal waveforms, and $\mathbf{n}(t) \in \mathbb{C}^{m \times 1}$ is an additive noise vector that is independent of the signal term. The array steering matrix is defined as

$$\mathbf{A}(\boldsymbol{\theta}, \boldsymbol{\rho}) = \begin{bmatrix} \mathbf{a}(\boldsymbol{\theta}_1, \boldsymbol{\rho}) & \cdots & \mathbf{a}(\boldsymbol{\theta}_d, \boldsymbol{\rho}) \end{bmatrix},$$

where $\mathbf{a}(\boldsymbol{\theta}_i, \boldsymbol{\rho}) \in \mathbb{C}^{m \times 1}$ denotes the array response to a unit waveform associated with the signal parameter $\boldsymbol{\theta}_i \in \mathbb{R}^{p \times 1}$. We will refer to $\boldsymbol{\theta}_i$ as the DOA of the ith signal. The vector $\boldsymbol{\rho} \in \mathbb{R}^{n \times 1}$ contains all additional parameters of the array steering matrix that may be unknown. Examples of typical models for the vector $\boldsymbol{\rho}$ will be considered later. When the array response is a function of the signal parameters only, we simply omit the dependency on $\boldsymbol{\rho}$ and write $\mathbf{A}(\boldsymbol{\theta})$. It is assumed that the array is unambiguous, so that the columns in $\mathbf{A}(\boldsymbol{\theta})$ are linearly independent as long as $\boldsymbol{\theta}_i \neq \boldsymbol{\theta}_j$, $i \neq j$.

The signal $\mathbf{s}(t)$ and the noise $\mathbf{n}(t)$ are modeled as zero-mean circular (Gaussian) random vectors with covariances

$$E\{\mathbf{s}(t)\mathbf{s}^H(s)\} = \mathbf{P}\,\delta_{t,s},$$
$$E\{\mathbf{n}(t)\mathbf{n}^H(s)\} = \sigma^2 \mathbf{I}\,\delta_{t,s},$$

where $(\cdot)^H$ denotes complex conjugate transpose, and $\delta_{t,s}$ is the Kronecker delta. Let d' denote the rank of the signal covariance matrix $\mathbf{P}$. Note that $d' < d$ when some of the signals are fully correlated or coherent.

Assuming the signals and the noise to be uncorrelated, the array output covariance matrix is given by

$$\mathbf{R} = E\{\mathbf{x}(t)\mathbf{x}^H(t)\} = \mathbf{A}(\boldsymbol{\theta}, \boldsymbol{\rho})\mathbf{P}\mathbf{A}^H(\boldsymbol{\theta}, \boldsymbol{\rho}) + \sigma^2 \mathbf{I}.$$

The eigendecomposition of $\mathbf{R}$ is

$$\mathbf{R} = \sum_{k=1}^{m} \lambda_k \mathbf{e}_k \mathbf{e}_k^H,$$

where $\lambda_1 \geq \cdots \geq \lambda_{d'} > \lambda_{d'+1} = \cdots = \lambda_m = \sigma^2$ are the eigenvalues and $\mathbf{e}_k$ the corresponding eigenvectors. Let $\mathbf{E}_s = \begin{bmatrix} \mathbf{e}_1 & \cdots & \mathbf{e}_{d'} \end{bmatrix}$ be the matrix of the *signal* eigenvectors and $\mathbf{E}_n = \begin{bmatrix} \mathbf{e}_{d'+1} & \cdots & \mathbf{e}_m \end{bmatrix}$ the matrix of *noise* eigenvectors. The range of $\mathbf{E}_s$ is called the *signal subspace*, while the range of $\mathbf{E}_n$ is the *noise subspace*.

From the structure of the covariance matrix model it is clear that

$$\mathcal{R}(\mathbf{E}_s) \subseteq \mathcal{R}(\mathbf{A}(\boldsymbol{\theta}, \boldsymbol{\rho})), \tag{13.1}$$

where $\mathcal{R}(\mathbf{A})$ denotes the range of $\mathbf{A}$. For noncoherent cases, when $d' = d$, there is equality in (13.1), and we also have

$$\mathbf{E}_n^H \mathbf{A}(\boldsymbol{\theta}, \boldsymbol{\rho}) = \mathbf{0}. \tag{13.2}$$

However, note that this does *not* hold when $d' < d$.

The two geometrical facts (13.1)–(13.2) form the basis for all subspace estimation techniques.

From (13.1) it is clear that there exists a full-rank matrix $\mathbf{T} \in \mathbb{C}^{d \times d'}$ such that

$$\mathbf{E}_s = \mathbf{A}(\boldsymbol{\theta}, \boldsymbol{\rho})\mathbf{T}. \tag{13.3}$$

For the parameters to be identifiable when using the subspace approach, they have to be uniquely determined from the subspace equation (13.3). In particular, identifiability is guaranteed if

$$\mathbf{A}(\boldsymbol{\theta}_1)\mathbf{T}_1 = \mathbf{A}(\boldsymbol{\theta}_2)\mathbf{T}_2$$

for any two full-rank matrices $\mathbf{T}_i \in \mathbb{C}^{d \times d'}$, $i = 1, 2$, implies $\boldsymbol{\theta}_1 = \boldsymbol{\theta}_2$ (with some convention for the ordering of the elements in $\boldsymbol{\theta}_i$) (Wax and Ziskind, 1989). If the parameters in $\boldsymbol{\rho}$ also need to be estimated, they should naturally also be included in the identifiability condition above.

13.3 Array model errors

When discussing array model errors, we will follow the approach mentioned in the introduction, which assumes $\boldsymbol{\rho}$ to be a random vector drawn from some distribution. More specifically, the perturbation parameter vector $\boldsymbol{\rho}$ is modeled as a random vector with mean $\mathrm{E}\{\boldsymbol{\rho}\} = \boldsymbol{\rho}_0$ and covariance

$$\mathrm{E}\{(\boldsymbol{\rho} - \boldsymbol{\rho}_0)(\boldsymbol{\rho} - \boldsymbol{\rho}_0)^T\} = \boldsymbol{\Omega}. \tag{13.4}$$

We assume that both $\boldsymbol{\rho}_0$ and $\boldsymbol{\Omega}$ are known. Similar to Viberg and Swindlehurst (1994a,b), Jansson *et al.* (1998), we assume that the elements in $\boldsymbol{\Omega}$ are "small" and, hence, consider only small perturbations in $\boldsymbol{\rho}$ around $\boldsymbol{\rho}_0$, to allow a first-order perturbation analysis.

Some examples of common array perturbation models are outlined below.

13.3.1 Sensor position errors

Assuming an array composed of identical sensors lying in the same two-dimensional plane as the signals of interest, a general model for the ith element of the array response vector is

$$[\mathbf{a}(\boldsymbol{\theta}_k)]_i = g_i(\boldsymbol{\theta}_k) \exp\left(j\frac{2\pi}{\lambda} \left(x_i \cos(\boldsymbol{\theta}_k) + y_i \sin(\boldsymbol{\theta}_k)\right)\right),$$

where λ is the carrier wavelength, $g_i(\cdot)$ denotes the gain pattern, and (x_i, y_i) denotes the position coordinates of the ith sensor. If the sensor positions are not precisely known, we set

$$\boldsymbol{\rho} = [x_1, y_1, \ldots, x_m, y_m]^T,$$

letting $\boldsymbol{\rho}_0$ represent the nominal (assumed) vector of sensor positions, and $\boldsymbol{\Omega}$ the covariance matrix of the position errors.

13.3.2 Receiver gain and phase variations

The gain and phase of each sensor's RF receiver front end vary due to a variety of factors, including differences in cable lengths, nonidentical components, and temperature fluctuations. While to some extent these variations can be calibrated out, there always remain some differences between receivers. The following model is a simple way of representing these effects:

$$\mathbf{A}(\boldsymbol{\theta}, \boldsymbol{\rho}) = \text{diag}(\boldsymbol{\rho})\mathbf{A}(\boldsymbol{\theta}),$$

where $\mathbf{A}(\boldsymbol{\theta})$ is the nominal (calibrated) array response, and $\text{diag}(\boldsymbol{\rho})$ represents a diagonal matrix whose nonzero elements are given by $\rho_i \exp(j\rho_{i+m})$, $i = 1, 2, \ldots, m$. Here, the first m elements of $\boldsymbol{\rho}$ model the gain and the remaining m elements model the phase. Again, $\boldsymbol{\rho}_0$ contains the nominal gain and phase values, and $\boldsymbol{\Omega}$ models the expected variation in the receiver gain and phase. Mutual coupling effects can be modeled using a similar approach, where instead of a diagonal matrix, $\boldsymbol{\rho}$ is used to specify a more complicated matrix with off-diagonal elements that capture the element-to-element coupling.

13.3.3 Generic array perturbations

In many cases, the causes of the array model errors may be many; they may be too complex to model using physical reasoning, as in the two cases described above. One approach that can be employed in such cases is to model $\boldsymbol{\rho}$ implicitly, using

$$\tilde{\mathbf{A}} = \mathbf{A}(\boldsymbol{\theta}, \boldsymbol{\rho}) - \mathbf{A}(\boldsymbol{\theta}, \boldsymbol{\rho}_0),$$

with the statistics of $\tilde{\mathbf{A}}$ directly quantifying the array perturbation. For example, Swindlehurst and ·Kailath (1993) and Viberg and Swindlehurst (1994a) assumed $\tilde{\mathbf{A}}$ to be zero mean with covariances given by

$$E\{\text{vec}(\tilde{\mathbf{A}}) \, \text{vec}^H(\tilde{\mathbf{A}})\} = \boldsymbol{\Psi} \otimes \boldsymbol{\Gamma}, \tag{13.5}$$

$$E\{\text{vec}(\tilde{\mathbf{A}}) \, \text{vec}^T(\tilde{\mathbf{A}})\} = \mathbf{0}. \tag{13.6}$$

Looking at the ijth block in (13.5), $E\{\tilde{\mathbf{a}}(\boldsymbol{\theta}_i)\tilde{\mathbf{a}}^H(\boldsymbol{\theta}_j)\} = \boldsymbol{\Psi}_{ij}\boldsymbol{\Gamma}$, it can be seen that this error model assumes the same spatial error covariance matrix $\boldsymbol{\Gamma}$ for all sensors, with a possible DOA dependency modeled via $\boldsymbol{\Psi}_{ij}$.

13.4 DOA estimation and subspace fitting

The subspace relations discussed in Section 13.2 can be used in various ways for the estimation of the model parameters, and in particular for the DOAs. It has been shown that most approaches can be put quite conveniently into a common subspace fitting framework (Viberg and Ottersten, 1991). The idea of the signal subspace fitting methods is to minimize a suitable norm of the error between a sample estimate $\hat{\mathbf{E}}_s$ of the signal eigenvector matrix $\mathbf{E}_s$ and the model $\mathbf{A}(\boldsymbol{\theta}, \boldsymbol{\rho})\mathbf{T}$. More precisely, the basic signal subspace fitting criterion is (Viberg and Ottersten, 1991; Ottersten *et al.*, 1993)

$$V(\boldsymbol{\theta}, \boldsymbol{\rho}) = \min_{\mathbf{T}} \|\hat{\mathbf{E}}_s - \mathbf{A}(\boldsymbol{\theta}, \boldsymbol{\rho})\mathbf{T}\|_{\mathbf{W}}^2, \tag{13.7}$$

where $\mathbf{W}$ denotes a positive definite Hermitian weighting matrix, and $\|\mathbf{X}\|_{\mathbf{W}}^2 = \text{Tr}(\mathbf{X}\mathbf{W}\mathbf{X}^H)$. The estimates of the parameters are obtained as the minimizing argument of the criterion, i.e.,

$$\hat{\boldsymbol{\theta}}, \hat{\boldsymbol{\rho}} = \arg\min_{\boldsymbol{\theta}, \boldsymbol{\rho}} V(\boldsymbol{\theta}, \boldsymbol{\rho}).$$

Here, the possibility of including estimation of the array model parameter vector $\boldsymbol{\rho}$ is explicitly indicated (see the discussion about autocalibration in Section 13.1). However, in the following the focus will be on the estimation of the DOA parameters only.

For certain error models, the weighted norm (13.7) needs to be generalized, so that each residual element gets its own weight relative to all the others.

This can be achieved with the more general criterion

$$V(\boldsymbol{\theta}) = \min_{\mathbf{T}} \ \bar{\boldsymbol{\varepsilon}}^H(\boldsymbol{\theta}) \mathbf{W} \bar{\boldsymbol{\varepsilon}}(\boldsymbol{\theta}), \tag{13.8}$$

$$\text{with} \quad \bar{\boldsymbol{\varepsilon}} = \begin{bmatrix} \boldsymbol{\varepsilon} \\ \boldsymbol{\varepsilon}^* \end{bmatrix}, \quad \boldsymbol{\varepsilon} = \text{vec}(\hat{\mathbf{E}}_s - \mathbf{A}(\boldsymbol{\theta})\mathbf{T}),$$

where $\mathbf{W}$ is again a positive definite weighting matrix (we use $\mathbf{W}$ as a generic notation for weighting matrices throughout). An extended residual vector $\bar{\boldsymbol{\varepsilon}}$ is obtained by combining the residual $\boldsymbol{\varepsilon}$ and its complex conjugate $\boldsymbol{\varepsilon}^*$. Alternatively, the extended residual vector could have been formed from the real and imaginary parts of $\boldsymbol{\varepsilon}$. Clearly, these two representations yield equivalent results (with an appropriate $\mathbf{W}$), since there exists an invertible transformation between them.

An alternative DOA estimation criterion can be formulated by utilizing the geometrical relation (13.1) as follows: let $\mathbf{B}(\boldsymbol{\theta}) \in \mathbb{C}^{m \times (m-d)}$ be a full-rank matrix whose columns span the null space of $\mathbf{A}^H(\boldsymbol{\theta})$. This implies that $\mathbf{B}^H(\boldsymbol{\theta})\mathbf{A}(\boldsymbol{\theta}) = \mathbf{0}$ and $\mathbf{B}^H(\boldsymbol{\theta}_0)\mathbf{E}_s = \mathbf{0}$, where $\boldsymbol{\theta}_0$ denotes the true DOA. Hence, assuming parameter identifiability, the equations $\mathbf{B}^H(\boldsymbol{\theta})\mathbf{E}_s = \mathbf{0}$ uniquely determine the true DOA. Given an estimate $\hat{\mathbf{E}}_s$ of $\mathbf{E}_s$, these equations will not be fulfilled exactly for any $\boldsymbol{\theta}$, and it is reasonable to consider the minimization of a suitable norm of $\mathbf{B}^H(\boldsymbol{\theta})\hat{\mathbf{E}}_s$. Similar to the above, consider

$$V(\boldsymbol{\theta}) = \bar{\boldsymbol{\varepsilon}}^H(\boldsymbol{\theta}) \mathbf{W} \bar{\boldsymbol{\varepsilon}}(\boldsymbol{\theta}), \tag{13.9}$$

$$\text{with} \quad \bar{\boldsymbol{\varepsilon}} = \begin{bmatrix} \boldsymbol{\varepsilon} \\ \boldsymbol{\varepsilon}^* \end{bmatrix}, \quad \boldsymbol{\varepsilon} = \text{vec}(\mathbf{B}^H(\boldsymbol{\theta})\hat{\mathbf{E}}_s).$$

It can be shown that the estimates obtained by the minimization of the subspace fitting criteria (13.8) and (13.9) are asymptotically equivalent (Stoica *et al.*, 1997) (see Cardoso and Moulines (2000)). In other words, for each subspace fitting weighting matrix in (13.8) there is a weighting matrix in (13.9) leading to asymptotically equivalent DOA estimates and vice versa. When studying asymptotic equivalence, it is therefore sufficient to study one of these formulations.

However, when attempting to minimize these criteria, some parameterizations can lead to efficient optimization algorithms. Also, the weighting matrices are often parameter dependent as well as data dependent, and must be estimated. Estimating the weighting matrices is not equivalent for the two formulations.

Henceforth, we will refer to the above methods as the generalized weighted subspace fitting (GWSF) method, and in particular focus on the second formulation (13.9).

13.5 Special cases of GWSF

The GWSF formulation is in fact very general, and many existing subspace DOA estimation methods can be related to GWSF by a proper choice of the weighting matrix.

13.5.1 GWSF for combined noise and array errors

It is well known that, within the class of estimators based on $\bar{\varepsilon}$ in (13.9), the optimal choice of the weighting to minimize the parameter estimation error variance is

$$\mathbf{W} = \mathbf{C}_{\bar{\varepsilon}}^{-1}, \tag{13.10}$$

where $\mathbf{C}_{\bar{\varepsilon}}$ is the asymptotic covariance matrix of the residual vector $\bar{\varepsilon}$ at the true DOA $\boldsymbol{\theta}_0$.

For the case of combined noise and small array perturbation (13.4), $\mathbf{C}_{\bar{\varepsilon}}$ can be shown to be (Jansson *et al.*, 1998)

$$\mathbf{C}_{\bar{\varepsilon}} = \bar{\mathbf{L}} + \bar{\mathbf{G}}\bar{\mathbf{G}}^H, \tag{13.11}$$

where

$$\bar{\mathbf{L}} = \begin{bmatrix} \mathbf{L} & \mathbf{0} \\ \mathbf{0} & \mathbf{L}^* \end{bmatrix}, \qquad \mathbf{L} = \left(\sigma^2 \tilde{\boldsymbol{\Lambda}}^{-2}\boldsymbol{\Lambda}_s \otimes \mathbf{B}^H\mathbf{B}\right),$$

$$\bar{\mathbf{G}} = \begin{bmatrix} \mathbf{G} \\ \mathbf{G}^* \end{bmatrix}, \qquad \mathbf{G} = \left(\mathbf{T}^T \otimes \mathbf{B}^H\right)\mathbf{D}_\rho \bar{\boldsymbol{\Omega}}^{1/2}. \tag{13.12}$$

Here, $\boldsymbol{\Lambda}_s$ is a diagonal matrix containing the d' largest signal eigenvalues of $\mathbf{R}$, $\tilde{\boldsymbol{\Lambda}} = \boldsymbol{\Lambda}_s - \sigma^2\mathbf{I}$, $\bar{\boldsymbol{\Omega}}^{1/2}$ is a (symmetric) square root of $\bar{\boldsymbol{\Omega}} = N\boldsymbol{\Omega}$, $\mathbf{T} = \mathbf{A}^\dagger\mathbf{E}_s$, where $(\cdot)^\dagger$ denotes the Moore–Penrose pseudoinverse, and

$$\mathbf{D}_\rho = \left[\frac{\partial\,\mathrm{vec}(\mathbf{A}(\boldsymbol{\theta},\boldsymbol{\rho}))}{\partial\rho_1} \quad \cdots \quad \frac{\partial\,\mathrm{vec}(\mathbf{A}(\boldsymbol{\theta},\boldsymbol{\rho}))}{\partial\rho_n}\right]\Bigg|_{\boldsymbol{\theta}=\boldsymbol{\theta}_0,\boldsymbol{\rho}=\boldsymbol{\rho}_0}.$$

The covariance matrix (13.11) contains two terms, one due to the noise and another one that accounts for the array perturbations. Thus, with some knowledge of the relation between errors causes by the measurement noise and perturbations in the array model, an optimal trade-off is obtained when forming the weighting matrix. Also, note that the weighting matrix depends on the unknown parameters in $\mathbf{B}$ and $\mathbf{D}_\rho$. Fortunately, it is possible to replace the weighting matrix with an estimate thereof without affecting the asymptotic properties of the DOA estimate. More details regarding GWSF for the combined effects of noise and model errors, including implementation issues, can be found in the paper by Jansson *et al.* (1998).

Zhu and Wang (1988), Wahlberg *et al.* (1991), and Viberg and Swindlehurst (1994b) derive an asymptotically valid Cramér–Rao lower bound (CRB) for the problem of interest. Below we give the lower bound on the signal parameters only. Assuming that $\hat{\boldsymbol{\theta}}$ is an asymptotically unbiased estimate of $\boldsymbol{\theta}_0$ and that $\boldsymbol{\rho}$ is Gaussian, then for large N and small $\boldsymbol{\Omega}$,

$$\mathrm{E}\{(\hat{\boldsymbol{\theta}} - \boldsymbol{\theta}_0)(\hat{\boldsymbol{\theta}} - \boldsymbol{\theta}_0)^T\} \geq \mathbf{CRB}_{\boldsymbol{\theta}} \triangleq \frac{\sigma^2}{2N} \left[\mathbf{C} - \mathbf{F}_{\boldsymbol{\theta}}^T \boldsymbol{\Upsilon}^{-1} \mathbf{F}_{\boldsymbol{\theta}}\right]^{-1}, \qquad (13.13)$$

where

$$\mathbf{C} = \mathrm{Re}\{\mathbf{D}_{\boldsymbol{\theta}}^H \mathbf{M} \mathbf{D}_{\boldsymbol{\theta}}\},$$

$$\mathbf{M} = \mathbf{U}^T \otimes \boldsymbol{\Pi}_{\mathbf{A}}^{\perp},$$

$$\mathbf{U} = \mathbf{A}^\dagger \mathbf{E}_s \tilde{\boldsymbol{\Lambda}}^2 \boldsymbol{\Lambda}_s^{-1} \mathbf{E}_s^H \mathbf{A}^{\dagger H},$$

$$\mathbf{D}_{\boldsymbol{\theta}} = \left[\frac{\partial\,\mathrm{vec}(\mathbf{A}(\boldsymbol{\theta}, \boldsymbol{\rho}))}{\partial\theta_1} \cdots \frac{\partial\,\mathrm{vec}(\mathbf{A}(\boldsymbol{\theta}, \boldsymbol{\rho}))}{\partial\theta_d}\right],$$

$$\mathbf{F}_{\boldsymbol{\theta}} = \mathrm{Re}\{\mathbf{D}_{\boldsymbol{\rho}}^H \mathbf{M} \mathbf{D}_{\boldsymbol{\theta}}\},$$

$$\boldsymbol{\Upsilon} = \mathrm{Re}\left\{\mathbf{D}_{\boldsymbol{\rho}}^H \mathbf{M} \mathbf{D}_{\boldsymbol{\rho}} + \frac{\sigma^2}{2} \bar{\boldsymbol{\Omega}}^{-1}\right\},$$

and $\boldsymbol{\Pi}_{\mathbf{A}}^{\perp} = \mathbf{I} - \boldsymbol{\Pi}_{\mathbf{A}}$, where $\boldsymbol{\Pi}_{\mathbf{A}} = \mathbf{A} \mathbf{A}^\dagger$. The above expressions are evaluated at $\boldsymbol{\theta}_0$ and $\boldsymbol{\rho}_0$.

It is interesting to notice that the CRB for the case with only measurement noise and no calibration errors is $\sigma^2 \mathbf{C}^{-1}/2N$. Clearly, this is a lower bound for the combined CRB in (13.13), since $\mathbf{F}_{\boldsymbol{\theta}}^T \boldsymbol{\Upsilon}^{-1} \mathbf{F}_{\boldsymbol{\theta}}$ is positive semidefinite.

As shown by Jansson *et al.* (1998), the optimally weighted GWSF DOA estimator is consistent and has a limiting zero-mean Gaussian distribution with a covariance equal to the CRB matrix in (13.13). Hence, GWSF is a statistically efficient estimator for this quite general estimation problem. The MAP-NSF (Viberg and Swindlehurst, 1994b) and MAPprox (Wahlberg *et al.*, 1991; Jansson *et al.*, 1998) estimators also attain the CRB given above. However, GWSF has some advantages compared with those methods, especially for scenarios with coherent or highly correlated emitters. GWSF also allows for an efficient polynomial rooting based implementation, similar to that of IQML (Bresler and Macovski, 1986) and MODE (Stoica and Sharman, 1990b) for the estimation of the DOAs when the nominal array is uniform and linear (see the paper by Jansson *et al.* (1998) for details).

In the following sections, two special cases will be studied; namely, when either the measurement noise or the model errors dominate. It will be shown that in these cases, the generic estimator reduces to well-known methods.

13.5.2 GWSF for model errors only

In the previous section, it was assumed that the DOA estimation errors are significantly influenced by both noise and array model perturbations. Next, we study GWSF for the case when the model errors dominate, and neglect the noise. In other words, we will study the GWSF criterion when $\sigma^2 \to 0$ (or when $N \to \infty$).

In particular, we will study GWSF for the generic array perturbation model discussed in Section 13.3. In the notation from Section 13.5.1, (13.5)–(13.6) become

$$\mathbf{D}_\rho \bar{\Omega} \mathbf{D}_\rho^H = \mathbf{\Psi} \otimes \mathbf{\Gamma},$$
$$\mathbf{D}_\rho \bar{\Omega} \mathbf{D}_\rho^T = \mathbf{0}.$$

The GWSF optimal weighting matrix (13.10), (13.11) reduces to

$$\mathbf{W} = \begin{bmatrix} \mathbf{G}\mathbf{G}^H & \mathbf{0} \\ \mathbf{0} & (\mathbf{G}\mathbf{G}^H)^* \end{bmatrix}^{-1},$$

where $\mathbf{G} = (\mathbf{T}^T \otimes \mathbf{B}^H) \mathbf{D}_\rho \bar{\Omega}^{1/2}$. Hence, the GWSF criterion (13.9), (13.10) simplifies to

$$\begin{aligned}
V(\boldsymbol{\theta}) &= 2 \operatorname{vec}^H (\mathbf{B}^H \hat{\mathbf{E}}_s) \left[(\mathbf{T}^T \mathbf{\Psi} \mathbf{T}^* \otimes \mathbf{B}^H \mathbf{\Gamma} \mathbf{B}) \right]^{-1} \operatorname{vec}(\mathbf{B}^H \hat{\mathbf{E}}_s) \\
&= 2 \operatorname{Tr} \left\{ \hat{\mathbf{E}}_s^H \mathbf{B} (\mathbf{B}^H \mathbf{\Gamma} \mathbf{B})^{-1} \mathbf{B}^H \hat{\mathbf{E}}_s (\mathbf{T}^H \mathbf{\Psi}^T \mathbf{T})^{-1} \right\} \\
&= 2 \operatorname{Tr} \left\{ \mathbf{\Pi}_{\mathbf{\Gamma}^{-1/2}\mathbf{A}}^\perp \mathbf{\Gamma}^{-1/2} \hat{\mathbf{E}}_s (\mathbf{T}^H \mathbf{\Psi}^T \mathbf{T})^{-1} \hat{\mathbf{E}}_s^H \mathbf{\Gamma}^{-1/2} \right\},
\end{aligned}$$

where in the last equality we used that $\mathbf{B}^H \mathbf{\Gamma}^{1/2} \mathbf{\Gamma}^{-1/2} \mathbf{A} = \mathbf{0}$, and hence that

$$\mathbf{\Pi}_{\mathbf{\Gamma}^{1/2}\mathbf{B}} = \mathbf{\Pi}_{\mathbf{\Gamma}^{-1/2}\mathbf{A}}^\perp.$$

The above simplified expression of the GWSF criterion function corresponds exactly to the criterion function used by the robust subspace fitting (RSF) method by Swindlehurst and Kailath (1993). As shown by Swindlehurst and Kailath (1993), RSF is an optimally weighted subspace fitting method for the studied generic array response error model.

The RSF perturbation model is in a sense "nonparametric," and is probably most suitable for a case where the array response is measured by a calibration procedure. The GWSF formulation, on the other hand allows for general parameterizations of the model error, e.g., in terms of physical quantities.

13.5.3 GWSF with no model errors

Here, it is shown that the optimally weighted GWSF simplifies to the WSF method (Viberg and Ottersten, 1991) when no model errors are taken into account. For this case, $\bar{\mathbf{\Omega}} = \mathbf{0}$ in (13.12), and the GWSF criterion (13.9), (13.10), (13.11) can be rewritten as

$$
\begin{aligned}
V(\boldsymbol{\theta}) &= \bar{\boldsymbol{\varepsilon}}^H(\boldsymbol{\theta}) \begin{bmatrix} \mathbf{L}^{-1} & \mathbf{0} \\ \mathbf{0} & \mathbf{L}^{-*} \end{bmatrix} \bar{\boldsymbol{\varepsilon}}(\boldsymbol{\theta}) = 2\operatorname{vec}^H(\mathbf{B}^H\hat{\mathbf{E}}_s)\mathbf{L}^{-1}\operatorname{vec}(\mathbf{B}^H\hat{\mathbf{E}}_s) \\
&= 2\operatorname{vec}^H(\mathbf{B}^H\hat{\mathbf{E}}_s)\left(\sigma^{-2}\widetilde{\mathbf{\Lambda}}^2\mathbf{\Lambda}_s^{-1} \otimes (\mathbf{B}^H\mathbf{B})^{-1}\right)\operatorname{vec}(\mathbf{B}^H\hat{\mathbf{E}}_s) \\
&= 2\operatorname{vec}^H(\mathbf{B}^H\hat{\mathbf{E}}_s)\operatorname{vec}\left((\mathbf{B}^H\mathbf{B})^{-1}\mathbf{B}^H\hat{\mathbf{E}}_s\sigma^{-2}\widetilde{\mathbf{\Lambda}}^2\mathbf{\Lambda}_s^{-1}\right) \\
&= 2\operatorname{Tr}\left\{\hat{\mathbf{E}}_s^H\mathbf{B}(\mathbf{B}^H\mathbf{B})^{-1}\mathbf{B}^H\hat{\mathbf{E}}_s\sigma^{-2}\widetilde{\mathbf{\Lambda}}^2\mathbf{\Lambda}_s^{-1}\right\} \\
&= \frac{2}{\sigma^2}\operatorname{Tr}\left\{\mathbf{\Pi}_{\mathbf{A}}^{\perp}\hat{\mathbf{E}}_s\mathbf{W}_{\mathrm{WSF}}\hat{\mathbf{E}}_s^H\right\},
\end{aligned} \tag{13.14}
$$

where $\mathbf{W}_{\mathrm{WSF}} = \widetilde{\mathbf{\Lambda}}^2\mathbf{\Lambda}_s^{-1}$. Above, we again used that $\mathbf{B}$ spans the null space of $\mathbf{A}$ and, hence, $\mathbf{B}(\mathbf{B}^H\mathbf{B})^{-1}\mathbf{B}^H = \mathbf{\Pi}_{\mathbf{B}} = \mathbf{\Pi}_{\mathbf{A}}^{\perp}$. Equation (13.14) is exactly $2/\sigma^2$ times the WSF criterion as introduced by Viberg and Ottersten (1991). It is known that WSF is asymptotically statistically efficient. Consequently, GWSF will also be efficient.

Viberg and Ottersten (1991) showed that several known methods can be viewed as special cases of the subspace fitting formulation. In particular, it was shown that DML (Böhme, 1984), MD-MUSIC (Schmidt, 1981; Roy, 1987; Cadzow, 1988), TLS-ESPRIT (Paulraj *et al.*, 1986; Roy *et al.*, 1986; Roy and Kailath, 1989), and ML-ESPRIT (Roy, 1987) all have the same asymptotic performance as certain members of the signal subspace fitting (SSF) family of methods [obtained by choosing different weighting matrices in the SSF criterion (13.7)]. Since SSF is a special case of GWSF, weighting matrices can be chosen so that GWSF is asymptotically equivalent to the above-mentioned methods as well.

In order to see the connection between (one-dimensional spectral) MUSIC (Schmidt, 1979, 1981) and subspace fitting, consider the following subspace fitting criterion

$$
V(\boldsymbol{\theta}) = \min_{\mathbf{T}}\|\hat{\mathbf{E}}_s - \mathbf{a}(\theta)\mathbf{T}\|_F^2 = \operatorname{Tr}\left\{(\mathbf{I} - \mathbf{a}(\mathbf{a}^H\mathbf{a})^{-1}\mathbf{a}^H)\hat{\mathbf{E}}_s\hat{\mathbf{E}}_s^H\right\},
$$

where the dependence on θ was omitted for simplicity, and where $\|\mathbf{X}\|_F^2 =$

$\text{Tr}\{\mathbf{X}\mathbf{X}^H\}$ is the squared Frobenius norm. Next, notice that

$$\min_{\theta} \text{Tr}\Big\{(\mathbf{I} - \mathbf{a}(\mathbf{a}^H\mathbf{a})^{-1}\mathbf{a}^H)\hat{\mathbf{E}}_s\hat{\mathbf{E}}_s^H\Big\}$$

$$= \max_{\theta} \text{Tr}\Big\{\mathbf{a}(\mathbf{a}^H\mathbf{a})^{-1}\mathbf{a}^H\hat{\mathbf{E}}_s\hat{\mathbf{E}}_s^H\Big\} = \min_{\theta} \text{Tr}\Big\{\mathbf{a}(\mathbf{a}^H\mathbf{a})^{-1}\mathbf{a}^H\hat{\mathbf{E}}_n\hat{\mathbf{E}}_n^H\Big\}$$

$$= \min_{\theta} \frac{\mathbf{a}^H\hat{\mathbf{E}}_n\hat{\mathbf{E}}_n^H\mathbf{a}}{\mathbf{a}^H\mathbf{a}} = \max_{\theta} \frac{\mathbf{a}^H\mathbf{a}}{\mathbf{a}^H\hat{\mathbf{E}}_n\hat{\mathbf{E}}_n^H\mathbf{a}},$$

since $\hat{\mathbf{E}}_s\hat{\mathbf{E}}_s^H = \mathbf{I} - \hat{\mathbf{E}}_n\hat{\mathbf{E}}_n^H$. The last expression involves the well-known MU-SIC pseudospectrum. Clearly, MUSIC can be viewed as an (unweighted) one-dimensional subspace fitting method where the different DOAs are found by locating the d smallest local minima.

13.6 Numerical example

In this section, the performance of different DOA estimators are illustrated by means of a simulation example involving DOA estimation errors due to both noise and array model imperfections. Consider a uniform linear array of $m = 6$ sensors, separated by half a wavelength. Two signals impinge from the directions $0°$ and $10°$ relative to broadside. The signals are uncorrelated and of equal power. The sample size is fixed to $N = 200$, while the signal-to-noise ratio (SNR) is varied. The nominal unit gain sensors are perturbed by additive Gaussian random variables with a standard deviation of 5%. The nominal phases of the sensors are also perturbed by adding uncorrelated Gaussian random variables with standard deviation 0.05. This phase error corresponds approximately to a direction error of $1°$ around broadside.

Fig. 13.1 shows the root-mean-square (RMS) error versus the SNR. Only the RMS errors for the DOA angle θ_1 are displayed; the results corresponding to θ_2 are similar. The empirical RMS values are computed from 1000 independent Monte Carlo trials. The DOA estimators included in the comparison are: Root-MUSIC (Schmidt, 1979; Barabell, 1983), ESPRIT (using maximum overlap subarrays) (Paulraj *et al.*, 1986; Roy *et al.*, 1986; Roy and Kailath, 1989), WSF (Viberg and Ottersten, 1991) and GWSF (Jansson *et al.*, 1998) (see Section 13.5.1). The WSF and GWSF methods are implemented in their "rooting versions" (Stoica and Sharman, 1990b; Jansson *et al.*, 1998).

The approximate CRB (13.13) that accounts both for array model errors and the noise is denoted MAP-CRB in Fig. 13.1. The CRB for the ideal nominal case without array model errors is denoted NOM-CRB (i.e., the CRB accounting only for the additive noise, see, e.g., Stoica and Nehorai,

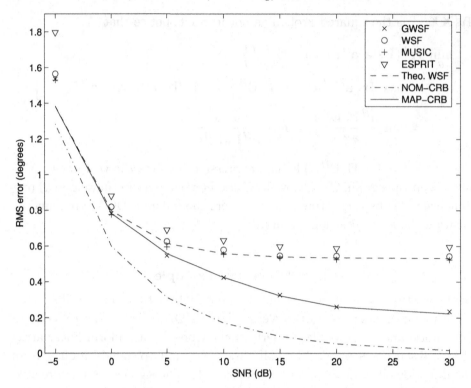

Fig. 13.1. The RMS errors for θ_1 versus the SNR.

1990). Also included in Fig. 13.1 is a curve resulting from a theoretical performance analysis of WSF for the case under study (i.e., for small model errors and large N or high SNR) (Viberg and Swindlehurst, 1994b; Jansson *et al.*, 1998).

It can be seen that for lower SNRs, when the measurement noise is the dominating error source, all methods have a very similar performance in this scenario, since the emitter signals are uncorrelated. For higher SNRs, the gain and phase errors dominate, and, as expected, the GWSF method outperforms the other methods that do not take the model errors into account. It can also be seen that the RMS error of GWSF attains the MAP-CRB as predicted by the theory. Similarly, the theoretical RMS curve for WSF predicts the corresponding empirical RMS values very well.

13.7 Concluding remarks

The subspace approach to signal parameter estimation has been successful in providing high accuracy at a reasonable cost. This chapter has reviewed a class of optimal subspace methods that maintains the asymptotic performance of maximum likelihood and similar techniques. The price for the statistical efficiency is in general that the complexity approaches that of maximum-likelihood methods. However, in special cases, computationally more attractive implementations are available, see, e.g., Stoica and Sharman (1990a), Jansson et al. (1998). We have also shown how the original MUSIC algorithm can be obtained from the general GWSF formulation. The GWSF framework, therefore, provides a clear link between classical statistical estimation and a wide class of practically useful subspace-based methods. Similar ideas as those presented here for the direction estimation problem can also be used in related applications, e.g., frequency estimation (Eriksson et al., 1994; Kristensson et al., 2001), subspace-based system identification (Verhaegen, 1994; Van Overschee and De Moor, 1996; Viberg et al., 1997), and blind channel identification (Moulines et al., 1995; Kristensson and Ottersten, 1998).

References

Barabell, A. J. (1983). Improving the resolution performance of eigenstructure-based direction-finding algorithms. In *Proc. IEEE Int. Conf. Acoust., Speech, Signal Process.*, 336–339.

Böhme, J. F. (1984). Estimation of source parameters by maximum likelihood and nonlinear regression. In *Proc. IEEE Int. Conf. Acoust., Speech, Signal Process.*, 7.3.1–7.3.4.

Bresler, Y. and Macovski, A. (1986). Exact maximum likelihood parameter estimation of superimposed exponential signals in noise. *IEEE Trans. Acoust., Speech, Signal Process.*, **34** (5), 1081–1089.

Cadzow, J. A. (1988). A high resolution direction-of-arrival algorithm for narrowband coherent and incoherent sources. *IEEE Trans. Acoust., Speech, Signal Process.*, **36** (7), 965–979.

Cardoso, J.-F. and Moulines, E. (2000). In-variance of subspace based estimators. *IEEE Trans. Signal Process.*, **48** (9), 2495–2505.

Compton, R. T. (1982). The effect of random steering vector errors in the Applebaum adaptive array. *IEEE Trans. Aerosp. Electron. Syst.*, **18** (5), 392–400.

Eriksson, A., Stoica, P., and Söderström, T. (1994). Markov-based eigenanalysis method for frequency estimation. *IEEE Trans. Signal Process.*, **42** (3), 586–594.

Flieller, A., Ferreol, A., Larzabal, P., and Clergeot, H. (1995). Robust bearing estimation in the presence of direction-dependent modelling errors: Identifiability and treatment. In *Proc. IEEE Int. Conf. Acoust., Speech, Signal Process.*, 1884–1887.

Friedlander, B. (1990a). A sensitivity analysis of the MUSIC algorithm. *IEEE Trans. Acoust., Speech, Signal Process.*, **38** (10), 1740–1751.

—— (1990b). Sensitivity of the maximum likelihood direction finding algorithm. *IEEE Trans. Aerosp. Electron. Syst.*, **26** (6), 953–968.

Friedlander, B. and Weiss, A. J. (1994). Effects of model errors on waveform estimation using the MUSIC algorithm. *IEEE Trans. Signal Process.*, **42** (1), 147–155.

Gustafsson, K., McCarthy, F., and Paulraj, A. (1996). Mitigation of wing flexure induced errors for airborne direction-finding applications. *IEEE Trans. Signal Process.*, **44** (2), 296–304.

Jansson, M., Swindlehurst, A. L., and Ottersten, B. (1998). Weighted subspace fitting for general array error models. *IEEE Trans. Signal Process.*, **46** (9), 2484–2498.

Kangas, A., Stoica, P., and Söderström, T. (1994). Finite sample and modeling error effects on ESPRIT and MUSIC direction estimators. *IEE Proc. Radar, Sonar, Navig.*, **141** (5), 249–255.

—— (1996). Large sample analysis of MUSIC and min-norm direction estimators in the presence of model errors. *Circ., Sys., Signal Process.*, **15** (3), 377–393.

Koerber, M. and Fuhrmann, D. (1993). Array calibration by Fourier series parameterization: Scaled principle components method. In *Proc. IEEE Int. Conf. Acoust., Speech, Signal Process.*, IV340–IV343.

Krim, H. and Viberg, M. (1996). Two decades of array signal processing research: The parametric approach. *IEEE Signal Process. Mag.*, **13** (4), 67–94.

Kristensson, M., Jansson, M., and Ottersten, B. (2001). Further results and insights on subspace based sinusoidal frequency estimation. *IEEE Trans. Signal Process.*, **49** (12), 2962–2974.

Kristensson, M. and Ottersten, B. (1998). A statistical approach to subspace based blind identification. *IEEE Trans. Signal Process.*, **46** (6), 1612–1623.

Li, F. and Vaccaro, R. (1992). Sensitivity analysis of DOA estimation algorithms to sensor errors. *IEEE Trans. Aerosp. Electron. Syst.*, **28** (3), 708–717.

Lo, J. and Marple, S. (1987). Eigenstructure methods for array sensor localization. In *Proc. IEEE Int. Conf. Acoust., Speech, Signal Process.*, 2260–2263.

McArthur, D. and Reilly, J. (1994). A computationally efficient self-calibrating direction of arrival estimator. In *Proc. IEEE Int. Conf. Acoust., Speech, Signal Process.*, IV–201 – IV–205.

Moulines, E., Duhamel, P., Cardoso, J.-F., and Mayrargue, S. (1995). Subspace methods for the blind identification of multichannel FIR filters. *IEEE Trans. Signal Process.*, **43** (2), 516–525.

Ng, B. and Nehorai, A. (1993). Active array sensor location calibration. In *Proc. IEEE Int. Conf. Acoust., Speech, Signal Process.*, IV21–IV24.

Ottersten, B., Viberg, M., Stoica, P., and Nehorai, A. (1993). Exact and large sample ML techniques for parameter estimation and detection in array processing. In Haykin, Litva, and Shepherd (eds.), *Radar Array Processing*, 99–151 (Springer-Verlag, Berlin).

Paulraj, A. and Kailath, T. (1985). Direction-of-arrival estimation by eigenstructure methods with unknown sensor gain and phase. In *Proc. IEEE Int. Conf. Acoust., Speech, Signal Process.*, 17.7.1–17.7.4.

Paulraj, A., Roy, R., and Kailath, T. (1986). A subspace rotation approach to signal parameter estimation. *Proc. IEEE*, **74** (7), 1044–1046.

Quazi, A. H. (1982). Array beam response in the presence of amplitude and phase fluctuations. *J. Acoust. Soc. Amer.*, **72** (1), 171–180.

Ramsdale, D. J. and Howerton, R. A. (1980). Effect of element failure and random errors in amplitude and phase on the sidelobe level attainable with a linear array. *J. Acoust. Soc. Amer.*, **68** (3), 901–906.

Rockah, Y. and Schultheiss, P. M. (1987a). Array shape calibration using sources in unknown locations – Part I: Far-field sources. *IEEE Trans. Acoust., Speech, Signal Process.*, **35** (3), 286–299.

—— (1987b). Array shape calibration using sources in unknown locations – Part II: Near-field sources and estimator implementation. *IEEE Trans. Acoust., Speech, Signal Process.*, **35** (3), 724–735.

Roy, R. and Kailath, T. (1989). ESPRIT – Estimation of signal parameters via rotational invariance techniques. *IEEE Trans. Acoust., Speech, Signal Process.*, **37** (7), 984–995.

Roy, R., Paulraj, A., and Kailath, T. (1986). ESPRIT – A subspace rotation approach to estimation of parameters of cisoids in noise. *IEEE Trans. Acoust., Speech, Signal Process.*, **34** (4), 1340–1342.

Roy, R. H. (1987). *ESPRIT, Estimation of Signal Parameters via Rotational Invariance Techniques*. Ph.D. thesis, Stanford Univ., Stanford, CA.

Schmidt, R. O. (1979). Multiple emitter location and signal parameter estimation. In *Proc. RADC Spectrum Estimation Workshop*, 243–258.

—— (1981). *A Signal Subspace Approach to Multiple Emitter Location and Spectral Estimation*. Ph.D. thesis, Stanford Univ., Stanford, CA.

Soon, V. C. and Huang, Y. F. (1992). An analysis of ESPRIT under random sensor uncertainties. *IEEE Trans. Signal Process.*, **40** (9), 2353–2358.

Stoica, P. and Nehorai, A. (1990). Performance study of conditional and unconditional direction-of-arrival estimation. *IEEE Trans. Acoust., Speech, Signal Process.*, **38** (10), 1783–1795.

Stoica, P. and Sharman, K. (1990a). Maximum likelihood methods for direction-of-arrival estimation. *IEEE Trans. Acoust., Speech, Signal Process.*, **38** (7), 1132–1143.

—— (1990b). Novel eigenanalysis method for direction estimation. *Radar and Signal Process., IEE Proc. F*, **137** (1), 19–26.

Stoica, P., Viberg, M., Wong, M., and Wu, Q. (1997). A unified instrumental variable approach to direction finding in colored noise fields. In Madisetti, V. and Williams, D. (eds.), *Digital Signal Processing Handbook*, 64.1–64.18 (CRC Press and IEEE Press).

Swindlehurst, A. (1996). A maximum a posteriori approach to beamforming in the presence of calibration errors. In *Proc. 8th Workshop Stat. Signal and Array Process.*, 82–85.

Swindlehurst, A. and Kailath, T. (1990). On the sensitivity of the ESPRIT algorithm to non-identical subarrays. *Sādhanā, Academy Proc. in Eng. Sciences*, **15** (3), 197–212.

—— (1992). A performance analysis of subspace-based methods in the presence of model errors – Part 1: The MUSIC algorithm. *IEEE Trans. Signal Process.*, **40** (7), 1758–1774.

—— (1993). A performance analysis of subspace-based methods in the presence of model errors – Part 2: Multidimensional algorithms. *IEEE Trans. Signal Process.*, **41** (9), 2882–2890.

Van Overschee, P. and De Moor, B. (1996). *Subspace Identification for Linear Systems: Theory–Implementation–Applications* (Kluwer Academic Publishers).

Van Trees, H. L. (2002). *Optimum Array Processing (Detection, Estimation, and Modulation Theory, Part IV)* (Wiley-Interscience).

Verhaegen, M. (1994). Identification of the deterministic part of MIMO state space models given in innovations form from input-output data. *Automatica*, **30** (1),

61–74.

Viberg, M. and Ottersten, B. (1991). Sensor array processing based on subspace fitting. *IEEE Trans. Signal Process.*, **39** (5), 1110–1121.

Viberg, M. and Swindlehurst, A. L. (1994a). Analysis of the combined effects of finite samples and model errors on array processing performance. *IEEE Trans. Signal Process.*, **42** (11), 3073–3083.

—— (1994b). A Bayesian approach to auto-calibration for parametric array signal processing. *IEEE Trans. Signal Process.*, **42** (12), 3495–3507.

Viberg, M., Wahlberg, B., and Ottersten, B. (1997). Analysis of state space system identification methods based on instrumental variables and subspace fitting. *Automatica*, **33** (9), 1603–1616.

Wahlberg, B., Ottersten, B., and Viberg, M. (1991). Robust signal parameter estimation in the presence of array perturbations. In *Proc. IEEE Int. Conf. Acoust., Speech, Signal Process.*, 3277–3280.

Wax, M. and Ziskind, I. (1989). On unique localization of multiple sources by passive sensor arrays. *IEEE Trans. Acoust., Speech, Signal Process.*, **37** (7), 996–1000.

Weiss, A. J. and Friedlander, B. (1989). Array shape calibration using sources in unknown locations - A maximum likelihood approach. *IEEE Trans. Acoust., Speech, Signal Process.*, **37** (12), 1958–1966.

Wong, K. M., Walker, R. S., and Niezgoda, G. (1988). Effects of random sensor motion on bearing estimation by the MUSIC algorithm. *IEE Proc., Part F*, **135** (3), 233–250.

Wylie, M., Roy, S., and Messer, H. (1994). Joint DOA estimation and phase calibration of linear equispaced (LES) arrays. *IEEE Trans. Signal Process.*, **42** (12), 3449–3459.

Yang, J. and Swindlehurst, A. (1995). The effects of array calibration errors on DF-based signal copy performance. *IEEE Trans. Signal Process.*, **43** (11), 2724–2732.

Yip, P. and Zhou, Y. (1995). A self-calibration algorithm for cyclostationary signals and its uniqueness analysis. In *Proc. IEEE Int. Conf. Acoust., Speech, Signal Process.*, 1892–1895.

Zhu, J. X. and Wang, H. (1988). Effects of sensor position and pattern perturbations on CRLB for direction finding of multiple narrowband sources. In *Proc. 4th ASSP Workshop Spectral Estimation and Modeling*, 98–102.

14

Blind and semiblind MIMO channel estimation

Dirk Slock

Institut Eurécom

Abdelkader Medles

Bell Laboratories, Lucent Technologies

The goal of this chapter is to expose a number of key ideas in blind and, especially, semiblind (SB) channel estimation (CE), and attract attention to various considerations that should be kept in mind in this context. As will become clear, the topic considered is vast. Due to space limitations, the inclusion and discussion of references is far from exhaustive. See also de Carvalho and Slock (2001) for an overview of semiblind single-input multiple-output (SIMO) channel estimation approaches. The use of blind information in digital communications is motivated by a desire to limit capacity loss due to training. Such capacity loss potentially increases with increasing time variation, occupied bandwidth, and number of transmitters. In other applications, blind techniques may be the only option (e.g., acoustic dereverberation). A particularity of digital communications, however, is that the sources are discrete time, white, and finite alphabet.

14.1 Signal model

In a first instance, the nonblind information considered will be provided by training or pilot information. As for terminology, the term *training sequence* (TS) tends to be used for a limited consecutive sequence of known symbols, whereas pilot symbols are typically isolated known symbols. A pilot signal is a continuous stream of known symbols, superimposed on the data signal.

Consider an (in a first instance time-invariant) discrete-time multiple-input multiple-output (MIMO) system with N_t inputs and N_r outputs,

$$\underbrace{\mathbf{y}_k}_{N_r \times 1} - \underbrace{\mathbf{v}_k}_{N_r \times 1} = \sum_{i=1}^{N_t} \sum_{l=0}^{L_i} \underbrace{\mathbf{h}_{i,l}}_{N_r \times 1} \underbrace{x_{i,k-l}}_{1 \times 1} = \sum_{l=0}^{L} \underbrace{\mathbf{H}_l}_{N_r \times N_t} \underbrace{\mathbf{x}_{k-l}}_{N_t \times 1} = \underbrace{\mathbf{H}(q)}_{N_r \times N_t} \underbrace{\mathbf{x}_k}_{N_t \times 1}, \quad (14.1)$$

where $\mathbf{H}(q) = \sum_{l=0}^{L} \mathbf{H}_l\, q^{-l}$, $L = \max\{L_i, i = 1, \ldots, N_t\}$, and we introduced

279

the one-sample delay operator: $q^{-1}\mathbf{x}_k = \mathbf{x}_{k-1}$; $\mathbf{v}_k$ is the additive noise. For the case of an (orthogonal frequency-division multiplexing (OFDM) or single-carrier) cyclic prefix (CP) block transmission system with N samples per block, the introduction of a cyclic prefix of $K \geq L$ samples means that the last K samples of the current block (of N samples) are repeated before the actual block. If we assume without loss of generality that the current block starts at time 0, then samples $\mathbf{x}_{N-K}\cdots\mathbf{x}_{N-1}$ are repeated at time instants $-K, \ldots, -1$. This means that the output at sample periods $0, \ldots, N-1$ can be written in matrix form as

$$\begin{bmatrix} \mathbf{y}_0 \\ \vdots \\ \mathbf{y}_{N-1} \end{bmatrix} = \mathbf{Y}_0 = \mathcal{C}(\mathbf{h})\,\mathbf{X}_0 + \mathbf{V}_0, \tag{14.2}$$

where the matrix $\mathcal{C}(\mathbf{h})$ is not only (block) Toeplitz, but even (block) circulant: each (block) row is obtained by a (block) cyclic shift to the right of the previous row. Consider now applying an N-point FFT to both sides of (14.2) at block m:

$$F_{N,N_r}\mathbf{Y}_m = F_{N,N_r}\mathcal{C}(\mathbf{h})F_{N,N_t}^{-1}\quad F_{N,N_t}\mathbf{X}_m \quad + \quad F_{N,N_r}\mathbf{V}_m, \tag{14.3}$$

or with new notation:

$$\mathbf{Y}_m = \mathcal{H} \qquad\qquad \mathbf{X}_m \qquad + \mathbf{V}_m, \tag{14.4}$$

where $F_{N,p} = F_N \otimes I_p$ (Kronecker product: $A \otimes B = [a_{ij}B]$), F_N is the N-point $N \times N$ DFT matrix, $\mathcal{H} = \mathrm{diag}\{\mathbf{H}_0, \ldots, \mathbf{H}_{M-1}\}$ is a block-diagonal matrix with diagonal blocks $\mathbf{H}_n = \sum_{l=0}^{L}\mathbf{H}_l e^{-j2\pi ln/N}$, the $N_r \times N_t$ channel transfer function at tone (subcarrier) n (frequency $= n/N$ times the sample frequency). In OFDM, the transmitted symbols are in $\mathbf{X}_m$ and, hence, are in the frequency domain. The corresponding time domain samples are in $\mathbf{X}_m$. The OFDM symbol period index is m. In single-carrier (SC) CP systems, the transmitted symbols are in $\mathbf{X}_m$ and, hence, are in the time domain. The corresponding frequency domain data are in $\mathbf{X}_m$. The components of $\mathbf{V}_m$ are assumed to be white noise; hence, the components of $\mathbf{V}_m$ are also white. At tone $n \in \{0, \ldots, N-1\}$, we get the following input-output relation

$$\underbrace{\mathbf{y}_n[m]}_{N_r \times 1} = \underbrace{\mathbf{H}_n}_{N_r \times N_t}\,\underbrace{\mathbf{x}_n[m]}_{N_t \times 1} + \underbrace{\mathbf{v}_n[m]}_{N_r \times 1}, \tag{14.5}$$

where the elements (symbols) of $\mathbf{x}_n[m]$ belong to some finite alphabet (constellation) in the case of OFDM.

14.2 Structured deterministic and stochastic channel models

The MIMO channel for the spatial multiplexing case is a special form of the multiuser channel, one in which the multiple users are colocated. As a result, the channel lengths L_i are usually equal. This remains the case even if sensors with different polarizations are used, but not necessarily if pattern (beam) diversity is used. Also refer to Chapter 1 for channel models.

14.2.1 Pathwise channel models

To go beyond the finite impulse response (FIR) channel model, more structured channel models, such as infinite impulse response (IIR) models, could be useful for applications in which the output may return to the input in some sense, creating natural modes, as in acoustic applications in an enclosed medium, or for transmission lines. For wireless communications, however, structured channel models can be based on the physics of the propagation mechanism. Since attenuation increases rapidly after a few reflections or diffractions, direct input-to-output transfer models are most appropriate. So, apart from the nonparametric FIR model (in which knowledge of the pulse shape may also be expressed, see Section 14.6.1), parametric pathwise channel models may be considered. The specular time-varying MIMO channel impulse response $\mathbf{H}(t, \tau)$ is of the form

$$\mathbf{H}(t, kT) = \sum_{i=1}^{N_p} A_i(t)\, e^{j2\pi f_i t}\, \mathbf{a}_R(\phi_i)\mathbf{a}_T^T(\theta_i)\, p(kT - \tau_i),$$

to which each path contributes a rank-one component in three dimensions: delay, direction of arrival (DOA), and direction of departure (DOD). The N_p pathwise contributions involve: complex attenuation A_i; Doppler shift $f_i \in (-f_d, f_d)$ (where f_d is the Doppler frequency); angle of departure θ_i; angle of arrival ϕ_i; path delay τ_i; $\mathbf{a}_R(\cdot)$, $\mathbf{a}_T(\cdot)$ are the Rx/Tx[†] antenna array responses, $p(\cdot)$ is the pulse shape (Tx filter), and T is the symbol period. Note that the DOA and DOD may involve more than one angle parameter. Consider stacking the columns of the consecutive impulse response matrix coefficients to obtain

$$\mathbf{h}(t) = \text{vec}\{\mathbf{H}(t, kT)\} = \sum_{i=1}^{N_p} \mathbf{h}^{(i)} e^{j2\pi f_i t} A_i(t) = \mathbf{PA}(t), \qquad (14.6)$$

[†]For the sake of brevity, Rx may alternatively refer to receiver(s)/receive/receiving etc., Tx to transmitter(s)/transmit/transmitting and similarly for other abbreviations.

where $\mathbf{h}$ is $N \times 1$ with $N = N_t N_r L$, $\mathbf{P} = [\mathbf{h}^{(1)} \cdots \mathbf{h}^{(N_p)}]$, $\mathbf{A}(t)$ is $N_p \times 1$, containing the $e^{j2\pi f_i t} A_i(t)$. The "fast" parameters $A_i(t)$ are band-limited (often to much less than f_d), so that $e^{j2\pi f_i t} A_i(t)$ is a modulated lowpass signal. Fast fading is essentially due to the Doppler shifts f_i, and leads to much faster variation of the channel coefficients with a more complex Doppler spectrum (compared with the $A_i(t)$), due to a superposition of path contributions. The "slow" parameters f_i, τ_i, θ_i, ϕ_i vary much more slowly, according to the slow fading (varying multipath structure).

14.2.2 Deterministic and stochastic time-varying channel models

Two approaches can be introduced for time variation: modeling $\mathbf{h}_k = \mathbf{h}(kT)$ as a stationary vector process, called the stationary model for short, or using a basis expansion model (BEM), in which the time-varying channel coefficients are expanded into known time-varying basis functions, and the unknown channel parameters are now no longer the channel coefficients but the combination coefficients in the BEM. The BEM model was introduced by Y. Grenier around 1980 for time-varying filtering, by E. Karlsson in the early 1990s for time-varying channel modeling, and by M. Tsatsanis and G. Giannakis in 1996 for blind time-varying channel estimation. The BEM has also been revived and generalized in the canonical coordinates concept of A. Sayeed, in which the basis functions are signal independent (nonparametric).

The choice of the channel model interacts with the design of the modulation format. For instance, a stationary model may be more appealing for the case of long transmission packets (e.g., corresponding to a packet of data within one convolutional coding operation), whereas the BEM model might be more appealing in the case of shorter packets (e.g., OFDM symbols), the length of which would correspond to a potential subsampling period of the channel variation (related to maximum Doppler spread) appearing in the BEM. In the case of the stationary model, the stationarity suggests Wiener filtering of brute channel estimates, but the transients at both edges of the packet may be more properly treated with a Kalman filter or smoother. For the BEM model, the question arises whether to model potential correlations between basis expansion coefficients, or just their variances (stationary and BEM models are equivalent if all correlations are accounted for in the BEM model). A (nonparametric) BEM is well designed if the expansion coefficients are fairly uncorrelated. A parametric BEM approach is obtained when working with the Karhunen–Loève expansion of the channel temporal variation correlation. Nonparametric BEMs typically correspond to subsampling and interpolation of lowpass signals. The *block-fading* model is a BEM with rectangular basis functions, leading to a subsampling of the channel variation. In Tong *et al.* (2004), it is mentioned that a stationary channel estimation error is obtained when the downsampled version of the channel satisfies the Nyquist criterion (i.e., allows reconstruction of the continuous-time lowpass channel variation). The block-fading and stationary models can be combined (subsampling first). BEMs can be applied to $\mathbf{h}(t)$ directly or to $\mathbf{A}(t)$. Modeling the $A_i(kT)$ (and hence the $e^{j2\pi f_i kT}A_i(kT)$) as independent autoregressive (AR) processes leads to a subspace AR model (Slock, 2004a) for $\mathbf{h}_k$ with spectrum $S_{\mathbf{hh}}(z) = \mathbf{P}S_{\mathbf{AA}}(z)\mathbf{P}^H$. So the channel may be quite predictable (especially in the wideband MIMO case), since $S_{\mathbf{hh}}(z)$ may be doubly singular, owing to limited bandwidth of $S_{\mathbf{AA}}(z)$ and limited rank, $N_p < N$. The block-fading model applied to the $A_i(t)$ leads to a four-dimensional rank-one contribution per path to $S_{\mathbf{hh}}(z)$. Although it would rarely make sense in practice, a four-dimensional separable correlation model of the form $S_{\mathbf{hh}}(f) = R_\tau \otimes R_T \otimes R_R\, S_d(f)$ may be considered for the purpose of performance analysis, where R_τ is the correlation matrix (R) between delays (typically diagonal with power delay profile), R_T is the Tx side R, R_R is the Rx side R, and $S_d(f)$ is the scalar common Doppler spectrum (shape) of all channel coefficients. For the estimation of structured channels, an unstructured version may be estimated first, with the structure being imposed in a second stage.

14.3 Performance indicators

14.3.1 Channel capacity

See also Chapters 4 and 6 for channel capacity and its dependence on channel state information at the receiver (CSIR), or a relevant discussion and references in Tong *et al.* (2004). However, in those references, mostly the gap to channel capacity with CSIR of schemes based on training only is studied.

Consider a possibly time-selective frequency-flat channel $\mathbf{y}_k = \mathbf{H}_k\,\mathbf{x}_k + \mathbf{v}_k$. Within the block duration $T = N_{TS} + N_B$, the pilot symbols are contained in $\mathbf{X}^{TS}$, whereas the data (blind part) is contained in $\mathbf{X}^B$, where we define $\mathbf{X}_i^k = [\mathbf{x}_i, \mathbf{x}_{i+1}, \ldots, \mathbf{x}_k]$.

Mutual information (MI) decomposition

The MI is a crucial quantity, since the channel capacity is the MI when an optimal input distribution is used. Assuming the Rx has no side information about the channel, apart from the Rx signal, the MI between the Tx and

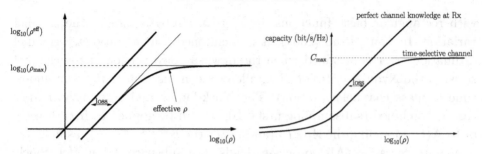

Fig. 14.1. Effective SNR and capacity vs. SNR behavior for a time selective channel.

Rx signals is then

$$I(\mathbf{Y}^{TS}, \mathbf{Y}^B; \mathbf{X}^B | \mathbf{X}^{TS}) = I(\mathbf{Y}^B; \mathbf{X}^B | \mathbf{X}^{TS}, \mathbf{Y}^{TS}).$$

The sequential expansion of this expression (Medles and Slock, 2003b) leads to

$$I(\mathbf{Y}^B; \mathbf{X}^B | \mathbf{X}^{TS}, \mathbf{Y}^{TS}) = \sum_{i=1}^{N_B} I(\mathbf{y}_i; \mathbf{x}_i | \mathbf{X}^{TS}, \mathbf{x}_1^{i-1}, \overline{\mathbf{Y}}_i),$$

where $\overline{\mathbf{Y}}_i = [\mathbf{Y}^{TS}, \mathbf{y}_1^{i-1}, \mathbf{y}_{i+1}^{N_B}]$ contains all the Rx signal apart from $\mathbf{y}_i$. From this expression, we conclude that an optimal way of processing is to use the past detected symbols as training, and the future (not detected) symbols as blind information for the CE, hence SB CE. This is consistent with the DFE canonical Rx concept: not only Rx but also CE is based on all Rx signal plus past symbols. Furthermore, in Medles and Slock (2003b), bounds on the MI are obtained, assuming white Gaussian inputs, corresponding to modeling CE error induced noise as an additive independent Gaussian noise. These bounds lead to a decrease in effective SNR ρ and ensuing MI loss as depicted in Fig. 14.1. Upper and lower bounds are obtained by considering CE based on different types of information:

> TS-type channel estimate based on past symbols (as TS)
> $\leq$ channel capacity without CSIR
> $=$ semiblind channel estimate based on known past symbols
>   and unknown Gaussian future symbols
> $\leq$ TS-type channel estimate based on past and future symbols

Asymptotic behavior of the capacity for block-fading channels

In this case, $\mathbf{H}_k = \mathbf{H}$ for $k = 1, \ldots, T$. The average MI is defined as

$$I_{\text{avg}}(T) = \frac{1}{T} I(\mathbf{Y}^B; \mathbf{X}^B | \mathbf{X}^{TS}, \mathbf{Y}^{TS}) = \frac{1}{T} \sum_{k=1}^{N_B} I(\mathbf{y}_k; \mathbf{x}_k | \mathbf{X}^{TS}, \mathbf{x}_1^{k-1}, \overline{\mathbf{Y}}_k).$$

(14.7)

For infinite blocks, we obtain the following limit (Medles and Slock, 2003b)

$$\lim_{T \to \infty} I_{\text{avg}}(T) = I(\mathbf{y}; \mathbf{x} | \mathbf{H}),$$

where $I(\mathbf{y}; \mathbf{x} | \mathbf{H})$ is the average MI with perfect CSIR. For block-fading channels, there is no loss in MI for large blocks when optimal SB CE algorithms are used.

Remark 1 As T grows, the use of detected data only to estimate the channel allows us to achieve asymptotically the average MI I_{avg} of the system. But for finite T, it is also necessary to use future undetected symbols to reach it.

Remark 2 The MI expression of (14.7) does not differentiate between training and past detected symbols. Then, for a fixed T, and when all the entries of $\mathbf{X}$ (training and data) are i.i.d., it is easy to see that the average MI I_{avg} of the system is maximized when the number of TS symbols N_{TS} is as small as possible (i.e., allows SB identifiability of the channel).

14.3.2 Other performance criteria

Once the Tx signal structure has been fixed (including the pilot structure), performance may be characterized in terms of the channel estimation quality, via, e.g., the Cramér–Rao bound (CRB), or in terms of the bit error rate (BER). A BER upper bound is obtained by considering the channel estimation error induced noise as additional additive independent Gaussian. This BER upper bound is, hence, described by this noise variance increase and ensuing shift in SNR. More discussion on performance criteria can be found in Tong *et al.* (2004).

14.4 Basic semiblind techniques

Basic refers here to the only nonblind information coming from time-multiplexed (TM) pilots, the channel being deterministic (block fading), the

noise being spatio-temporally (ST) white Gaussian, the model for the unknown symbols being deterministic symbols (DSB) or i.i.d. Gaussian symbols (GSB). Consider the blind case first (DB/GB).

In the DB case, the compressed likelihood, after eliminating the unknown
symbols via $\mathbf{x}_k = (\mathbf{H}^\dagger(q)\mathbf{H}(q))^{-1}\mathbf{H}^\dagger(q)\,\mathbf{y}_k$ (MMSE-ZF equalizer), is a function of $P_{\mathbf{H}(z)} = \mathbf{H}(z)(\mathbf{H}^\dagger(z)\mathbf{H}(z))^{-1}\mathbf{H}^\dagger(z)$, or the column space of $\mathbf{H}(z)$.
Obviously, $N_r > N_t$ is required, and if $\mathbf{H}(z)$ is irreducible (no zeros) and
column reduced, with columns ordered in nonincreasing length, then the re-
lation between $\mathbf{H}(z)$ and a $\widehat{\mathbf{H}}(z)$ that can be deterministically identified is
$\widehat{\mathbf{H}}(z) = \mathbf{H}(z)\mathbf{L}(z)$, where $\mathbf{L}(z)$ is block lower triangular, the diagonal blocks
have dimensions commensurate with the groups of columns of $\mathbf{H}(z)$ that
have identical length L_i (order L_i-1); the diagonal blocks are instantaneous
mixtures, and the lower triangular entries (i,j) are FIR of order $L_j - L_i$. So,
if all L_i are identical, then $\mathbf{L}$ is a square instantaneous mixture.

In the GB case, we get for the Rx spectrum $S_{\mathbf{yy}}(z) = \sigma_x^2\,\mathbf{H}(z)\,\mathbf{H}^\dagger(z) +$
$S_{\mathbf{vv}}(z)$, with the usual assumption $S_{\mathbf{vv}}(z) = \sigma_v^2\,I_{N_r}$ (ST white), although
$S_{\mathbf{vv}}(z) = S_{vv}(z)\,I_{N_r}$ (spatially white) with scalar $S_{vv}(z) = \lambda_{\min}(S_{\mathbf{yy}}(z))$ is
sufficient with $N_r > N_t$. With columns of $\mathbf{H}(z)$ arranged in nonincreasing
length, we get

$$\mathbf{H}(z) = (S_{\mathbf{yy}}(z) - S_{vv}(z)\,I_{N_r})^{\frac{1}{2}}\,\mathbf{L},$$

where $(\cdot)^{1/2}$ denotes a minimum-phase spectral factor (with columns in nonincreasing length), and $\mathbf{L}$ is block diagonal (same structure as the $\mathbf{L}(z)$ of
the DB case) with unitary diagonal blocks. So, when all L_i are identical,
the unidentifiable part $\mathbf{L}$ is a unitary instantaneous mixture. Although GB
reduces the number of unidentifiable parameters, DB leads to consistency
in SNR (zero error in absence of noise) for the DB-identifiable part (while
GB does not for the GB extra identifiable part). Note that GB provides
blind information even when $N_t \geq N_r$. Going from blind to semiblind, a
proper design should introduce enough nonblind information to allow complete channel identifiability, regardless of the channel realization. The nonblind information thus required is much less in the GB case than in the DB
case.

14.4.1 Frequency-flat MIMO channels

In this case, the channel impulse response in (14.1) is limited to $\mathbf{H} = \mathbf{H}_0$,
and there is no intersymbol interference (ISI). The TS has a length of N_{TS}
pilot symbol vectors $\mathbf{x}_k$, and the blind data part is composed of N_B data
symbol vectors. The second-order statistics (SOS) of the Rx signal are given

by

$$\widehat{\mathbf{R}} = \frac{1}{N_B} \sum_{k=1}^{N_B} \mathbf{y}_k \mathbf{y}_k^H = \mathbf{U}_e \mathbf{S}_e \mathbf{U}_e^H.$$

Using the singular value decomposition (SVD) of the channel

$$\mathbf{H} = \mathbf{UDQ} = \mathbf{WQ},$$

it is easy to see that only $\mathbf{W}$ can be identified blindly (GB), whereas $\mathbf{Q}$ has to be identified using TS. However, this does not mean that estimation of $\mathbf{W}$ and $\mathbf{Q}$ is decoupled. The optimal technique is the Maximum Likelihood (ML) approach, involving the complete Rx signal. However, in Medles and Slock (2003a) it was shown that for $N_B \gg (\sigma_x^2/\sigma_v^2)\, N_{TS}$, ML can be simplified by first estimating $\mathbf{W}$ from $\widehat{\mathbf{R}}$ and then estimating $\mathbf{Q}$ from TS as follows:

(i) Estimation of $\mathbf{W}$:

- $\widehat{\mathbf{U}}$ corresponds to the p dominant eigenvectors in $\mathbf{U}_e$ (where $p = \min\{N_t, N_r\}$)
- $\widehat{\mathbf{D}}$ matches the p dominant eigenvalues of $(1/\sigma_x)\left(\lfloor \mathbf{S}_e - \sigma_v^2 \mathbf{I}_{N_r}\rfloor_+\right)^{1/2}$
- $\widehat{\mathbf{W}} = \widehat{\mathbf{U}}\widehat{\mathbf{D}}$

(ii) Estimation of $\mathbf{Q}$:
$\widehat{\mathbf{Q}} = \mathbf{V}\,\mathbf{S}^H$, where $\mathbf{S}$ and $\mathbf{V}$ denote the unitary factors in the SVD of $\sum_{k=1}^{N_{TS}} \mathbf{x}_k^{TS} \mathbf{y}_k^{TSH} \mathbf{W} = \mathbf{S}\mathbf{\Sigma}\mathbf{V}^H$, assuming $\sum_{k=1}^{N_{TS}} \mathbf{x}_k^{TS} \mathbf{x}_k^{TSH} \sim I_{N_t}$.

The evaluation of the performance shows that this technique achieves the GSB CRB for $N_B \gg (\sigma_x^2/\sigma_v^2)\, N_{TS}$; see Fig. 14.2 for an illustration, and Medles and Slock (2005) for more details.

14.4.2 Frequency-selective MIMO channels

Time domain approaches

In this case, there is ISI and nonorthogonality between pilot and data symbols, as depicted in Fig. 14.3 for a single burst of pilots. If we regroup all pilots in $\mathbf{X}_{TS}$, and hence $\mathbf{X}_U$ collects all unknown symbols, we can write $\mathcal{T}(\mathbf{h})\,\mathbf{X} = \mathcal{T}_K(\mathbf{h})\,\mathbf{X}_{TS} + \mathcal{T}_U(\mathbf{h})\,\mathbf{X}_U$, where $\mathcal{T}_K$ produces the channel output due to known TS. The Gaussian SB (GSB) ML criterion becomes

$$\ln \det(C_U) + (\mathbf{Y} - \mathcal{T}_K(\mathbf{h})\,\mathbf{X}_{TS})^H\, C_U^{-1}\, (\mathbf{Y} - \mathcal{T}_K(\mathbf{h})\,\mathbf{X}_{TS}),$$

where $C_U = \sigma_x^2\, \mathcal{T}_U(\mathbf{h})\, \mathcal{T}_U^H(\mathbf{h}) + C_{\mathbf{VV}}$. This GSB ML criterion often gets simplified to the following augmented TS (ATS) criterion (information from the mean only) (Medles *et al.*, 2001): $(\mathbf{Y}_{\mathrm{TS}} - \mathcal{T}_{\mathrm{TS}}(\mathbf{h})\mathbf{X}_{TS})^H C_{\mathrm{TS}}^{-1}(\mathbf{Y}_{\mathrm{TS}} -$

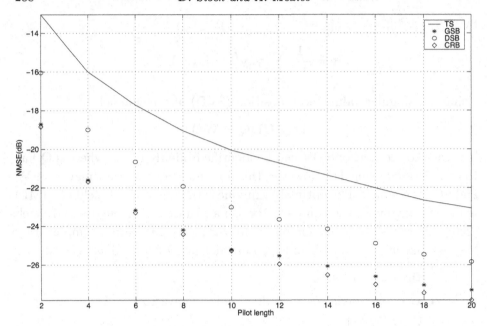

Fig. 14.2. Normalized channel estimation MSE vs. pilot length N_{TS}: frequency-flat channel, $N_t = 2$, $N_r = 4$, $N_B = 400$, SNR = 10 dB.

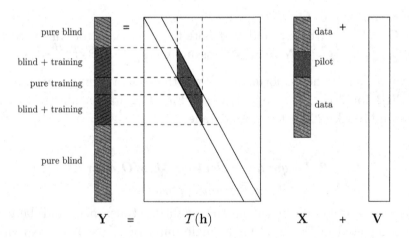

Fig. 14.3. Received signal structure for a frequency-selective channel.

$\mathcal{T}_{\mathrm{TS}}(\mathbf{h})\mathbf{X}_{\mathrm{TS}})$, where $\mathbf{Y}_{\mathrm{TS}}$ is the part of $\mathbf{Y}$ containing at least one pilot symbol. ATS can be solved iteratively as a weighted LS problem. See Pladdy *et al.* (2004) for a simplified ATS algorithmic approach, whereas in Rousseaux and Leus (2004) the interference from unknown data on pilots is canceled blindly in an iterative approach. In Rousseaux *et al.* (2003b), ATS is treated as a simplified ML criterion for the case in which $C_{\mathbf{VV}}$ is considered totally unknown (and unstructured), in which case C_{TS} is built from a sample covariance matrix. In Rousseaux *et al.* (2003a), distributed training bursts are treated. Though a significant improvement over TS, ATS only makes limited use of the blind information. Although the GSB ML problem is meaningful, and becomes straightforward to solve when the pilots are isolated, in some unpublished work we have found that grouped pilots lead to better performance here as well (when the channel does not vary fast).

To exploit the blind information, a general technique is to use a weighted least-squares (WLS) combination of TS and blind information:

$$\min_{\widehat{\mathbf{h}}}\left\{\left\|\mathbf{Y}_{\mathrm{TS}} - \mathcal{X}_{\mathrm{TS}}\,\widehat{\mathbf{h}}\right\|_{C_{\mathrm{TS}}^{-1}(\widehat{\mathbf{h}})}^{2} + \left\|\mathcal{B}\widehat{\mathbf{h}}\right\|_{C^{\#}(\widehat{\mathbf{h}})}^{2}\right\}, \qquad (14.8)$$

where $\|\mathbf{Y}\|_C^2 = \mathbf{Y}^H C \mathbf{Y}$, $\mathcal{X}_{\mathrm{TS}}\mathbf{h} = \mathcal{T}_{\mathrm{TS}}(\mathbf{h})\mathbf{X}_{\mathrm{TS}}$, $C(\mathbf{h}) = \mathrm{E}\big[(\mathcal{B}\mathbf{h})(\mathcal{B}\mathbf{h})^H\big]$, and correlation between blind and TS parts is neglected. $\mathcal{B}$ is a matrix that (soft-) constrains the channel and expresses the blind information by parameterizing the noise subspace, or capturing the whiteness of the transmitted signal. $\mathcal{B}$ can be appropriately parameterized in terms of the prediction error filter $\mathbf{P}_K(q) = \mathbf{I} + \sum_{i=1}^{K-1}\mathbf{P}_i q^{-i}$ for the noise-free signal SOS: $\mathbf{P}_K(q)\mathbf{H}(q) = \mathbf{H}_0$. For $N_r \geq N_t$, the channel predictor generically exists and is FIR for $N_r > N_t$ with $K \geq \lceil (L - N_t)/(N_r - N_t)\rceil$; it can be evaluated from the Rx signal SOS $\widehat{\mathbf{R}}_{\mathbf{yy}}(k)$, $k = 0,\ldots,K$, and leads to a parameterization of the channel: $\mathbf{H}(z) = \mathbf{P}^{-1}(z)\mathbf{W}\mathbf{Q}$, $(S_{\mathbf{yy}}(z) - \sigma_v^2 I_{N_r})\mathbf{P}^{\dagger}(z) = \mathbf{H}(z)\sigma_x^2 \mathbf{Q}^H \mathbf{W}^H$, and $\mathbf{W}$ is obtained from $\mathbf{P}(z)(S_{\mathbf{yy}}(z) - \sigma_v^2 I_{N_r})\mathbf{P}^{\dagger}(z) = \sigma_x^2 \mathbf{W}\mathbf{W}^H$. However, unlike the flat channel case, there is no simple ML variation to estimate $\mathbf{Q}$. If we reduce the exploitation of $\mathbf{P}(q)\mathbf{H}_k = \mathbf{H}_0\delta_{k0}$ to $\mathbf{W}^{\perp H}\mathbf{H}_0 = 0$ and $\mathbf{P}(q)\mathbf{H}_k = 0$, $k > 0$, and combine it with the TS part in a WLS approach, then the result is the quadratic criterion of (14.8). Further results on this approach can be found in Medles *et al.* (2001) and Medles and Slock (2005).

It is interesting to mention that the more conventional subspace-based techniques only exploit the set of equations $\mathbf{W}^{\perp H}\mathbf{P}(q)\mathbf{H}_k = 0$, which corresponds to the (DB) information in the noise subspace (here parameterized by $\mathbf{W}^{\perp H}\mathbf{P}(q)$). The full criterion enhances the performance by further exploiting the temporal whiteness of the Tx symbols (GB), which is characterized by $\mathbf{P}(q)\mathbf{H}_k = 0$, $k > 0$. In Fig. 14.4, an illustration is given of the performance of basic TS and ATS, and DSB and GSB with basic TS or ATS (DSBA, GSBA). The number of pilots leads to unidentifiability for TS, ATS or DSB, and identifiability for GSB, DSBA and GSBA. This example illustrates the usefulness of blind information and the extra information present in GSB (be it not consistent in SNR).

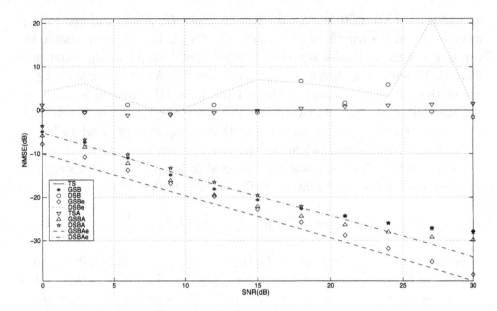

Fig. 14.4. Normalized channel estimation MSE vs. SNR: frequency-selective channel, $N_t = 2$, $N_r = 4$, channel lengths $[3, 1]$, $N_B = 300$, $N_{TS} = 5$.

Frequency domain approaches

Note that time-domain-multiplexed pilots become embedded pilots (see further) in the frequency domain, and vice versa. In the CP case, we get for the SOS from (14.4)

$$\mathbf{R_{YY}} = \sigma_x^2 \mathcal{H}\mathcal{H}^H + \sigma_v^2 I_{N_r N},$$

which shows the decoupling between tones, or from (14.5) per tone

$$\mathbf{R}_n = \mathbf{R}_{y_n y_n} = \sigma_x^2 H_n H_n^H + \sigma_{v_n}^2 I_{N_r} = V_{\mathcal{S},n}\Lambda_{\mathcal{S},n}V_{\mathcal{S},n}^H + \sigma_{v_n}^2 V_{\mathcal{N},n}V_{\mathcal{N},n}^H,$$

where $\sigma_{v_n}^2$ can vary between tones for a temporally colored but spatially white noise (which hence becomes straightforward to handle, approximating or assuming $\mathbf{R_{VV}}$ to be block circulant). The coupling between tones comes from the FIR channel response. Let $h_n = \text{vec}(H_n)$; then, the FFT relation leads to $h_n = G_n \mathbf{h}$ for some matrix G_n. Now, if at tone n we have a cost function of the form $h_n^H Q_k h_n$, then this induces a cost function for $\mathbf{h}$ of the form $\mathbf{h}^H \left[\sum_{n=0}^{N-1} G_n^H Q_n G_n\right]\mathbf{h}$, and similarly for Fisher information matrices. So, one can just concentrate on the cost function for a given tone. For instance, if $\widehat{\mathbf{R}}_n = \widehat{\mathbb{E}}[u_n u_n^H] = \widehat{V}_{\mathcal{S},n}\widehat{\Lambda}_{\mathcal{S},n}\widehat{V}_{\mathcal{S},n}^H + \widehat{V}_{\mathcal{N},n}\widehat{\Lambda}_{\mathcal{N},n}\widehat{V}_{\mathcal{N},n}^H$, then we get the signal subspace fitting cost function: $\min_{\mathbf{h}} \sum_{n=0}^{N-1} \|H_n^H \widehat{V}_{\mathcal{N},n}\|_F^2$, requiring only SVDs of small matrices.

Due to the decoupling between tones, one may also envisage the introduction of optimal weighting: $\max_{\mathbf{h}} \sum_{n=0}^{N-1} \text{tr}\{P_{\boldsymbol{H}_n} \widehat{V}_{\mathcal{S},n} \widehat{\widetilde{\Lambda}}_{\mathcal{S},n}^2 \widehat{\Lambda}_{\mathcal{S},n}^{-1} \widehat{V}_{\mathcal{S},n}^H\}$. More discussion can be found in Slock (2004b). See also Liu *et al.* (2001) for a per-tone application of the Constant Modulus Algorithm (CMA). In Zeng and Ng (2004), a subspace method (and semiblind version) is proposed based on zero padding, which leads to more robust channel estimates (compared with CP), but is computationally complex (SVDs are not per tone but of OFDM symbol size). A simpler version is discussed in Slock (2004b).

14.5 Bayesian semiblind (BSB) channel estimation

BSB CE has been introduced in Slock (2004a), where more details can be found. Separation property: BSB can be organized as basic SB followed by Bayesian filtering, especially if enough training for basic SB is available to lead to full identifiability. The basic semiblind CE leads to a measurement equation $\widehat{\mathbf{h}}_k = \mathbf{h}_k + \widetilde{\mathbf{h}}_k$, with $\widetilde{\mathbf{h}}_k \sim \mathcal{CN}(b(\mathbf{h}_k), C(\mathbf{h}_k))$, temporally decorrelated. The measurement equation then gets combined with the prior channel model (e.g., Rayleigh fading) in a Bayesian filtering operation. In the TS case, $b \equiv 0$, $C = \sigma_{\tilde{h}}^2 I$, leading to Wiener filtering/smoothing. In the SB case, $b(\mathbf{h}_k), C(\mathbf{h}_k) \longrightarrow b(\widehat{\mathbf{h}}_k), C(\widehat{\mathbf{h}}_k)$, the measurement equation being possibly time-varying, Kalman filtering/smoothing may be required. The full Bayesian approach requires joint estimation of Bayesian channel and deterministic prior hyper-parameters, which may be done with a variety of approaches, including EM. The appropriate CRB here is for joint deterministic slow and Bayesian fast parameters. See also Hassan *et al.* (2004) (and references therein) for a combination of Kalman filtering and CMA, and Haykin *et al.* (2004) (and references therein) for an overview on the use of Kalman and particle filtering. Whereas most applications of Wiener/Kalman filtering to channel tracking proposed in the literature assume the state space model to be known, the estimation of the channel variation statistics is incorporated in Lenardi and Slock (2002), Montalbano and Slock (2003).

14.6 Other forms of side information

With perfect CSIR (and no CSIT, and i.i.d. channel elements), the optimal input signal is zero-mean ST white Gaussian noise. Any deviation from this (side information) will lower the perfect CSIR channel capacity. But of course, there usually is no CSIR, so any such deviation may allow channel estimation, hence leading to an increase in actual channel capacity (see

Zheng and Tse (2002) for optimal input distributions in the absence of CSIR).
Possible forms of side information are:

- higher-order statistics of data symbols (Cardoso, 1998; Cichocki and Amari, 2002);
- finite alphabet (FA) of unknown symbols, exploited through iterative channel estimation and data detection, see, e.g., Talwar and Paulraj (1997), Li and Yang (2003), Zhu *et al.* (2003), Scaglione and Vosoughi (2004), Souza *et al.* (2004), or Yue *et al.* (2004) with two-level Kalman filtering. In Sadler *et al.* (2001), it is shown that when constraints such as FA constraints on the symbols only leave a discrete ambiguity, then the CRB (which is a local bound) for channel estimation is the same as for the case when the unknown symbols were known;
- channel coding in unknown symbols, exploited through turbo detection and estimation; in Scherb *et al.* (2004) a channel estimation CRB is provided when data symbol channel coding is exploited, involving the minimum distance amplification introduced by the channel code. As SNR increases from low to high values, this CRB moves from the case of the data symbols being unknown Gaussian to being known as a TS;
- partial FA knowledge: constant modulus (8-PSK in EDGE) (Safavi and Abed-Meraim, 2003; Hassan *et al.*, 2004; Liu *et al.*, 2001);
- some training/pilot symbols, only enough to allow iterative joint data detection/channel estimation to converge;
- symbol modulus variation pattern (a particular form of Tx-induced cyclostationarity); some of the techniques proposed here lead to wide-sense cyclostationarity (Tsatsanis and Giannakis, 1997), without consistency in SNR. The technique proposed in Leus *et al.* (2001), though, is deterministic;
- space-time coding redundancies through reduced rate linear precoding, introducing subspaces in the transmitted signal covariance, e.g., Alamouti or other orthogonal ST coding schemes, or Choi (2004) and Liu *et al.* (2001);
- guard intervals in time or frequency, see Scaglione *et al.* (1999), Zeng and Ng (2004), cyclic prefix structure;
- symbol stream color (see below);
- known pulse shape (see below);
- CDMA spreading code(s) (see below);
- Tx induced nonzero mean (see superimposed pilots below).

Spatial multiplexing schemes that achieve the optimal rate-diversity trade-off (see Chapter 8) typically do not introduce any blind information (other than GB) for the channel estimation. In Medles and Slock (2004), for instance, a previously introduced linear prefiltering scheme was shown to attain this optimal trade-off. Since the prefilter is a MIMO allpass filter, it leaves the white vector input white. However, perturbations of optimal trade-off achieving schemes can be derived that introduce side information (see Section 14.6.1). So, questions that so far are only partially answered are: what is the optimal amount of side information to maximize capacity, as more side information reduces capacity with CSIR, but also reduces channel estimation error and hence increases capacity? More importantly, what is the optimal distribution of side information over the various forms? Note that only DB and GB are, strictly speaking, blind approaches. The exploitation of any form of side information mentioned above should be called a semiblind approach.

14.6.1 Coloring linear precoding

In Hua and Tugnait (2000), it is shown that colored inputs can be separated if their spectra are linearly independent. Correlation can be introduced by linear convolutive precoding, which corresponds to MIMO prefiltering of $\mathbf{x}_k$ with a MIMO prefilter $\mathbf{T}(z)$, such that the Tx vector signal becomes $\mathbf{a}_k = \mathbf{T}(q)\,\mathbf{x}_k$. We consider full rate linear precoding, so that $\mathbf{T}(z)$ is square $(N_t \times N_t)$ (in Leus *et al.* (2001), an example of low rate precoding appears since the same symbol sequence gets distributed over all Tx antennas). We get for the Rx signal spectrum $S_{\mathbf{yy}}(z) = \sigma_x^2\,\mathbf{H}(z)\,\mathbf{T}(z)\,\mathbf{T}^\dagger(z)\,\mathbf{H}^\dagger(z) + \sigma_v^2 I_m$. An appropriate $\mathbf{T}(z)$ may reduce the nonidentifiability to a phase factor per source, or even to a global phase factor. Consider a generic reducible channel that can be factored as $\mathbf{H}(z) = \mathbf{G}(z)\mathbf{C}(z)$, where $\mathbf{G}(z)$ is irreducible and column reduced with columns in order of., e.g., nonincreasing degree. If r is the (generic) rank of $\mathbf{H}(z)$, then $\mathbf{G}(z)$ is $N_r \times r$, whereas $\mathbf{C}(z)$ is $r \times N_t$. For $r \leq N_r - 1$, we can DB identify $\mathbf{G}(z)$. $\mathbf{G}(z)$ is unique up to a factor $\mathbf{L}(z)$ mentioned earlier. For whichever $\mathbf{G}(z)$ in this equivalence class, it remains to identify $\mathbf{C}(z)$ from

$$\mathbf{S}(z) = \mathbf{G}^{\#}(z)\,(S_{\mathbf{yy}}(z) - \sigma_v^2 I_m)\mathbf{G}^{\#\dagger}(z) = \mathbf{C}(z)\mathbf{S}_{\mathbf{aa}}(z)\mathbf{C}^\dagger(z), \qquad (14.9)$$

where $\mathbf{G}^{\#}(z)$ is a MMSE-ZF equalizer for $\mathbf{G}(z)$: $\mathbf{G}^{\#}(z)\mathbf{G}(z) = I_r$. The degree of $\mathbf{C}_j(z)$ is unpredictable and can be up to $L_j - 1$, the degree of the corresponding column $\mathbf{h}_j(z)$ of $\mathbf{H}(z)$. Two scenarios may be distinguished:

Noncooperative scenario

This scenario typically corresponds to the multiuser case (on the Tx side) without cooperation between users. Consider the simple case with users with a single Tx antenna. In this scenario, $\mathbf{H}(z)$ has no structure other than possibly being FIR, and $\mathbf{T}(z)$ and $S_{\mathbf{aa}}(z)$ are diagonal. This scenario has been considered in Abed-Meraim *et al.* (2001), Hua and Xiang (2001), Xavier *et al.* (2001). In Medles and Slock (2002), two approaches have been proposed for the identification of $\mathbf{C}(z)$ from (14.9).

Frequency-domain approach The idea here is to introduce zeros into the diagonal elements of $\mathbf{T}(z)$, or hence $\mathbf{S}_{\mathbf{aa}}(z)$, such that all other elements other than diagonal element j share N_j zeros

$$\mathbf{T}_{jj}(z) = \prod_{i=1,\neq j}^{p} \prod_{k=1}^{L_i}(1 - z_{i,k}z^{-1}).$$

This allows identifiability of $\mathbf{C}_j(z)$ from $\mathbf{S}(z)$ up to a phase, since

$$\mathbf{S}(z_{j,k}) = \mathbf{C}_j(z_{j,k})\mathbf{S}_{a_j a_j}(z_{j,k})\mathbf{C}_j^\dagger(z_{j,k}), \qquad k = 1,\ldots,L_j,$$

where $\mathbf{S}_{a_j a_j}(z) = \sigma_x^2 T_{jj}(z)T_{jj}^\dagger(z)$.

Time-domain approach The idea here is to introduce delay in the pre-filter, so that the correlations of each $\mathbf{C}_j(z)$ appear separately in certain delay portions of the correlation sequence of $\mathbf{S}(z)$. This can be obtained, for instance, with

$$\mathbf{T}_{jj}(z) = 1 - \alpha_j z^{-d_j}, \qquad d_j = \sum_{i=1}^{j-1} L_i.$$

Identification can be done with a correlation sequence peeling approach that starts with the last column $\mathbf{C}_{N_t}(z)$, of which the (single-sided) correlation sequence appears in an isolated fashion in the last L_{N_t} correlations of $\mathbf{S}(z)$. Identification of $\mathbf{C}_{N_t}(z)$ from its correlation sequence can be done up to a phase factor $e^{j\theta_{N_t}}$ (and up to the phase of zeros if $\mathbf{C}_{N_t}(z)$ has zeros). We can then subtract $\mathbf{S}_{a_j a_j}(z)\mathbf{C}_{N_t}(z)\mathbf{C}_{N_t}^\dagger(z)$ (which does not require $\mathbf{C}_{N_t}(z)$, but only its correlation sequence) from $\mathbf{S}(z)$, which will then reveal the correlation sequence of $\mathbf{C}_{N_t-1}(z)$ in its last L_{N_t-1} correlations, etc. The degree of $\mathbf{S}_{\mathbf{aa}}(z)$ is in this case the degree d_{N_t} of $\mathbf{S}_{a_{N_t} a_{N_t}}(z)$, which, in the case of all equal L_j, is again $(N_t-1)L_1$, which leads to a degree of $N_t L_1 - 1$ for $\mathbf{S}(z)$, or hence $N_t L_1$ correlations. Such a degree for $\mathbf{S}_{\mathbf{aa}}(z)$ is not only sufficient but also necessary, since when $r = 1$, there are $N_t L_1$ parameters to be identified, for which, indeed, at least $N_t L_1$ correlations are needed.

Cooperative scenario

This is the single-user spatial multiplexing scenario. $\mathbf{S_{aa}}(z)$ is allowed to be nondiagonal. Noncooperative approaches can, of course, also be applied here. However, that would lead to at least an unknown phase per Tx antenna, and hence requires either differential encoding or training symbols per Tx antenna. By applying full prefiltering, such that $\mathbf{S_{aa}}(z)$ is not block diagonal (in which case it is said to be fully diverse), $\mathbf{H}(z)$ can be identified up to a global phase factor only under certain conditions on $\mathbf{S_{aa}}(z)$ (see Medles and Slock (2002)). Since this case results in better identifiability, better estimation quality may be another consequence.

Precoder optimization

We consider here the optimization of the ergodic capacity w.r.t. the precoder. As discussed earlier, the optimized prefilter is the result of a compromise, and is expected to be a perturbation of a paraunitary filter, transforming the white $\mathbf{x}_k$ into slightly colored $\mathbf{a}_k$, allowing channel identification. An example of this optimization for a frequency-flat channel is provided in Medles and Slock (2002).

Oversampling, known pulse shapes, and CDMA

So far, the multitude of outputs was assumed to stem from a multitude of sensors. Another output dimension may be added by oversampling the output w.r.t. the discrete-time input. If now also the Tx (and Rx) pulse shape is known, then it can be represented as an $N_{os} \times 1$ vector prefilter $\mathbf{T}(z)$ per input, with N_{os} the oversampling factor. The channel $\mathbf{h}_i(z)$ for input i now also becomes a so-called pseudocirculant $N_r N_{os} \times N_{os}$ matrix filter. A known pulse shape is treated in, e.g., Ghauri and Slock (2000). The knowledge of the pulse shape helps to improve the channel estimation accuracy by reducing the remaining delay spread and capturing the ill-conditioning in time (tapering) and frequency domain (limited bandwidth). But it will not help in resolving inputs if they all use the same pulse shape.

Direct sequence spectrum spreading (or DS-CDMA) is a special case, in which the oversampling factor corresponds to the spreading factor. A sample is called a chip, and the column prefilter $\mathbf{T}(z)$ is static, corresponding to an instantaneous multiplication with the spreading code (which can be time-varying in the case of long/aperiodic/pseudorandom codes, or time-invariant as in the case of short/periodic/deterministic codes). Of course, CDMA can be combined with oversampling w.r.t. the chip rate, and exploitation of a chip pulse shape. The use of different spreading codes for different inputs allows for fairly robust blind source separation and channel estimation, see Ghauri and Slock (1998, 1999), and also Hochwald *et al.* (2001) and Liu *et al.* (2001). In Sung and Tong (2004), long-code CDMA and fast fading channels are considered.

14.7 Pilot structure optimization

Most existing work on pilot structure optimization considers channel estimation based on training only, see Chapter 17. Basic work on TS based MIMO CE appears in Hassibi and Hochwald (2003). In Barhumi *et al.* (2003), Ma *et al.* (2003), and Yang *et al.* (2004), TS-based CE in doubly selective MIMO OFDM systems is considered. See Dong *et al.* (2004) for Bayesian pilot based estimation of frequency-flat AR single-input single-output (SISO) channels, and Tong *et al.* (2004) for a tutorial. A (not so) recent twist on the training paradigm is, besides the usual time-multiplexed (TM) pilots, the appearance of superimposed pilots (SI, also called embedded). SI pilots are actually classical in CDMA standards, which use a pilot signal, sometimes combined with TM pilots. In Zhu *et al.* (2003), SI-pilot-based channel estimates are used to initialize an iterative receiver. In Berriche *et al.* (2004), optimization of a mixture of TM and SI pilots is considered. The continuous SI pilots actually form a pilot signal, and their large duration leads to quasi-orthogonality with the data. It is found that for large enough and equivalent pilot power, both pilot forms lead to similar performance. Only the channel estimation (CRB) is considered, though, as a performance indicator. In Vosoughi and Scaglione (2004b), the effect of both types of pilots on the throughput is considered, and TM pilots appear to be favored. Indeed, pilots not only allow channel estimation, but also influence the data detection. The presence of TM pilots leads to reduced ISI in frequency-selective channels with time-domain Tx. Semiblind channel estimation and detection with SI pilots is considered in Meng and Tugnait (2004). So, an important question here is: is orthogonality of pilots and data desirable? The answer may depend on how mixed information (pilot/data) is used.

14.8 Other research avenues ahead

We already mentioned the optimization of the side information mix.

Multiuser case In this case, the number of unknowns per received sample increases further. Whereas spatial multiplexing is the cooperative case of multiple-input, the multiuser case corresponds to the noncooperative version. Differentiation of users at the level of SOS can be obtained through coloring (e.g., CDMA) as mentioned earlier. In Zeng and Ng (2004), a semiblind multiuser scenario is considered.

Noncoherent approaches See also Chapter 10. In the comparison between noncoherent approaches (no CSIR) and coherent approaches based on channel estimates, the current trend in improving noncoherent approaches involves exploiting the Doppler structure (predictability) of the channel. This may be one indication that coherent approaches based on channel estimation should work better.

Semiblind direct receiver estimation Training optimization may depend on the receiver architecture (Vosoughi and Scaglione, 2004a). See Scaglione *et al.* (1999) for the blind determination of linear equalizers and Bugallo *et al.* (2002) for the semiblind determination of a MIMO DFE.

Channel estimation for the transmitter The availability of channel state information at the Tx (CSIT) allows us to improve transmission through adaptive modulation, see, e.g., Xia *et al.* (2004) and Chapter 6. Questions that arise here involve not only channel estimation, but also its possible quantization and (digital or analog) retransmission. A key issue here is also the degree of reciprocity of the channel, or, e.g., its pathwise parameters (direction, delay, Doppler shift, power). Another issue is the effect of sensor array design on channel estimation and reciprocity, e.g., beamspace (beam selection should be reciprocal).

Description of channel variation in terms of user mobility Such an approach would possibly allow a more compact description of temporal variation, lead to better channel predictability, and allow the separation of users with less side information and mobile localization applications (Amar and Weiss, 2004). See Bug and Jakoby (2004) for an approach in this direction. Variability of the environment also needs to be taken into account, however.

References

Abed-Meraim, K., Xiang, Y., Manton, J. H., and Hua, Y. (2001). Blind source-separation using second-order cyclostationary statistics. *IEEE Trans. Signal Process.*, **49** (4), 694–701.

Amar, A. and Weiss, A. J. (2004). Direct position determination of narrowband radio transmitters. In *Proc. IEEE Int. Conf. Acoust., Speech, Signal Process.*, vol. 2, 81–84.

Barhumi, I., Leus, G., and Moonen, M. (2003). Optimal training design for MIMO OFDM systems in mobile wireless channels. *IEEE Trans. Signal Process.*, **51** (6), 1615–1624.

Berriche, L., Abed-Meraim, K., and Belfiore, J.-C. (2004). Cramer-Rao bounds for MIMO channel estimation. In *Proc. IEEE Int. Conf. Acoust., Speech, Signal Process.*, vol. 4, 397–400.

Bug, S. and Jakoby, R. (2004). Modeling of the mobile radio channel using theory of dynamics—first derivations and results. In *Proc. Eur. Wireless Conf.*

Bugallo, M. F., Miguez, J., and Castedo, L. (2002). Decision-feedback semiblind channel equalization in space-time coded systems. In *Proc. IEEE Int. Conf. Acoust., Speech, Signal Process.*, vol. 3, 2425–2428.

Cardoso, J.-F. (1998). Blind signal separation: Statistical principles. *Proc. IEEE*, **86** (10), 2009–2025. Special Issue on Blind Identification and Estimation.

de Carvalho, E. and Slock, D. T. M. (2001). Semi-blind methods for FIR multichannel estimation. In Giannakis, G. B., Hua, Y., Stoica, P., and Tong, L. (eds.), *Signal Processing Advances in Wireless & Mobile Communications* (Prentice Hall).

Choi, J. (2004). Equalization and semi-blind channel estimation for space-time block coded signals over a frequency-selective fading channel. *IEEE Trans. Signal Process.*, **52** (3), 774–785.

Cichocki, A. and Amari, S. (2002). *Adaptive Blind Signal and Image Processing, Learning Algorithms, and Applications* (Wiley).

Dong, M., Tong, L., and Sadler, B. (2004). Optimal insertion of pilot symbols for transmissions over time-varying flat fading channels. *IEEE Trans. Signal Process.*, **52** (5), 1403–1418.

Ghauri, I. and Slock, D. T. M. (1998). Blind and semi-blind single user receiver techniques for asynchronous CDMA in multipath channels. In *Proc. IEEE Global Telecommun. Conf.*, vol. 6, 3572–3577.

—— (1999). MMSE-ZF receiver and blind adaptation for multirate CDMA. In *Proc. IEEE Veh. Technol. Conf. (VTC Fall)*, vol. 1, 628–632.

—— (2000). Structured estimation of sparse channels in quasi-synchronous DS-CDMA. In *Proc. IEEE Int. Conf. Acoust., Speech, Signal Process.*, vol. 5, 2873–2876.

Hassan, M. A. S., Sharif, B. S., Woo, W. L., and Jimaa, S. (2004). Semiblind estimation of time varying STFBC-OFDM channels using Kalman filter and CMA. In *Proc. IEEE Int. Symp. Comp. Commun.*, vol. 2, 594–599.

Hassibi, B. and Hochwald, B. M. (2003). How much training is needed in multiple-antenna wireless links? *IEEE Trans. Inf. Theory*, **49** (10), 951–964.

Haykin, S., Huber, K., and Chen, Z. (2004). Bayesian sequential state estimation for MIMO wireless communications. *Proc. IEEE*, **92** (3), 439–454.

Hochwald, B., Marzetta, T. L., and Papadias, C. B. (2001). A transmitter diversity scheme for wideband CDMA systems based on space-time spreading. *IEEE J. Sel. Areas Commun.*, **19** (1), 48–60.

Hua, Y. and Tugnait, J. K. (2000). Blind identifiability of FIR-MIMO systems with colored input using second-order statistics. *IEEE Signal Process. Lett.*, **7** (12), 348–350.

Hua, Y. and Xiang, Y. (2001). The BID for blind equalization of FIR MIMO channels. In *Proc. IEEE Int. Conf. Acoust., Speech, Signal Process.*, vol. 1, 415–424.

Lenardi, M. and Slock, D. T. M. (2002). Estimation of time-varying wireless channels and application to the UMTS W-CDMA FDD downlink. In *Proc. Eur. Wireless Conf.*, 204–207.

Leus, G., Vandaele, P., and Moonen, M. (2001). Deterministic blind modulation-induced source separation for digital wireless communications. *IEEE Trans. Signal Process.*, **49** (1), 219–227.

Li, Y. and Yang, L. (2003). Semi-blind MIMO channel identification based on error adjustment. In *Proc. IEEE Conf. Neural Netw. Signal Process.*, vol. 2, 1429–1432.

Liu, Z., Giannakis, G. B., Barbarossa, S., and Scaglione, A. (2001). Transmit-antennae space-time block coding for generalized OFDM in the presence of unknown multipath. *IEEE J. Sel. Areas Commun.*, **19** (7), 1352–1364.

Ma, X., Giannakis, G. B., and Ohno, S. (2003). Optimal training for block transmission over doubly selective wireless fading channels. *IEEE Trans. Signal Process.*, **51** (5), 1351–1366.

Medles, A., de Carvalho, E., and Slock, D. T. M. (2001). Linear prediction based semi-blind estimation of MIMO FIR channels. In *Proc. IEEE Workshop Signal Process. Advances Wireless Commun.*, 58–61.

Medles, A. and Slock, D. (2004). Achieving the optimal diversity-vs-multiplexing tradeoff for MIMO flat channels with QAM space-time spreading and DFE equalization. *IEEE Trans. Inf. Theory*. Submitted.

—— (2005). Semiblind estimation of FIR MIMO channels in single-carrier systems. *IEEE Trans. Signal Process.* Submitted.

Medles, A. and Slock, D. T. M. (2002). Linear precoding for spatial multiplexing MIMO systems: Blind channel estimation aspects. In *Proc. IEEE Int. Conf. Commun.*, vol. 1, 401–405.

—— (2003a). Augmenting the training sequence part in semiblind estimation for MIMO channels. In *Proc. 37th Asilomar Conf. Signals, Syst., Comput.*, vol. 2, 1825–1829.

—— (2003b). Mutual information without channel knowledge at the receiver. In *Proc. IEEE Workshop Signal Process. Advances Wireless Commun.*, 398–402.

Meng, X. and Tugnait, J. K. (2004). Semi-blind channel estimation and detection using superimposed training. In *Proc. IEEE Int. Conf. Acoust., Speech, Signal Process.*, vol. 4, 417–420.

Montalbano, G. and Slock, D. T. M. (2003). Joint common-dedicated pilots based estimation of time-varying channels for W-CDMA receivers. In *Proc. IEEE Veh. Technol. Conf. (VTC Fall)*, vol. 2, 1253–1257.

Pladdy, C., Özen, S., Nerayanuru, S. M., Fimoff, M. J., and Zoltowski, M. (2004). Taylor series approximation of semi-blind best linear unbiased channel estimates for the general linear model. In *Proc. Asilomar Conf. Signals, Syst., Comput.*, vol. 2, 2208–2212.

Rousseaux, O. and Leus, G. (2004). An iterative method for improved training-based estimation of doubly selective channels. In *Proc. IEEE Int. Conf. Acoust., Speech, Signal Process.*, vol. 4, 889–892.

Rousseaux, O., Leus, G., Stoica, P., and Moonen, M. (2003a). Generalized training based channel identification. In *Proc. IEEE Global Telecommun. Conf.*, vol. 5, 2432–2436.

—— (2003b). Training based maximum likelihood channel identification. In *Proc.*

IEEE Workshop Signal Process. Advances Wireless Commun., 334–338.

Sadler, B. M., Kozik, M. J., and Moore, T. (2001). Bounds on bearing and symbol estimation with side information. *IEEE Trans. Signal Process.*, **49** (4), 822–834.

Safavi, A. and Abed-Meraim, K. (2003). Blind channel identification robust to order overestimation: a constant modulus approach. In *Proc. IEEE Int. Conf. Acoust., Speech, Signal Process.*, vol. 2, 1248–1251.

Scaglione, A., Giannakis, G. B., and Barbarossa, S. (1999). Redundant filterbank precoders and equalizers part II: Blind channel estimation, synchronization, and direct equalization. *IEEE Trans. Signal Process.*, **47** (7), 2007–2022.

Scaglione, A. and Vosoughi, A. (2004). Turbo estimation of channel and symbols in precoded MIMO systems. In *Proc. IEEE Int. Conf. Acoust., Speech, Signal Process.*, vol. 4, 413–416.

Scherb, A., Kühn, V., and Kammeyer, K.-D. (2004). Cramer-Rao lower bound for semiblind channel estimation with respect to coded and uncoded finite-alphabet signals. In *Proc. Asilomar Conf. Signals, Syst., Comput.*, vol. 2, 2193–2197.

Slock, D. T. M. (2004a). Bayesian blind and semiblind channel estimation. In *Proc. IEEE Workshop Sensor, Array, Multich. Signal Process.*

—— (2004b). Blind FIR channel estimation in multichannel cyclic prefix systems. In *Proc. IEEE Workshop Sensor, Array, Multich. Signal Process.*

Souza, R. D., Garcia-Frias, J., and Haimovich, A. M. (2004). A semi-blind receiver for iterative data detection and decoding of space-time coded data. In *Proc. IEEE Wireless Commun. Netw. Conf.*, vol. 3, 1626–1630.

Sung, Y. and Tong, L. (2004). Tracking of fast-fading channels in long-code CDMA. *IEEE Trans. Signal Process.*, **52** (3), 786–795.

Talwar, S. and Paulraj, A. (1997). Blind separation of synchronous co-channel digital signals using an antenna array—part II: Performance analysis. *IEEE Trans. Signal Process.*, **45** (3), 706–718.

Tong, L., Sadler, B. M., and Dong, M. (2004). Pilot-assisted wireless transmission. *IEEE Signal Process. Mag.*, **21** (6), 12–25.

Tsatsanis, M. K. and Giannakis, G. B. (1997). Transmitter induced cyclostationarity for blind channel estimation. *IEEE Trans. Signal Process.*, **45** (7), 1785–1794.

Vosoughi, A. and Scaglione, A. (2004a). The best training depends on the receiver architecture. In *Proc. IEEE Int. Conf. Acoust., Speech, Signal Process.*, vol. 4, 409–412.

—— (2004b). On the effect of channel estimation error with superimposed training upon information rates. In *Proc. IEEE Int. Symp. Inf. Theory*, 316.

Xavier, J., Barroso, V. A. N., and Moura, J. M. F. (2001). Closed-form correlative coding (cfc_2) blind identification of MIMO channels: Isometry fitting to second order statistics. *IEEE Trans. Signal Process.*, **49** (5), 1073–1086.

Xia, P., Zhou, S., and Giannakis, G. B. (2004). Adaptive MIMO-OFDM based on partial channel state information. *IEEE Trans. Signal Process.*, **52** (1), 202–213.

Yang, L., Ma, X., and Giannakis, G. B. (2004). Optimal training for MIMO fading channels with time- and frequency-selectivity. In *Proc. IEEE Int. Conf. Acoust., Speech, Signal Process.*, vol. 3, 821–824.

Yue, J., Kim, K. J., Reid, T., and Gibson, J. D. (2004). Joint semi-blind channel estimation and data detection for MIMO-OFDM systems. In *Proc. IEEE Symp.*

Emerging Tech.: Mobile and Wireless Commun., vol. 2, 709–712.

Zeng, Y. and Ng, T.-S. (2004). A semi-blind channel estimation method for multiuser multiantenna OFDM systems. *IEEE Trans. Signal Process.*, **52** (5), 1419–1429.

Zheng, L. and Tse, D. (2002). Communication on the Grassmann manifold: A geometric approach to the noncoherent multiple antenna channel. *IEEE Trans. Inf. Theory*, **48** (2), 359–383.

Zhu, H., Farhang-Boroujeny, B., and Schlegel, C. (2003). Pilot embedding for joint channel estimation and data detection in MIMO communication systems. *IEEE Trans. Signal Process.*, **7** (1), 30–32.

15

MIMO receive algorithms

Thomas Kailath

Stanford University

Haris Vikalo and Babak Hassibi

California Institute of Technology

In multiantenna wireless communication systems, data is transmitted across channels that can often be modeled as linear and time-invariant. The received signal in such systems is given by a linear combination of the transmitted data symbols, corrupted by an additive Gaussian noise,

$$\mathbf{x} = \mathbf{H}\mathbf{s} + \mathbf{v}, \tag{15.1}$$

where $\mathbf{H}$ is an $N \times M$ complex valued channel whose realization is known to the receiver (and is estimated, for instance, by means of sending a known training sequence), $\mathbf{s}$ is an M-dimensional transmitted symbol, and $\mathbf{v}$ is an N-dimensional noise with $\mathcal{C}(0, \sigma^2)$ Gaussian entries. Furthermore, we will assume that the entries in the transmitted symbol vector $\mathbf{s}$ in (15.1) are points in a QAM constellation.

For computational reasons, we shall replace the complex-valued model (15.1) by its real-valued equivalent in the usual way. To this end, we define the $m = 2M$ dimensional vector s, and the $n = 2N$ dimensional vectors x and v, composed of the real and imaginary parts of $\mathbf{s}$, $\mathbf{x}$, and $\mathbf{v}$, respectively, as

$$s = \begin{bmatrix} \mathcal{R}(\mathbf{s})^T & \mathcal{I}(\mathbf{s})^T \end{bmatrix}^T, \quad x = \begin{bmatrix} \mathcal{R}(\mathbf{x})^T & \mathcal{I}(\mathbf{x})^T \end{bmatrix}^T, \quad v = \begin{bmatrix} \mathcal{R}(\mathbf{v})^T & \mathcal{I}(\mathbf{v})^T \end{bmatrix}^T,$$

and the $n \times m$ matrix H

$$H = \begin{bmatrix} \mathcal{R}(\mathbf{H}) & -\mathcal{I}(\mathbf{H}) \\ \mathcal{I}(\mathbf{H}) & \mathcal{R}(\mathbf{H}) \end{bmatrix}.$$

Then the real-valued equivalent of the model (15.1) is given by

$$x = Hs + v. \tag{15.2}$$

This work is supported in part by the NSF under grant no. CCR-0133818, by the ONR under grant no. N00014-02-1-0578, and by Caltech's Lee Center for Advanced Networking.

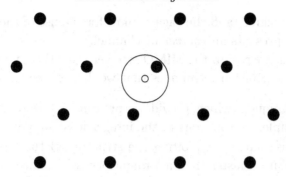

Fig. 15.1. Geometric interpretation of the integer least-squares problem.

At a receiver, a detector forms an estimate of the transmitted symbol, $\hat{s}$. The optimal detector minimizes the average probability of error, i.e., it minimizes $P(\hat{s} \neq s)$. This is achieved by the maximum-likelihood (ML) design, which, under the previous assumptions, performs the nonlinear optimization

$$\min_{s \in \mathcal{D}_L^m} \|x - Hs\|^2, \tag{15.3}$$

where $\mathcal{D}_L^m$ denotes the m-dimensional square lattice spanned by an L-PAM constellation in each dimension. Furthermore, to obtain the soft decisions required by iterative decoding schemes in systems employing space-time or error-correcting codes, MIMO soft-decoding algorithms also often have to solve (15.3) or its modifications.

Problem (15.3), typically referred to as an *integer least-squares* problem, has a simple geometric interpretation. As the entries of s run over the points in the L-PAM constellation, s spans the "rectangular" m-dimensional lattice, $\mathcal{D}_L^m$. However, for any given *lattice-generating matrix H*, the n-dimensional vector Hs spans a "skewed" lattice. Thus, given the skewed lattice Hs and the vector $x \in \mathcal{R}^n$, the integer least-squares problem is to find the "closest" lattice point (in a Euclidean sense) to x, as illustrated in Fig. 15.1.

Problem (15.3) is, for a general H, known to be exponentially complex both in the worst-case sense (Grotschel *et al.*, 1993) as well as in the average sense (Ajtai, 1998).

Optimal detection reduces to solving an integer least-squares problem not only for the simple uncoded transmission problem modeled by (15.2), but also in the context of lattice codes (Banihashemi and Khandani, 1998; Agrell *et al.*, 2002), CDMA systems (Brutel and Boutros, 1999; Viterbo and Boutros, 2000), multiantenna systems employing space-time codes (Foschini, 1996; Damen *et al.*, 2000; Hassibi and Hochwald, 2002), etc. Many of these applications are characterized by an affine mapping between the transmitted and the received signal, and thus again allow for the use of the model (15.2), where H now represents an *equivalent* channel.

In this chapter, we review the MIMO receiver algorithms for solving (15.3). In particular, the solution techniques that we discuss are the following:

- heuristic techniques, which provide approximate but readily implementable low-complexity solutions to the integer least-squares problem, and
- exact methods that, by exploiting the structure of the lattice, generally obtain the solution faster than a straightforward exhaustive search.

It is a pleasantly surprising fact that the exact techniques turn out to have complexity comparable to that of the heuristic techniques over a useful range of channel signal to noise ratios (SNR) (see Section 15.2).

Finally, we should point out potential connection between the MIMO detection problem and the multiuser detection (Verdu, 1998). However, space limitation precludes further discussion on multiuser detection and we refer the interested reader to Verdu (1998) and the references therein.

15.1 Heuristic techniques

Finding the exact solution of (15.3) is, in general, NP hard. Therefore, many wireless communication systems employ some approximations, heuristics or combinations thereof, often with manageable computational complexity. We briefly discuss some of these techniques.

Zero-forcing

Solve the unconstrained least-squares problem to obtain $\hat{s} = H^\dagger x$, where $H^\dagger$ denotes the pseudoinverse of H. Since the entries of $\hat{s}$ will not necessarily be integers, round them off to the closest integer (a process referred to as slicing) to obtain

$$\hat{s}_B = \left[H^\dagger x \right]_{\mathcal{Z}}. \tag{15.4}$$

The above $\hat{s}_B$ is often called a Babai estimate (Grotschel *et al.*, 1993). In the communications literature, this procedure is referred to as *zero-forcing equalization*.

The complexity of finding the Babai estimate is essentially determined by the complexity of finding the pseudoinverse of the matrix H in (15.4). The simplest way of calculating the pseudoinverse is by means of QR factorization, $H = QR$. It can also be calculated in a more stable way (which avoids inverting the upper triangular matrix R) by means of singular value decomposition (SVD) of H. In either case, assuming that H is square (i.e., $n = m$), the complexity of finding the Babai estimate is of cubic order, $O(m^3)$.

Nulling and canceling

In this method, the Babai estimate is used for only one of the entries of s, say the first. Then this entry, s_1, is assumed to be known and its effect is canceled out to obtain a reduced-order integer least-squares problem with $m - 1$ unknowns. The process is repeated to find s_2, etc. In communications parlance this is known as *decision-feedback equalization*.

We shall find it convenient to denote the partition of the channel matrix H into rows and columns as

$$H = \begin{bmatrix} \underline{h}_1 & \underline{h}_2 & \cdots & \underline{h}_m \end{bmatrix} = \begin{bmatrix} H_1 \\ \vdots \\ H_n \end{bmatrix}.$$

The nulling and canceling algorithm can be stated by the following pseudocode:

$$y_1 := x$$
$$\text{for } k = 0 \text{ to } m - 1$$
$$\quad \text{find weighting vector } w_{m-k}$$
$$\quad \hat{s}_{m-k} := \text{slice}(w_{m-k}y_{k+1})$$
$$\quad y_{k+2} = y_{k+1} - \underline{h}_{m-k}\hat{s}_{m-k}$$
$$\text{end}$$

In the algorithm, for each value of the index k, the entries of the auxiliary vector y_{k+1} are weighted by the components of the weight vector w_{m-k} and linearly combined to account for the effect of the interference. Depending on the criterion chosen for the design of w_{m-k} (i.e., for performing the nulling operation), we can distinguish between the following cases:

(i) *Zero-forcing (ZF) nulling*

In this case, interference from the yet undetected symbols is nulled. Denoting

$$H_{m-k} = \begin{bmatrix} \underline{h}_1 & \underline{h}_2 & \cdots & \underline{h}_{m-k} \end{bmatrix},$$

this condition can be stated as

$$H^*_{m-k}w_{m-k} = e_{m-k},$$

where e_{m-k} is an $(m - k) \times 1$ column vector that consists of all zeros except for the $(m-k)$-th entry, whose value is 1. The weighting vector is then given by the least-norm solution of the form

$$w_{m-k} = H_{m-k}^\dagger e_{m-k},$$

where $(\cdot)^\dagger$ denotes the pseudoinverse, i.e.,

$$H_{m-k}^\dagger = H_{m-k}(H_{m-k}^* H_{m-k})^{-1}.$$

(ii) *Minimum mean-square error (MMSE) nulling*

The objective in MMSE nulling is to minimize the expected mean-square error between the receiver's estimate and the transmitted symbol. This can be expressed as

$$E\left[s_{m-k} y_{k+1}^*\right] = w_{m-k}^* E\left[y_{k+1} y_{k+1}^*\right].$$

Furthermore, we shall assume that the previous decisions made by the detector were correct, i.e., $\hat{s}_{m-k} = s_{m-k}$. Defining

$$s_{1:m-k} = \begin{bmatrix} s_1 & s_2 & \cdots & s_{m-k} \end{bmatrix},$$

we can write

$$y_{k+1} = H_{m-k} s_{1:m-k} + v,$$

where v is the noise vector in (15.2). Furthermore, assuming that the transmitted symbol sequence is spatially white and has variance $\mathcal{E}_s$, we readily find that the MMSE nulling vector is given by

$$w_{m-k} = \left(H_{m-k}^* H_{m-k} + \frac{1}{\rho} I\right)^{-1} \underline{h}_{m-k},$$

where $\rho = \mathcal{E}_s/\sigma^2$ denotes the signal-to-noise ratio (SNR).

The computational complexity is again determined by the complexity of solving the underlying unconstrained least-squares problem, i.e., calculating the pseudoinverse at each step of the algorithm. When $m = n$, we need to evaluate the pseudoinverse of a series of matrices with dimensions $m \times (m-k)$, where $k = m, m-1, \ldots, 1$. The computational complexity of performing this series of operations is clearly of the fourth order, i.e., $O(m^4)$, an order of magnitude higher than the complexity of finding the Babai estimate.

Nulling and canceling with optimal ordering

The nulling and canceling algorithm can suffer from *error-propagation*: if s_1 is estimated incorrectly it can have an adverse effect on estimation of the remaining unknowns s_2, s_3, etc. To minimize the effects of error propagation, it is advantageous to perform nulling and canceling from the "strongest" to the "weakest" signal. This is the method proposed for V-BLAST (Foschini, 1996).

Consider, for instance, the MMSE nulling and canceling algorithms. To perform optimal ordering, we consider the covariance matrix of the estimation error $s - \hat{s}$,

$$P = E(s - \hat{s})(s - \hat{s})^* = \left(H^*H + \frac{1}{\rho}I \right)^{-1}.$$

Consider the entries of the estimated symbol $\hat{s}$, i.e., $\{\hat{s}_i, i = 1, 2, \ldots, m\}$. The "strongest" signal, corresponding to the "best" estimate, is the one with the smallest variance, i.e., s_i for which P_{ii} is the smallest. If we reorder the entries of s so that the strongest signal is s_m, then the estimate $\hat{s}_m$ is going to be better than it would be for any other ordering of the entries in s. This ordering we perform at each step of the nulling and canceling algorithm.

The computational complexity of the algorithm is the same as the complexity of the standard nulling and canceling algorithm, namely $O(m^4)$, augmented by the complexity of the ordering operation, which, for the set of k elements, is $O(k^3)$.

Square-root algorithm for nulling and canceling

The increased complexity of the nulling and canceling algorithm as compared with the zero-forcing algorithm is that the former requires repeated evaluation of the pseudoinverse for each deflated channel matrix. It is of interest to seek cost effective implementations of the algorithm so that a pseudoinverse at a particular stage can be efficiently deduced from the pseudoinverse computed at the previous stage. To this end, using the array algorithm ideas of linear estimation theory (see, e.g., Kailath *et al.* (2000)), a so-called square-root algorithm was proposed in Hassibi (1999).

Such algorithms are characterized by numerical stability and robustness achieved by increasing the dynamic range of the quantities involved, their condition numbers, etc. In particular, this is obtained by insisting to

- avoid squaring objects, such as in the computation of H^*H.
- avoid inverting objects.
- make as much use as possible of unitary transformations.

For the following discussion, it will be convenient to write the basic linear least-mean-squares estimate of s, given the observation $x = Hs + v$, as

$$\hat{s} = \left(H^*H + \frac{1}{\rho}I\right)^{-1} H^*x = \begin{bmatrix} H \\ \frac{1}{\sqrt{\rho}}I_m \end{bmatrix}^\dagger \begin{bmatrix} x \\ 0 \end{bmatrix} = H_1^\dagger x, \qquad (15.5)$$

where $H_1^\dagger$ denotes the first n columns of the pseudoinverse of the augmented channel matrix in (15.5).

To start the use of unitary transformations, and to avoid squaring H, consider the QR decomposition of the augmented channel matrix

$$\begin{bmatrix} H \\ \frac{1}{\sqrt{\rho}}I_m \end{bmatrix} = QR = \begin{bmatrix} Q_1 \\ Q_2 \end{bmatrix} R,$$

where Q is an $(n + m) \times m$ matrix with orthonormal columns, and R is $m \times m$ nonsingular and upper-triangular. Note then that

$$P = \left(H^*H + \frac{1}{\rho}I_m\right)^{-1} = (R^*R)^{-1} = R^{-1}R^{-*}.$$

Thus we can identify R^{-1} as a square-root of P, say,

$$R^{-1} = P^{1/2}, \quad P^{1/2}P^{*/2} = P.$$

The pseudoinverse of the augmented channel matrix now becomes

$$\begin{bmatrix} H \\ \frac{1}{\sqrt{\rho}}I_m \end{bmatrix}^\dagger = R^{-1}Q^* = P^{1/2}Q^*,$$

and thus

$$H_1^\dagger = P^{1/2}Q_1^*.$$

Therefore, given $P^{1/2}$ and Q_1, we can compute both the pseudoinverse and the error covariance matrix, required for the nulling operation and the optimal ordering. The problem becomes one of finding the best way to compute $P^{1/2}$ and Q_1.

Recall that the optimal ordering is performed according to the values of the diagonal entries of P. This information can also be deduced from $P^{1/2}$. Since the diagonal entries of P are simply the squared length of the rows of $P^{1/2}$, the minimum diagonal entry of P corresponds to the minimum length row of $P^{1/2}$.

Now assume that the entries of the transmitted signal s have been re-ordered so that the mth diagonal entry is the smallest. Consider a unitary transformation Σ that rotates (or reflects) the mth row of $P^{1/2}$ to lie along the direction of the mth unit vector, i.e.,

$$P^{1/2}\Sigma = \begin{bmatrix} P^{(m-1)/2} & P_m^{(m-1)/2} \\ 0 & p_m^{1/2} \end{bmatrix}, \tag{15.6}$$

where $p_m^{1/2}$ is a scalar. It was shown in Hassibi (1999) that the block upper triangular square-root factor of P in (15.6), $P^{(m-1)/2}$, is a square-root factor of P^{m-1}. Therefore, to find the square-root factor of $P^{(m-1)}$, one needs to make P block upper triangular. The next signal to be detected is selected by finding the minimum length row of $P^{(m-1)/2}$. The rows of $P^{(m-1)/2}$ are then reordered so that this minimum length row corresponds to the last $(m-1)$-th row, and the upper block triangularization of $P^{(m-1)/2}$ gives the next square-root factor, $P^{(m-2)/2}$, and so on.

The process described above results in upper triangularization of the square root matrix $P^{1/2}$. Let $\underline{q}_{1,i}$, $i = 1, \ldots, m$, denote the resulting columns of Q_1, i.e.,

$$Q_1 = \begin{bmatrix} \underline{q}_{1,1} \cdots \underline{q}_{1,m} \end{bmatrix}.$$

It was shown in Hassibi (1999) that the nulling vectors for the signals s_1 to s_m are given by

$$H_{1,i}^{\dagger} = p_i^{1/2} \underline{q}_{1,i}^*,$$

where $p_i^{1/2}$ is the ith diagonal entry of $P^{1/2}$. Therefore, the nulling vectors are simply found by scaling the columns of Q_1 by the diagonals of $P^{1/2}$. Moreover, there is no need for recomputing $P^{1/2}$ and Q_1 for the deflated matrices $H^{(m-k)}$, $k = 1, \ldots, m-1$. The information needed for the optimal ordering and finding nulling vectors is already implicitly contained in $P^{1/2}$ and Q_1.

What remains to be specified is the computation of $P^{1/2}$ and Q_1. Note that we can write

$$P = \left(\sum_{j=1}^{n} H_j^* H_j + \frac{1}{\rho} I \right)^{-1}.$$

Denoting

$$P_{|i} \triangleq \left(\sum_{j=1}^{i} H_j^* H_j + \frac{1}{\rho} I \right)^{-1}, \quad P_{|n} = P,$$

and using a matrix inversion lemma, we obtain the so-called Riccati recursion of the RLS (recursive-least-squares) algorithm (Kailath *et al.*, 2000,

Section 2.6),

$$P_{|i} = P_{|i-1} - \frac{P_{|i-1} H_i^* H_i P_{|i-1}}{r_{e,i}}, \quad r_{e,i} = 1 + H_i P_{|i-1} H_i^*, \quad P_{|0} = \rho I. \quad (15.7)$$

On the other hand, to find a recursion for $H_1^\dagger$, note that the least-mean-square estimate of the signal, $\hat{s} = P^{1/2} Q_1^* x = H_1^\dagger x$, satisfies the recursion (Kailath *et al.*, 2000, Lemma 2.6.1)

$$\hat{s}_{|i} = \hat{s}_{|i-1} + \bar{K}_{p,i} r_{e,i}^{-1/2} \left(x_i - H_i \hat{s}_{|i-1} \right), \quad \bar{K}_{p,i} = P_{|i-1} H_i^* r_{e,i}^{-*/2}, \quad \hat{s}_{|0} = 0.$$

Then the recursion for the pseudoinverse $H_1^\dagger = P^{1/2} Q_1$ can be written as

$$H_{1|i}^\dagger = H_{1|i-1}^\dagger + \bar{K}_{p,i} r_{e,i}^{1/2} \left(e_i^* - H_i H_{1|i-1}^\dagger \right), \quad \bar{K}_{p,i} = P_{|i-1} H_i^* r_{e,i}^{-*/2},$$

$$H_{1|0}^\dagger = 0_{m \times n}. \quad (15.8)$$

Note that $H_1^\dagger = H_{1|n}^\dagger$.

One can further improve (15.7) and (15.8) by ensuring direct propagation of $P^{1/2}$. Incorporating these improvements, the algorithm of Hassibi (1999) can be summarized as follows:

(i) Compute $P^{1/2}$ and Q_1.

Propagate a square-root algorithm of the following form:

$$\begin{bmatrix} 1 & H_i P_{|i-1}^{1/2} \\ 0 & P_{|i-1}^{1/2} \\ -e_i & B_{i-1} \end{bmatrix} \Theta_i = \begin{bmatrix} r_{e,i}^{1/2} & 0 \\ \bar{K}_{p,i} & P_{|i}^{1/2} \\ A_i & B_i \end{bmatrix}, \quad P_{|0}^{1/2} = \sqrt{\rho} I, B_0 = 0_{n \times m},$$

where e_i is the ith unit vector of dimension n, and Θ_i is any unitary transformation that block lower triangularizes the prearray. After n steps, we obtain

$$P^{1/2} = P_{|n}^{1/2} \text{ and } Q_1 = B_n.$$

(ii) Find the minimum length row of $P^{1/2}$ and permute it to be the last (mth) row. Permute s accordingly.

(iii) Find a unitary Σ that makes $P^{1/2}\Sigma$ block upper triangular,

$$P^{1/2}\Sigma = \begin{bmatrix} P^{(m-1)/2} & P_m^{(m-1)/2} \\ 0 & p_m^{1/2} \end{bmatrix}.$$

(iv) Update Q_1 to $Q_1\Sigma$.

(v) The nulling vector for the mth signal is given by $p_m^{1/2} q_{1,m}^*$, where $q_{1,m}$ denotes the mth column of Q_1.

(vi) Go back to step 3, but now with $P^{(m-1)/2}$ and $Q_1^{(m-1)}$, (the first $m-1$ columns of Q_1).

The square-root algorithm achieves all the desired computational objectives. In particular, it avoids computing the pseudoinverse for each deflated channel matrix, avoids squaring or inverting any quantities, and makes extensive use of unitary transformations. Using the algorithm, the computational complexity of the nulling and canceling receiver can be reduced from $O(m^4)$ to $O(m^3)$, which is the complexity of the simple zero-forcing algorithm discussed earlier.

Solving relaxed convex optimization problems

Another heuristic approach to maximum-likelihood detection is via convex optimization techniques. The integer least-squares problem is essentially transformed into an optimization problem with both objective and constraint being convex functions. To illustrate the technique, we consider the detection problem where the entries in the symbol s are chosen from 4-QAM constellations, i.e., for each entry in the symbol vector s it holds that $s_i^2 = 1$.

Since

$$\|x - Hs\|^2 = s^T H^T H s - 2x^T H^T s + x^T x$$
$$= \operatorname{Tr} H^T H S - 2x^T H^T s + x^T x,$$

where $S = ss^T$, the integer least-squares problem can be expressed as

$$\min \left(\operatorname{Tr} H^T H S - 2x^T H^T s + x^T x \right)$$
$$\text{subject to } S_{ii} = 1, \ S \succeq ss^T, \ \operatorname{rank}(S) = 1.$$

Using

$$S \succeq ss^T \Leftrightarrow \begin{bmatrix} S & s \\ s^T & 1 \end{bmatrix} \succeq 0$$

and relaxing the rank one constraint, one can obtain a semidefinite program (SDP) with variables S and s of the form

$$\min \left(\operatorname{Tr} H^T H S - 2x^T H^T s + x^T x \right)$$
$$\text{subject to } S_{ii} = 1, \ \begin{bmatrix} S & s \\ s^T & 1 \end{bmatrix} \succeq 0.$$

Solving this SDP for s, an approximate solution to the detection problem can be found as $\hat{s} = \operatorname{sgn}(s)$ (recall that the algorithm is only for $s_i^2 = 1$). The complexity of solving the SDP is roughly cubic, $O(m^3)$.

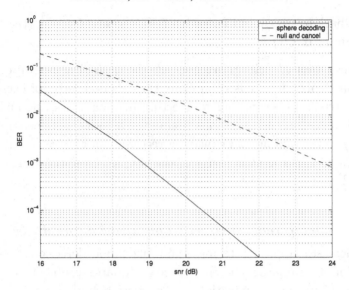

Fig. 15.2. Bit error performance of a sphere decoding vs. nulling and canceling with optimal ordering, $M = 8$, $N = 12$, 16-QAM.

15.2 Exact methods: sphere decoding

With an abundance of heuristic methods presented in the previous section, it is natural to ask how close they come to the optimal solution: in Fig. 15.2, the bit-error rate (BER) performance of an exact solution is compared with the ordered nulling and canceling (N/C) for a multiantenna system with $M = 8$ transmit and $N = 12$ receive antennas employing 16-QAM modulation scheme. Clearly, the ML receiver significantly outperforms N/C; thus, there is merit in studying exact solutions. The most obvious one is to search over the entire lattice, which invariably requires an exponential search. There do, however, exist exact methods that are more sophisticated than exhaustive search and can be employed for an arbitrary H. Such are Kannan's algorithm (Kannan, 1983) (which searches only over restricted parallelograms), the KZ algorithm (Lagarias *et al.*, 1990) (based on the Korkin–Zolotarev reduced basis (Korkin and Zolotarev, 1873)) and the sphere decoding algorithm of Fincke and Pohst (Pohst, 1981; Fincke and Pohst, 1985). We will focus on the latter, i.e., on solving (15.3) with the sphere decoding algorithm. (For the system in Fig. 15.2, finding the exact solution by means of exhaustive search requires testing 4.3×10^9 points and is thus practically infeasible. The exact performance curve in Fig. 15.2 is obtained with the sphere decoding algorithm, which on the other hand requires computational effort implementable in practice.)

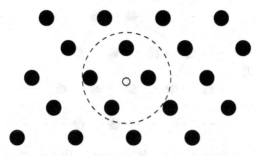

Fig. 15.3. Idea behind the sphere decoder.

The basic premise in sphere decoding is rather simple: attempt to search over only lattice points $s \in \mathcal{D}_L^m$ that lie in a certain sphere of radius d around the given vector x, thereby reducing the search space and hence the required computational effort (see Fig. 15.3). Clearly, the closest lattice point inside the sphere will also be the closest lattice point for the whole lattice. However, closer scrutiny of this basic idea leads to two key questions.

(i) *How to choose d?* Clearly, if d is too large, we may obtain too many points and the search may remain exponential in size, whereas if d is too small, we may obtain no points inside the sphere.

A natural candidate for d is the *covering radius* of the lattice, defined to be the smallest radius of spheres centered at the lattice points that cover the entire space. This is clearly the smallest radius that guarantees the existence of a point inside the sphere for any vector x. The problem with this choice of d is that determining the covering radius for a given lattice is itself NP hard (Conway and Sloane, 1993).

Another choice is to use d as the distance between the Babai estimate and the vector x, i.e., $d = \|x - H\hat{s}_B\|$, since this radius guarantees the existence of at least one lattice point (here the Babai estimate) inside the sphere. However, it may happen that this choice of radius will yield too many lattice points lying inside the sphere.

(ii) *How can we tell which lattice points are inside the sphere?* If this requires testing the distance of each lattice point from x (to determine whether it is less than d), then there is no point in sphere decoding as we shall still need an exhaustive search.

Sphere decoding does not really address the first question (we shall address it later by exploiting statistical assumptions in our model). However, it does propose an efficient way to answer the second one. The basic observation is the following. Although it is difficult to determine the lattice points inside a

T. Kailath, H. Vikalo, and B. Hassibi

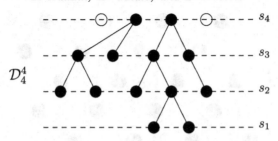

Fig. 15.4. Sample tree generated to determine lattice points in a 4-dimensional sphere.

general m-dimensional sphere, it is trivial to do so in the (one-dimensional) case of $m = 1$. The reason is that a one-dimensional sphere reduces to the endpoints of an interval and so the desired lattice points will be the integer values that lie in this interval. We can use this observation to go from dimension k to dimension $k + 1$. Suppose we have determined all k-dimensional lattice points that lie in a sphere of radius d. Then for any such k-dimensional point, the set of admissible values of the $(k+1)$th dimensional coordinate that lie in the higher dimensional sphere of the *same* radius d forms an interval.

The above means that we can determine all lattice points in a sphere of dimension m and radius d by successively determining all lattice points in spheres of lower dimensions $1, 2, \ldots, m$ and the same radius d. Such an algorithm for determining the lattice points in an m-dimensional sphere essentially constructs a tree where the branches in the kth level of the tree correspond to the lattice points inside the sphere of radius d and dimension k—see Fig. 15.4. Moreover, the complexity of such an algorithm will depend on the *size* of the tree, i.e., on the number of lattice points visited by the algorithm in different dimensions.

With this brief discussion we can now be more specific about the problem at hand. To this end, we shall assume that $n \geq m$, i.e., that there are at least as many equations as unknowns in $x \approx Hs$. Note that the lattice point Hs lies inside a sphere of radius d centered at x if, and only if,

$$d^2 \geq \|x - Hs\|^2 . \tag{15.9}$$

To break the problem into the subproblems described above, it is useful to introduce the QR factorization of the matrix H

$$H = Q \begin{bmatrix} R \\ 0_{(n-m) \times m} \end{bmatrix}, \tag{15.10}$$

where R is an $m \times m$ upper triangular matrix and $Q = [Q_1 \quad Q_2]$ is an $n \times n$ orthogonal matrix. The condition (15.9) can then be written as

$$d^2 \geq \left\| x - [Q_1 \quad Q_2] \begin{bmatrix} R \\ 0 \end{bmatrix} s \right\|^2 = \|Q_1^* x - Rs\|^2 + \|Q_2^* x\|^2,$$

where $(\cdot)^*$ here denotes Hermitian matrix transposition. Or in other words,

$$d^2 - \|Q_2^* x\|^2 \geq \|Q_1^* x - Rs\|^2. \tag{15.11}$$

Defining $y = Q_1^* x$ and $d'^2 = d^2 - \|Q_2^* x\|^2$ allows us to rewrite this as

$$d'^2 \geq \sum_{i=1}^{m} \left(y_i - \sum_{j=i}^{m} r_{i,j} s_j \right)^2, \tag{15.12}$$

where $r_{i,j}$ denotes an (i, j) entry of R. Here is where the upper triangular property of R comes in handy. The right-hand side (RHS) of the above inequality can be expanded as

$$d'^2 \geq (y_m - r_{m,m} s_m)^2 + (y_{m-1} - r_{m-1,m} s_m - r_{m-1,m-1} s_{m-1})^2 + \cdots \tag{15.13}$$

where the first term depends only on s_m, the second term on $\{s_m, s_{m-1}\}$ and so on. Therefore, a necessary condition for Hs to lie inside the sphere is that $d'^2 \geq (y_m - r_{m,m} s_m)^2$. This condition is equivalent to s_m belonging to the interval

$$\left\lceil \frac{-d' + y_m}{r_{m,m}} \right\rceil \leq s_m \leq \left\lfloor \frac{d' + y_m}{r_{m,m}} \right\rfloor, \tag{15.14}$$

where $\lceil \cdot \rceil$ denotes rounding to the nearest larger element in the L-PAM constellation that spans the lattice. Similarly, $\lfloor \cdot \rfloor$ denotes rounding to the nearest smaller element in the L-PAM constellation that spans the lattice.

Of course, (15.14) is by no means sufficient. For every s_m satisfying (15.14), defining $d_{m-1}'^2 = d'^2 - (y_m - r_{m,m} s_m)^2$ and $y_{m-1|m} = y_{m-1} - r_{m-1,m} s_m$, a stronger necessary condition can be found by looking at the first two terms in (15.13), which leads to s_{m-1} belonging to the interval

$$\left\lceil \frac{-d'_{m-1} + y_{m-1|m}}{r_{m-1,m-1}} \right\rceil \leq s_{m-1} \leq \left\lfloor \frac{d'_{m-1} + y_{m-1|m}}{r_{m-1,m-1}} \right\rfloor. \tag{15.15}$$

One can continue in a similar fashion for s_{m-2}, and so on until s_1, thereby obtaining all lattice points belonging to (15.9).

We can now formalize the algorithm.

> *Input:* $Q = \begin{bmatrix} Q_1 & Q_2 \end{bmatrix}$, R, x, $y = Q_1^* x$, d.

1. Set $k = m$, $d_m'^2 = d^2 - \|Q_2^* x\|^2$, $y_{m|m+1} = y_m$

2. (Bounds for s_k) Set $UB(s_k) = \lfloor \frac{d_k' + y_{k|k+1}}{r_{k,k}} \rfloor$, $s_k = \lceil \frac{-d_k' + y_{k|k+1}}{r_{k,k}} \rceil - 1$

3. (Increase s_k) $s_k = s_k + 1$. If $s_k \le UB(s_k)$, go to 5, else go to 4.

4. (Increase k) $k = k + 1$; if $k = m + 1$, terminate algorithm, else go to 3.

5. (Decrease k) If $k = 1$, go to 6. Else $k = k - 1$, $y_{k|k+1} = y_k - \sum_{j=k+1}^{m} r_{k,j} s_j$, $d_k'^2 = d_{k+1}'^2 - (y_{k+1|k+2} - r_{k+1,k+1} s_{k+1})^2$, and go to 2.

6. Solution found. Save s and its distance from x, $d_m'^2 - d_1'^2 + (y_1 - r_{1,1} s_1)^2$, and go to 3.

Note that the subscript $k|k+1$ in $y_{k|k+1}$ above is used to denote the received signal y_k adjusted with the already estimated symbol components $s_{k+1}, \dots, s_m$.

We also need a method to determine the desired radius d. Here is where our statistical model of the communication system helps. Note that $(1/\sigma^2) \cdot \|v\|^2 = (1/\sigma^2) \cdot \|x - Hs\|^2$ is a χ^2 random variable with n degrees of freedom. Thus we may choose the radius to be a scaled variance of the noise,

$$d^2 = \alpha n \sigma^2,$$

in such a way that with a high probability we find a lattice point inside the sphere,

$$\int_0^{\alpha n/2} \frac{\lambda^{n/2-1}}{\Gamma(n/2)} e^{-\lambda} d\lambda = 1 - \epsilon,$$

where the integrand is the probability density function of the χ^2 random variable with n degrees of freedom, and where $1 - \epsilon$ is set to a value close to 1, say, $1 - \epsilon = 0.99$. (If the point is not found, we can increase the probability $1 - \epsilon$, adjust the radius, and search again.)

With the above choice of the radius, and because of the random nature of H and v, the computational complexity of the sphere decoding algorithm is clearly a random variable. Moreover, and rather strikingly, we can compute the mean and the variance of the complexity. We omit the details, but note that in Hassibi and Vikalo (2005), the mean value is calculated

(i) for a 2-PAM constellation to be

$$C(m, \rho, d^2) = \sum_{k=1}^{m} f_p(k) \sum_{l=0}^{k} \binom{k}{l} \gamma\left(\frac{\alpha n}{2(1 + \frac{12\rho l}{m(L^2-1)})}, \frac{n-m+k}{2}\right)$$

(15.16)

(ii) for a 4-PAM constellation to be

$$C(m, \rho, d^2) = \sum_{k=1}^{m} f_p(k) \sum_{q} \frac{1}{2^k} \sum_{l=0}^{k} \binom{k}{l} g_{kl}(q)$$

(15.17)

$$\times \gamma\left(\frac{\alpha n}{2(1 + \frac{12\rho q}{m(L^2-1)})}, \frac{n-m+k}{2}\right),$$

where $g_{kl}(q)$ is the coefficient of x^q in the polynomial

$$(1 + x + x^4 + x^9)^l (1 + 2x + x^4)^{k-l}.$$

The number of elementary operations per visited point in (15.16) and (15.17) is $f_p(k) = 2k + 9 + 2L$, and $\gamma(\cdot, \cdot)$ denotes an incomplete gamma function.

Similar expressions can be obtained for 8-PAM, 16-PAM, etc., constellations.

Let $C(m, \rho)$ denote the expected complexity of actually finding the solution, i.e., the expected complexity of the search where we keep increasing radii until finding a lattice point. Fig. 15.5 shows the expected complexity exponent defined as $e_c = \log_m(C(m, \rho))$. For a wide range of SNR, $e_c \le 4$, and thus in such SNR regions the expected complexity of the sphere decoding is comparable with the complexity of the heuristic techniques.

Fig. 15.5 shows the complexity as a function of SNR for $m = 10$ and L^2-QAM constellations with $L = 2, 4, 8, 16$. A particular modulation scheme can be used only in the range of SNRs that supports transmission at the rate corresponding to that modulation scheme, i.e., the rate has to be smaller than the ergodic capacity of the MIMO channel,

$$C_{\text{erg}} = E\{\log \det(I_M + \mathbf{H}^*\mathbf{H})\}.$$

On the other hand, the complexity of the sphere decoding algorithm for such SNRs is practically feasible (as noted above, it is often $e_c \le 4$). For instance, although the complexity for $L = 16$ appears to be high over a wide range of SNR, it is only for $\rho > \rho_{40} = 27.9$ dB that this modulation scheme can be employed (ρ_{40} is the SNR for which the capacity $C_{\text{erg}} = 40 = R_4(L = 16)$). The complexity exponent at ρ_{40} and $L = 16$ is $e_c \approx 4.4$. The other

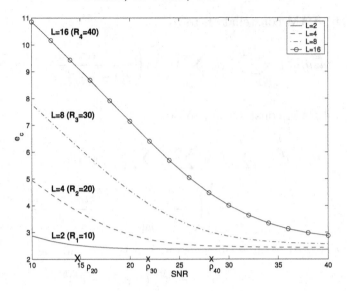

Fig. 15.5. The complexity exponent as a function of ρ for $m = n = 10$ and $L = 2, 4, 8, 16$.

SNRs marked on Fig. 15.5, $\rho_{30} = 21.6$ dB, and $\rho_{20} = 14.9$ dB, have similar meanings (only for $L = 8$ and $L = 4$, respectively).

On another note, the expected complexity above accounts for finding all the lattice points in the sphere. The point among those found that is closest to x is the solution to (15.3). There are some more efficient variations on the basic sphere decoding algorithm that potentially avoid having to search over all the lattice points inside the sphere. We briefly mention two of them here. In both cases, of course, the expected complexity will be no greater than that of the basic sphere decoding algorithm; however, exact calculation of the expected complexity appears to be difficult.

Sphere decoding with radius update Whenever the algorithm finds a point s_{in} inside the sphere (note that Hs_{in} is not necessarily the closest point to x), we set the new radius of the sphere $d^2 = \|x - Hs_{\text{in}}\|^2$ and restart the algorithm. Such radius update may be particularly useful at lower SNRs, where the number of points in the initial sphere is relatively large.

Schnorr–Euchner version of sphere decoding This strategy was proposed in Schnorr and Euchner (1994). The likelihood that the point will be found early is maximized if the search at each dimension k is performed from the middle of the allowed interval for s_k, and if the radius update strategy (as described above) is used. More details about the Schnorr–Euchner version of the sphere decoding, and some improvements thereof, can be found in Agrell *et al.* (2002). The connection between the Schnorr–Euchner strategy and the traditional stack sequential decoding algorithm is discussed in Damen *et al.* (2003).

15.3 Soft MIMO receive algorithms

Multiantenna wireless communication systems that protect transmitted data by either imposing error-correcting or space-time codes require probabilistic (soft) information at the MIMO receiver. This soft information is typically used to iterate between the receiver and the inner decoder (which recovers information from the error-correcting or the space-time encoder).

In Stefanov and Duman (2001), turbo-coded modulation for multiantenna systems was studied, and heuristics based on N/C were employed to obtain soft channel information. It was also noted there that if the soft information is obtained by means of an exhaustive search, the computational complexity grows exponentially in the number of transmit antennas and in the size of the constellation. Hence, for high-rate systems with large number of antennas, the exhaustive search proves to be practically infeasible.

In Vikalo *et al.* (2004), Hochwald and ten Brink (2003), two variations of the sphere decoding algorithm were proposed for obtaining the soft information. Both variations reduce the complexity of estimating the soft information by employing sphere decoding ideas to constrain the number of lattice points used for computing the required likelihood ratios. In Hochwald and ten Brink (2003), sphere decoding was employed to obtain a list of bit sequences that are "good" in a likelihood sense. This list is then used to generate soft information, which is subsequently updated by iterative channel decoder decisions. In Vikalo *et al.* (2004), a MIMO detector based on a modification of the original Fincke–Pohst algorithm was proposed to efficiently obtain soft information for the transmitted bit sequence. This modified Fincke–Pohst algorithm essentially performs a maximum a posteriori (MAP) search, i.e., it solves

$$\min_{s \in \mathcal{D}_L^m} \left[\|x - Hs\|^2 - \sum_{k=1}^{m} \log p(s_k) \right],$$

where $p(s_k)$ are *a priori* information for each symbol in the transmitted sequence. The MAP search is used to obtain a set of lattice points that contribute significantly to the likelihood ratios (for more details, see Vikalo *et al.*, 2004). These likelihood ratios (i.e., the required soft information) are then passed onto the channel decoder. The channel decoder's output is then fed back to the Fincke–Pohst MAP (FP-MAP) for the next iteration.

As discussed above, to obtain computationally efficient receiver schemes, the MIMO communication systems utilizing soft information may require modifications of the basic sphere decoding algorithm. Other MIMO systems may require such modifications as well. For FIR channels, the sphere decoding algorithm does not at all exploit the Markovian property of the channel, which is precisely what the Viterbi algorithm does. Practical algorithms that combine both structures (the lattice and the Markovian property) are highly desirable. On the other hand, when error-correcting codes are coupled with analog channels (through some modulation scheme) problems of joint detection and decoding arise. Some preliminary work addressing both these issues can be found in Vikalo (2003).

References

Agrell, E., Eriksson, T., Vardy, A., and Zeger, K. (2002). Closest point search in lattices. *IEEE Trans. Inf. Theory*, **48** (8), 2001–2214.

Ajtai, M. (1998). The shortest vector problem in L_2 is NP-hard for randomized reductions. In *Proc. 30th Ann. ACM Symp. Theory of Comput.*, 10–19.

Banihashemi, A. H. and Khandani, A. K. (1998). On the complexity of decoding lattices using the Korkin-Zolotarev reduced basis. *IEEE Trans. Inf. Theory*, **44** (2), 162–171.

Brutel, C. and Boutros, J. (1999). Euclidean space lattice decoding for joint detection in CDMA systems. In *Proc. 1999 IEEE Inf. Theory and Commun. Workshop*, 129.

Conway, J. H. and Sloane, N. J. (1993). *Sphere Packings, Lattices and Graphs* (Springer-Verlag).

Damen, M. O., Chkeif, A., and Belfiore, J. C. (2000). Lattice codes decoder for space-time codes. *IEEE Commun. Lett.*, 4, 161–163.

Damen, M. O., Gamal, H. E., and Caire, G. (2003). On maximum-likelihood detection and the search for the closest lattice point. *IEEE Trans. Inf. Theory*, **49** (10), 2389–2402.

Fincke, U. and Pohst, M. (1985). Improved methods for calculating vectors of short length in a lattice, including a complexity analysis. *Math. Comp.*, **44** (170), 463–471.

Foschini, G. J. (1996). Layered space-time architecture for wireless communication in a fading environment when using multi-element antennas. *Bell Labs Tech. J.*, **1** (2), 41–59.

Grotschel, M., Lovász, L., and Schriver, A. (1993). *Geometric Algorithms and Combinatorial Optimization* (Springer Verlag), second edn.

Hassibi, B. (1999). An efficient square-root algorithm for BLAST. *Submitted to IEEE Trans. Signal Process.* URL http://mars.bell-labs.com.

Hassibi, B. and Hochwald, B. (2002). High-rate codes that are linear in space and time. *IEEE Trans. Inf. Theory*, **48** (7), 1804–1824.

Hassibi, B. and Vikalo, H. (2005). On the sphere decoding algorithm I. Expected complexity. *IEEE Trans. Signal Process.*, **53**, 2806–2818.

Hochwald, B. M. and ten Brink, S. (2003). Achieving near-capacity on a multiple-antenna channel. *IEEE Trans. Commun.*, **51** (3), 389–399.

Kailath, T., Sayed, A. H., and Hassibi, B. (2000). *Linear Estimation* (Prentice-Hall, Englewood Cliffs, NJ).

Kannan, R. (1983). Improved algorithms on integer programming and related lattice problems. In *Proc. 15th Ann. ACM Symp. Theory of Comput.*, 193–206.

Korkin, A. and Zolotarev, G. (1873). Sur les formes quadratiques. *Math. Ann.*, **6**, 366–389.

Lagarias, J. C., Lenstra, H. W., and Schnorr, C. P. (1990). Korkin-Zolotarev bases and successive minima of a lattice and its reciprocal. *Combinatorica*, **10**, 333–348.

Pohst, M. (1981). On the computation of lattice vectors of minimal length, successive minima and reduced basis with applications. *ACM SIGSAM Bull.*, **15**, 37–44.

Schnorr, C. P. and Euchner, M. (1994). Lattice basis reduction: improved practical algorithms and solving subset sum problems. *Math. Program.*, **66** (2), 181–199.

Stefanov, A. and Duman, T. M. (2001). Turbo-coded modulation for systems with transmit and receive antenna diversity over block fading channels: system model, decoding approaches, and practical considerations. *IEEE J. Sel. Areas Commun.*, **19** (5).

Verdu, S. (1998). *Multiuser Detection* (Cambridge Univ. Press).

Vikalo, H. (2003). *Sphere Decoding Algorithms for Digital Communications.* Ph.D. thesis, Stanford University.

Vikalo, H., Hassibi, B., and Kailath, T. (2004). Iterative decoding for MIMO channels via modified sphere decoder. *IEEE Trans. Wireless Commun.*, **3** (6).

Viterbo, E. and Boutros, J. (2000). A universal lattice code decoder for fading channels. *IEEE Trans. Inf. Theory*, **45**, 1639–1642.

16

Space-time turbo coding

Stephan ten Brink

Realtek Semiconductors

16.1 Introduction

In digital communication systems, error correcting coding is used to combat channel impairments such as noise or fading. The discovery of an iterative "turbo" decoding strategy (Berrou *et al.*, 1993) started a new era in error correcting coding. Turbo codes were quickly adopted for wireless cellular standards like CDMA2000 and UMTS. Basic building blocks are soft in/soft out decoders connected through interleavers. With each decoding iteration, reliability information is exchanged, and *a priori* knowledge is updated by new, or *extrinsic* information—a mechanism similar to a turbo engine.

The advance of silicon technology facilitates the implementation of more sophisticated algorithms at the receiver, enabling iterative processing not only within the channel decoder, but also over the channel interface, such as the detector of a multiple input/multiple output (MIMO) antenna communication scheme. MIMO techniques (e.g., Winters *et al.* (1994)) allow to increase the data rate while keeping the bandwidth unchanged, thus making better use of the scarce spectral resources. They have recently found their way into a number of wireless communication standards, like IEEE 802.11n wireless LAN and 802.16 wireless MAN.

In this chapter, we apply turbo processing to the detection and decoding of signals transmitted over MIMO channels. We first outline several variants of coding over space and time, and determine the ultimate capacity limits of MIMO channels. We then study the properties of iterative processing structures, explain the exchange of reliability information and discuss the convergence behavior of iterative decoding in the MIMO context. Finally, we illustrate how low-density parity-check (LDPC) codes can approach capacity very closely by appropriate degree profile design matched to the MIMO detector.

322

16.2 Coding over space and time

In *open loop transmission* the communication channel is known at the receiver but not at the transmitter. This is commonly achieved by embedding pilot symbols or training sequences into the transmitted signal, to allow for channel estimation at the receiver. In *closed loop transmission* the channel is known at both transmitter and receiver, e.g., by feeding back channel state information from the receiver to the transmitter. This facilitates the application of beamforming techniques based on the singular value decomposition, which, however, are more difficult to implement in practice. We only consider open loop techniques in the sequel.

Essentially, there are two different ways of using multiple antennas at both sides, transmitter and receiver: *space-time coding*, and *spatial multiplexing*. Space-time codes are further divided into *space-time block codes* (Alamouti, 1998; Tarokh *et al.*, 1999), and *space-time trellis codes* (Tarokh *et al.*, 1998). Space-time block codes can be viewed as diversity mappings, or repetition codes over space and time providing increased robustness and range extension by transmitting the same data from different antennas. Typically, more transmit than receive antennas are applied. Optimal detection for orthogonal space-time block codes is simple, similar to maximum ratio combining. Due to its low implementation complexity, this technique has found prominent applications, e.g., in wireless cellular systems. Space-time trellis codes are an extension of trellis coded modulation to the multiple antenna case, offering an additional coding gain.

In spatial multiplexing (Foschini, 1996), different data streams are transmitted over different antennas, to increase the data rate at shorter range. Sometimes it is also referred to as "direct transmission", or simply "MIMO" (multiple input/multiple output). Optimal detection to recover the individual data streams tends to be complex, and different detection schemes have been devised, with a complexity/performance trade-off ranging from simple zero-forcing, ordered nulling-cancellation as pioneered in Golden *et al.* (1999), to complex maximum-likelihood and soft *a posteriori* probability (APP) detection, e.g., Tonello (2000).

We focus on spatial multiplexing (MIMO communications) in this chapter and study its combination with error correcting coding (channel coding). Particularly, we assume bit-interleaved coded modulation (Caire *et al.*, 1998) extended to the multiple antenna case, where a single channel encoder is connected to a multidimensional MIMO mapping through a (pseudo) random bit interleaver. At the receiver, MIMO demapping (detection) is performed. Next, we establish some capacity references for MIMO communications.

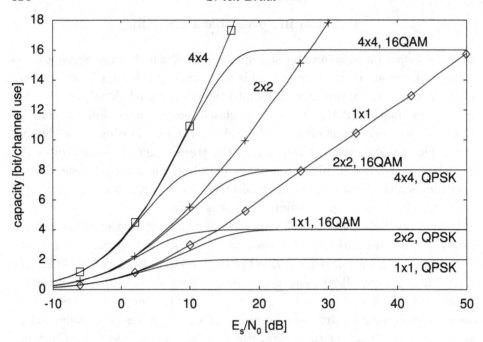

Fig. 16.1. Capacity and mutual information of ergodic MIMO channels.

16.3 Channel model and capacity limits

We assume a configuration with M transmit and N receive antennas. The transmitted symbols are $M \times 1$ vectors $\mathbf{s} = [s_1, \ldots, s_M]^T$ with entries taken from some complex constellation (e.g., QPSK, 16-QAM) of size 2^{M_c} signal points. Each vector symbol carries MM_c bits. The total transmitted power is E_s, with energy constraint $\mathrm{E}\left[|s_m|^2\right] = E_s/M$ per component. The received $N \times 1$ vector is $\mathbf{y}$, with

$$\mathbf{y} = \mathbf{H}\mathbf{s} + \mathbf{n}, \tag{16.1}$$

where $\mathbf{H}$ is the $N \times M$ complex channel matrix, known perfectly to the receiver, and $\mathbf{n}$ is an $N \times 1$ vector of independent zero-mean complex Gaussian noise entries with variance $\sigma^2 = N_0/2$ per real component. The normalized signal-to-noise ratio is defined as

$$\left.\frac{E_b}{N_0}\right|_{dB} = \left.\frac{E_s}{N_0}\right|_{dB} + 10\log_{10}\frac{N}{RMM_c}. \tag{16.2}$$

By convention, we use the value N in (16.2) to keep the $E_b/N_0|_{dB}$ at capacity close to each other for different N. We assume that $\mathbf{H}$ is known to the receiver only, and consider a Rayleigh fading channel, with entries of $\mathbf{H}$ being independent complex zero-mean Gaussian random variables with independent real and imaginary parts each having variance $1/2$. For a *quasi-static* channel the matrix $\mathbf{H}$ remains unchanged over long time intervals, while for an *ergodic* channel $\mathbf{H}$ changes for every symbol $\mathbf{s}$. An ergodic channel is approximated in practice by coding over many realizations of $\mathbf{H}$, like in multicarrier modulation employing time/frequency interleaving. We do not further consider a quasi-static model, where a comparison with an "outage" capacity would be more appropriate.

The ergodic MIMO channel capacity is (Foschini, 1996; Telatar, 1999)

$$ C = E\left[\log_2 \det\left(\mathbf{I} + \frac{E_s}{N_0}\frac{1}{M}\mathbf{H}\mathbf{H}^*\right)\right], \tag{16.3} $$

where $\mathbf{I}$ is the identity matrix and $\mathbf{H}^*$ is the complex-conjugate transpose of $\mathbf{H}$. Capacity is achieved by using continuous Gaussian distributed input symbols $\mathbf{s}$. For constrained input constellations (e.g., QPSK, 16-QAM) the maximal achievable rate is the mutual information between channel input $\mathbf{s}$ and output $\mathbf{y}$, $I(\mathbf{S}; \mathbf{Y})$. Generally, all signal points $s_1, \ldots, s_M$ are assumed to be equally likely. We do not further distinguish between capacity C and mutual information $I(\mathbf{S}; \mathbf{Y})$ in the following. Some capacity curves are given in Fig. 16.1. For error-free (reliable) transmission, only those points $(E_s/N_0, C)$ are attainable in the capacity plane for which channel code rate R and vector constellation size 2^{MM_c} fulfill $RMM_c \leq C$ (channel coding theorem (Shannon, 1948)). The corresponding E_s/N_0-value at $RMM_c = C$ is referred to as *Shannon limit* and serves as the ultimate performance limit.

16.4 System structures of MIMO turbo processing

A brief overview of possible iterative "turbo" processing structures for MIMO communications is depicted in Fig. 16.2. Two or more component detectors/decoders exchange *extrinsic* information in an iterative fashion, simply referred to as *iterative detection*. We start with a qualitative discussion before presenting numerical results in the later sections.

- *Structure (a): demapper, channel decoder.* This is a straightforward implementation of bit-interleaved coded modulation with iterative decoding over MIMO channels. The outer channel encoder and inner MIMO mapper can be regarded as a serially concatenated coding scheme (Benedetto *et al.*, 1998), allowing iterative demapping and decoding at the receiver. It is the most common configuration in practical MIMO systems.
- *Structure (b): demapper, decoder of a parallel concatenated code.* A serial concatenation of an outer parallel concatenated "turbo" code and a

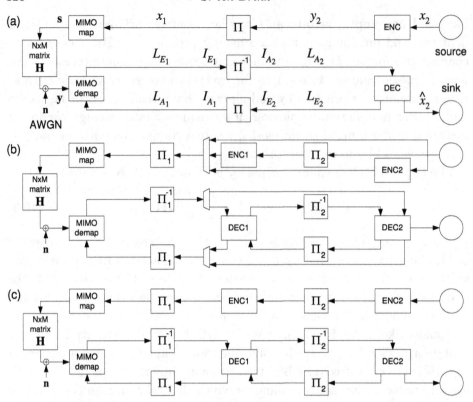

Fig. 16.2. Concatenated coding and MIMO mapping with corresponding iterative MIMO demapping and decoding structures: (a) demapper, decoder; (b) demapper, decoder of parallel concatenated code; (c) demapper, decoder of serially concatenated code.

MIMO mapper, with iterative demapping at the receiver, including iterative decoding of the turbo code. Extrinsic information with regard to the information bits is fed back within the turbo code iterations, while extrinsic information with regard to the coded bits is fed back for the demapper iterations. Receivers for the wireless cellular system UMTS exhibit this structure.

- *Structure (c): demapper, decoder of a serially concatenated code.* A serial concatenation of an outer code with an inner MIMO mapper, whereby the outer code by itself is a serial concatenation of two component codes, resulting in a doubly serially concatenated coding structure with two iteration loops.
- *Structure (c_a): combined demapper/inner decoder, outer decoder of a serially concatenated code.* Omitting the interleaver between inner encoder and mapper of *structure (c)* allows to include the MIMO demapping met-

ric into the inner decoding trellis, thus reducing complexity. The second iteration loop is no longer required, and we obtain *structure (a)*, with a combined MIMO demapper/inner decoder processing block.

The channel code introduces redundancies and dependencies between the coded bits that are mapped onto a multidimensional MIMO constellation. At the receiver, it is suboptimal yet pragmatic to operate MIMO detector and channel decoder separately. The optimal but more complex *joint detection* is approximated by iteratively exchanging soft information between detector and decoder. Next we describe details of soft iterative processing by focusing on *structure (a)*, *iterative MIMO demapping and decoding*.

16.5 Iterative exchange of reliability information

The channel code and the MIMO mapper with matrix channel $\mathbf{H}$ can be interpreted as a serially concatenated coding scheme. The outer encoder (e.g., a convolutional code) forwards outer coded bits through the bit interleaver to the inner space-time constellation mapping; the MIMO channel implicitly performs a block encoding with matrix $\mathbf{H}$, introducing memory, thus connecting the bits of the vector constellation across transmit antennas. Optimal decoding at the receiver corresponds to jointly detecting the MIMO constellation and decoding the outer code, which is prohibitively complex. The bit interleaver at the transmitter allows to separate this task into MIMO demapping and channel decoding, approximating *joint detection* using iterative exchange of reliability information between the processing elements.

The iterative algorithm is illustrated in Fig. 16.2. The detector takes channel observations $\mathbf{y}$ and *a priori* knowledge L_{A_1} on the outer coded bits and computes *extrinsic* information L_{E_1} for each of the MM_c coded bits per vector channel symbol $\mathbf{y}$. L_{E_1} is deinterleaved to become the *a priori* input L_{A_2} to the outer soft in/soft out decoder (MAP, APP, BCJR algorithm (Bahl *et al.*, 1974)), which calculates *extrinsic* information L_{E_2} on the outer coded bits. L_{E_2} is reinterleaved and fed back as *a priori* knowledge L_{A_1} to the inner detector, completing one iteration.

For iterative decoding it is essential that the MIMO demapper accepts *a priori* information as fed back from the outer channel decoder. In the next section we describe how to compute the various *a priori* and *extrinsic* quantities.

16.5.1 Inner MIMO detection

The complexity of the optimal *a posteriori* probability detector grows exponentially with the number of bits in the constellation, i.e., with the number of transmit antennas M and the size of the QAM constellation M_c. We assume optimal *a posteriori* probability processing blocks. Suboptimal blocks can also be used, but at the cost of performance degradation. Maximizing the *a posteriori* probability (MAP, or APP) for a given bit minimizes the probability of making an error on that bit. The *a posteriori* probability is usually expressed as a log-likelihood ratio value (L-value (Hagenauer *et al.*, 1996)). Equivalently, probabilities could be used. However, with L-values, simple add/subtract operations are sufficient to separate *a priori* or old information from new *extrinsic* information obtained from APP detection/decoding. The sign of the L-value determines whether the bit is a one or zero. The absolute value indicates the reliability of the decision; L-values near zero correspond to unreliable bits. By convention, we represent the logical zero for a bit by the amplitude level $x_k = -1$, and logical one by $x_k = +1$. The *a posteriori* L-value of the bit $x_{1,k}, k = 0, \ldots, MM_c - 1$, conditioned on the received vector channel symbol $\mathbf{y}$ is given as

$$L_{D_1}(x_{1,k} \,|\mathbf{y}) = \ln \frac{P\left[x_{1,k} = +1 \,|\mathbf{y}\right]}{P\left[x_{1,k} = -1 \,|\mathbf{y}\right]}. \tag{16.4}$$

Owing to the bit interleaver between channel encoder and MIMO mapper, the bits of $\mathbf{x}_1$ can be considered as approximately statistically independent. Using Bayes' theorem, and splitting up joint probabilities into products (independence assumption on the $x_{1,k}$), we can write the soft output as

$$L_{D_1}(x_{1,k} \,|\mathbf{y}) = L_{A_1}(x_{1,k}) + \underbrace{\ln \frac{\displaystyle\sum_{\hat{\mathbf{x}}_1 \in \mathbb{X}_{k,+1}} p\left(\mathbf{y} \,|\hat{\mathbf{x}}_1\right) \cdot \exp\left(\tfrac{1}{2}\hat{\mathbf{x}}_{1,[k]}^T \cdot \mathbf{L}_{A_1,[k]}\right)}{\displaystyle\sum_{\hat{\mathbf{x}}_1 \in \mathbb{X}_{k,-1}} p\left(\mathbf{y} \,|\hat{\mathbf{x}}_1\right) \cdot \exp\left(\tfrac{1}{2}\hat{\mathbf{x}}_{1,[k]}^T \cdot \mathbf{L}_{A_1,[k]}\right)}}_{L_{E_1}\left(x_{1,k}|\mathbf{y}\right)},$$

$$\tag{16.5}$$

where $\mathbb{X}_{k,+1}$ is the set of 2^{MM_c-1} bit vectors $\mathbf{x}_1$ having $x_{1,k} = +1$,

$$\mathbb{X}_{k,+1} = \{\mathbf{x} \,|\, x_k = +1\}, \quad \mathbb{X}_{k,-1} = \{\mathbf{x} \,|\, x_k = -1\} \quad \text{respectively,} \tag{16.6}$$

and *a priori* L-value

$$L_{A_1}(x_{1,j}) = \ln \frac{P[x_{1,j} = +1]}{P[x_{1,j} = -1]}. \tag{16.7}$$

The vector $\mathbf{x}_{1,[k]}$ denotes the subvector of $\mathbf{x}_1$ obtained by omitting its kth element $x_{1,k}$, and $\mathbf{L}_{A_1,[k]}$ denotes the vector of L_{A_1}-values omitting $L_{A_1}(x_{1,k})$. Thus, L_{D_1} can be written as a sum of *a priori* L-value L_{A_1} and *extrinsic* L-value L_{E_1}. From the linear channel model (16.1) and with constellation mapping $\hat{\mathbf{s}} = \text{map}(\hat{\mathbf{x}}_1)$ we find the *likelihood function* $p(\mathbf{y}|\hat{\mathbf{x}}_1)$ as

$$p\left(\mathbf{y}\,|\hat{\mathbf{s}} = \text{map}(\hat{\mathbf{x}}_1)\right) = \frac{\exp\left[-\frac{1}{2\sigma^2}\cdot\|\mathbf{y} - \mathbf{H}\cdot\hat{\mathbf{s}}\|^2\right]}{(2\pi\sigma^2)^N}. \qquad (16.8)$$

For the L-value calculation in (16.5) only the term in the exponent is relevant and the constant factor outside the exponent can be omitted. Equation (16.5) serves as the soft input metric to the outer channel decoder.

16.5.2 Outer channel decoding

We may also apply (16.4) to the channel (error-correcting) code. Equation (16.4) then becomes the *a posteriori* L-value obtained from APP decoding of the outer channel code. Thus, the channel decoder processing can also be decomposed into *a priori* and *extrinsic* components. The resulting equation for the channel code is

$$L_{D_2}(x_{2,k}|\mathbf{L}_{A_2}) = L_{A_2}(x_{2,k}) + \underbrace{\ln\frac{\displaystyle\sum_{\hat{\mathbf{x}}_2\in\mathbb{X}_{k,+1}}\exp\left(\frac{1}{2}\hat{\mathbf{x}}_{2,[k]}^T\cdot\mathbf{L}_{A_2,[k]}\right)}{\displaystyle\sum_{\hat{\mathbf{x}}_2\in\mathbb{X}_{k,-1}}\exp\left(\frac{1}{2}\hat{\mathbf{x}}_{2,[k]}^T\cdot\mathbf{L}_{A_2,[k]}\right)}}_{L_{E_2}\left(x_{2,k}|\mathbf{L}_{A_2,[k]}\right)}, \qquad (16.9)$$

with the data bits from the source denoted as $\mathbf{x}_2$. For channel codes that can be decoded using a trellis, (16.9) can be advantageously implemented using a forward and backward recursion (Bahl *et al.*, 1974).

16.6 Studying convergence using mutual information

We now turn to analyzing the dynamics of the iterative algorithm. The convergence behavior can be conveniently described by *transfer curves* (or *characteristics*) based on mutual information. We show how to compute extrinsic information transfer curves for the MIMO detector and the outer channel decoder. Then we predict the convergence of the iterative scheme for arbitrary detector/decoder combinations by plotting both curves into a single diagram, referred to as *extrinsic information transfer (EXIT) chart*.

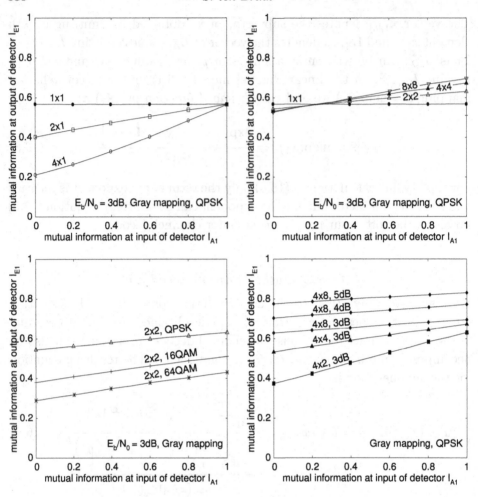

Fig. 16.3. Transfer curves of MIMO APP detection for different $M \times N$ antenna configurations and constellation sizes.

16.6.1 Inner detector transfer curves

To compute the mutual information transfer curve of the MIMO detector we assume L_{A_1} to be an L-value based on Gaussian distributions with variance $\sigma_{A_1}^2$ and mean value $\mu_{A_1} = \sigma_{A_1}^2/2$. The conditional probability density function is

$$p_{A_1}(\xi|X = x) = \frac{e^{-\left(\xi - \frac{\sigma_{A_1}^2}{2} \cdot x\right)^2 / \left(2\sigma_{A_1}^2\right)}}{\sqrt{2\pi}\,\sigma_{A_1}}. \qquad (16.10)$$

To measure the information content of the a priori knowledge over a block length of n, the average mutual information $I_{A_1} = \sum_k I(X_{1,k}; L_{A_1,k})/n$,

$0 \leq I_{A_1} \leq 1$, between transmitted inner information bits X_1 and the L-values L_{A_1} is used.

$$I_{A_1} = \frac{1}{2} \cdot \sum_{x_1=-1,1} \int_{-\infty}^{+\infty} p_{A_1}(\xi | X_1 = x_1)$$

$$\times \log_2 \frac{2 \cdot p_{A_1}(\xi | X_1 = x_1)}{p_{A_1}(\xi | X_1 = -1) + p_{A_1}(\xi | X_1 = 1)} \, d\xi. \quad (16.11)$$

With (16.10), equation (16.11) becomes

$$J(\sigma_{A_1}) := I_{A_1}(\sigma_{A_1}) = 1 - \int_{-\infty}^{+\infty} \frac{e^{-\frac{\left(\xi - \sigma_{A_1}^2/2\right)^2}{2\sigma_{A_1}^2}}}{\sqrt{2\pi}\,\sigma_{A_1}} \cdot \log_2\left[1 + e^{-\xi}\right] d\xi. \quad (16.12)$$

The function $J(\sigma)$ can be evaluated using numerical integration.

Mutual information is also used to quantify the extrinsic output $I_{E_1} = \sum_k I(X_{1,k}; L_{E_1,k})/n$. It is computed according to (16.11), using probability density functions p_{E_1} obtained from histogram measurements at the detector output. The inner extrinsic information transfer characteristic plots I_{E_1} over I_{A_1} for the E_b/N_0 value of interest. In the following, we study the basic dependencies of the detector transfer behavior while changing different parameters of the MIMO system (Fig. 16.3). We use Gaussian *a priori* knowledge and assume an ergodic channel where coding is applied over many realizations of **H**. The observations can be summarized as follows:

- *Increasing the number of transmit antennas M* results in steeper detector transfer curves, as the bits get more and more mutually dependent: initial detection without *a priori* knowledge ($I_{A_1} = 0$) is poor, but improves with higher values of I_{A_1} (1×1, 2×1 and 4×1 QPSK curve).
- *Increasing both M and N* results in a moderate steepening of the detector curve (1×1, 2×2, 4×4, and 8×8 QPSK curves).
- *Increasing the constellation size* results in a vertical shift down (for fixed E_b/N_0): more bits are involved in the detection of constellation symbols and thus a higher E_b/N_0 value is required for reliable communication. For the same reason, the slope becomes slightly steeper (2×2, QPSK, 16-QAM and 64-QAM curves).
- *With more receive antennas N*, the receive diversity is increased, resulting in a gain in detector transfer behavior (4×2, 4×4, 4×8 QPSK curve).
- *An increase in E_b/N_0* merely relates to a vertical shift of the detector transfer characteristic (4×8 curves) toward higher extrinsic output.

Most curves resemble straight lines. Moreover, it can be shown that any

$1 \times N$ curve for QPSK is a horizontal line. The $M \times 1$ curves meet at $I_{A_1} = 1$, and the curves decrease with M when $N = 1$. Similarly, one can show analytically that the $M \times N$ curves meet the $1 \times N$ curve at $I_{A_1} = 1$, and that they decrease with M for fixed I_{A_1} and N.

Note that for a nonergodic, quasi-static channel model the transfer curves would change with each new instance of **H**.

The mutual information transfer behavior of various detectors turns out to be quite different, depending on the number of transmit and receive antennas, and the constellation size. As a consequence, the performance of iterative detection will strongly depend on how well the outer channel code and the inner detector interact in exchanging extrinsic information.

16.6.2 Outer decoder transfer curves

We consider the *a priori* information $I_{A_2} = \sum_k I(Y_{2,k}; L_{A_2,k})/n$ and the *extrinsic* information $I_{E_2} = \sum_k I(Y_{2,k}; L_{E_2,k})/n$ at the decoder output. For the computation of outer transfer curves we again assume L_{A_2} to be Gaussian distributed and determine I_{E_2} by simulation (Fig. 16.4). For convolutional codes, there is a smooth transition from low to high extrinsic output I_{E_2} with increasing input I_{A_2}. The bigger the code memory, the more we obtain a step-function like behavior. For the parallel concatenated code, this transition is much more abrupt, once the input mutual information I_{A_2} exceeds the rate of the code (here $1/2$), resulting in a pronounced *threshold effect*. This behavior is typical for concatenated codes with iterative decoders. Note also that the parallel concatenated code *after no internal iteration* performs worse than both convolutional codes over the entire range of I_{A_2}-values. After eight iterations, however, the parallel concatenated code outperforms the convolutional codes for higher I_{A_2}-input. The effect of these transfer curves on the iterative decoding dynamics is studied next.

16.6.3 Extrinsic information transfer chart

We obtain an EXIT chart (Fig. 16.5) by plotting both inner and outer transfer curve into a single diagram, with the axes for the outer curve swapped. The *decoding trajectory* visualizes the exchange of extrinsic information. It is a simulation result of the iterative decoder. For sufficiently long codes and random bit interleavers, the independence assumption of *a priori* knowledge holds over many iterations, and the trajectory matches with the transfer curves of the individual components quite well. The iteration starts at the origin with zero *a priori* knowledge $I_{A_1} = 0$, and progresses until both trans-

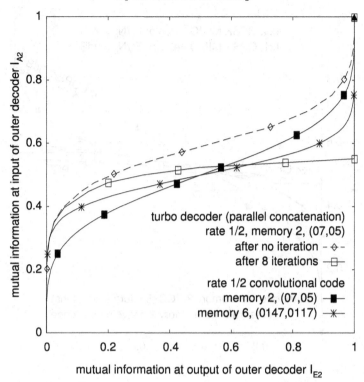

Fig. 16.4. Outer decoder transfer curves; rate 1/2 convolutional code of different memory, parallel concatenated code after up to eight internal iterations; axes swapped.

fer curves intersect ($I_{E_2} = I_{E_1}$), from high bit error rate (BER) in the lower left corner, to low BER in the upper right corner. For successful convergence to low BER, the inner curve must lie above the outer curve over the entire course of the iteration.

Representatively of many possible detector/decoder combinations, we shall study four cases: *structure (a)* with inner 4×1 and 4×4 MIMO detector, combined with an outer convolutional code; and *structure (b)* with the same two detectors, but an outer decoder for a parallel concatenated code. Fig. 16.5 shows decoding trajectories for two such combinations to verify our convergence predictions.

- *Structure (a), 4×1 MIMO detector at 9 dB, convolutional code:* the slope of inner and outer transfer curve are quite similar; the decoding trajectory indicates convergence to a reasonably low BER floor, which we measured to be $5 \cdot 10^{-4}$. The distance to capacity is about 2.5 dB (capacity at 6.6 dB).

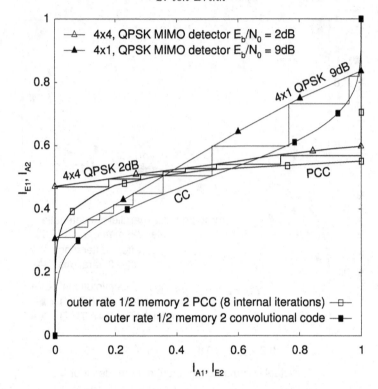

Fig. 16.5. EXIT chart: inner demapper and outer decoder transfer curves with decoding trajectories; 4×4 demapper and parallel concatenated code (PCC) at $E_b/N_0 = 2$ dB, 4×1 demapper and convolutional code (CC) at $E_b/N_0 = 9$ dB.

The error floor cannot be lowered by a longer interleaver, since both inner and outer curves intersect.

• *Structure (a)*, 4×4 *MIMO demapper at* 2 *dB, convolutional code:* convergence is possible, but hardly any improvement can be achieved by iterations, resulting in a rather high BER at 10^{-1}.

• *Structure (b)*, 4×1 *MIMO demapper at* 9 *dB, parallel concatenated code:* early intersection of both curves, no convergence. The 4×1 curve has to be raised up to 13 dB to achieve convergence, resulting in a 4 dB penalty over the convolutional code. Obviously, the parallel concatenated code is not a good match for detector curves with steep slopes.

• *Structure (b)*, 4×4 *MIMO demapper at* 2 *dB, parallel concatenated code:* the trajectory shows that convergence is possible for 2 dB, which is very close to capacity (Fig. 16.1, capacity at about 1.5 dB). The BER floor, if any, can be lowered by increasing the interleaver length.

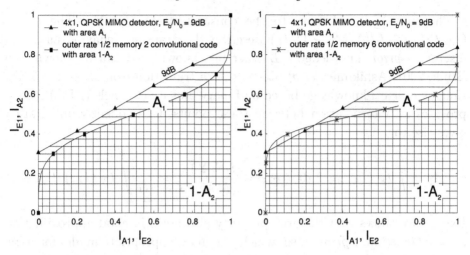

Fig. 16.6. Illustration of area property of inner demapper and outer decoder.

16.6.4 Design guidelines based on area property

In this section we motivate an *area property* for transfer curves based on mutual information. The MIMO constellation/labeling maps the bit vector $\mathbf{X} = (X_0, \ldots, X_{M-1})$ into the vector symbol $\mathbf{S}$. With the *chain rule of mutual information* we can decompose the *vector-wise* mutual information $I(\mathbf{X}; \mathbf{Y}) = I(\mathbf{S}; \mathbf{Y})$ into a sum of MM_c *bitwise* mutual informations I_L,

$$I(\mathbf{X}; \mathbf{Y}) = \sum_{L=0}^{MM_c-1} I_L. \tag{16.13}$$

I_L is a shorthand notation of

$$I_L = \overline{I(X_m; \mathbf{Y} \,|\, L \text{ other bits known})}, \tag{16.14}$$

$$0 \leq m < MM_c; \; 0 \leq I_L \leq 1. \tag{16.15}$$

A number of L "other" unmapped bits $X_n, n \neq m$ of the vector constellation are perfectly known, whereas no knowledge is available for the remaining $MM_c - 1 - L$ bits. The bar in (16.14) indicates that I_L is averaged a) over bitwise mutual information with respect to all bits of the constellation, $X_m, 0 \leq m < MM_c$, b) over all possible $\binom{MM_c-1}{L}$ combinations to choose L known bits out of the total of $MM_c - 1$ other bits of the constellation, and c) over all 2^L bit vector realizations thereof. The vector channel with mutual information $0 \leq I(\mathbf{X}; \mathbf{Y}) \leq MM_c$ carrying MM_c bits can be viewed as being composed of MM_c parallel subchannels with mutual information $0 \leq I_L \leq 1$, each carrying a single bit.

The sum $\sum I_L$ adds up to the constant value $I(\mathbf{X};\mathbf{Y})$. The points $(I_A, I_E) = (L/(MM_c - 1), I_L)$ describe the transfer behavior based on discrete a priori knowledge ("L other bits known"). It can be shown (ten Brink, 2001; Ashikhmin et al., 2004) that a transition from discrete to continuous a priori knowledge based on binary erasure channels (BEC) merely provides an interpolation between these points. Consequently, integration gives rise to

$$A_1 = \int_0^1 I_{E_1}(I_{A_1,\mathrm{BEC}})\,dI_{A_1,\mathrm{BEC}} = \frac{I(\mathbf{S};\mathbf{Y})}{MM_c} \underset{\substack{\mathrm{Gaussian} \\ \mathrm{a\ priori}}}{\approx} \int_0^1 I_{E_1}(I_{A_1})\,dI_{A_1}.$$

Experience shows that the area property also holds in good approximation for the Gaussian a priori case, which is a more appropriate model for iterative decoding schemes exchanging L-values. For the outer code, the area is connected to the rate of the code (Ashikhmin et al., 2004)

$$A_2 = \int_0^1 I_{E_2}(I_{A_2,\mathrm{BEC}})\,dI_{A_2,\mathrm{BEC}} = 1 - R \underset{\substack{\mathrm{Gaussian} \\ \mathrm{a\ priori}}}{\approx} \int_0^1 I_{E_2}(I_{A_2})\,dI_{A_2}.$$

This is illustrated in Fig. 16.6. *Horizontal lines* indicate the area A_1 under the inner detector curve, while *vertical lines* correspond to the area $1 - A_2$ under the outer decoder curve. Convergence is possible for $A_1 > 1 - A_2$, i.e., $I(\mathbf{S};\mathbf{Y}) > R$, which is just another way of writing Shannon's channel coding theorem (Shannon, 1948). The area property restricts the shape of the transfer curves. An area gap directly relates to a rate loss: both curves should be *matched* to each other to minimize this gap; code design reduces to *curve fitting*. Irregular codes offer the flexibility to shape the outer decoder curve as desired, to account for any detector, as discussed in the next section.

16.7 Irregular codes for approaching capacity

Irregular low-density parity-check codes can be designed using a curve fitting approach to minimize the area gap between inner detector and outer decoder transfer curve (Ashikhmin et al., 2004; ten Brink et al., 2004). Alternatively, punctured convolutional codes can be used (Tuechler, 2004). In the following, we consider an LDPC code of length n and rate $R = k/n$ (Richardson and Urbanke, 2001). An iterative decoder for this code is essentially identical to that one of the serially concatenated *structure (a)*. It is composed of n variable nodes, an edge interleaver, and $n - k$ check nodes. The ith variable node represents the ith bit of the codeword. This bit is involved in $d_v^{(i)}$ parity checks, and thus its node has $d_v^{(i)}$ edges going into the edge interleaver. The edge interleaver connects the variable nodes to the check nodes, each of which representing a parity-check equation. The ith check node checks $d_c^{(i)}$ bits and thus has $d_c^{(i)}$ edges. The sets of variable and check nodes are referred to as the *inner variable node decoder* (VND) and *outer check node decoder* (CND), respectively. The MIMO detector is used as the channel interface; it is part of the inner decoder, providing L-values as input to the VND, and accepting extrinsic messages from the VND as *a priori* input. Owing to the edge interleaver, no separate interleaver between detector and VND is required.

16.7.1 Transfer curve of the inner variable node decoder

A variable node of degree d_v has $d_v + 1$ incoming messages, d_v from the edge interleaver and one from the channel. The variable node decoder just performs an addition of L-values

$$L_{i,\text{out}} = L_{\text{ch}} + \sum_{j \neq i} L_{j,\text{in}}, \qquad (16.16)$$

where $L_{j,\text{in}}$ is the jth *a priori* L-value going into the variable node, $L_{i,\text{out}}$ is the ith extrinsic L-value coming out of the variable node, and L_{ch} is the L-value from the channel interface, e.g., the MIMO detector. For an AWGN channel with BPSK (± 1) modulation and noise variance σ_n^2, we define $E_b/N_0 = 1/(2R\sigma_n^2)$. The channel L-value is

$$L_{\text{ch}} := \log \frac{p(y|x = +1)}{p(y|x = -1)} = \frac{2}{\sigma_n^2} y, \qquad (16.17)$$

where $p(y|x)$ is the Gaussian probability density function evaluated at the channel output y given the input x. Let X and Y be random variables representing the respective channel input and output. The variance of L_{ch} conditioned on X is $\sigma_{\text{ch}}^2 = 4/\sigma_n^2 = 8R \cdot E_b/N_0$. To compute an extrinsic information transfer curve for (16.16), we model $L_{j,\text{in}}$ as the output L-value of an AWGN channel whose input is the jth interleaver bit transmitted using BPSK. The transfer function of a degree-d_v variable node is

$$I_{E,\text{VND}}(I_{A,\text{VND}}, d_v) = J\left(\sqrt{(d_v - 1)\left[J^{-1}(I_{A,\text{VND}})\right]^2 + \sigma_{\text{ch}}^2} \right), \qquad (16.18)$$

with inverse function $J^{-1}(\cdot)$, and $J(\cdot)$ as defined in (16.12). Instead of an AWGN channel, we assume a MIMO channel interface with detector transfer curve $I_{E,\mathrm{DET}}(I_{A,\mathrm{DET}}, E_b/N_0)$, see examples in Fig. 16.3, and set

$$\sigma_{\mathrm{ch}} = J^{-1}(I_{E,\mathrm{DET}}(I_{A,\mathrm{DET}}, E_b/N_0)),$$

with *a priori* knowledge at the MIMO detector

$$I_{A,\mathrm{DET}}(I_{A,\mathrm{VND}}, d_v) = J(\sqrt{d_v} \cdot J^{-1}(I_{A,\mathrm{VND}})).$$

16.7.2 Transfer curve of the outer check node decoder

The decoding of a degree d_c check node corresponds to the decoding of a length d_c (or rate $(d_c - 1)/d_c$) single parity check code. The output L-values are thus (box-plus operation (Hagenauer *et al.*, 1996))

$$L_{i,\mathrm{out}} = \ln \frac{1 - \prod_{j \neq i} \frac{1 - e^{L_{j,\mathrm{in}}}}{1 + e^{L_{j,\mathrm{in}}}}}{1 + \prod_{j \neq i} \frac{1 - e^{L_{j,\mathrm{in}}}}{1 + e^{L_{j,\mathrm{in}}}}} = \sum_{j \neq i} \boxplus L_{j,\mathrm{in}}. \qquad (16.19)$$

Again, $L_{j,\mathrm{in}}$ is modeled as the output L-value of an AWGN channel whose input is the jth interleaver bit transmitted using BPSK. The check node transfer curves can be computed by simulation. Alternatively, for the binary erasure channel a duality property exists (Ashikhmin *et al.*, 2004) that gives the transfer curve $I_{E,\mathrm{SPC}}(\cdot)$ of the length d_c single parity check code in terms of the transfer curve $I_{E,\mathrm{REP}}(\cdot)$ of the length d_c (or rate $1/d_c$) repetition code, i.e., $I_{E,\mathrm{SPC}}(I_A, d_c) = 1 - I_{E,\mathrm{REP}}(1 - I_A, d_c)$. Simulations show that this property is an accurate approximation for BPSK/Gaussian *a priori* inputs $L_{j,\mathrm{in}}$. The check node transfer curve is approximated as

$$I_{E,\mathrm{CND}}(I_{A,\mathrm{CND}}, d_c) \approx 1 - J\left(\sqrt{d_c - 1} \cdot J^{-1}(1 - I_{A,\mathrm{CND}})\right). \qquad (16.20)$$

16.7.3 Transfer curves for code mixtures

We assume a *check-regular* LDPC code with check node degree d_c. The LDPC code is designed by specifying the variable node degrees $d_v^{(i)}$, $i = 1, \ldots, n$. Let D be the number of different variable node degrees, and denote these degrees by $d_{v,i}$, $i = 1, \ldots, D$. The average variable node degree is $\bar{d}_v = \sum_{i=1}^{D} a_i \cdot d_{v,i}$ where a_i is the fraction of *nodes* having degree $d_{v,i}$. The a_i must satisfy $\sum_i a_i = 1$. The number of edges at the VND and CND are the same, and thus we have $n\,\bar{d}_v = (n - k)\,d_c$ or

$$\bar{d}_v = (1 - R) \cdot d_c. \qquad (16.21)$$

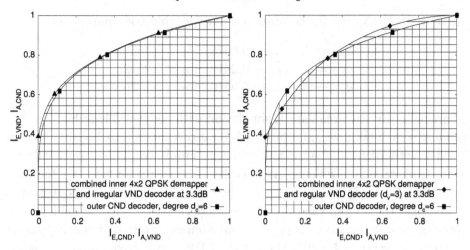

Fig. 16.7. Transfer curves of inner demapper/variable node decoder and outer check node decoder for *irregular* (left) and *regular* (right) variable node decoder.

Let b_i be the fraction of *edges* connected to variable nodes of degree $d_{v,i}$. There are in total $(n\,a_i)\,d_{v,i}$ edges involved with such nodes, so we have

$$b_i = \frac{n\,a_i\,d_{v,i}}{n\,\overline{d_v}} = \frac{d_{v,i}}{(1-R)\,d_c} \cdot a_i\,.$$

The b_i must satisfy $\sum_i b_i = 1$. It can be shown analytically that the transfer curve of a mixture of codes is an average of the component EXIT curves. The *edges* carry the extrinsic messages, and thus averaging must be performed using the b_i. The effective VND transfer curve is given by

$$I_{E,\mathrm{VND}}(I_{A,\mathrm{VND}}) = \sum_{i=1}^{D} b_i \cdot I_{E,\mathrm{VND}}(I_{A,\mathrm{VND}}, d_{v,i})\,, \qquad (16.22)$$

i.e., (16.22) is a weighted sum of the VND curves of (16.18). Only $D-2$ edge fractions can be adjusted because we must enforce (16.21) and $\sum_i b_i = 1$. $D = 3$ already provides good results, as shown next.

16.7.4 Design example

We design a code with rate $R = 1/2$ to be used with a 4×2 MIMO detector. For simplicity we restrict the VND to have only three different variable node degrees ($D = 3$). Only one b_i can be chosen freely. Fig. 16.7 (left) shows a curve fit at $E_b/N_0 = 3.3$ dB for $d_c = 6$ and variable node parameters

$$d_{v,1} = 2, \quad a_1 = 0.6978, \quad b_1 = 0.4652,$$
$$d_{v,2} = 3, \quad a_2 = 0.2612, \quad b_2 = 0.2612,$$
$$d_{v,3} = 20, \quad a_3 = 0.0410, \quad b_3 = 0.2736.$$

There is hardly any area gap. A simulation with codeword length $n = 10^5$, a random edge interleaver, and hundred iterations shows a turbo cliff at about 3.6 dB, with BER 10^{-4} after fourty blocks simulated. This is only 0.65 dB away from capacity, which is at 2.95 dB. The corresponding *regular* code with $d_v = 3$ and $d_c = 6$ is shown on the right of Fig. 16.7. No convergence is possible; the E_b/N_0-value has to be increased to 5.8 dB to achieve convergence. For comparison, a parallel concatenated code with 4×2 detector starts to converge at 5.4 dB. Making the LDPC code *irregular* by adjusting the variable node degrees using curve fitting thus avoided this 2 dB penalty.

16.8 Conclusions

Turbo methods have evolved from improving error correcting coding to applications in receiver design like iterative detection and decoding. While the performance enhancement through iterative detection is limited by the conventional channel codes used, recent developments in code design open up new opportunities to jointly optimize detector and channel code. In future systems, adjusting code properties such as degree profiles of low-density parity-check codes offers the potential of bridging the final gap to capacity.

Acknowledgments

The material of this chapter has been compiled from contributions of many researchers in the field. In particular, the author was privileged to work with A. Ashikhmin, B. Hochwald, G. Kramer, R. Mahadevappa, J. Salz, and the members of the wireless communications research department of Bell Laboratories, Crawford Hill, on topics of multiple antenna communications.

References

Alamouti, S. M. (1998). A simple transmit diversity technique for wireless communications. *IEEE J. Sel. Areas Commun.*, **16** (8), 1451–1458.

Ashikhmin, A., Kramer, G., and ten Brink, S. (2004). Extrinsic information transfer functions: model and erasure channel properties. *IEEE Trans. Inf. Theory*, **50** (11), 2657–2673.

Bahl, L., Cocke, J., Jelinek, F., and Raviv, J. (1974). Optimal decoding of linear codes for minimizing symbol error rate. *IEEE Trans. Inf. Theory*, **20** (2), 284–287.

Benedetto, S., Divsalar, D., Montorsi, G., and Pollara, F. (1998). Serial concatenation of interleaved codes: Performance analysis, design and iterative decoding. *IEEE Trans. Inf. Theory*, **44** (3), 909–926.

Berrou, C., Glavieux, A., and Thitimajshima, P. (1993). Near Shannon limit error–correcting coding and decoding: Turbo–codes. In *Proc. IEEE Conf. Commun.*, 1064–1070.

ten Brink, S. (2001). Exploiting the chain rule of mutual information for the design of iterative decoding schemes. In *Proc. Allerton Conf. Contr., Commun., Comp.*, 293–300.

ten Brink, S., Kramer, G., and Ashikhmin, A. (2004). Design of low-density parity-check codes for modulation and detection. *IEEE Trans. Commun.*, **52** (4), 670–678.

Caire, G., Taricco, G., and Biglieri, E. (1998). Bit-interleaved coded modulation. *IEEE Trans. Inf. Theory*, **44** (3), 927–946.

Foschini, G. J. (1996). Layered space–time architecture for wireless communication in a fading environment when using multi–element antennas. *Bell Labs. Tech. J.*, **1** (2), 41–59.

Golden, G. D., Foschini, G. J., Valenzuela, R. A., and Wolniansky, P. W. (1999). Detection algorithm and initial laboratory results using V-BLAST space-time communication architecture. *Electron. Lett.*, **35** (1), 14–16.

Hagenauer, J., Offer, E., and Papke, L. (1996). Iterative decoding of binary block and convolutional codes. *IEEE Trans. Inf. Theory*, **42** (2), 429–445.

Richardson, T. J. and Urbanke, R. L. (2001). The capacity of low–density parity-check codes under message–passing decoding. *IEEE Trans. Inf. Theory*, **47** (2), 599–618.

Shannon, C. E. (1948). A mathematical theory of communication. *Bell Syst. Tech. J.*, **27**, 379–423, 623–656.

Tarokh, V., Jafarkani, H., and Calderbank, A. R. (1999). Space–time block codes from orthogonal designs. *IEEE Trans. Inf. Theory*, **45** (5), 1456–1467.

Tarokh, V., Seshadri, N., and Calderbank, A. R. (1998). Space-time codes for high data rate wireless communication: Performance criterion and code construction. *IEEE Trans. Inf. Theory*, **44** (2), 744–765.

Telatar, I. E. (1999). Capacity of multi–antenna Gaussian channels. *Eur. Trans. Telecommun.*, **10** (6), 585–595.

Tonello, A. M. (2000). Space–time bit–interleaved coded modulation with an iterative decoding strategy. In *Proc. IEEE Veh. Technol. Conf.*, vol. 1, 473–478.

Tuechler, M. (2004). Design of serially concatenated systems depending on the block length. *IEEE Trans. Commun.*, **52**, 209–218.

Winters, J. H., Salz, J., and Gitlin, R. D. (1994). The impact of antenna diversity on the capacity of wireless communication systems. *IEEE Trans. Commun.*, **42** (2/3/4), 1740–1751.

17

Training for MIMO communications

Youngchul Sung and Tae Eung Sung

Cornell University

Brian M. Sadler

U.S. Army Research Laboratory

Lang Tong

Cornell University

In this chapter, an overview of training signal design for multiple-input multiple-output (MIMO) systems is provided with basic theoretical frameworks related to parameter estimation and information theory, as well as generalization and practical issues.

17.1 Introduction

Many MIMO communication systems and space-time techniques, for example, BLAST (Foschini, 1996), are designed for coherent detection, which requires channel state information for successful decoding. To facilitate channel estimation and synchronization in such systems, training or pilot signals are usually embedded in transmitted data streams. The design of these training signals can affect significantly the overall performance of a wireless system.

Since the use of training signals reduces effective data throughput, it is natural to seek optimal design of these embedded signals; one may ask "how many training symbols are necessary?" or "what is the optimal pilot sequence and its placement within data streams?"

Optimal training design for MIMO systems is a challenging task since the number of channel parameters to estimate increases rapidly as the number of transmitting and receiving antennas increases. Optimality of design depends on various factors such as receiver implementation, channel model, and design criteria. Although receiver architecture must be taken into account, training design is primarily a transmitter technique. Once a training scheme is chosen, it may be standardized for a specific application. It is therefore important that a training scheme is optimal or near optimal for a wide range of channel conditions and receiver implementations. Furthermore, since designers may have different design constraints and objectives, it is preferable that the training scheme is optimal for different design criteria.

The problem of optimal training design has been investigated extensively for various systems in recent years, and a substantial literature is available (see Tong *et al.*, 2004, and references therein). Typical design criteria include the information-theoretic metrics such as the cut-off rate, information (ergodic) capacity, and outage probability (Foschini, 1996; Telatar, 1999). For example, training signals for MIMO systems are optimized by maximizing data throughput under outage probability constraint (Marzetta, 1999), and by maximizing a lower-bound on training-based information capacity as a function of training design parameters (Baltersee *et al.*, 2001; Ma *et al.*, 2002; Hassibi and Hochwald, 2003; Yang *et al.*, 2004). Typically, these information-theoretic approaches to training design have been applied to a class of systems constrained to using training in a specific way, where channel is estimated using classical estimators based on training signal only, and the channel estimate is subsequently used for data decoding.

The information-theoretic metrics are global measures of a communication system, yielding good insights into the trade-off between the quality of channel estimate and data throughput. Often, we are interested in the optimality of specific system components. For example, one may be interested in the training scheme that minimizes channel estimation error for a given amount of training. In this case, a sensible measure is the mean square error (MSE) of the channel estimator. Classical training-based estimators form a channel estimate based only on the observations of training signals. For these channel estimators the problem of training placement is straightforward; training symbols should be clustered in a single block since corrupted observations by unknown data are discarded. (In the case of a flat-fading channel, the symbol-time location of the training symbols does not matter.) Hence, only the design of optimal pilot sequence remains and has been investigated by minimizing the MSE for flat-fading MIMO channels (Silverstein, 1997; Marzetta, 1999; Guey *et al.*, 1999; Wong and Park, 2004) and for frequency-selective MIMO channels (Yang and Wu, 2002), by minimizing auto- and cross-correlations (Fan and Mow, 2004), and by maximizing a lower bound on the training-based capacity (Hassibi and Hochwald, 2003). These works confirm the merit of orthogonal training sequences, e.g., Hadamard codes, between multiple transmitting antennas.

On the other hand, a more sophisticated class of estimators, referred to as semiblind, use all available observations (i.e., training and data) to estimate channel. The semiblind methods treat unknown data as nuisance parameters that can either be marginalized or estimated jointly with a channel.

These semiblind approaches improve the estimation performance for a given amount of training. Because it is desirable that the design of an optimal training scheme does not depend on the specific algorithm used at the receiver, the Cramér–Rao bound (CRB) (Van Trees, 1968) is a natural choice as a figure of merit in this case. Semiblind CRBs have been developed under deterministic and stochastic data assumption (de Carvalho and Slock, 1997; Dong and Tong, 2002), and for MIMO channels (Sadler *et al.*, 2001b; Kozick *et al.*, 2003). For semiblind channel estimation, the position of training symbols makes a difference since both training and data are modeled in estimation. Using the CRB as metric, training design has been investigated for linear finite impulse response (FIR) MIMO channels (Dong and Tong, 2002), for orthogonal space-time coded systems (Budianu and Tong, 2002), and for FIR MIMO systems using affine precoding (Vosoughi and Scaglione, 2003, 2004).

MIMO systems are also considered for multicarrier systems (Bölcskei *et al.*, 2002). Using training-based estimators and their MSEs, optimal design of a training signal for MIMO orthogonal frequency-division multiplexing (OFDM) systems is considered (Li *et al.*, 1999; Tung *et al.*, 2001; Barhumi *et al.*, 2003).

Although frameworks and methods for training design have been vigorously investigated in recent years, there are still many open problems and practical issues such as superimposed versus multiplexed training, optimal design for tracking of MIMO channels, the trade-off between optimality and complexity, etc.

The remainder of this chapter is organized as follows. Section 17.2 describes MIMO system models. Section 17.3 provides fundamental frameworks based on parameter estimation and information-theoretic approaches. Some generalization and practical issues are given in Section 17.4, followed by the conclusion in Section 17.5.

17.2 System model

17.2.1 FIR MIMO channel

Channel models are crucial for the further development of theoretical frameworks. We first consider the block fading channel model (Biglieri *et al.*, 1998), which provides a good approximation to slowly varying channels. In this model, channels are assumed to be time invariant within a block of length B symbol intervals, and then to change to a different fading state. The value of B is designed to be approximately less than or equal to the coherence time of the fading channel. Fig. 17.1 illustrates a typical model of a MIMO system

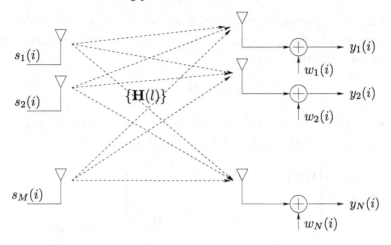

Fig. 17.1. A MIMO system with M transmitting and N receiving antennas.

with M transmitting and N receiving antennas, where the channel between each pair of transmitting and receiving antennas can be viewed as a single-input single-output (SISO) channel. When the bandwidth of a transmitted signal is wide and intersymbol interference occurs due to multipaths, it is usually assumed that the channel has a linear finite impulse response of order L, which captures the delay spread of the channel. Thus, in complex baseband representation the output of a typical discrete-time FIR MIMO channel is described by

$$\mathbf{y}(i) = \sum_{l=0}^{L} \mathbf{H}(l)\mathbf{s}(i-l) + \mathbf{w}(i), \tag{17.1}$$

where the input M-vector at symbol interval i, $\mathbf{s}(i) \triangleq \begin{bmatrix} s_1(i) & \cdots & s_M(i) \end{bmatrix}^T$, the output N-vector $\mathbf{y}(i) = \begin{bmatrix} y_1(i) & \cdots & y_N(i) \end{bmatrix}^T$, and the receiver noise vector $\mathbf{w}(i) \triangleq \begin{bmatrix} w_1(i) & \cdots & w_N(i) \end{bmatrix}^T$, where $(\cdot)^T$ denotes the matrix transpose. The noise $\mathbf{w}(i)$ is usually modeled as complex Gaussian $\mathcal{N}(\mathbf{0}, \sigma_w^2 \mathbf{I})$, independent over i. The channel is represented by a sequence of $M \times N$ matrices, $\{\mathbf{H}(l), l = 0, 1, \ldots, L\}$. (Here, the block index for the channel matrices is omitted.) For antenna pair (m, n), the channel impulse response is given by

$$\mathbf{h}^{(mn)} = \begin{bmatrix} H_{mn}(0) & H_{mn}(1) & \cdots & H_{mn}(L) \end{bmatrix}^T,$$

where $H_{mn}(l)$ is the (m, n) element of $\mathbf{H}(l)$. Depending on the approach, there are several choices for the assumption on the elements of $\{\mathbf{H}(l)\}$; one may view the elements as deterministic but unknown quantities, or as realizations of random variables with a known joint probability distribution.

Stacking the observation vectors for an entire block, except the observations corrupted by the previous block, we can write the received vector for the block in a filtering form:

$$\mathbf{y} = \mathcal{T}(\mathbf{H})\mathbf{s} + \mathbf{w}, \tag{17.2}$$

where $\mathbf{y} = \left[\mathbf{y}(B)^T\ \mathbf{y}(B-1)^T\ \cdots\ \mathbf{y}(L+1)^T\right]$, $\mathbf{w} = \left[\mathbf{w}(B)^T\ \mathbf{w}(B-1)^T\ \cdots\ \mathbf{w}(L+1)^T\right]$, $\mathbf{s} = \left[\mathbf{s}(B)^T\ \mathbf{s}(B-1)^T\ \cdots\ \mathbf{s}(1)^T\right]$, and

$$\mathcal{T}(\mathbf{H}) = \begin{bmatrix} \mathbf{H}(0) & \cdots & \mathbf{H}(L) & & \\ & \ddots & & \ddots & \\ & & \mathbf{H}(0) & \cdots & \mathbf{H}(L) \end{bmatrix}_{(B-L)N \times BM}. \tag{17.3}$$

The matrix in (17.3) is known as a generalized Sylvester matrix. In many cases, it is useful to rewrite the block observation vector $\mathbf{y}$ in a data matrix form:

$$\mathbf{y} = \mathcal{H}(\mathbf{s})\mathbf{h} + \mathbf{w}, \tag{17.4}$$

where $\mathbf{h} = \mathrm{vec}\{\left[\mathbf{H}(0)\ \mathbf{H}(1)\ \cdots\ \mathbf{H}(L)\right]\}$, and

$$\mathcal{H}(\mathbf{s}) = \begin{bmatrix} \mathbf{s}^T(B) & \cdots & \mathbf{s}^T(B-L) \\ \mathbf{s}^T(B-1) & \cdots & \mathbf{s}^T(B-L-1) \\ \vdots & & \vdots \\ \mathbf{s}^T(L+1) & \cdots & \mathbf{s}^T(1) \end{bmatrix} \otimes \mathbf{I}_N, \tag{17.5}$$

which is a block Hankel matrix of size $(B-L)N$ by $(L+1)MN$.

When the bandwidth of the transmitted signal is narrow so that all spectral components of the signal experience almost the same fading magnitude and phase, the flat-fading MIMO channel model can be applied as a special case of (17.1), and the channel output is given by

$$\mathbf{y}(i) = \mathbf{H}\mathbf{s}(i) + \mathbf{w}(i), \ i = 1,\ldots,B, \tag{17.6}$$

where the channel between a pair of transmitting and receiving antennas is described by a single scalar coefficient. In the flat-fading case we can write the observation block of length B in a matrix form:

$$\mathbf{Y} = \mathbf{H}\mathbf{S} + \mathbf{W}, \tag{17.7}$$

where $\mathbf{Y} \triangleq \left[\mathbf{y}(1)\ \cdots\ \mathbf{y}(B)\right]$, $\mathbf{S} \triangleq \left[\mathbf{s}(1)\ \cdots\ \mathbf{s}(B)\right]$, and $\mathbf{W} \triangleq \left[\mathbf{w}(1)\cdots\mathbf{w}(B)\right]$.

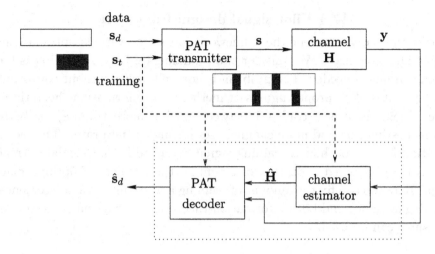

Fig. 17.2. A general training-based transceiver structure ($\mathbf{s}_t$: training signal, $\mathbf{s}_d$: data signal, $\mathbf{s}$: total signal, $\mathbf{y}$: entire observation corresponding to $\mathbf{s}$).

17.2.2 Pilot-assisted transceiver structure

While Shannon theory does not mandate channel estimation, the idea of acquiring channel state before decoding using training signals is widely employed in practice and has been proposed for many space-time systems (Foschini, 1996; Wolniansky *et al.*, 1998; Alamouti, 1998; Marzetta, 1999). So-called pilot-assisted transmissions (PAT) embed pilot signals into data streams. The presence of training signals implies that they will be used at the receiver explicitly or implicitly. Explicit approaches, as illustrated in Fig. 17.2, estimate channel parameters and subsequently use the channel estimate for demodulation and decoding. The channel estimator takes the training signal $\mathbf{s}_t$ (and possibly the entire observation $\mathbf{y}$), produces a channel estimate $\hat{\mathbf{H}}$, and feeds the estimate to the decoder. Practical decoders may assume that the estimated channel parameters are perfect. Such an assumption is of course not valid in a strict sense, and the corresponding scheme is referred to as a mismatched decoder (Merhav *et al.*, 1994; Lapidoth and Narayan, 1998). An alternative is to treat the estimated channel parameters as part of the observation. The decoder exploits the joint statistics of $(\hat{\mathbf{H}}, \mathbf{y})$.

Implicit approaches, on the other hand, treat the training signal as side information. The channel estimator in Fig. 17.2 is bypassed, and the training signal is used to tune the receiver directly. In adaptive equalization techniques, for example, channel estimates might not be obtained explicitly, rather the training is used to adaptively update an equalizer.

17.3 Pilot signal design: framework

In this section, we present the frameworks on which various optimal training designs are based. We consider two fundamental issues. The first is how much training is needed. The problem is not well-posed without constraints on data rates. The proper figures of merit are of information-theoretic nature, where there is a trade-off between having more training for better channel estimation and more channel uses for higher data rates. The second question focuses on how a training signal is embedded into a data stream when there is a fixed allocation of training—the problem of optimal placement. Here, we can fix the amount of training and optimize the placement of training symbols, and both information-theoretic and parameter-estimation measures can be used.

17.3.1 *Information-theoretic approaches*

The use of information-theoretic metrics is crucial to revealing trade-offs among training designs. In these settings, a training scheme provides side information about unknown channels. Along this line, there is an extensive literature on reliable communications under channel uncertainty (see Lapidoth and Narayan, 1998, and references therein), which provides useful tools for obtaining achievable rate expressions that may be optimized with respect to training design parameters. Specifically, when channel estimates are available to the receiver, bounds on mutual information (Lapidoth and Narayan, 1998; Caire and Shamai, 1999; Médard, 2000; Lapidoth and Shamai, 2002) are the primary equations used in many of the optimal training designs. These expressions, however, do not incorporate the resources required to obtain channel estimates. For the analysis of training-based schemes, one must take into account the resources allocated for pilot transmission.

For MIMO systems, the problem of training design for the BLAST system was first investigated by Marzetta (1999). Based on the nonergodic channel model (i.e., one codeword is contained within one block length of a quasi-static channel) and on the outage probability due to the error of a pure training-based channel estimator, the required training period was obtained by maximizing data throughput for the block fading flat MIMO channel (17.6), assuming that H_{mn} is i.i.d. $\mathcal{N}(0,1)$. It is shown by Marzetta (1999) that for maximum data throughput, the required training period is half of the block length and no less than the number of transmitting antennas.

For the class of ergodic block fading[†] MIMO channels, the problem of

---

[†]In this case, one codeword is spread over many blocks, which may be a reasonable scenario for a system employing an interleaver after the channel encoder.

training design was investigated using the information capacity as metric with the flat fading model (17.6) (Hassibi and Hochwald, 2003). Assuming time-division multiplexed (TDM) training within a block, Hassibi and Hochwald maximized a lower bound on training-based channel capacity with respect to the number of pilots used in a block and the power allocated to pilots. Their work provides a useful framework for analyzing the capacity achievable by training-based schemes in general, and leads to several interesting insights. They identified that, in the low SNR regime and when the coherence time (block length) was short, training-based schemes incurred a substantial penalty; training can lead to bad channel estimates, and no training may be preferable. On the other hand, training-based schemes were close to being optimal in high SNR and long coherence time regimes. This is consistent with the intuition that, with a negligible price paid to obtaining high quality estimates, we can assume that the channel is approximately known at the receiver. Later, similar approaches have been applied for the training design for frequency-selective and time- and frequency-selective MIMO channels (Ma *et al.*, 2002; Yang *et al.*, 2004).

Information-theoretic formulation

We illustrate an example of information-theoretic formulation for MIMO training design by Hassibi and Hochwald (2003), which is germane to many scenarios considered in the literature. A typical information-theoretic setup assumes the explicit transceiver structure described in Fig. 17.2, where the channel estimate is formed by a classical estimator based on training signals only. In this section, we also assume the transceiver structure in which the transmitter does not know the channel and the receiver learns the channel using the minimum mean squared error (MMSE) channel estimator. The presentation follows that of Hassibi and Hochwald (2003).

Consider the blockwise time-varying flat MIMO channel (17.7) with normalization such that the elements of $\underline{\mathbf{H}}$ and $\underline{\mathbf{W}}$ have unit mean-square:

$$\mathbf{Y} = \sqrt{\frac{\rho}{M}}\underline{\mathbf{H}}\mathbf{S} + \underline{\mathbf{W}}, \tag{17.8}$$

where $\mathbf{Y}$ is the $N \times B$ channel output observed for the block, $\mathbf{S}$ is the $M \times B$ channel input matrix consisting of the training subblock $\mathbf{S}_t$ and the data subblock $\mathbf{S}_d$ (i.e., $\mathbf{S} = [\mathbf{S}_t \ \mathbf{S}_d]$), $\underline{\mathbf{W}}$ is additive white Gaussian noise with zero mean and unit variance, and ρ is the expected SNR at each receiving antenna. For simplicity, we assume that the channel matrix $\underline{\mathbf{H}}$ is Gaussian, with independent entries of zero mean and unit variance, independent over each block. We assume that B_t symbol intervals are used for transmitting training within a block, and ρ_t and ρ_d are the received SNR at each receiving antenna for the training period and the data period, respectively. Thus, we have

$$B = B_t + B_d, \quad \text{and} \quad \rho B = \rho_t B_t + \rho_d B_d.$$

Notice that for given B and ρ, $\{B_t, \rho_t\}$ is sufficient to specify the training parameters together with the values of $\mathbf{S}_t$. Assuming that the training-based MMSE channel estimate $\hat{\mathbf{H}}$ is available at the decoder, the model in (17.8) can be rewritten for the data period as

$$\mathbf{Y}_d = \sqrt{\frac{\rho_d}{M}}\hat{\mathbf{H}}\mathbf{S}_d + \mathbf{Z}, \quad \mathbf{Z} = \sqrt{\frac{\rho_d}{M}}\tilde{\mathbf{H}}\mathbf{S}_d + \mathbf{W}_d, \quad (17.9)$$

where $\mathbf{S}_d$ is the unknown data with power constraint $\text{tr}\,\mathbb{E}\{\mathbf{S}_d\mathbf{S}_d^H\} = MB_d$, and $\tilde{\mathbf{H}} = \underline{\mathbf{H}} - \hat{\mathbf{H}}$ is the estimation error, which is uncorrelated (independent for the Gaussian assumption) to $\hat{\mathbf{H}}$ by the orthogonality principle for MMSE estimation. The difference between (17.8) and (17.9) is that $\hat{\mathbf{H}}$ of (17.9) is known to the receiver, whereas $\underline{\mathbf{H}}$ of (17.8) is unknown. In addition, $\mathbf{Z}$ is not independent of $\mathbf{S}_d$ in (17.9), and its normalized variance is given by

$$\sigma_{\mathbf{Z}}^2 \triangleq \frac{1}{NB_d}\,\text{tr}\,\mathbb{E}\{\mathbf{Z}\mathbf{Z}^H\},$$

$$= \frac{\rho_d}{MNB_d}\,\text{tr}\left[\mathbb{E}\{\tilde{\mathbf{H}}\tilde{\mathbf{H}}^H\}\mathbb{E}\{\mathbf{S}_d^H\mathbf{S}_d\}\right] + 1.$$

If the decoder takes $\hat{\mathbf{H}}$ as part of the observation and uses it along with $\mathbf{Y}_d$ to decode $\mathbf{S}_d$, the capacity of such a scheme is lower bounded by

$$C \geq \inf_{p_{\mathbf{z}};\,\text{tr}\,\mathbb{E}\{\mathbf{Z}\mathbf{Z}^H\}=NB_d\sigma_{\mathbf{Z}}^2}\left(\sup_{p_{\mathbf{S}_d};\,\text{tr}\,\mathbb{E}\{\mathbf{S}_d^H\mathbf{S}_d\}=MB_d}\frac{B-B_t}{B}I(\mathbf{S}_d;\mathbf{Y}_d,\hat{\mathbf{H}})\right)$$

$$= \inf_{p_{\mathbf{z}};\,\text{tr}\,\mathbb{E}\{\mathbf{Z}\mathbf{Z}^H\}=NB_d\sigma_{\mathbf{Z}}^2}\left(\sup_{p_{\mathbf{S}_d};\,\text{tr}\,\mathbb{E}\{\mathbf{S}_d^H\mathbf{S}_d\}=MB_d}\frac{B-B_t}{B}I(\mathbf{S}_d;\mathbf{Y}_d|\hat{\mathbf{H}})\right), \quad (17.10)$$

where the equality is the result of applying the chain rule under the assumption that the data $\mathbf{S}_d$ is independent of the channel estimate $\hat{\mathbf{H}}(\mathbf{S}_t, \mathbf{Y}_t)$ based on the training signals only. The right-hand side of (17.10) is simply the channel capacity of the scheme added by the worst-case noise. If we choose $\mathbf{S}_d$ to be zero mean Gaussian with $\mathbb{E}\{\mathbf{S}_d^H\mathbf{S}_d\} = B_d\mathbf{I}_M$ and Gaussian noise with the power constraint in (17.10), the capacity is lower bounded by

$$C \geq C_{\text{LB}} \triangleq \mathbb{E}_{\hat{\mathbf{H}}}\left\{\frac{B-B_t}{B}\log\det\left(\mathbf{I}_N + \frac{\rho_d\sigma_{\hat{\mathbf{H}}}^2}{1+\rho_d\sigma_{\hat{\mathbf{H}}}^2}\frac{\hat{\mathbf{H}}^H\hat{\mathbf{H}}}{M}\right)\right\}, \quad (17.11)$$

where $\sigma_{\bar{\mathbf{H}}}^2 = 1-\sigma_{\hat{\mathbf{H}}}^2$, $\sigma_{\hat{\mathbf{H}}}^2 = (1/NM)\,\mathrm{tr}\,\mathbb{E}\{\hat{\mathbf{H}}^H\hat{\mathbf{H}}\}$, and the normalized estimate $\bar{\mathbf{H}} = (1/\sigma_{\hat{\mathbf{H}}})\hat{\mathbf{H}}$. The proof can be found in Hassibi and Hochwald (2003). The fact that Gaussian noise is the worst uncorrelated additive noise for the Gaussian model can also be found in Médard (2000), Diggavi and Cover (2001), Lapidoth and Shamai (2002).

Notice that the lower bound is a function of training design parameters. The problem of training design can be formulated as maximizing the lower bound on capacity (17.11). Let $\mathcal{P}$ denote the set of training parameters that specify B_t, ρ_t, and the values and placement of pilot symbols. For the considered training scheme, the optimal training scheme $\mathcal{P}^*$ is given by

$$\mathcal{P}^* = \sup_{\mathcal{P}} C_{\mathrm{LB}}.$$

The above optimization is performed with respect to the training percentage B_t/B, power allocation ρ_t, and training symbol values and placement. This framework can be extended to channel estimators with some constraint other than the MMSE estimator, and to more general channel models with frequency-selectivity and time-variation (Baltersee *et al.*, 2001; Ma *et al.*, 2002; Yang *et al.*, 2004).

17.3.2 Signal processing perspectives on MIMO training design

We now consider MIMO training design, when there is a fixed allocation of training, using performance measures related to parameter estimation. We focus here on the FIR MIMO channel in Section 17.2 and the corresponding training design, especially on the optimal placement of training symbols.

Multidimensional training placement models

Training symbols are traditionally time-division multiplexed. The use of an antenna array in MIMO systems extends multiplexing to the spatial dimension. In addition to interleaving training and data in the time and spatial domains, we may consider superimposing pilots and data. The combination of these factors multiplies possible scenarios to examine.

The key to unifying various schemes is to view the problem of training design as one of power allocation. Specifically, in each design dimension— time, space, frequency, etc.—a pair of power allocation parameters is used to describe the design. In the case of a single-carrier MIMO system with M transmitting antenna as illustrated in Fig. 17.3(a), each transmitted symbol $s_m(i)$ is specified by the time index i and the transmitting antenna index m.

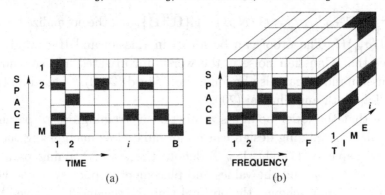

Fig. 17.3. An illustration of the multidimensional training placement model (a partially shaded square indicates a pilot symbol superimposed onto a data symbol).

If we assume that pilots may be superimposed in any position, we have

$$s_m(i) = \sqrt{\phi_{mi}}\, s_m^t(i) + \sqrt{\gamma_{mi}}\, s_m^d(i), \quad m = 1, \ldots, M, \; i = 1, \ldots, B,$$

where the $s_m^t(i)$ satisfy $|s_m^t(i)| = 1$, are known, and have allocated power $\phi_{mi} \geq 0$, and $s_m^d(i)$ are unknown data with zero mean, unit variance, and with $\gamma_{mi} \geq 0$. For the transmission of a block of size B, the training scheme is parameterized by an $M \times B$ training matrix $\mathbf{S}_t = [s_m^t(i)]_{(m,i)}$, and by nonnegative power allocation matrices $\mathbf{\Phi} = [\phi_{mi}]$ and $\mathbf{\Gamma} = [\gamma_{mi}]$. The formulation can be extended to higher dimensions illustrated in Fig. 17.3(b).

Constraints can be imposed on training design in several ways. For example, a constraint on the total power of training can be given by

$$\frac{1}{MB} \sum_{i=1}^{B} \sum_{m=1}^{M} \mathbb{E}\{|\sqrt{\phi_{mi}} s_m^t(i)|^2\} = \frac{1}{MB} \sum_{i=1}^{B} \sum_{m=1}^{M} \phi_{mi} = P. \qquad (17.12)$$

Training design for block fading channels

For block fading MIMO channel models, training symbols are typically inserted in each block, and channel estimation and symbol detection are performed within each block with the help of training signals. These training signals can be exploited in different ways (see Fig. 17.2). A classical training-based estimator forms a channel estimate $\hat{\mathbf{H}}$ based only on the observations $\mathbf{y}_t$ corresponding to the training signal $\mathbf{s}_t$, whereas semiblind estimators use all available observations $\mathbf{y}$ (i.e., training and data) to obtain a channel estimate.

When training-based estimators are used for single carrier systems over MIMO channels with flat or frequency-selective fading, the design of a given amount of training is straightforward, and reduces to the problem of optimal sequence design (Silverstein, 1997; Marzetta, 1999; Guey *et al.*, 1999; Yang and Wu, 2002; Hassibi and Hochwald, 2003; Wong and Park, 2004; Fan and Mow, 2004). Hence, we focus here on semiblind estimation of FIR MIMO channels with the CRB on MSE as design metric. Specifically, the MSE of any unbiased estimator $\hat{\mathbf{H}}$, under regularity conditions, is lower bounded by

$$\mathbb{E}\{\|\hat{\mathbf{H}} - \mathbf{H}\|_F^2\} \geq \mathrm{tr}\{\mathbf{F}^{-1}(\mathbf{\Gamma}, \mathbf{\Phi}, \mathbf{S}_t)\}, \tag{17.13}$$

where $\mathbf{F}(\mathbf{\Gamma}, \mathbf{\Phi}, \mathbf{S}_t)$ is the Fisher information matrix (FIM), which is a function of the training design parameters to be optimized. Thus, optimal pilot placement is found by minimizing the CRB over all feasible allocations $\{\mathbf{\Gamma}, \mathbf{\Phi}, \mathbf{S}_t\}$ satisfying given constraints. For the parameter estimation formulation the CRB serves as the counterpart of the lower bound C_{LB} of (17.11) in the information-theoretic setup.

The CRB may be formulated with both random and deterministic parameter models; random models lead to useful insights into ensemble behavior, whereas deterministic models provide means for assessing specific realizations of channels and sources. The channel CRB may also be a function of the unknown data transmitted simultaneously with the training (de Carvalho and Slock, 1997). When the unknown data are treated as random parameters, they need to be marginalized to obtain the likelihood function, and the CRB is a function of the (prior) data distribution. If, on the other hand, these unknown data are treated as deterministic unknown parameters, then the data may be viewed as nuisance parameters that affect the CRB of the channel estimator. Note that although the CRB can be achieved with finite data samples in some estimation problems such as in the case of a linear system model, the achievability is not always guaranteed with finite data samples. However, the existence of asymptotically efficient algorithms (e.g., the maximum-likelihood estimator (MLE) in many cases) justifies the use of the CRB as a design criterion.

Using the CRB as design metric and a random channel model, optimal placement of training symbols and power allocations is derived for FIR SISO and MIMO channels by Dong and Tong (2002). It is shown that the optimal training placement scheme that minimizes the CRB is independent of the prior channel distribution and the SNR. The placement depends on the power allocation and on the number and placement of training symbols in a block. Using the CRB as metric, the design of affine precoder is optimized as well by Vosoughi and Scaglione (2003, 2004), where it is shown that different assumptions about unknown data lead to different affine precoder designs.

The CRB has also been employed to study the impact of side information, including training, with MIMO channels by Kozick *et al.* (2003). Identifiability of the unknown channel(s) may also be studied by considering the minimum conditions under which the CRB exists (that is, the minimum conditions under which the FIM becomes full rank, a condition that is referred to as the FIM identifiability). Necessary and sufficient conditions for this to occur are provided by Moore and Sadler (2004), covering cases from SISO to FIR MIMO channels. Minimum conditions include the length of the training.

Superimposed training schemes are also considered for MIMO systems. A superimposed pilot scheme for space-time coded transmission over flat block fading MIMO channels was considered by Budianu and Tong (2002), where the problem setting is general but the optimal placement was not obtained. However, the analysis revealed the weakness of superimposed training in the block stationary case, showing that TDM training scheme had smaller CRB than that of superimposed training. On the other hand, if training should be included in every block and the channel estimation is accurate, the superimposed scheme provides higher mutual information.

Block fading channels: a design example

In this section, we present a particular example by Dong and Tong (2002). We consider optimal training placement of the block-fading FIR MIMO channel in Section 17.2.1 with the semiblind CRB with random-parametric formulation as design metric. (The notations and definitions follow those in Section 17.2.1.)

For the placement problem, each transmitted symbol is either training or data, i.e., we have

$$s_m(i) = \sqrt{\phi_{mi}}\, s_m^t(i) + \sqrt{\gamma_{mi}}\, s_m^d(i), \quad m = 1, \ldots, M, \ i = 1, \ldots, B, \quad (17.14)$$

where ϕ_{mi} is one for the training locations in space-time and $\gamma_{mi} = 1 - \phi_{mi}$. Thus, the set $\mathcal{P}$ of all possible placements can be specified as follows. Let $B_t^{(m)}$ and $B_d^{(m)}$ be the number of training and data symbols, respectively, within a block of length B for transmitting antenna m. For transmitting antenna m we define $\boldsymbol{\mu}^{(m)}$ and $\boldsymbol{\nu}^{(m)}$; $\boldsymbol{\mu}^{(m)} = \left[\mu_1^{(m)} \cdots \mu_{\beta_m+1}^{(m)} \right]$ a data sub-block length vector, $\boldsymbol{\nu}^{(m)} = \left[\nu_1^{(m)} \cdots \nu_{\beta_m}^{(m)} \right]$ a training cluster length vector, where β_m is the number of training clusters in a transmitted block at antenna m, as illustrated in Fig. 17.4. Constraining the total number of data and pilot symbols, we have $\sum_{j=1}^{\beta_m+1} \mu_j^{(m)} = B_d^{(m)}$ and $\sum_{j=1}^{\beta_m} \nu_j^{(m)} = B_t^{(m)}$. Thus, the set of all possible placements is given by $\mathcal{P} = \{ (\boldsymbol{\mu}, \boldsymbol{\nu}) : \boldsymbol{\mu} = \left[\boldsymbol{\mu}^{(1)} \cdots \boldsymbol{\mu}^{(M)} \right], \boldsymbol{\nu} = \left[\boldsymbol{\nu}^{(1)} \cdots \boldsymbol{\nu}^{(M)} \right] \}$ satisfying the power constraint.

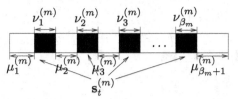

Fig. 17.4. Training placement within a block for transmitting antenna m (dark areas represent the training clusters).

We make the following assumptions:

A1) The data symbols $\{s_m^d(i), i = 1, \ldots, B\}$ for transmitting antenna m form an i.i.d. sequence (independent over m) drawn from the probability density function (pdf) $p_s(\cdot)$ with zero mean and variance $\sigma_{d;m}^2$. The power of training symbols for antenna m is defined as $\sigma_{t;m}^2 \triangleq (1/B_t^{(m)}) \sum_{i=1}^{B_t^{(m)}} |s_m^t(i)|^2$.

A2) The $(L+1)MN$ coefficients of the channel matrices $\{\mathbf{H}(l), l = 0, \ldots, L\}$ are i.i.d. random variables with pdf $p_h(\cdot)$.

A3) The data $\mathbf{s}$, channel $\mathbf{h}$, and noise $\mathbf{w}$ are jointly independent, where $\{w_n(i)\}$ are i.i.d. circular complex Gaussian noise with zero mean and variance σ_w^2.

A4) $B_t^{(m)} \sigma_{t;m}^2 > \sigma_{d;m}^2$ for $m = 1, \ldots, M$. That is, the total power of training symbols is at least larger than that of one data symbol for all transmitting antennas.

From (17.14) the data MB-vector $\mathbf{s}$ of (17.2) for the entire block is decomposed into two MB-vectors:

$$\mathbf{s} = \mathbf{s}_t + \mathbf{s}_d, \tag{17.15}$$

where $\mathbf{s}_t$ is made by putting the known symbols at the training locations in space-time and zero padding elsewhere, i.e., at the data locations. Similarly, the data matrix $\mathcal{H}(\mathbf{s})$ of (17.4) is decomposed by

$$\mathcal{H}(\mathbf{s}) = \mathcal{H}(\mathbf{s}_t) + \mathcal{H}(\mathbf{s}_d),$$

where $\mathcal{H}(\mathbf{s}_t)$ and $\mathcal{H}(\mathbf{s}_d)$ are formed by (17.5) using $\mathbf{s}_t$ and $\mathbf{s}_d$, respectively. We define the following matrices:

$$\mathcal{R}_{\mathbf{s}} \triangleq \mathcal{H}^H(\mathbf{s})\mathcal{H}(\mathbf{s}) \text{ and } \mathcal{R}_{\mathbf{s}_t} \triangleq \mathcal{H}^H(\mathbf{s}_t)\mathcal{H}(\mathbf{s}_t).$$

Theorem 17.1 *(Dong and Tong, 2002) Under the assumptions A1–A4 and appropriate regularity conditions (Van Trees, 1968; Weinstein and Weiss,*

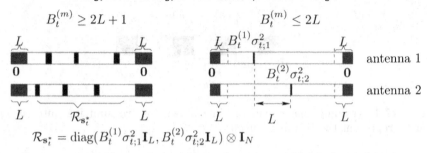

$$\mathcal{R}_{\mathbf{s}_t^*} = \mathrm{diag}(B_t^{(1)}\sigma_{t;1}^2\mathbf{I}_L, B_t^{(2)}\sigma_{t;2}^2\mathbf{I}_L) \otimes \mathbf{I}_N$$

Fig. 17.5. Optimal placements of training symbols for a MIMO channel with two transmitting antennas.

1988), the MSE matrix of any unbiased estimator $\hat{\mathbf{h}}(\mathbf{y}, \mathbf{s}_t)$, *defined as*

$$\mathcal{M}(\hat{\mathbf{h}}) \triangleq \mathbb{E}\{[\hat{\mathbf{h}}(\mathbf{y}, \mathbf{s}_t) - \mathbf{h}][\hat{\mathbf{h}}(\mathbf{y}, \mathbf{s}_t) - \mathbf{h}]^H\},$$

satisfies the inequality

$$\mathcal{M}(\hat{\mathbf{h}}) \geq \mathbf{\Lambda}(\mathcal{P}, \mathbf{s}_t) \triangleq \left[\frac{1}{\sigma_w^2}\mathbb{E}\{\mathcal{R}_{\mathbf{s}}\} + \rho_h^2\mathbf{I}_{(L+1)MN}\right]^{-1}, \qquad (17.16)$$

where $\mathbf{\Lambda}(\mathcal{P}, \mathbf{s}_t)$ *is the complex CRB, the expectation in* (17.16) *is taken with respect to the unknown data symbols, and* $\rho_h^2 = \mathbb{E}\{|\partial \ln p_\rho(h)/\partial h^*|^2\}$, *with the expectation taken with respect to* $p_h(h)$.

Thus, the CRB is a function of the placement parameters $\mathcal{P}$, the training symbol values $\mathbf{s}_t$ via $\mathbb{E}\{\mathcal{R}_{\mathbf{s}}\}$, and the channel distribution through ρ_h^2. The CRB is optimized over $(\mathcal{P}, \mathbf{s}_t)$ to obtain the best training scheme.

It is shown by Dong and Tong (2002) that if the number of training symbols is large enough to cover the block edges corrupted by interblock interference, it is optimal to place (known) zeros wasting no power at the two block edges, and to place the remaining known symbols in the middle of the block (not corrupted by interblock interference) such that these symbols satisfy a certain orthogonal condition between transmitting antennas. On the other hand, if the known symbols cannot cover the block edges, it is optimal to place roughly half of the known symbols of value zero at each edge and one symbol with all available training power in the middle of the block such that this symbol does not interfere with the middle symbol of different transmitting antenna streams through intersymbol interference. Fig. 17.5 illustrates such a placement for the case of two transmitting antennas. The difference between training-based and semiblind estimations can be clearly seen. For training-based estimators, the locations of known symbols within a transmitted stream should be clustered as a single subblock, and the single cluster of one transmitting antenna should be aligned with the training cluster of the other transmitting antenna stream.

Generalizations to correlated channel taps, symbol-by-symbol power constraints, etc., are available in the reference. The framework illustrated here can be generalized by modifying the signal decomposition (17.15). For example, superimposed training schemes can be modeled by allowing ϕ_{mi} to take arbitrary values between 0 and 1. Also, the combination of linear precoding and training embedding (possibly including superimposing) can be considered (Vosoughi and Scaglione, 2003, 2004). In this case, (17.15) can be modified as

$$\mathbf{s} = \mathbf{F}\mathbf{s}_I + \mathbf{s}_t,$$

with a precoding matrix $\mathbf{F}$ that has full column rank, known symbols $\mathbf{s}_t$ and information symbols $\mathbf{s}_I$. The precoder $\mathbf{F}$ can be optimized along with $\mathbf{s}_t$ to obtain minimum CRB.

17.4 Generalization and other issues

17.4.1 MIMO-OFDM systems

In the previous sections, we have considered training signal design for single carrier systems over MIMO channels. Multiple transmitting and receiving antennas can also be used in multicarrier systems using OFDM, which has been standardized for many applications such as digital television broadcasting, digital audio broadcasting, wireless local area networks (WLANs), etc. A typical configuration is to apply the OFDM processing at each transmitting and receiving antenna. For OFDM systems, each OFDM symbol is preceded by a cyclic prefix (CP) or guard interval, and intersymbol interference in time is irrelevant if the guard interval is larger than the maximum channel dispersion, whereas the interference between subcarriers may occur due to the mobility of transmitter or receiver and the frequency offset of local oscillators.

For MIMO-OFDM systems, assuming a TDM training period, optimal design has been considered for training-based schemes. In the TDM case, the problem reduces to a two-dimensional sequence design in the space and subcarrier domains. Optimal sequence was obtained by minimizing the MSE of a training-based channel estimator (Li *et al.*, 1999; Tung *et al.*, 2001; Larsson and Li, 2001), showing the optimality of orthogonal sequences (in the subcarrier domain) between different transmitting antennas. On the other hand, the optimal placement of pilot tones in single or multiple OFDM symbols was considered when the training period contains both pilot tones and data tones (Barhumi *et al.*, 2003). Barhumi *et al.* (2003) show that the optimal training requires some orthogonality between different transmitting antenna signals in addition to the well-known conditions of equal power and equal spacing for pilot tones for SISO-OFDM systems.

17.4.2 Fast-fading MIMO channels

In the previous sections, we considered block fading channels where the channel is time-invariant for a block. However, for fast fading channels this assumption may not be valid, i.e., the channel varies from symbol to symbol.

In this case, we typically assume that channel variations are highly correlated, at least for a short time, which is consistent with mobile channel measurements (Jakes, 1974). Indeed, if the channel process behaves independently from sample to sample, then training is required for every sample and training placement is not an issue. A practical model is the first order autoregressive (AR) model of the channel process $\{\mathbf{h}_i\}$ that leads to a state space representation

$$\mathbf{h}_{i+1} = \mathbf{A}\mathbf{h}_i + \mathbf{B}\mathbf{u}_i, \tag{17.17}$$

where the channel vector $\mathbf{h}_i$ at time i is made by concatenating the MIMO channel matrix $\mathbf{H}_i$, $\mathbf{A}$ characterizes the fading rate, and $\mathbf{u}_i$ is the driving noise. Higher order AR models have also been employed for mobile channel modeling. AR models provide a reasonable fit to the widely used Jakes model, which characterizes the power spectral density of the channel process $\mathbf{h}_i$.

Optimal training design for the fast-fading MIMO channels is a challenging task that provides many interesting research topics, such as the comparison of the superimposed scheme with TDM training, the investigation of optimal training amount as a function of fading rate, etc. From a signal processing perspective, one may consider the Kalman filter for MIMO channel tracking (Komninakis *et al.*, 2002) and optimal training design. Since the CRB is not tractable for fast fading channels, one possible design metric is the steady-state MSE of the training-based optimal estimator, Kalman filter, as considered by Dong *et al.* (2004). While the CRB in the block fading model gives a lower bound that can be achieved by the MLE at least asymptotically, the steady-state MSE of the optimal filter gives a good figure of merit for the fast fading model (17.17). Another useful technique for fast fading channels is a basis expansion model (Giannakis and Tepedelenlioglu, 1998). This method absorbs the time variation of a channel into a set of known basis functions and converts a general time-varying channel model to a block frequency-selective fading model.

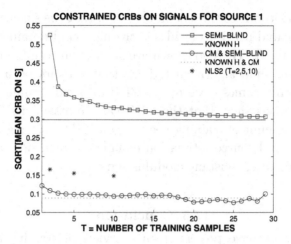

Fig. 17.6. CRBs and symbol estimation against training size for a 2×2 system (block fading, block size = 30) (Kozick *et al.*, 2003).

17.4.3 Semiblind approach with side information

In Section 17.3.2, we have noted how training can be treated as side information and incorporated into semiblind channel estimation and decoding. Often, the receiver has additional side information such as constant modulus signals (or a more generally known constellation), known power levels, known angles of arrival, space-time coding, precoding, etc. Exploiting this additional side information can result in enhanced channel estimates and cochannel interference rejection. Alternatively, the side information can be exploited to decrease the required amount of training for a given performance level. The constant modulus (CM) property is particularly useful in this regard, and it leads to tractable algorithms. The impact of many forms of side information can be analyzed using the constrained CRB (Stoica and Ng, 1998; Sadler *et al.*, 2001a). Sadler and Kozick (2000) demonstrate that reduced training sizes are needed when the CM property is exploited.

An example from Kozick *et al.* (2003) is shown in Fig. 17.6, showing CRBs and symbol estimation against training size for a 2×2 system (block fading, with block size = 30). Independent 8-PSK data streams come from each transmit antenna, and results are shown for recovering one of these streams. A 3-ray flat channel is used, with significant spatial angle overlap. The SNR is 20 dB. The top curve bounds the performance of a semiblind system that exploits both training and the unknown symbols, and indicates that the value of training is most significant in the first few training samples, approaching the known channel bound as the training size grows. The bottom curve incorporates the side information that the signals are constant modulus; this bound nearly coincides with the bound obtained assuming a known channel matrix $\mathbf{H}$. Also shown are simulation results for a nonlinear least squares semiblind algorithm (NLS2) that exploits both training and CM. The algorithm comes close to the CRB for small training size. (Cases with better channels show that the CRB can be attained.) This example shows how the amount of training needed in a MIMO setting can be significantly reduced, and source estimation enhanced, when side information is exploited (in this case, constant modulus signaling).

17.5 Conclusion

In this chapter, we have presented an overview of training signal design for MIMO systems. General MIMO channel models have been given, and common design criteria have been reviewed. Also, information-theoretic and signal processing frameworks have been discussed with particular examples.

Acknowledgments

This work was supported in part by the Multidisciplinary University Research Initiative (MURI) under the Office of Naval Research (ONR) Contract N00014-00-1-0564, Army Research Laboratory (ARL) CTA on Communication and Networks under Grant DAAD19-01-2-0011, and National Science Foundation under Contract CCR-0311055.

References

Alamouti, S. M. (1998). A simple transmit diversity technique for wireless communications. *IEEE J. Sel. Areas Commun.*, **16** (8), 1451–1458.

Baltersee, J., Fock, G., and Meyr, H. (2001). Achievable rate of MIMO channels with data-aided channel estimation and perfect interleaving. *IEEE J. Sel. Areas Commun.*, **19** (12), 2358–2368.

Barhumi, I., Leus, G., and Moonen, M. (2003). Optimal training design for MIMO OFDM systems in mobile wireless channels. *IEEE Trans. Signal Process.*, **51** (6), 1615–1624.

Biglieri, E., Proakis, J., and Shamai, S. (1998). Fading channels: Information-theoretic and communications aspects. *IEEE Trans. Inf. Theory*, **44** (6), 2619–2692.

Bölcskei, H., Gesbert, D., and Paulraj, A. J. (2002). On the capacity of OFDM-based spatial multiplexing systems. *IEEE Trans. Commun.*, **50** (2), 225–234.

Budianu, C. and Tong, L. (2002). Channel estimation for space-time block coding systems. *IEEE Trans. Signal Process.*, **50** (10), 2515–2528.

Caire, G. and Shamai, S. (1999). On the capacity of some channels with channel state information. *IEEE Trans. Inf. Theory*, **45** (6), 2007–2019.

de Carvalho, E. and Slock, D. T. M. (1997). Cramer-Rao bounds for semi-blind, blind and training sequence based channel estimation. In *Proc. IEEE Workshop Signal Process. Advances Wireless Commun.*, 129–132.

Diggavi, S. and Cover, T. (2001). The worst additive noise under a covariance constraint. *IEEE Trans. Inf. Theory*, **47** (7), 3072–3081.

Dong, M. and Tong, L. (2002). Optimal design and placement of pilot symbols for channel estimation. *IEEE Trans. Signal Process.*, **50** (6), 3055–3069.

Dong, M., Tong, L., and Sadler, B. (2004). Optimal insertion of pilot symbols for transmissions over time-varying flat fading channels. *IEEE Trans. Signal Process.*, **52** (5), 1403–1418.

Fan, P. and Mow, W. (2004). On optimal training sequence design for multiple-antenna systems over dispersive fading channels and its extensions. *IEEE Trans. Veh. Technol.*, **53** (5), 1623–1626.

Foschini, G. J. (1996). Layered space-time architecture for wireless communication in a fading environment when using multi-element antennas. *Bell Labs Tech. J.*, **1** (2), 41–59.

Giannakis, G. and Tepedelenlioglu, C. (1998). Basis expansion models and diversity techniques for blind identification and equalization of time-varying channels. *Proc. IEEE*, **86** (10), 1969–1986.

Guey, J., Fitz, M., Bell, M., and Kuo, W. (1999). Signal design for transmitter diversity wireless communication systems over Rayleigh fading channels. *IEEE Trans. Commun.*, **47** (4), 527–537.

Hassibi, B. and Hochwald, B. (2003). How much training is needed in multiple-antenna wireless links. *IEEE Trans. Inf. Theory*, **49** (4), 951–963.

Jakes, W. (1974). *Microwave Mobile Communications* (Wiley, New York).

Komninakis, C., Fragouli, C., Sayed, A., and Wesel, R. D. (2002). Multi-input multi-output fading channel tracking and equalization using Kalman estimation. *IEEE Trans. Signal Process.*, **50** (5), 1065–1076.

Kozick, R. J., Sadler, B. M., and Moore, T. (2003). Performance of MIMO: CM and semi-blind cases. In *Proc. IEEE Workshop Signal Process. Advances Wireless Commun.*, 309–313.

Lapidoth, A. and Narayan, P. (1998). Reliable communication under channel uncertainty. *IEEE Trans. Inf. Theory*, **44** (6), 2148–2177.

Lapidoth, A. and Shamai, S. (2002). Fading channels: How perfect need 'perfect side information' be. *IEEE Trans. Inf. Theory*, **48** (5), 1118–1134.

Larsson, E. and Li, J. (2001). Preamble design considerations for multiple-antenna OFDM based WLANs. *IEEE Signal Process. Lett.*, **8** (11), 285–288.

Li, Y., Seshadri, N., and Ariyavisitakul, S. (1999). Channel estimation for OFDM systems with transmitter diversity in mobile wireless channels. *IEEE J. Sel. Areas Commun.*, **17** (3), 461–471.

Ma, X., Yang, L., and Giannakis, G. B. (2002). Optimal training for MIMO frequency-selective fading channels. In *Proc. Asilomar Conf. Signals, Syst., Comput.*, vol. 2, 1107–1111.

Marzetta, T. (1999). BLAST training: Estimating channel characteristics for high capacity space-time wireless. In *Proc. Ann. Allerton Conf. Commun., Contr., Comput.*, 958–966.

Médard, M. (2000). The effect upon channel capacity in wireless communication of perfect and imperfect knowledge of the channel. *IEEE Trans. Inf. Theory*, **46** (3), 933–946.

Merhav, N., Kaplan, G., Lapidoth, A., and Shamai, S. (1994). On information rates for mismatched decoders. *IEEE Trans. Inf. Theory*, **40** (6), 1953–1967.

Moore, T. and Sadler, B. M. (2004). Sufficient conditions for regularity and strict identifiability in MIMO systems. *IEEE Trans. Signal Process.*, **52** (9), 2650–2655.

Sadler, B. M. and Kozick, R. J. (2000). Bounds on uncalibrated array signal processing. In *Proc. IEEE Workshop Statistical Signal and Array Process.*, 73–77.

Sadler, B. M., Kozick, R. J., and Moore, T. (2001a). Bounds on bearing and symbol estimation with side information. *IEEE Trans. Signal Process.*, **49** (4), 822–834.

—— (2001b). Bounds on MIMO channel estimation and equalization with side information. In *Proc. IEEE Int. Conf. Acoust., Speech, Signal Process.*, vol. 4, 2145–2148.

Silverstein, S. D. (1997). Application of orthogonal codes to the calibration of active phased array antennas for communication satellites. *IEEE Trans. Signal Process.*, **45** (1), 206–218.

Stoica, P. and Ng, B. (1998). On the Cramer-Rao bound under parametric constraints. *IEEE Signal Process. Lett.*, **5** (7), 177–179.

Telatar, İ. E. (1999). Capacity of multi-antenna Gaussian channels. *Eur. Trans. Telecomm*, **10** (6), 585–596.

Tong, L., Sadler, B. M., and Dong, M. (2004). Pilot-assisted wireless transmissions. *IEEE Signal Process. Mag.*, **21** (6), 12–25.

Tung, T. L., Yao, K., and Hudson, R. E. (2001). Channel estimation and adaptive power allocation for performance and capacity improvement of multiple-antenna OFDM systems. In *Proc. IEEE Workshop Signal Process. Advances Wireless Commun.*, 82–85.

Van Trees, H. L. (1968). *Detection, Estimation and Modulation Theory*, vol. 1 (Wiley, New York).

Vosoughi, A. and Scaglione, A. (2003). Channel estimation for precoded MIMO systems. In *Proc. IEEE Workshop Statistical Signal Process.*, 442 – 445.

—— (2004). The best training depends on the receiver architecture. In *Proc. IEEE Int. Conf. Acoust., Speech, Signal Process.*, vol. 4, 409–412.

Weinstein, E. and Weiss, A. (1988). A general class of lower bounds in parameter estimation. *IEEE Trans. Inf. Theory*, **34** (2), 338–342.

Wolniansky, P., Foschini, G., Golden, G., and Valenzuela, R. (1998). V-BLAST: An architecture for realizing very high data rates over the rich-scattering wireless channel. In *Proc. URSI Int. Symp. Signals, Syst., Electron.*, 295–300.

Wong, T. and Park, B. (2004). Training sequence optimization in MIMO systems with colored interference. *IEEE Trans. Commun.*, **52** (11), 1939–1947.

Yang, L., Ma, X., and Giannakis, G. B. (2004). Optimal training for MIMO fading channels with time- and frequency-selectivity. In *Proc. IEEE Int. Conf. Acoust., Speech, Signal Process.*, vol. 3, 821–824.

Yang, S. and Wu, J. (2002). Optimal binary training sequence design for multiple-antenna systems over dispersive fading channels. *IEEE Trans. Veh. Technol.*, **51** (5), 1271–1276.

Part IV

System-level issues of multiantenna systems

18

MIMO Gaussian multiple access channels

Mehdi Mohseni, Mark Brady, and John Cioffi

Stanford University

From the earliest days of multiuser information theory (Ahlswede, 1971), multiple access channels (MACs) have assumed a prominent role and profited from the development of a rich theory. Recent years have seen these theoretical results find manifest application in MIMO fixed wireless, cellular voice/data uplink, "vectored" upstream DSL, and wireless LAN standards, to name only a few. The rapid pace of research and development continues in earnest, where both techniques to implement classical results as well as developing extensions to novel channel scenarios are of interest.

This chapter surveys classical and modern theoretical aspects of multiple access channels that are germane to the design of MIMO communications systems. The emphasis is upon design principles that arise naturally from information-theoretic considerations; to this end, it is fitting to focus on a class of MACs termed Gaussian MACs, for which a diverse body of literature exists under numerous channel scenarios. Detailed proofs of the results may be found in the references contained herein.

Channel models are considered in a paradigm of space, frequency, and time. The "space" dimension refers to the number of users and number of channel inputs and outputs (e.g., transmit and receive antennas), and their correlation. The "frequency" dimension refers to frequency-selectivity present in the channel. Finally, "time" refers to time-dependence of the channel process (i.e., fading). As one might expect, the qualitative results vary considerably based on the scenario being studied; more surprisingly, there are many commonalities that arise.

This chapter is organized as follows. In the framework of space, time, and frequency, the simplest case corresponds to channels that are fixed in time, have single inputs (per user) and outputs, and are frequency-flat. Section 18.1 introduces MACs by examining this setting, and proceeds to vary the "frequency" and "space" dimensions with "time" held fixed. In Section 18.2, the "time" dimension is modified to examine ergodic capacities of fading channels under different "time" and "space" parameters. Common results and conclusions are given in Section 18.3.

Throughout the chapter, the following notation is employed: $[X]_{i,j}$ denotes the element in the ith row and jth column of the matrix X. $X \succeq 0$ denotes that the matrix X is positive semidefinite, and accordingly, $X \succ 0$ for X positive definite. For a vector $\mathbf{x}$, let $\mathbf{x}_n$ denote its nth element. For vectors $\mathbf{a}, \mathbf{b} \in \mathbb{R}^N$, the notation $\mathbf{a} \succeq \mathbf{b}$ means that $\mathbf{a}_n \geq \mathbf{b}_n$ for each $n = 1, \ldots, N$. The probability distribution function of a random variable X with support $\mathfrak{X}$ is denoted $p(x)$, where $p : \mathfrak{X} \mapsto [0, 1]$. Logarithms are taken to be natural logarithms.

18.1 Deterministic multiple access channels

This section considers MACs that are fixed in the "time" dimension; as such, the channel is presumed to be constant and known perfectly by the receiver (full CSIR). In practice, receiver channel knowledge can be acquired for stable channels (such as in DSL upstream) though channel training or sounding. The case of perfect transmitter channel information (full CSIT) is discussed first, followed by channels that are unknown at the transmitter but known to be members of a certain class.

A real, discrete-time Gaussian multiple access channel is defined by the input-output relation:

$$\mathbf{y} = \sum_{k=1}^{K} H_k \mathbf{x}_k + \mathbf{z}, \tag{18.1}$$

where $H_k \in \mathbb{R}^{r \times t_k}$, $\mathbf{x}_k \in \mathbb{R}^{t_k}$, $k = 1, \ldots, K$, $\mathbf{y} \in \mathbb{R}^r$, and $\mathbf{z} \in \mathbb{R}^r$, $p(\mathbf{z}) \sim \mathcal{N}(0, S_z)$ is i.i.d. additive white Gaussian noise with zero mean and covariance S_z. There are a total of K users, with r channel outputs, and t_k inputs for user k. For the special case of $r = 1$, S_z is denoted as σ^2.

The MAC capacity region is defined to be the convex closure of the set of achievable rate pairs for which messages may be conveyed with vanishingly small probability of error.[†] It is convenient to now consider the two-user MAC for concreteness, as the K-user results generalize readily. The rate pair (R_1, R_2) is said to be *achievable* if there exists a sequence of $((2^{nR_1}, 2^{nR_2}), n)$ codes defining encoding functions $X_1^n : W_1^n \mapsto \mathbb{R}^{t_1 \times n}$ and $X_2^n : W_2^n \mapsto \mathbb{R}^{t_2 \times n}$ (where $W_1^n = \{1, \ldots, 2^{nR_1}\}$ and $W_2^n = \{1, \ldots, 2^{nR_2}\}$), as well as a decoding function $\widehat{X} : \mathbb{R}^{r \times n} \mapsto W_1^n \times W_2^n$ with probability of error $P_e^{(n)} =$

[†]This chapter exclusively considers asymptotically vanishing *average* probability of error.

$\Pr\{\widehat{X}(y) \neq (W_1^n, W_2^n)\}$ satisfying $\lim_{n\to\infty} P_e^{(n)} = 0$. Furthermore, users 1 and 2 have power constraints P_1 and P_2 respectively, which require that $(1/n)\sum_{i=1}^{t_1}\sum_{j=1}^{n}|[X_1^n(w_1)]_{i,j}|^2 \leq P_1$, and $(1/n)\sum_{i=1}^{t_2}\sum_{j=1}^{n}|[X_2^n(w_2)]_{i,j}|^2 \leq P_2$ for each $w_1 \in W_1^n$, $w_2 \in W_2^n$, $n \in \mathbb{Z}_{++}$.

A seminal result is that the MAC capacity region takes on a compact form (Cover and Thomas, 1991):

$$\mathcal{C} = \bigcup_{\substack{p(x_1,x_2,\ldots,x_K)= \\ p(x_1)\cdot p(x_2)\cdots p(x_K), \\ \mathbf{E}\ X_k^2 \leq P_k,\ k=1,\ldots,K}} \left\{ R \in \mathbb{R}_+^K : \sum_{s\in S} R_s \leq I(X_S; Y | X_{S^c}), \forall S \subseteq E \right\},$$

(18.2)

where $E = \{1,\ldots,K\}$ is an index set on a total of K users. This general region is simplified further in the following sections by restricting consideration to Gaussian MACs.

18.1.1 Scalar Gaussian MAC

When $H_k \in \mathbb{R}^{1\times 1}$, $k = 1,\ldots,K$, the MAC is said to be a *scalar* MAC. For the scalar Gaussian MAC, it can be shown that (independent) zero-mean Gaussian input distributions $(p(x_k) \sim \mathcal{N}(0, P_k)$, $k = 1,\ldots,K)$ are optimal in the sense of simultaneously maximizing $I(X_S; Y | X_{S^c})$ for each $S \subseteq E$ (subject to the power constraints). For the two-user case, the capacity region $\mathcal{C}$ reduces to the Cover–Wyner pentagon:

$$\mathcal{C} = \left\{ (R_1, R_2) \in \mathbb{R}_+^2 : R_1 + R_2 \leq \frac{1}{2}\log\left(1 + \frac{P_1 + P_2}{\sigma^2}\right), \right.$$

$$\left. R_1 \leq \frac{1}{2}\log\left(1 + \frac{P_1}{\sigma^2}\right),\ R_2 \leq \frac{1}{2}\log\left(1 + \frac{P_2}{\sigma^2}\right) \right\}.$$

The capacity region for two users $(E = \{1,2\})$ is the pentagon $ABDEF$, as illustrated in Fig. 18.1(a). In general, the K-user capacity region is a polyhedron; additionally, $\mathcal{C}$ possesses a *polymatroid* structure that will be of subsequent importance. A polymatroid P

$$P = \left\{ \mathbf{x} \in \mathbb{R}_+^n : \sum_{s\in S} \mathbf{x}_s \leq f(S), \forall S \subseteq E \right\}$$

is defined (Tse and Hanly, 1998) by a set function $f : 2^E \mapsto \mathbb{R}_+$ satisfying the following properties for all $S, T \subseteq E$: (1) $f(\emptyset) = 0$, (2) $S \subseteq T \Rightarrow f(S) \leq f(T)$, and (3) $f(S) + f(T) \leq f(S\bigcap T) + f(S\bigcup T)$. It can be shown that

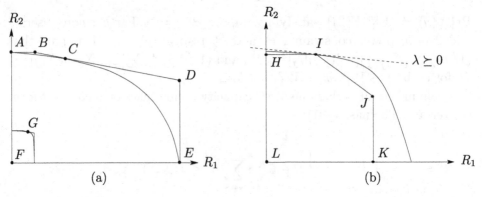

Fig. 18.1. Capacity and orthogonal signaling regions of (a) the scalar MAC $H = [1\ 3]$, $\sigma^2 = 1$, $P_1 = P_2 = \{1/2, 1/10\}$, and (b) canonical vector MAC showing supporting hyperplane λ.

$f(S) = I(X_S; Y|X_{S^c})$ is a such a set function, and consequently that $\mathcal{C}$ is a polymatroid (Tse and Hanly, 1998).

Superposition and orthogonal signaling

To achieve all points in $\mathcal{C}$, more than one user may need to transmit at the same time. For example, to achieve the Pareto-optimal[†] point B in Fig. 18.1(a), it can be readily verified that it is necessary that $p(x_1) = \mathcal{N}(0, P_1)$ and $p(x_2) = \mathcal{N}(0, P_2)$. Thus, both users transmit concurrently, the codewords are scaled and superimposed by the MAC, and the decoding function $\widehat{X}$ jointly detects the messages W_1, W_2.

In practical systems, the implementation of such a joint decoding function $\widehat{X}$ may be highly complex.[‡] It is of practical as well as theoretic interest to consider orthogonal signaling, whereby different users transmit on orthogonal dimensions (in time or frequency). The function $\widehat{X}$ may then be substantially simplified because each user may be decoded independently. The region $\mathcal{R}_T \subset \mathcal{C}$ is achievable using only orthogonal transmission (time- or frequency-division):

$$\mathcal{R}_T = \left\{ R \in \mathbb{R}^n_+ : \lambda \in \mathbb{R}^n_+, \mathbf{1}^T \lambda = 1, R_k \leq \frac{\lambda_k}{2} \log\left(1 + \frac{P_k H_k^2}{\lambda_k \sigma^2}\right) \right\}.$$

This achievable region is also shown as the curved inner lines in Fig. 18.1(a). A simple and intuitive interpretation of this result is to consider a scheme

---

[†]A rate vector $R \in \mathcal{C}$ is *Pareto-optimal* if $\{R' \in \mathcal{C} : R' \succeq R,\ R' \neq R\} = \emptyset$.

[‡]BLAST (Foschini, 1996) and generalized decision-feedback equalization (GDFE) (Cioffi and Forney, 1997) (Ginis and Cioffi, 2001), alternatively termed MMSE-SIC, are examples of reduced-complexity decoding structures; further details are given by, e.g., Paulraj *et al.* (2003).

whereby user k transmits a fraction λ_k of the time at power P_k/λ_k, while all other users do not transmit. Clearly, such a scheme satisfies the power constraints, and achievability follows readily from (18.2). Generally, the difference between C and $\mathcal{R}_T$ is most pronounced for disparate channels (i.e., when the near-far effect is predominant).

The *sum capacity*, alternatively termed the *total capacity*, is the maximum achievable sum of each user's rate. Surprisingly, the sum capacity may be achieved using orthogonal signaling (Cover and Thomas, 1991).[†] This fact is illustrated in Fig. 18.1(a), where TDMA achieves the sum capacity at points C and G.

Orthogonal vs. superposition coding; wideband regime

A natural question is the amount of inefficiency induced by using simpler orthogonal coding as opposed to superposition coding. It has been shown that, as the user power constraints converge to 0, $\mathcal{R}_T$ approaches C in the sense that the ratio of their areas converges to 1. This phenomenon is illustrated in Fig. 18.1(a) where the inner pentagon (containing G) corresponds to the same channel with lower power constraints. It can be seen from the figure that $C \setminus \mathcal{R}_T$ is a smaller fraction of the capacity region for the lower power constraints. To formalize this notion, consider an orthogonal coding scheme (henceforth, TDMA for specificity) whereby user 1 is allocated a fraction λ, $0 < \lambda < 1$, of the time for transmission, and accordingly, user 2 is allocated a fraction $1 - \lambda$ of the total time. Suppose that $\sigma^2 = 1$. Certainly, user 1 cannot hope to achieve higher rate than $R_1 = (1/2)\log(1 + H_1^2\alpha_1 P_1)$, and similarly for user 2. It has been shown that (Caire et al., 2004):

$$\lim_{\alpha_1 \downarrow 0} \frac{\lambda \log(1 + H_1^2\alpha_1/\lambda)}{\log(1 + H_1^2\alpha_1 P_1)} = \lim_{\alpha_2 \downarrow 0} \frac{\log(1 + H_2^2\alpha_2 P_2)}{(1 - \lambda)\log(1 + H_2^2\alpha_2/(1 - \lambda))} = 1.$$

Although it is tempting to conclude from this limiting result that TDMA is optimal in the low-rate[‡] regime, this is not the case; Caire et al. (2004), Verdú (2002) have shown that while TDMA is able to achieve the preceding optimality criterion and the minimum theoretical energy per bit ($E_b^m/N_0 = -1.59$ dB), the TDMA spectral efficiency (bits per channel use) for $E_b/N_0 \gtrsim E_b^m/N_0$ is in general less than that attained by superposition schemes. Consequently, orthogonal coding schemes are suboptimal, even in the low-rate regime.[§]

[†]The maximum is achieved because C and $\mathcal{R}_t$ are closed and bounded, hence compact.

[‡]The low-rate regime in discrete channels corresponds to the wideband regime in continuous channels; both consider transmission with few bits per dimension.

[§]Caire et al. (2004) define the MAC *slope regions* in analogy with the capacity region and shows that they are unequal for TDMA and superposition coding.

18.1.2 Frequency-selective scalar MACs

Having developed several results for frequency-flat channels, this section considers generalizing the "frequency" dimension to admit a class of frequency-selective scalar MACs. It is convenient to the present analysis to consider frequency-selective channels in terms of their time-domain impulse responses. In particular, consider the class of all scalar MACs having *finite* impulse response as defined by the following channel model:

$$\mathbf{y}_t = \sum_{j=0}^{D} \sum_{k=1}^{K} H_k(\mathbf{x}_k)_{t-j} + \mathbf{z}_t,$$

where t is the sample index, k is the user number, and $D < \infty$ is the length of the channel memory. The noise process z is postulated to be stationary and satisfy $p(z_t) \sim \mathcal{N}(0, \sigma^2)$, and $\mathbf{E}[z_t z_{t+\tau}] = 0$ for $|\tau| > D_z$, where $D_z < \infty$. Denote the discrete Fourier transform of the sequence $x_1, x_2, \ldots$ as $X(\omega) = \sum_{n=1}^{\infty} x_n \exp(-j\omega n)$. The two-user capacity region of this class of frequency-selective MACs is given by (Cheng and Verdú, 1993):

$$\mathcal{C} = \bigcup_{\substack{S_k(\omega) \geq 0, \\ \frac{1}{\pi} \int_0^{\pi} S_k(\omega) d\omega \leq P_k, \\ k=1,2.}} \left\{ (R_1, R_2) \in \mathbb{R}_+^2 : \right.$$

$$R_1 \leq \frac{1}{2\pi} \int_0^{\pi} \log \left(1 + \frac{S_1(\omega)|H_1(\omega)|^2}{Z(\omega)} \right) d\omega,$$

$$R_2 \leq \frac{1}{2\pi} \int_0^{\pi} \log \left(1 + \frac{S_2(\omega)|H_2(\omega)|^2}{Z(\omega)} \right) d\omega,$$

$$R_1 + R_2 \leq \frac{1}{2\pi} \int_0^{\pi} \log \left(1 + \frac{S_1(\omega)|H_1(\omega)|^2 + S_2(\omega)|H_2(\omega)|^2}{Z(\omega)} \right) d\omega \left. \right\}.$$

Unlike frequency-flat MACs, the capacity region of such channels does not have a simple formulation, nor is it (in general) a pentagon. Nevertheless, certain qualitative properties may be deduced. The most relevant of these pertains to achieving the sum capacity of the ISI MAC.

A key result is that the K-user scalar ISI MAC sum capacity may be achieved using an FDMA solution between users, whereby each user transmits on disjoint frequencies: $S_i(\omega) > 0 \Longrightarrow S_j(\omega) = 0$ for all $j \neq i$. An intuitive motivation of this result is that at each frequency, it is reasonable that the sum rate is maximized by having the user with the "best" channel transmitting. Note that this statement strengthens the prior result that orthogonal multiplexing is sum-capacity optimal in frequency-flat scalar MACs. Furthermore, if each user's channel is equal ($H_1(\omega) = H_2(\omega) = \cdots = H_K(\omega)$)

for all $\omega \in [-\pi, \pi]$, then the capacity region is again a pentagon (for the two-user case), and a polymatroid in the general case.

18.1.3 Vector channels

Scalar MACs may be generalized to consider multiple channel outputs and multiple inputs per user. These channels are defined here as vector Gaussian MACs. In the space-time-frequency paradigm, the use of multiple transmit and/or receive antennas corresponds to exploiting the "space" dimension of transmission.

Vector Gaussian MAC capacity region

Again, consider the two-user channel for specificity. We return to the setting of channels that are "flat" in the frequency dimension. The capacity region of the two-user vector Gaussian MAC (18.2) may be evaluated as:

$$
\mathcal{C} = \bigcup_{\substack{S_1 \succeq 0, S_2 \succeq 0, \\ \mathrm{Tr}(S_1) \leq P_1, \mathrm{Tr}(S_2) \leq P_2}} \left\{ (R_1, R_2) \in \mathbb{R}_+^2 : \right.
$$

$$
R_1 + R_2 \leq (1/2) \log |S_z + H_1 S_1 H_1^T + H_2 S_2 H_2^T| / |S_z|,
$$

$$
\left. R_1 \leq \frac{1}{2} \log \frac{|S_z + H_1 S_1 H_1^T|}{|S_z|}, \quad R_2 \leq \frac{1}{2} \log \frac{|S_z + H_2 S_2 H_2^T|}{|S_z|} \right\}, \quad (18.3)
$$

where the matrices S_k, $k = 1, \ldots, K$, are each user's transmit covariance. As in the scalar case, it can be shown that Gaussian inputs are optimal and achieve each point in the capacity region, which is shown in Fig. 18.1(b). Observe that for *given* values of $\{S_k\}$, the achievable region is a pentagon (in general, a polymatroid) as shown in Fig. 18.1b. While all points on the segment IJ achieve the same rate sum, as few as one point may be a Pareto-optimal point of $\mathcal{C}$ (point I).

A scenario of particular importance to cellular uplinks is when the receiver (e.g., base station) has multiple antennas, while the (mobile) users have only one antenna each ($t_1 = t_2 = \cdots = t_K = 1$, $r > 1$). In this case, the capacity region is again a polymatroid. However, it can be readily verified that unlike the scalar Gaussian MAC, orthogonal signaling does *not* in general achieve the sum capacity. Hence, it also does not achieve sum capacity in the general vector Gaussian MAC (18.3).

18.1.4 Numerical solution of the capacity region

It is possible to generalize the preceding results on sum capacity by considering the maximization of a (nonnegative) weighted sum of users' rates $\sum_k \lambda_k R_k$ as illustrated by the hyperplane in Fig. 18.1b. Finding the optimal inputs $p(x_1), \ldots, p(x_K)$ that achieve this maximization is in general nontrivial. However, for the deterministic Gaussian MAC, Gaussian inputs are optimal, and this may be expressed as a convex optimization problem (Boyd and Vandenberghe, 2004):

$$\max \quad \sum_{k=1}^{K} (\lambda_{\pi_k} - \lambda_{\pi_{k-1}}) \sum_{j=k}^{K} R_{\pi_j}$$

$$\text{subject to} \quad \sum_{j=k}^{K} R_{\pi_j} \leq \frac{1}{2} \log \frac{\left| S_z + \sum_{j=k}^{K} H_{\pi_j} S_{\pi_j} H_{\pi_j}^T \right|}{|S_z|}, \quad k = 1, \ldots, K,$$

$$\text{Tr}(S_k) \leq P_k, \quad k = 1, \ldots, K,$$

$$S_k \succeq 0, \quad k = 1, \ldots, K, \tag{18.4}$$

where π is a permutation on $\{1, \ldots, K\}$ such that $\lambda_{\pi_1} \geq \cdots \geq \lambda_{\pi_K}$ and we define $\lambda_{\pi_0} = 0$. The convexity of the constraint functions follows from the fact that $f : \mathbb{S}_+^N \mapsto \mathbb{R}$, $f(X) = -\log \det X$ is convex in the set of positive definite matrices. This optimization may be solved numerically using efficient interior-point algorithms (Boyd and Vandenberghe, 2004).

In the case of scalar MACs, the optimization simplifies because $\mathcal{C}$ is a polymatroid and the objective is maximized at one of its $K!$ Pareto-optimal vertices.[†] Moreover, it is not necessary to search each vertex to find the maximum; the optimal rate vector $(R_1^*, \ldots, R_K^*)$ satisfies:

$$R_{\pi_k}^* = \frac{1}{2} \log \left(1 + \frac{H_{\pi_k}^2 P_{\pi_k}}{\sum_{l=k+1}^{K} H_{\pi_l}^2 P_{\pi_l} + \sigma^2} \right).$$

Iterative waterfilling for sum-rate maximization

In the event that $\lambda = (1/N)\mathbf{1}$, a simplification of the preceding optimization problem (18.4) is possible; finding the inputs that achieve the sum capacity can be accomplished by the following simple procedure (Yu *et al.*, 2004):

(i) Initialize $S_k = 0$, $k = 1, \ldots, K$.
(ii) For user $k = 1$ to K

    (a) Compute $S_z' = S_z + \sum_{j \neq k} H_j S_j H_j^T$.

[†]In general, there may be as many as $K!$ unique Pareto-optimal vertices, with fewer if the problem is degenerate.

(b) Set $S_k = \arg\max_S \log|H_k S H_k^T + S_z'|$.

(iii) Return to step (ii).

Whereby each user successively optimizes its own transmit spectrum (S_k) as though the interference from all other users were additional additive noise (S_z'). The maximization step may be readily computed as water-pouring over the eigenvalues of $H_k^T H_k$ (Cover and Thomas, 1991). It can be shown that this algorithm reaches a feasible point whose sum rate is within $(K-1)r/2$ nat of optimum after only one round of iteration, and always converges.

18.1.5 Capacity of unknown Gaussian MACs

It is not possible in all situations to have perfect channel knowledge available at each transmitter. In a scenario where the channel is unknown to the transmitters yet constant, the capacity region may be considered in a game-theoretic sense. Suppose that transmitting users select input distributions $p(x_1),\ldots,p(x_K)$, and in response, an adversary selects the values of each user's channel $(H_1,\ldots,H_K)$ from a set $\mathcal{H}$ so as to most penalize the users. In particular, define the capacity region $\mathcal{C}(\mathcal{H})$ to be the convex closure of the set of rates achievable *regardless* of the adversary's choice of $(H_1,\ldots,H_K) \in \mathcal{H}$.

In general, $\mathcal{C}(\mathcal{H})$ is not known to possess a simple structure, nor is its computation straightforward. However, this is not the case if a certain condition holds in the "space" dimension. Define $\mathcal{H} \subset \times_{k=1}^K \mathbb{R}^{r \times t_k}$ to be *isotropic* if $(H_1,\ldots,H_K) \in \mathcal{H}$, $U_k \in \mathbb{R}^{t_k \times t_k}$, and $U_k^T U_k = I$ for each $k = 1,\ldots,K$ implies that $(H_1 U_1,\ldots,H_K U_K) \in \mathcal{H}$. Observe that the isotropy condition holds vacuously for scalar MACs. In the event that the isotropy condition holds on $\mathcal{H}$, it has been shown (Palomar *et al.*, 2003) that:

$$\mathcal{C}(\mathcal{H}) = \Big\{ (R_1,\ldots,R_K) \in \mathbb{R}_+^K :$$

$$\sum_{k \in S} R_k \leq \inf_{(H_1,\ldots,H_K) \in \mathcal{H}} \frac{1}{2} \log \Big| I + \sum_{k \in S} \frac{P_k}{t_k} H_k H_k^T \Big|, \forall S \subseteq E \Big\}, \quad (18.5)$$

which is a polymatroid. Furthermore, *each* $(R_1,\ldots,R_K) \in \mathcal{C}(\mathcal{H})$ is achieved by $p(x_k) \sim \mathcal{N}(0, S_k)$ for each $k = 1,\ldots,K$, with the uniform power allocation $S_k = (P_k/t_k)I$, $k = 1,\ldots,K$. Depending on the structure of $\mathcal{H}$, (18.5) may be simple or difficult to evaluate numerically. Irrespective of the analytical tractability of $\mathcal{C}(\mathcal{H})$, the uniform power allocation is optimal in the broad sense of achieving each point in $\mathcal{C}(\mathcal{H})$.

18.2 Fading MACs

This section considers the channels that vary in the "time" dimension. Such channels are of particular interest in mobile wireless communication where users experience varying channel conditions due to multipath, shadowing, and interference effects. Thus, for mobile users, the channel is dependent on position, and hence time.[†] We concentrate exclusively on *ergodic* capacity, which represents the long-term average of achievable rates. In particular, by carefully examining the role of multiuser *dimensions*, or *degrees of freedom*, in achieving sum capacity, numerous insights are possible.

In single-user fading channels, multiple transmit dimensions (such as space, frequency, and time) can generally be utilized in two different ways (Yu and Rhee, 2004). First, redundant copies of the same data stream can be transmitted through multiple dimensions; if these dimensions are chosen properly, it is unlikely that all of them will experience deep fades concurrently, leading to an increase in transmission reliability. This concept is termed *diversity*. Alternatively, multiple dimensions may be used to transmit additional information, thereby achieving higher rates; this concept is termed *multiplexing*. In multiuser fading channels, an additional notion of diversity may be identified: that of multiuser diversity. Each user in a multiuser system may be viewed as an available transmission dimension. Since users scattered geographically experience different fading states, it is unlikely that many users simultaneously experience poor channels. Therefore, the total system throughput can be made resilient to fading by choosing some "best" set of users for communication. This concept is termed *multiuser diversity*.[‡]

18.2.1 Ergodic MAC capacity with CSIT

Availability of instantaneous channel state information at the transmitter side (CSIT) enables the transmitters to dynamically allocate power among various transmit dimensions (i.e., space, frequency and time), thereby adapting the transmission scheme to channel states. In practice, channel state may be estimated at the receiver and sent back to the transmitter via a feedback channel. This section considers fading Gaussian MACs with full CSIT. Depending on the fading coherence time and the delay constraint of the application, there are two principal types of power constraints that may

---

[†]The interested reader may refer to Paulraj *et al.* (2003) for a survey of results on single-user fading channels.

[‡]Recent results (Tse *et al.*, 2004) have characterized the fundamental trade-offs between diversity, multiplexing, and multiuser diversity of MACs in the asymptotic SNR regime.

be considered: long-term power constraints (corresponding to fast-fading channels) and short-term power constraint (corresponding to slow-fading channels). Throughout this section, the following channel model is adopted:

$$\mathbf{y}_n = \sum_{k=1}^{K} H_k(\nu_n)(\mathbf{x}_k)_n + \mathbf{z}_n, \tag{18.6}$$

where $\mathbf{y}_n \in \mathbb{C}^r$ is the received signal, $H_k(\nu_n) \in \mathbb{C}^{r \times t_k}$, $k = 1, \ldots, K$ is the channel of the kth user as a function of the fading state ν at time n (ν_n), and $\mathbf{z}_n$ is i.i.d. circularly-symmetric complex Gaussian noise $p(z_n) \sim \mathcal{CN}(0, S_z)$. In this section, optimal transmit strategies to achieve the *sum capacity* are investigated; by carefully inspecting the optimality conditions of the throughput maximization problem, insights on the trade-off between spatial multiplexing and multiuser diversity are obtained.

Long-term power constraints

When the application delay constraint is much longer than the coherence time of the fading process, the code block-length can be chosen long enough to cover all fading states without violating the delay constraint. In this case, the mutual information terms averaged over all fading states determine the ergodic capacity region.

It can be shown that for the model (18.6) with long-term power constraints, the use of zero-mean Gaussian input signals is optimal. Thus, for each ν it is sufficient to determine the input covariance of each user $S_k(\nu)$, $k = 1, \ldots, K$, where $S_k(\nu) = \mathbf{E}\{\mathbf{x}_k(\nu)\mathbf{x}_k(\nu)^*\}$. Note that the expectation for the input covariance matrix in state ν is taken over the codebook of user k. Then the capacity region can be expressed as:

$$\mathcal{C} = \bigcup_{\substack{S_k(\nu) \succeq 0, \\ \mathbf{E}_H\{\text{Tr}(S_k)\} \leq P_k, \\ \forall \nu, k}} \left\{ R \in \mathbb{R}_+^K : \sum_{s \in S} R_s \leq \mathbf{E}_H I(X_S; Y | X_{S^c}, H), \forall S \subset E \right\}. \tag{18.7}$$

In (18.7), the expectations are with respect to the joint channel distribution of $H(\nu)$. Therefore, the mutual information term may be evaluated as:

$$I(X_S; Y | X_{S^c}, H(\nu)) = \log \frac{\left| S_z + \sum_{k \in S} H_k(\nu) S_k(\nu) H_k(\nu)^* \right|}{|S_z|}.$$

Ergodic sum capacity of fading vector Gaussian MACs

Considering the sum capacity as the performance criterion of MACs, this capacity may be expressed as the solution to the optimization problem:

$$\max \quad \int_\nu \log \frac{\left|S_z + \sum_{k=1}^K H_k(\nu)S_k(\nu)H_k(\nu)^*\right|}{|S_z|} dp(\nu)$$

$$\text{such that} \quad \int_\nu \text{Tr}(S_k(\nu))dp(\nu) \le P_k, \quad k = 1, \ldots, K,$$

$$S_k(\nu) \succeq 0, \quad \forall \nu, k. \tag{18.8}$$

The structure of the optimal solution can be characterized by exploring the Karush–Kuhn–Tucker (KKT) optimality conditions associated with (18.8) (Boyd and Vandenberghe, 2004). Associate the scalar dual variables $\lambda_k \ge 0$, $k = 1, \ldots, K$, to the corresponding power constraints, and $t_k \times t_k$ matrices $\Phi_k(\nu) \succeq 0$ to each positive semidefinite constraint. The resulting KKT conditions are:

$$\lambda_k I = H_k(\nu)^* \left(S_z + \sum_{i=1}^K H_i(\nu)S_i(\nu)H_i(\nu)^* \right)^{-1} H_k(\nu) + \Phi_k(\nu),$$

$$\int_\nu \text{Tr}(S_k(\nu))d\rho(\nu) \le P_k,$$

$$\text{Tr}(S_k(\nu)\Phi_k(\nu)) = 0,$$

$$\Phi_k(\nu), S_k(\nu) \succeq 0, \quad \lambda_k \ge 0. \tag{18.9}$$

For a fixed fading state ν and fixed user k, equations (18.9) have exactly the same form as the KKT conditions for the single-user rate maximization problem in a channel with an equivalent noise covariance matrix equal to $\tilde{S}_z(k, \nu) = \sum_{i \ne k} H_i(\nu)S_i(\nu)H_i(\nu)^* + S_z$. The equivalent noise contains the interference from other users as well as the Gaussian noise. For a single user channel, it is well known that the optimal transmit covariance matrix is obtained by waterfilling over the singular values of the equivalent channel $\tilde{H}_k(\nu) = \tilde{S}_z(k, \nu)^{-1/2} H_k(\nu)$, with water level equal to $1/\lambda_k$. Also, the transmit directions must align with the right eigenvectors of the equivalent channel. Formally, if $\tilde{H}_k(\nu) = F_k(\nu)\Gamma_k(\nu)M_k(\nu)^*$ denotes the singular value decomposition of the channel, then $S_k(\nu) = M_k(\nu)\Sigma_k(\nu)M_k(\nu)^*$, where $\Sigma_k(\nu)$ is a diagonal matrix with ith entry equal to $(1/\lambda_k - 1/\gamma_k^i(\nu)^2)^+$ (where γ^i is the ith diagonal element of Γ), for $i = 1, \ldots, t$.

This is true for each $k = 1, \ldots, K$; thus, at each state ν, all users must simultaneously waterfill with their own water levels $1/\lambda_k$ against combined interference and noises. However, the water levels are the same for each state, which implies that each user needs to waterfill over both space and time with the same water level. For each user, the water level can be obtained from the total power constraint:

$$\int_{\nu} \left(\sum_{i=1}^{t} (1/\lambda_k - 1/\gamma_k^i(\nu)^2)^+ \right) dp(\nu) = P_k, \quad k = 1, \ldots, K. \tag{18.10}$$

Using the similarity between single user and multiuser water filling solutions, and based on the fact that all users must simultaneously waterfill over all states, the iterative waterfilling procedure of Section 18.1.4 may be modified to obtain the optimal input covariance matrices: (Yu and Rhee, 2004).

(i) Initialize $S_k(\nu) = 0$, $k = 1, \ldots, K$, $\forall \nu$.
(ii) For user $k = 1$ to K
    For each state ν
    (a) Compute $\tilde{S}_z(k, \nu) = \sum_{i \neq k} H_i(\nu) S_i(\nu) H_i(\nu)^* + S_z$.
    (b) Compute SVD of $\tilde{S}_z(k, \nu)^{-1/2} H_k(\nu) = F_k(\nu) \Gamma_k(\nu) M_k(\nu)^*$.
    (c) Compute λ_k such that 18.10 is satisfied.
    (d) Compute the diagonal matrix $\Sigma_k(\nu)$ with elements $(1/\lambda_k - 1/\gamma_k^i(\nu)^2)^+$.
    (e) Set $S_k(\nu) = M_k(\nu) \Sigma_k(\nu) M_k(\nu)^*$.
(iii) Return to step (ii).

Sum capacity with long-term power constraints and Rayleigh fading

As mentioned previously, spatial dimensions can be distributed among the best set of users during each state to maximize the total system throughput. Since not all the users are in deep fade at each state, this also makes the throughput resilient to channel variations. Simultaneously, various spatial dimensions assigned to each active user can be employed to transmit independent data streams achieving higher data rates for the given user and maximizing the total data rate. Hence, to maximize the total rate-sum, spatial dimensions need to be distributed properly among users at each state. In this context, an important question is: how many users need to be simultaneously active at each time, and how ought dimensions be distributed among them? This question may be addressed by inspecting the KKT optimality conditions (Rhee *et al.*, 2004).

Consider an i.i.d. Rayleigh fading model, where the elements of H_k are i.i.d. $\sim \mathcal{CN}(0, 1)$. The subsequent analysis considers the special case where ν is an i.i.d. process. For simplicity of the analysis, assume all transmitters have t transmit antennas ($t_1 = \cdots = t_K = t$). Define a user k to be *active* in state ν if $S_k(\nu) \neq 0$. An upper bound on the number of simultaneously active users when achieving sum capacity can be obtained by counting the number of variables and unknowns in the set of KKT equations (18.9). For a given state ν, let $C(\nu)$ be equal to $(S_z + \sum_{i=1}^{K} H_i(\nu)S_i(\nu)H_i(\nu)^*)^{-1}$. As seen from (18.9), the term $C(\nu)$ is common among all the equations corresponding to different users. Therefore, the matrix $C(\nu)$ is an $r \times r$ Hermitian matrix intro-ducing r^2 variables to the system of equations for each state. For each user k, $H_k(\nu)^*C(\nu)H_k(\nu)$ and $\Phi_k(\nu)$ are Hermitian $t \times t$ matrices; hence, the equal-ity $H_k(\nu)^*C(\nu)H_k(\nu) + \Phi_k(\nu) = \lambda_k I$ represents t^2 linear equations, which are independent due to the i.i.d. Rayleigh fading assumption for different antennas and users.

A total of Kt^2 equations need to be satisfied for each state, and due to the i.i.d. Rayleigh fading process assumption, these sets of equations are independent with probability 1. The waterfilling levels λ_k are only a function of fading distribution and can be computed *a priori* from (18.10), making them constants in this set of equations. Let $\rho_k(\nu) = \text{rank}(S_k(\nu))$. Since $\text{Tr}(S_k(\nu)\Phi_k(\nu)) = 0$ and both $S_k(\nu)$ and $\Phi_k(\nu)$ are positive semidefinite matrices, it follows immediately that because $S_k(\nu)\Phi_k(\nu) = 0$, the rank of $\Phi_k(\nu)$ can be at most $t - \rho_k(\nu)$. Using this rank constraint on $\Phi_k(\nu)$, it can be shown that Φ_k can introduce at most $t^2 - \rho_k^2$ degrees of freedom to the set of equations. This is because a $t \times t$ Hermitian matrix of rank not exceeding $t - \rho_k(\nu)$ has at most $t^2 - \rho_k(\nu)^2$ independent variables. Considering the total number of equations and unknowns, with probability 1 the system (18.9) has a solution if and only if the number of variables is greater than or equal to the number of (independent) equations; thus, $Kt^2 \leq r^2 + \sum_{k=1}^{K}(t^2 - \rho_k(\nu)^2)$, which reduces to:

$$\sum_{k=1}^{K} \rho_k(\nu)^2 \leq r^2. \qquad (18.11)$$

This inequality, which was first obtained by Yu and Rhee (2004), suggests that each user can utilize at most t spatial dimensions, and the sum of the squared users' ranks cannot exceed r^2. Since for each active user k we have that $\rho_k \geq 1$, the total number of active users is at most r^2 at each state.

Inequality (18.11) is now applied to two special configurations and is shown to verify previously known results for fading and nonfading MACs. For the special case of scalar Gaussian MACs, (18.11) implies that at each fading state only one user may be active. Hence, the optimal multiuser scheduling (in the sense of achieving sum capacity) is to assign the channel to (at most) one "best" user at each state, which verifies the result previously established by Knopp and Humblet (1995). By choosing ν uniformly distributed over $[-\pi, \pi]$, the capacity region of a nonfading scalar MAC with ISI (as considered in Section 18.1.2) can be expressed by (18.7). Here the role of frequency and time is exchanged: only one user can be active at each frequency, which confirms the optimality of FDMA obtained formerly (Cheng and Verdú, 1993).

Short-term power constraints

When the communication system has a limited delay constraint or the fading coherence time is comparable to the delay constraint, the code block-length cannot be chosen arbitrary long to cover all fading states. In this case, the block length is chosen to only cover a single state, during which the channel is assumed to be constant. The term "block fading channels" also refers to this scenario, where average power constraints are applied to each block. The averaged mutual information over all fading states is the ergodic MAC capacity (Hanly and Tse, 1998):

$$\mathcal{C} = \bigcup_{\substack{S_k(\nu) \succeq 0, \\ \mathrm{Tr}(S_k(\nu)) \leq P_k, \ \forall k, \nu.}} \left\{ R \in \mathbb{R}_+^K : \right.$$

$$\left. \sum_{s \in S} R_s \leq \mathbf{E}_H \left\{ I(X_S; Y | X_{S^c}, H) \right\}, \forall S \subseteq E \right\}.$$

One may again consider maximizing $\sum_k \lambda_k R_k$ for $\lambda \succeq 0$ over all possible input distributions $\{S_k(\nu)\}$ that satisfy the power constraints. In block fading channels, power constraints are applied to each fading state independently; consequently, the mutual information terms corresponding to different states are not coupled together through power constraints. Hence, the corresponding capacity region of each state can be characterized independently, and by averaging over all fading states, one can obtain the total capacity region. Suppose $R_k(\nu)$, $k = 1, \ldots, K$, is the boundary point of the MAC channel at state ν associated with weights λ_k. Equivalently, it maximizes $\sum_k \lambda_k R_k(\nu)$ subject to the power constraints, and $\mathbf{E}_\nu \{R_k(\nu)\}$ will maximize $\sum_k \lambda_k R_k$ and is a boundary point of the capacity region of the block fading MAC associated with weights λ_k.

Sum capacity with short-term power constraints and Rayleigh fading

Parallel to the long-term power constraint scenario, the consideration of sum capacity with i.i.d. Rayleigh fading leads to additional insights. For each state ν, the input covariance matrices $S_k(\nu)$ that achieve sum capacity satisfy the KKT optimality conditions:

$$\lambda_k I = H_k^* \left(S_z + \sum_{i=1}^{K} H_i S_i H_i^* \right)^{-1} H_k + \Phi_k,$$

$$\text{Tr}(S_k) \leq P_k,$$

$$\text{Tr}(S_k \Phi_k) = 0,$$

$$\Phi_k, S_k \succeq 0, \quad \lambda_k \geq 0. \tag{18.12}$$

To simplify the expressions, the variable ν is omitted from the equations, but S_k, Φ_k, H_k, and λ_k all depend on and hold for each channel state ν. In contrast with the fast fading scenario, the water levels $1/\lambda_k$ vary from state to state and waterfilling may be thought of as performed over spatial dimensions independently on each state.

As with fast fading channels, it is of practical importance to know how many users may share the channel (for example, wireless media) simultaneously, and how the spatial dimensions are distributed between them. These questions can be answered by exploring KKT conditions (18.12). Unlike the fast fading case, it can be shown that the total rate sum can be maximized only if each user utilizes all its available power at each state. This is because each user is allowed to use some given amount of power at each state independent of channel quality and cannot trade it for occasions when the channel quality is better. Hence, at each state, all users are active and communicate with the base station. Another interesting question is how users share available spatial dimensions, especially how many users employ just one spatial dimension, i.e., perform beamforming. These questions are motivated by the simplicity of beamforming; a single data stream can be encoded by an AWGN code and multiplied by a beamforming vector for transmission without the need for complicated prefiltering or orthogonal block coding structures.

Consider the KKT conditions given in (18.12). The matrix $C = (S_z + \sum_{i=1}^{K} H_i S_i H_i^*)^{-1}$ appears in all equations corresponding to different users and introduces r^2 variables. As mentioned earlier, water levels are no longer fixed between fading states, and therefore comprise another K variables. Define $\rho_k = \text{rank}(S_k)$. By exactly the same reasoning as in the fast fading case, it follows that Φ_k has $t^2 - \rho_k^2$ degrees of freedom. Hence there are a total of $\sum_{k=1}^{K}(t^2 - \rho_k^2) + K + r^2$ variables. Each equality of the form $H_k^* C H_k + \Phi_k = \lambda_k I$ corresponds to t^2 equations, which are independent with probability 1 due to the independent Rayleigh fading across the users and antennas. Therefore, with probability 1, the KKT conditions can have a solution if and only if the number of equations is less than or equal to the number of variables: $\sum_{k=1}^{K} \rho_k^2 \leq K + r^2$. Clearly $\sum_k \rho_k \leq \sum_k \rho_k^2$ and since all users are active at each time, $\rho_k \geq 1$. Combining these inequalities yields $K \leq \sum_{k=1}^{K} \rho_k \leq K + r^2$. These bounds on $\sum_k \rho_k$ imply that at most r^2 users could concurrently transmit over more than one spatial dimension. Thus, when the number of users is much larger than the number of receive antennas ($K \gg r^2$), beamforming becomes an optimal (sum capacity achieving) transmission scheme for almost all users. This result was observed in Rhee *et al.* (2004).

18.2.2 Ergodic MAC capacity with CSIR only

This section considers the capacity region of a fading MAC when channel state information is available only at the receiver (CSIR). As in the previous analysis, the fading process is postulated to be frequency nonselective, leaving space and time available as transmit dimensions. Depending on the application delay constraint and fading coherence time, different notions of capacity may be considered. For fast fading channels, long block-length codes covering all states achieve the mutual information averaged over all fading states. With CSIR only, the channel inputs are not allowed to vary as a function of the state. This is referred to as ergodic capacity.

Alternatively, when the fading process is slow, codewords can only experience a few channel states. One sense of capacity is the smallest achievable rate over all fading states (see Section 18.1.5). However, for some important fading processes (such as i.i.d. Rayleigh fading), this minimum mutual information can be zero, and, hence, such a notion of capacity is vacuous. In such cases, it is reasonable to allow the transmission fail for a given outage fraction, and achieve a rate R otherwise.[†] Returning to the ergodic capacity with CSIR, it can be shown that Gaussian inputs are optimal for i.i.d. fading MACs, and that the ergodic capacity region is given by:

$$
\mathcal{C} = \bigcup_{\substack{S_k \succeq 0, \\ \mathrm{Tr}(S_k) \leq P_k, \, \forall k.}} \left\{ R \in \mathbb{R}_+^K : \right.
$$

$$
\left. \sum_{s \in S} R_s \leq \mathbf{E}_H \left\{ \log \left(\frac{|S_z + \sum_{k \in S} H_k S_k H_k^*|}{|S_z|} \right) \right\}, \forall S \subset E \right\}. \quad (18.13)
$$

For the case of i.i.d. Rayleigh fading between each pair of transmit and

---

[†]Outage capacity of Gaussian MACs with CSIR has been considered in several studies (e.g., Biglieri *et al.*, 1998; Rhee and Cioffi, 2003).

receive antennas, (18.13) has been evaluated by Telatar (1999) for the single-user case and generalized by Rhee and Cioffi (2003). For such *"unbiased"* channels, the capacity region is a polymatroid and all Pareto-optimal points are achievable with $S_k = P_k/t_k I$, $k = 1, \ldots, K$. That is, each user allocates power uniformly among t_k channel inputs (i.e., uniformly in space):

$$\mathcal{C} = \left\{ R \in \mathbb{R}_+^K : \sum_{s \in S} R_s \leq \mathbf{E}_H \left\{ \log \left(\frac{|S_z + \sum_{k \in S} \frac{P_k}{t_k} H_k H_k^*|}{|S_z|} \right) \right\}, \forall S \subseteq E \right\}.$$

This result is a straightforward generalization of single-user ergodic capacity and is based on the convexity of $-\log|X|$ in the set of positive definite matrices. This is first argued for the single-user case. Observe that if the matrix H has elements $[H]_{i,j}$ distributed as i.i.d. circularly-symmetric complex Gaussian variables, then for any unitary matrix Q, the matrix HQ has the same joint distribution as H. Consider the eigendecomposition of $S = Q\Lambda Q^*$, and observe that $\mathbf{E}\{\log|S_z + HQ\Lambda Q^* H^*|\} = \mathbf{E}\{\log|S_z + H\Lambda H^*|\}$. Based on this observation, it is sufficient to consider diagonal S_k in (18.13). Let $\mathrm{Tr}(\Lambda) = P$ and J_i denote a specific $t \times t$ permutation matrix (among all $t!$ such permutations). Evidently, $(1/t!) \sum_i J_i \Lambda J_i^* = (P/t) I_t$, and since each J_i is unitary, $\mathbf{E}\{\log|S_z + HJ_i\Lambda J_i^* H^*|\} = \mathbf{E}\{\log|S_z + H\Lambda H^*|\}$. From the concavity of the logarithm function and the linearity of expectation, $\mathbf{E}\{\log|S_z + H\Lambda H^*|\} \leq \mathbf{E}\{\log|S_z + (P/t)HH^*|\}$, which implies that the uniform power allocation maximizes the average mutual information. This result can be generalized to MACs easily. Considering other S_i, $i \neq k$ fixed, each $\mathbf{E}\{\log|S_z + \sum_{k \in S} H_k S_k H_k^*|\}$ term in (18.13) can be maximized by choosing $S_k = (P_k/t_k) I_{t_k}$.

As discussed earlier in this section, multiple spatial dimensions can be utilized to increase the total data rate by spatial multiplexing. It is of interest, however, to know how the ergodic capacity scales with the number of transmit and receive antennas. For single-user Gaussian fading channels with Rayleigh fading $H \in \mathbb{C}^{r \times t}$, $[H]_{i,j}$ i.i.d. circularly symmetric $\sim \mathcal{CN}(0,1)$, and fixed total power constraint P, let $C(r,t,P)$ denote the ergodic capacity. It can be shown that the ergodic capacity varies linearly with t in the case when $r = t$ or $C(t,t,P) \simeq tC(1,1,P)$, and for the case $r \neq t$, $C(r,t,P) \simeq \min(r,t)C(1,1,P) + \mathcal{O}(\log(1 + |t - r|))$ (Foschini, 1996). Additional increase in capacity is due to higher transmit or receive diversity gains when $t > r$ or $t < r$, respectively.

Considering the total rate sum as a figure of merit for MACs, we pursue a connection between the sum capacity of MACs and ergodic capacity of single-user channels. Let $C_{\mathrm{MAC}}(r,t,P,K)$ be the sum capacity of a MAC with K mobile users, each equipped with t transmit antennas, r receive antennas at the base station, i.i.d. Rayleigh fading (as described above), and power constraint $P_k = P/K$ for each user. Define $H = [H_1, \ldots, H_K] \in \mathbb{C}^{r \times tK}$. H therefore contains i.i.d. Gaussian elements and $C_{\text{MAC}}(r, t, P, K) = \mathbf{E}\{\log(|S_z + (P/tK)HH^*|/|S_z|)\} = C(r, tK, P)$. This equality is combined with the scaling results for single-user channels in Rhee and Cioffi (2003) to obtain the following three observations: (1) when $tK \leq r$, sum capacity increases linearly by increasing either K or t, (2) when $tK > r$ capacity increases at most logarithmically by increasing either K or t, (3) when $tK \gg r$ capacity does not change significantly by increasing either K or t.

Note that as K or t becomes large, $1/(tK)HH^* \to I_r$ (a.s.) and the ergodic sum capacity tends to an upper bound equal to $r \log(1 + P)$. Increasing the number of transmit antennas or the number of users in the system equivalently increases the spatial or multiuser diversity gains and helps to mitigate fading; consequently, the sum capacity converges to capacity of an equivalent AWGN channel. While increasing either K or t is sufficient to achieve the upper bound, multiantenna transmitters (large t) boost individual users' rates and provide spatial diversity gain (Visuri and Bölcskei, 2004).

18.3 Conclusions

This chapter's survey of recent results in Gaussian MACs from a viewpoint of space, time, and frequency has revealed numerous recurring themes. General expressions for channel capacity and capacity-achieving techniques have been discussed for different channel models. In scalar Gaussian MACs, the role of orthogonal signaling strategies was noted in a diverse number of settings to attain sum capacity. As such, orthogonal signaling can be a pragmatic technique when system throughput is of primary concern. In vector channels, the generality of inputs uniform over the "space" dimension was observed. It was seen that the uniform distribution in space is optimal when information about the spatial "directions" of the channel are not available at the transmitter. An analysis of ergodic capacity in fading MACs found that multiuser diversity gain may often be obtained by using only a small fraction of the available users or spatial dimensions. An exciting implication of these findings is that optimal transmission strategies can often be implemented with far less complexity one might presume *a priori*.

References

Ahlswede, R. (1971). Multi-way communication channels. In *Proc. IEEE Int. Symp. Inf. Theory*, 23–52 (Publishing House of the Hungarian Academy of Sciences).

Biglieri, E., Proakis, J., and Shamai (Shitz), S. (1998). Fading channels: Information-theoretic and communication aspects. *IEEE Trans. Inf. Theory,* **44** (6), 2619–2692.

Boyd, S. and Vandenberghe (2004). *Convex Optimization* (Cambridge Univ. Press).

Caire, G., Tuninetti, D., and Verdú, S. (2004). Suboptimality of TDMA in the Low-Power regime. *IEEE Trans. Inf. Theory,* **50** (4), 608–620.

Cheng, R. S. and Verdú, S. (1993). Gaussian multiaccess channels with ISI: capacity region and multiuser water-filling. *IEEE Trans. Inf. Theory,* **39** (3), 773–785.

Cioffi, J. M. and Forney, G. D. (1997). GDFE for packet transmission with ISI and Gaussian noise. In Paulraj, A., Roychowdhury, V., and Schaper, C. (eds.), *Communication, Computation, Control and Signal Processing,* 79–127 (Kluwer, Boston).

Cover, T. M. and Thomas, J. A. (1991). *Elements of Information Theory* (Wiley-Interscience, New York).

Foschini, G. (1996). Layered space-time arch. for wireless comm. in a fading environment when using multi-element antennas. *Bell Labs Tech. J.,* **1** (2), 41–59.

Ginis, G. and Cioffi, J. M. (2001). On the relation between V-BLAST and the GDFE. *IEEE Commun. Lett.,* **5** (9), 364–366.

Hanly, S. V. and Tse, D. N. C. (1998). Multiaccess fading channels—Part II: Delay-limited capacities. *IEEE Trans. Inf. Theory,* **44** (7), 2816–2831.

Knopp, R. and Humblet, P. A. (1995). Information capacity and power control in single-cell multiuser communications. In *Proc. IEEE Int. Conf. Commun.,* vol. 1, 331–335.

Palomar, D. P., Cioffi, J. M., and Lagunas, M. A. (2003). Uniform power allocation in MIMO channels: a game-theoretic approach. *IEEE Trans. Inf. Theory,* **49** (7), 1707–1727.

Paulraj, A., Nabar, R., and Gore, D. (2003). *Introduction to Space-Time Wireless Communications* (Cambridge Univ. Press).

Rhee, W. and Cioffi, J. M. (2003). On the capacity of multiuser wireless channels with multiple antennas. *IEEE Trans. Inf. Theory,* **49** (10), 2580–2595.

Rhee, W., Yu, W., and Cioffi, J. M. (2004). The optimality of beamforming in uplink multiuser wireless systems. *IEEE Trans. Wireless Commun.,* **3** (1), 86–96.

Telatar, E. (1999). Capacity of multi-antenna Gaussian channels. *Eur. Trans. Telecommun.,* **10** (6), 585–595.

Tse, D. N. C. and Hanly, S. V. (1998). Multiaccess fading channels—Part I: Polymatroid structure, optimal resource allocation and throughput capacities. *IEEE Trans. Inf. Theory,* **44** (7), 2796–2815.

Tse, D. N. C., Viswanath, P., and Zheng, L. (2004). Diversity-multiplexing tradeoff in multple-access channels. *IEEE Trans. Inf. Theory,* **50** (9), 1859–1874.

Verdú, S. (2002). Spectral efficiency in the wideband regime. *IEEE Trans. Inf. Theory,* **48** (6), 1319–1343.

Visuri, S. and Bölcskei, H. (2004). MIMO-OFDM multiple access with variable amount of collision. In *Proc. IEEE Int. Conf. Commun.,* vol. 1, 286–291.

Yu, W. and Rhee, W. (2004). Degrees of freedom in multi-user spatial multiplex systems with multiple antennas. *Submitted to IEEE Trans. Commun.*

Yu, W., Rhee, W., Boyd, S., and Cioffi, J. (2004). Iterative water-filling for Gaussian vector multiple access channels. *IEEE Trans. Inf. Theory,* **50** (1), 145–152.

19

On information-theoretic aspects of MIMO broadcast channels

Giuseppe Caire

University of Southern California

Shlomo Shamai (Shitz), Yossef Steinberg, and Hanan Weingarten

Technion

19.1 Introduction

The broadcast channel (BC) first introduced by Cover (1972) is now a standard channel model in information theory (Cover and Thomas, 1991), which has attracted massive attention, and yet the capacity region is not known in general. The BC models downlink communications, where a central hub (cell-site, for example) transmits to potentially multiple users. Multiple-input multiple-output (MIMO) channels have been identified as a central factor in significantly amplifying the capability of wireless communications (Goldsmith *et al.*, 2003; Tse and Viswanath, 2005). Motivated by practical high reliable rate demands of future wireless and wireline systems, well modeled by the MIMO Gaussian broadcast channel (MIMO GBC), such as cellular systems, wireless local area networks, and xDSL links, the ultimate potential of the MIMO GBC has been identified as a theoretical challenge, which carries evident practical implications.

The central information-theoretic hardship arises in the general MIMO GBC setting, even if all channel state information (CSI) is available at every node (transmitter as well as receivers). This is due to loss of degradation ordering in the general MIMO case, directly leading to the general BC, for which the capacity region is yet unknown. Evidently one may resort to bounding techniques, such as the well known Marton region (Marton, 1979), which constituted for decades the best achievable rate region for the general BC. Different upper bounds are also available, and some of the most interesting are the Marton–Körner (Marton, 1979) and Sato (Sato, 1978) outer bounds.

Fortunately, the MIMO GBC is one of the classes of nondegraded broadcast channels for which the capacity region of individual rates is fully characterized. This chapter briefly overviews the information-theoretic aspects of these results.

Standard approaches to communicate over the MIMO GBC are based on precoding, with the focus on linear precoding and a variety of different performance criteria such as signal-to-noise ratio admissibility, zero forcing, average (or worst-case) reliable rates, min-max strategies and the like (Wiesel et al., 2006, and references therein). The first paper addressing the information-theoretic implications of MIMO GBC is by Caire and Shamai (Shitz) (2003). The approach taken by Caire and Shamai hinges on what has been found as the central theoretical ingredient in the successful attack on MIMO GBCs, dirty paper coding (DPC). This immediately put in focus the Gelfand–Pinsker setting (Gelfand and Pinsker, 1980), and in particular the Gaussian Costa DPC model (Costa, 1983). The optimality of the basic approach of Costa (1983), which is a modification of what is named ranked known interference (RKI), in terms of throughput in a simple two transmit antenna two user case has been substantiated via the Sato bound (Sato, 1978).

The results of Caire and Shamai (Shitz) (2001) spurred an exploding interest in the problem. Yu et al. (2001) applied the DPC idea directly to a vector setting, which is also evident by considering scalar DPC with singular value decomposition. This, together with evident optimization by permutation, describes what is now known as the DPC achievable region for the two-user MIMO GBC. The concept of uplink-downlink duality, first introduced by Jindal et al. (2004), Viswanath and Tse (2003), and Vishwanath et al. (2003), happens to constitute a central tool that facilitated the significant extension and generalization of the results of Caire and Shamai (Shitz) (2003), establishing the optimality of the sum rate in a general MIMO GBC, when combined with the Sato bound. In Yu and Cioffi (2004), the sum-rate optimality of the DPC rate region of the MIMO GBC has been established under minor regularity conditions (removed in Yu and Lan, 2004, where also minimax duality is introduced; see also Yu, 2006) adhering to the notion of a canonic generalized decision-feedback equalizer. The sum capacity can also be interpreted as the saddle-point of a mutual information game, where a signal player chooses a transmit covariance to maximize the mutual information, and "Nature" chooses a fictitious noise correlation to minimize the mutual information. The DPC rate region of the MIMO GBC does admit many interpretations, including optimal standard beamforming with DPC (Schubert and Boche, 2002). Important observations were made in Vishwanath et al. (2002) and Tse and Viswanath (2002), where the DPC rate region is given in terms of the rate region of an associated degraded MIMO

GBC. This observation facilitates the replacement of the Sato upper bound on the sum rate with the degraded same marginals (DSM) bound. Thus, it has been concluded then that if, indeed, Gaussian coding is optimal for the MIMO degraded GBC (as conjectured in these works), then the DPC region is the capacity region. This conjecture has first been substantiated in Weingarten *et al.* (2004), which has introduced a novel and important concept of an enhanced broadcast channel. Another step forward reported in Weingarten *et al.* (2006) provides a fully cohesive view that hinges neither on the concepts of the uplink-downlink duality nor on the DSM bound. Rather, the capacity region of the MIMO GBC is established by first principles, relying on the new concept of enhanced broadcast channel. Efficient algorithms to determine the capacity region of the MIMO GBC, focusing especially on sum rate and exploiting convex programming and the duality principle, are now available (Jindal *et al.*, 2002).

The outline of this chapter is as follows. In Section 19.2, we provide a preliminary information-theoretic background on the BC and DPC. Section 19.3 contrasts some signal-processing-based techniques as applied for the MIMO GBC, and focuses on the original RKI and related approaches that invoke DPC. Section 19.4 addresses the important notion of uplink-downlink duality and its implication in determining the optimal sum rate. Section 19.5 focuses on the full capacity region, discussing the DSM bound, and the "enhanced channel" notion. Comparisons of the capacity regions for a simple two-user channel associated with different approaches are provided in Section 19.6. Concluding remarks are given in Section 19.7, followed by a short reference list.

19.2 Preliminaries

We consider a MIMO GBC with m users. Receivers $i = 1, \ldots, m$ have r_i antennas and the transmitter has t transmit antennas. A time sample at the receiver of user i is given by

$$\mathbf{y}_i = \mathbf{H}_i \mathbf{x} + \mathbf{n}_i, \qquad i = 1, 2, \ldots, m, \tag{19.1}$$

where $\mathbf{n}_i \sim \mathcal{CN}(0, \mathbf{I})$ are additive circularly symmetric complex Gaussian noise vectors with covariance matrices $\mathbf{N}_i = \mathbb{E}[\mathbf{n}\mathbf{n}^H] = \mathbf{I}_{r_i \times r_i}$ and $\mathbf{H}_i \in \mathbb{C}^{r_i \times t}$ are fading matrices that are fixed and perfectly known at both transmitter and receivers. In addition, we assume that there exists a total power constraint, $\mathbb{E}[\mathbf{x}^H \mathbf{x}] \leq P$. The focus here is to characterize the largest possible region of rates (capacity region) at which independent information messages can be sent to the users.

To better understand the difficulties imposed by this channel, let us first briefly overview some basic material on general BCs. A general two-user BC with an input X and outputs Y_1 and Y_2 is defined by the transition probabilities $P_{Y_i|X}$, $i = 1, 2$. Remarkably, there is no known solution for the capacity region of a general BC. The best known achievable rate region (inner bound of the capacity region) is given in Marton (1979). Nevertheless, for some special cases, such as the stochastically degraded BC, there exist single-letter formulas for the capacity region. A two user BC is said to be stochastically degraded (Cover and Thomas, 1991) if there exists a conditional probability function $P'_{Y_1,Y_2|X}$ with marginals $P'_{Y_i|X} = P_{Y_i|X}$, $i = 1, 2$ such that $X \to Y_1 \to Y_2$ forms a Markov chain.

It is easy to show that the single-input ($t = 1$) GBC is stochastically degraded. Therefore, there exists a single-letter formula for its capacity region. Yet the formula requires optimization over the joint distribution ($P_{U,X}$) of an auxiliary variable, U, and the channel's input, X. The optimality of Gaussian inputs was shown by Bergmans (1974) who, in effect, proved that encoding the messages using a superposition of Gaussian codes and applying successive decoding at the receivers is optimal. It is interesting to note that, even though he relied on the degradedness of the BC, Bergmans did not use the single-letter capacity formula in his proof of the converse, but rather the entropy power inequality (EPI).

On the other hand, the MIMO GBC is easily seen to be nondegraded in general. Furthermore, even in the special case where the MIMO GBC (with $t > 1$) is degraded, it was unknown until recently (Weingarten et al., 2004) whether Gaussian coding and successive decoding is optimal. The inability to use successive decoding in nondegraded MIMO GBCs required the introduction of new types of coding techniques that replaced standard superposition coding by random binning (Marton, 1979). The use of random binning is not specific to the BC. Gelfand and Pinsker (1980) used it in an encoding scheme that achieves the capacity of a channel controlled by random states that are noncausally known at the transmitter. In this application of random binning, information messages index subsets (bins) of a codebook. Based on the knowledge of the channel state sequence, the encoder finds in the bin corresponding to the message to be transmitted a codeword that is close, in some sense, to the state sequence. In this way, the effect of the channel state (unknown to the receiver) is mitigated. Costa (1983) applied the result of Gelfand and Pinsker (1980) to the specific model of a Gaussian channel with additive Gaussian interference that is noncausally known at the transmitter, $y_i = x_i + s_i + n_i$. Costa showed that by proper coding at the transmitter, the effect of the Gaussian interference can be canceled out and the capacity is the same as that of a channel without interference s_i and with the same power constraint. This encoding technique was termed dirty paper coding (DPC) and was later extended to vector Gaussian interference channels by Yu *et al.* (2001).

Caire and Shamai (Shitz) (2003) suggested using DPC for encoding in nondegraded MIMO GBCs. Consider a two-user MIMO GBC. We can use Gaussian coding to encode the information for user 2, and then encode for user 1 using DPC by treating the signal to the first user as noncausally known interference. User 2 decodes the Gaussian code and suffers from an additional Gaussian interference due to the signal sent to user 1. User 1 can decode his data without suffering from interference (from user 2's signal) due to dirty-paper successive encoding at the transmitter. Clearly, the roles of the two users can be reversed. Such a successive encoding at the transmitter allows us to encode over nondegraded BCs. By relying on the direct extension of DPC to vector channels (Yu *et al.*, 2001), we can write the following DPC region for the Gaussian MIMO BC in (19.1):

$$\mathcal{R}^{\mathrm{DPC}}(P, \mathbf{H}_{1\ldots m}) =$$

$$\mathbf{conv}\left\{ \bigcup_{\pi \in \Pi} \bigcup_{\mathbf{B}_1,\ldots,\mathbf{B}_m} \begin{pmatrix} R_1^{\mathrm{DPC}}(\pi, \mathbf{B}_{1\ldots m}, \mathbf{H}_{1\ldots m}), \ldots, \\ R_m^{\mathrm{DPC}}(\pi, \mathbf{B}_{1\ldots m}, \mathbf{H}_{1\ldots m}) \end{pmatrix} \right\}, \quad (19.2)$$

where **conv** is the convex hull operator, Π is the set of all permutations on $\{1, \ldots, m\}$, and the permutation $\pi \in \Pi$ corresponds to a reverse encoding order such that user π_{m+1-j} is the jth user in line to be encoded. The union is taken over all possible encoding orders and over all coding power allocation matrices $\mathbf{B}_i \succeq 0$, $\forall i$ that satisfy $\mathrm{tr}\{\sum_{i=1}^{m} \mathbf{B}_i\} \leq P$. R_i^{DPC} are defined such that for all $i = 1, \ldots, m$,

$$R_{\pi_i}^{\mathrm{DPC}}(\pi, \mathbf{B}_{1\ldots m}, \mathbf{H}_{1\ldots m}) = \log \frac{\left| \mathbf{H}_{\pi_i} \left(\sum_{j=1}^{i} \mathbf{B}_{\pi_j} \right) \mathbf{H}_{\pi_i}^{\mathsf{H}} + \mathbf{I}_{r_{\pi_i} \times r_{\pi_i}} \right|}{\left| \mathbf{H}_{\pi_i} \left(\sum_{j=1}^{i-1} \mathbf{B}_{\pi_j} \right) \mathbf{H}_{\pi_i}^{\mathsf{H}} + \mathbf{I}_{r_{\pi_i} \times r_{\pi_i}} \right|}. \quad (19.3)$$

19.3 Pre-equalization approaches

Consider the simpler case of the general MIMO GBC (19.1) where the transmitter has t antennas and the m receivers (users) have one antenna each. This channel, referred to as multiple-input single-output (MISO) GBC, is given by $\mathbf{y} = \mathbf{H}\mathbf{x} + \mathbf{n}$, where $\mathbf{H} \in \mathbb{C}^{m \times t}$ and each element of $\mathbf{y}$ corresponds to a different user.

Joint processing of the user signals at the transmitter is well established in digital communications. Given the analogy of the MISO GBC with an ISI channel with symbol-by-symbol detection (just replace the general matrix $\mathbf{H}$ with a Toeplitz matrix), known pre-equalization techniques can be applied almost verbatim to this case. We shall review linear and nonlinear pre-equalization techniques and point out the fact, discussed in the next sections, that optimal nonlinear pre-equalization used jointly with DPC for canceling known interference achieves the full capacity region.

Perhaps the most intuitive scheme is linear pre-equalization, which forms the transmitted signal as $\mathbf{x} = \mathbf{Gu}$, where $\mathbf{u} \in \mathbb{C}^m$ is a vector of (independent) code symbols. If the spatial signatures of the users are linearly independent, it is possible to create m noninterfering channels by letting $\mathbf{G} = \mathbf{H}^+ = \mathbf{H}^{\mathsf{H}}(\mathbf{HH}^{\mathsf{H}})^{-1}$, the Moore–Penrose pseudoinverse of $\mathbf{H}$. This approach is referred to as *channel inversion*, or *zero-forcing beamforming*. The transmit powers $q_i = \mathbb{E}[|u_i|^2]$ can be allocated according to different criteria, depending on the system performance desired target. In particular, if the objective is to maximize the achievable system throughput (sum rate) and if one assumes Gaussian capacity-achieving codes, the optimal power allocation vector $\mathbf{q}$ is found by waterfilling (Dimic and Sidiropoulos, 2004).

The design of *optimal* linear pre-equalization schemes is greatly simplified by using the fundamental result on uplink-downlink signal-to-interference-plus-noise (SINR) duality (Viswanath and Tse, 2003, and references therein). Namely, the uplink channel dual of the MISO GBC is given by $\mathbf{y}_{\mathrm{ul}} = \mathbf{H}^{\mathsf{H}}\mathbf{x}_{\mathrm{ul}} + \mathbf{n}_{\mathrm{ul}}$, where $\mathbf{n}_{\mathrm{ul}} \sim \mathcal{CN}(\mathbf{0}, \mathbf{I})$ and the inputs $x_{\mathrm{ul},i}$ must be encoded independently, subject to the same total power constraint as the original downlink channel, i.e., $\sum_i p_i \leq P$, where we let $p_i = \mathbb{E}[|x_{\mathrm{ul},i}|^2]$. It can be shown (Viswanath and Tse, 2003) that the set of achievable downlink SINRs under a given pre-equalization matrix $\mathbf{G}$ and power constraint P coincides with the set of achievable uplink SINRs under the given linear detection matrix $\mathbf{G}^{\mathsf{H}}$ and the same total power constraint P.

The joint optimization of the transmit powers $\mathbf{p}$ and of the receiver matrix $\mathbf{G}^{\mathsf{H}}$ for the uplink is well known (Ulukus and Yates, 1998). In fact, for any $\mathbf{p}$, the optimal receiving filters are proportional (via irrelevant nonzero scaling coefficients) to the linear minimum mean squared error (MMSE) estimators of the symbols $x_{\mathrm{ul},i}$ based on the observation $\mathbf{y}_{\mathrm{ul}}$. Then, the component-wise minimal uplink power allocation vector $\mathbf{p}$ can be computed via a standard power control iterative algorithm (Ulukus and Yates, 1998). Thanks to SINR duality, the downlink pre-equalization matrix achieving the same target SINRs is the same $\mathbf{G}$ obtained by uplink optimization, and the downlink powers are immediately obtained as the solution of a simple linear equation (Viswanath and Tse, 2003).

Like the case of ISI channels, better performance can be achieved by non-linear pre-equalization schemes, which can be seen as a generalization of the well-known Tomlinson–Harashima (TH) precoding (Windpassinger *et al.*, 2004). In the context of the MISO GBC, we have the freedom of "ranking" interference between users, i.e., choosing an arbitrary precoding order such that the signal destined to user i is precoded against the interference generated by users ranked before him. Moreover, simple presubtraction of known interference at the transmitter can be replaced by optimal DPC, which can be implemented by a multidimensional modulo-lattice generalization (Zamir *et al.*, 2002) of the one-dimensional modulo-integer operation of TH precoding, or by alternative techniques (Bennatan *et al.*, 2004, 2006, and references therein).

Assume that $\mathbf{H}$ has rank m (otherwise, we select a subset of $K \leq t$ linearly independent users). Then, the most intuitive form of modified TH precoding for the MISO GBC is based on the LQ factorization of the channel matrix $\mathbf{H} = \mathbf{LQ}$, such that $\mathbf{L}$ is lower triangular and $\mathbf{Q}$ is unitary. The transmitted signal is formed as $\mathbf{x} = \mathbf{Q}^H \mathbf{x}'$, where the components of $\mathbf{x}'$ are produced in sequence, from $i = 1$ to $i = m$. The signal seen by user i in the absence of precoding is given by

$$y_i = [\mathbf{L}]_{i,i} x_i' + \sum_{j<i} [\mathbf{L}]_{i,j} x_j' + n_i.$$

When producing the symbol x_i', the term $s_i = \sum_{j<i} [\mathbf{L}]_{i,j} x_j'$ can be treated as known interference, and DPC is applied so that the achievable rate for user i is the same as if s_i were not present. This scheme was proposed by Caire and Shamai (Shitz) (2003) under the name of ranked known interference (RKI). The user powers q_i can be allocated using waterfilling to maximize the sum rate. Results on the achievable throughput of the RKI scheme under DPC are provided in Caire and Shamai (Shitz) (2003), where it is shown that RKI is asymptotically throughput-wise optimal in the limits of small and large SNR.

The RKI pre-equalization strategy is analogous to zero-forcing decision-feedback equalization (ZF-DFE). Its performance can be further improved by replacing ZF-DFE by its minimum mean squared error counterpart, MMSE-DFE. SINR uplink-downlink duality is also established in the case of successive decoding (uplink) and successive dirty-paper encoding (down-link) by Viswanath and Tse (2003). The original modified RKI of Caire and Shamai (Shitz) (2003) can be viewed as a particularization of this for the two-antenna, two-user case. As for the linear case, it turns out that the optimal transmit matrix for a given encoding order can be obtained as the Hermitian transpose of the optimal receiver matrix of the dual uplink for the reverse decoding order. Moreover, a simple recursive (noniterative) algorithm computes the uplink filters explicitly (Schubert and Boche, 2004) in exactly m steps. The optimal downlink powers achieving the same target SINRs are obtained again as the solution of a linear equation. Finally, while in the linear case the power allocation and pre-equalization filter design that maximize the sum rate form an open problem, in the nonlinear case this is a convex optimization (Boyd and Vandenberghe, 2004) and is solved using duality and a simple iterative multiuser waterfilling algorithm (Vishwanath et al., 2004). As illustrated in the following sections, "MMSE-DFE" pre-equalization filter design with optimal DPC and Gaussian codes achieves the full capacity region of the MISO GBC.

19.4 Uplink-downlink duality and the sum capacity

We return to the DPC region expression (19.2) and note that there is no closed-form solution for its boundary points. Furthermore, as the expressions for the rates, $R_i^{\mathrm{DPC}}(\pi, \mathbf{B}_{1...m}, \mathbf{H}_{1...m})$, are neither concave nor convex, numerical analysis of the boundary of this region seems difficult. To overcome this problem we consider a dual uplink channel instead and exploit the duality relation between the uplink and downlink channel (19.1).

We define a dual multiple access channel (MAC) that has m transmitting users, each equipped with r_i, $i = 1, \ldots, m$ antennas, which transmit to a receiver that has t antennas. The fading matrix between each user and the receiver is given by the matrices $\mathbf{H}_i^{\mathsf{H}}$, $i = 1, \ldots, m$. In addition, the receiver is subjected to an additive white Gaussian noise vector, $\mathbf{w}$, with a covariance matrix $\mathbb{E}[\mathbf{w}\mathbf{w}^{\mathsf{H}}] = \mathbf{I}_{t \times t}$. A time sample of this channel is given by

$$\mathbf{y} = \sum_{j=1}^{m} \mathbf{H}_j^{\mathsf{H}} \mathbf{x}_j + \mathbf{w}. \tag{19.4}$$

Let $\mathcal{C}^{\mathrm{MAC}}(\mathbf{P}_{1...m}, \mathbf{H}_{1...m}^{\mathsf{H}})$ denote the capacity region of a MAC under a covariance matrix constraint, $\mathbb{E}[\mathbf{x}_i \mathbf{x}_i^{\mathsf{H}}] \preceq \mathbf{P}_i$, $i = 1, \ldots, m$, on each of the channel inputs. As the Gaussian distribution optimizes the entropy under a covariance matrix constraint, this capacity region is given by the polyhedron created by the intersection of all halfspaces defined by inequalities on the partial rate sums as follows:

$$
\mathcal{C}^{\mathrm{MAC}}(\mathbf{P}_{1\ldots m}, \mathbf{H}_{1\ldots m}^{\mathsf{H}})
$$
$$
= \left\{ \begin{array}{l} (R_1, \ldots, R_m): \\ \displaystyle\sum_{i \in S} R_i \leq \log \left| \mathbf{I} + \sum_{i \in S} \mathbf{H}_i^{\mathsf{H}} \mathbf{P}_i \mathbf{H}_i \right| \ \forall S \subseteq \{1, \ldots, m\} \end{array} \right\}. \tag{19.5}
$$

Each of the $m!$ vertices of the polyhedron is achieved using Gaussian codes and successive decoding. By modifying the decoding order we can choose between vertices. Thus, we may rewrite $\mathcal{C}^{\mathrm{MAC}}$ as follows:

$$
\mathcal{C}^{\mathrm{MAC}}(\mathbf{P}_{1\ldots m}, \mathbf{H}_{1\ldots m}^{\mathsf{H}})
$$
$$
= \mathbf{conv} \left\{ \bigcup_{\pi \in \Pi} \left\{ \begin{array}{l} (R_1, \ldots, R_m): \\ R_i = R_i^{\mathrm{MAC}}(\pi, \mathbf{P}_{1\ldots m}, \mathbf{H}_{1\ldots m}^{\mathsf{H}}) \ \forall i \end{array} \right\} \right\}, \tag{19.6}
$$

where π is the decoding order such that user π_j is the jth user to be decoded and where

$$
R_{\pi_i}^{\mathrm{MAC}}(\pi, \mathbf{P}_{1\ldots m}, \mathbf{H}_{1\ldots m}^{\mathsf{H}}) = \log \frac{\left| \sum_{j=i}^{m} \mathbf{H}_{\pi_j}^{\mathsf{H}} \mathbf{P}_{\pi_j} \mathbf{H}_{\pi_j} + \mathbf{I} \right|}{\left| \sum_{j=i+1}^{m} \mathbf{H}_{\pi_j}^{\mathsf{H}} \mathbf{P}_{\pi_j} \mathbf{H}_{\pi_j} + \mathbf{I} \right|}. \tag{19.7}
$$

We now consider a MAC with a total power constraint $\sum_{i=1}^{m} \mathrm{tr}\{\mathbb{E}[\mathbf{x}_i \mathbf{x}_i^{\mathsf{H}}]\} \leq P$, instead of a per-user power constraint. That is, the constraint is on the sum of the powers consumed by all users. The capacity region of the MAC for this case is given by the following union:

$$
\mathcal{C}^{\mathrm{Union}}(P, \mathbf{H}_{1\ldots m}^{\mathsf{H}}) = \bigcup_{\mathrm{tr}\{\sum_{i=1}^{m} \mathbf{P}_i\} \leq P} \mathcal{C}^{\mathrm{MAC}}(\mathbf{P}_{1\ldots m}, \mathbf{H}_{1\ldots m}^{\mathsf{H}}), \tag{19.8}
$$

where the union is taken over all matrices $\mathbf{P}_i \succeq 0 \ \forall i$ such that $\mathrm{tr}\{\sum_{i=1}^{m} \mathbf{P}_i\} \leq P$. We can now formally state a duality relation between the MAC (19.4) and the BC (19.1).

Theorem 19.1 (Viswanath and Tse, 2003; Vishwanath _et al._, 2003)
The DPC region of the MIMO BC in (19.1) with power constraint P is equal to the capacity region of the dual MIMO MAC in (19.4) with sum power constraint P,

$$
\mathcal{C}^{\mathrm{Union}}(P, \mathbf{H}_{1\ldots m}^{\mathsf{H}}) = \mathcal{R}^{\mathrm{DPC}}(P, \mathbf{H}_{1\ldots m}).
$$

Furthermore, there is sequence of transformations (Vishwanath _et al._, 2003) that allows us to calculate, for each set of matrices $\mathbf{P}_{1\ldots m}$ that achieve a given rate vector in the dual MAC with a given decoding order π, a set of matrices $\mathbf{B}_{1\ldots m}$ that satisfy $\sum \text{tr}\{\mathbf{B}_i\} \leq \sum \text{tr}\{\mathbf{P}_i\}$ and achieve the same rate vector in the BC with the reverse encoding order π, and vice versa. Consequently, $R_i^{\text{MAC}}(\pi, \mathbf{P}_{1\ldots m}, \mathbf{H}_{1\ldots m}^{\mathsf{H}}) = R_i^{\text{DPC}}(\pi, \mathbf{B}_{1\ldots m}, \mathbf{H}_{1\ldots m})$, $\forall i$.

Thus, instead of plotting the DPC rate region using expression (19.2), we can plot the union of regions of the dual MAC using expression (19.8). As this region is convex, its boundary is characterized by the supporting hyperplanes (Boyd and Vandenberghe, 2004) and is found by solving the set of optimization problems $\max_{R_{1\ldots m} \in \mathcal{C}^{\text{Union}}} \sum_{i=1}^{m} \mu_i R_i$ where $\mu_i \geq 0$ $\forall i$. However, it can be shown that for the union MAC these equations are convex and are given by

$$\max_{\mathbf{P}_i} \sum_{i=1}^{m} (\mu_{\pi_i^\mu} - \mu_{\pi_{i-1}^\mu}) \log \left| I + \sum_{j=i}^{m} \mathbf{H}_{\pi_j^\mu} \mathbf{P}_{\pi_j^\mu} \mathbf{H}_{\pi_j^\mu}^{\mathsf{H}} \right|, \qquad (19.9)$$

where the maximization is performed over the matrices $\mathbf{P}_i$ such that $\sum_{i=1}^{m} \text{tr}\{\mathbf{P}_i\} \leq P$ and where π^μ is such that if $\mu_i > \mu_j$ then $\pi_i^\mu > \pi_j^\mu$. Therefore, the boundary of $\mathcal{C}^{\text{Union}}(P, \mathbf{H}_{1\ldots m}^{\mathsf{H}})$ can be numerically calculated using interior-point methods (Boyd and Vandenberghe, 2004).

An additional benefit of the uplink-downlink duality is gained from the fact that the dimensionality of the optimization problem may be significantly reduced. In many practical cases, the number of transmit antennas in a BC is greater than the number of receive antennas of any of the receivers. In such a case, instead of optimizing over m matrices of size $t \times t$ in (19.2), we need to optimize over m matrices of sizes $r_i \times r_i$.

Interestingly, DPC achieves the sum capacity of the channel as stated bellow. Furthermore, as a result of the uplink-downlink duality, there exist efficient methods of calculating this rate (Jindal et al., 2002).

Theorem 19.2 (Caire and Shamai (Shitz), 2003; Viswanath and Tse, 2003; Vishwanath et al., 2003; Yu and Cioffi, 2004) *The sum rate capacity of the MIMO BC is achieved by the DPC strategy,*

$$\mathcal{C}_{\text{sum rate}}^{\text{BC}}(P, \mathbf{H}_{1\ldots m}) = \max_{R_{1\ldots m} \in R^{\text{DPC}}(P, \mathbf{H}_{1\ldots m})} \sum_{i=1}^{m} R_i.$$

Theorem 19.2 was proved under different constraints on the number of receive antennas and number of users and was finally proved in the most general form by Viswanath and Tse (2003), Vishwanath et al. (2003), and Yu and Cioffi (2004). The common theme in these papers is the use of the Sato upper bound (Sato, 1978), as originally suggested by Caire and Shamai (Shitz) (2003), to give an upper limit on the sum capacity and then show that this upper bound is achievable using DPC. Sato used the capacity of the cooperative channel, where the receivers may cooperate to receive a single message, as an upper bound on the sum rate of the BC. However, as there is no cooperation between the users in the BC, the sum rate in the GBC does not depend on the correlation between the noise vectors $\mathbf{n}_i$. Therefore, the sum capacity is bounded by the capacity of the cooperative channel with the "least favorable" noise vector $(\mathbf{n}_1^\mathsf{T}, \ldots, \mathbf{n}_m^\mathsf{T})^\mathsf{T}$ whose marginals $(\mathbf{n}_i)$ have identity (or smaller) covariance matrices.

Viswanath and Tse (2003) and Vishwanath *et al.* (2003) used the uplink-downlink duality and, instead of proving directly that using DPC can achieve the Sato upper bound, it was sufficient to show that the maximum sum rate in the dual MAC channel with the same total power constraint is equivalent to the Sato upper bound. As the sum-capacity problem in the dual MAC is convex, this could be done through the use of convex duality theory with positive semidefinite matrices (Boyd and Vandenberghe, 2004).

Viswanath and Tse (2003) and Vishwanath *et al.* (2003) addressed only the total power constraint, while Yu and Cioffi (2004) showed that there exists a coding scheme for the cooperative channel with the "least favorable" noise that both achieves the Sato upper bound and for which it is also possible to decipher the information at the receivers independently. As this idea does not rely on the uplink-downlink duality, the authors' proof holds for any convex power constraints, $\mathbb{E}[\mathbf{x}\mathbf{x}^\mathsf{H}] \in$ Convex Region, accounting also for a per-antenna power constraint.

19.5 The full rate region

Vishwanath *et al.* (2002) and Tse and Viswanath (2002) introduce the degraded same marginals (DSM) outer bound. Unlike the Sato upper bound, the DSM bound outer-bounds an entire region. Consider a degraded Gaussian MIMO BC that is based upon the original BC (19.1) by allowing each user to access all channel outputs of the following users (for a given ordering of the users). That is, for a given ordering π, the outputs of the new degraded BC $\mathbf{y}_i^{\mathrm{DSM}}$, gains $\mathbf{H}_i^{\mathrm{DSM}}$, and additive noise vectors are given by:

$$
\begin{aligned}
\mathbf{y}_i^{\mathrm{DSM}} &= [\mathbf{y}_{\pi(i)}^\mathsf{H}, \ldots, \mathbf{y}_{\pi(m)}^\mathsf{H}]^\mathsf{H}, \\
\mathbf{H}_i^{\mathrm{DSM}} &= [\mathbf{H}_{\pi(i)}^\mathsf{H}, \ldots, \mathbf{H}_{\pi(m)}^\mathsf{H}]^\mathsf{H}, \\
\mathbf{n}_i^{\mathrm{DSM}} &= [\mathbf{n}_{\pi(i)}^\mathsf{H}, \ldots, \mathbf{n}_{\pi(m)}^\mathsf{H}]^\mathsf{H}.
\end{aligned}
\tag{19.10}
$$

We term this new channel the DSM channel. As the users in the original BC decode the data independently, the correlation between the additive noise vectors of different users, $\mathbb{E}[\mathbf{n}_i \mathbf{n}_j^{\mathsf{H}}]$, has no effect on the capacity region. Therefore, the noise covariances in the DSM channel, $\mathbf{N}_i^{\mathrm{DSM}} = \mathbb{E}[\mathbf{n}_i^{\mathrm{DSM}} \mathbf{n}_i^{\mathrm{DSM}\,\mathsf{H}}]$, may be chosen arbitrarily, as long as the noise marginals remain identical to those of the original BC (i.e., $\mathbf{n}_i$). Furthermore, note that $\mathbf{N}_i^{\mathrm{DSM}}$ for $i \geq 2$ are submatrices of $\mathbf{N}_1^{\mathrm{DSM}}$ and, hence, in effect, only $\mathbf{N}_1^{\mathrm{DSM}}$ is chosen and the rest are set accordingly.

We can therefore outer-bound the capacity region of the original BC by the intersection of the capacity regions of DSM channels with different choices of $\mathbf{N}_i^{\mathrm{DSM}}$ and different user orderings. To formalize this result, denote by $\mathcal{C}_{\mathrm{DSM}}^{\mathrm{BC}}(P, \pi, \mathbf{H}_{1\ldots m}^{\mathrm{DSM}}, \mathbf{N}_{1\ldots m}^{\mathrm{DSM}})$ the capacity region of the DSM channel. Then,

$$\mathcal{C}^{\mathrm{BC}}(P, \mathbf{H}_{1\ldots m}) \subseteq \bigcap_{\pi \in \Pi} \bigcap_{\mathbf{N}_{1\ldots m}^{\mathrm{DSM}}} \mathcal{C}_{\mathrm{DSM}}^{\mathrm{BC}}(P, \pi, \mathbf{H}_{1\ldots m}^{\mathrm{DSM}}, \mathbf{N}_{1\ldots m}^{\mathrm{DSM}}), \qquad (19.11)$$

where the intersection is taken over all possible user orderings and all noise covariances $\mathbf{N}_{1\ldots m}^{\mathrm{DSM}}$ with the same marginals as in the original BC (i.e., choose $\mathbf{N}_1^{\mathrm{DSM}}$ such that $\mathbf{N}_{\pi_j}$, $j = 1, \ldots, m$, appear on the diagonal and $\mathbf{N}_i^{\mathrm{DSM}}$ for $i \geq 2$ are set as submatrices of $\mathbf{N}_1^{\mathrm{DSM}}$).

Note that unlike the original BC, the DSM channel is a degraded BC. Based on the scalar Gaussian BC, it was conjectured that Gaussian coding is optimal for the degraded MIMO BC as well. We shall denote the achievable rate region due to Gaussian coding in the DSM channel by $\mathcal{R}_{\mathrm{DSM_G}}^{\mathrm{BC}}(P, \pi, \mathbf{H}_{1\ldots m}^{\mathrm{DSM}}, \mathbf{N}_{1\ldots m}^{\mathrm{DSM}})$. The main result of Vishwanath et al. (2002) and Tse and Viswanath (2002) is stated in the following theorem.

Theorem 19.3 (Vishwanath et al., 2002, Tse and Viswanath, 2002)
If Gaussian coding is optimal for the DSM channel, then the capacity region of the nondegraded MIMO BC (19.1) under a power constraint P, is equal to the DPC region,

$$\mathcal{C}^{\mathrm{BC}}(P, \mathbf{H}_{1\ldots m}) = \mathcal{R}^{\mathrm{DPC}}(P, \mathbf{H}_{1\ldots m}) = \bigcap_{\pi \in \Pi} \bigcap_{\mathbf{N}_{1\ldots m}^{\mathrm{DSM}}} \mathcal{R}_{\mathrm{DSM_G}}^{\mathrm{BC}}(P, \pi, \mathbf{H}_{1\ldots m}^{\mathrm{DSM}}, \mathbf{N}_{1\ldots m}^{\mathrm{DSM}}),$$

where the intersection is calculated over the same set as in (19.11).

Therefore, in view of this theorem, it only remains to show that Gaussian coding is optimal for the DSM channel. This was finally proved by Weingarten et al. (2004) and thus the authors closed the gap left by Theorem 19.3 and proved that DPC is indeed optimal. The degraded MIMO BC discussed by Weingarten et al. (2004) is a generalization of the DSM channels discussed above and is given by the following recursive expression:

$$\mathbf{y}_1 = \mathbf{H}\mathbf{x} + \tilde{\mathbf{n}}_1 \quad \text{and} \quad \mathbf{y}_i = \mathbf{y}_{i-1}^{1\ldots r_i} + \tilde{\mathbf{n}}_i, \quad i = 2, \ldots, m, \qquad (19.12)$$

where $\tilde{\mathbf{n}}_i \sim \mathcal{CN}(0, \tilde{\mathbf{N}}_i)$ are additive Gaussian noise vectors of size $r_i \times 1$. The notation $\mathbf{y}_{i-1}^{1\cdots r_i}$ stands for the first r_i elements of the vector $\mathbf{y}_{i-1}$. The number of receive antennas of user i is r_i, and we assume that $r_1 \geq r_2 \geq \cdots \geq r_m$.

As this channel is degraded, user 1 can decode the messages intended for all other users and cancel them out before decoding his own message. Therefore, if a superposition of Gaussian codes is used with covariances $\mathbf{B}_i$, $i = 1, \ldots, m$, the rate achieved by user i is given by

$$R_i^{\mathrm{G}}(\mathbf{B}_{1\ldots m}, \mathbf{H}, \tilde{\mathbf{N}}_{1\ldots m}) = \log \frac{\left| \sum_{j=1}^{i} \tilde{\mathbf{N}}_i^{1\cdots r_i} + \sum_{j=1}^{i} \mathbf{H}^{1\cdots r_i} \mathbf{B}_j \mathbf{H}^{1\cdots r_i \mathsf{H}} \right|}{\left| \sum_{j=1}^{i} \tilde{\mathbf{N}}_i^{1\cdots r_i} + \sum_{j=1}^{i-1} \mathbf{H}^{1\cdots r_i} \mathbf{B}_j \mathbf{H}^{1\cdots r_i \mathsf{H}} \right|},$$

(19.13)

where the matrix $\mathbf{H}^{1\cdots r_i}$ contains the first r_i rows of $\mathbf{H}$ and $\tilde{\mathbf{N}}_i^{1\cdots r_i}$ contains the first r_i rows and columns of $\tilde{\mathbf{N}}_i$. The capacity region of (19.12) follows.

Theorem 19.4 (Weingarten *et al.*, 2004)   *The capacity region of the degraded Gaussian MIMO BC (19.12) under a total power constraint, P, is obtained by a superposition of Gaussian codes at the encoder and successive decoding at the receivers such that*

$$\mathcal{C}^{\mathrm{Deg}}(P, \mathbf{H}, \tilde{\mathbf{N}}_{1\ldots m}) = \bigcup_{\mathbf{B}_{1\ldots m}} \{ (R_1, \ldots, R_m) | \ R_i \leq R_i^{\mathrm{G}}(\mathbf{B}_{1\ldots m}, \mathbf{H}, \tilde{\mathbf{N}}_{1\ldots m}) \ \forall i \},$$

where the union is taken over all matrices $\mathbf{B}_i \succeq 0$, $\forall i$ such that $\mathrm{tr}\{\sum_{i=1}^{m} \mathbf{B}_i\} \leq P$.

To prove Theorem 19.4, we consider a degenerate case of the channel in (19.12) that is *aligned* such that $r_1 = r_2 = \cdots = r_m$ and $\mathbf{H} = \mathbf{I}$. In addition, only real vectors are considered. Hence, the channel is given by

$$\mathbf{y}_i = \mathbf{x} + \mathbf{w}_i, \qquad \forall i, \tag{19.14}$$

where $\mathbf{w}_i \sim \mathcal{N}(0, \mathbf{W}_i)$ is an additive real Gaussian noise vector such that $\mathbf{W}_1 \preceq \mathbf{W}_2 \preceq \cdots \mathbf{W}_m$. Furthermore, instead of considering a total power constraint, we will consider an input covariance constraint, $\mathbb{E}[\mathbf{x}\mathbf{x}^{\mathsf{H}}] \preceq \mathbf{S}$. It is shown by Weingarten *et al.* (2006) that if Gaussian coding is optimal for the GBC in (19.14) under an input covariance constraint, then it is optimal for the GBC in (19.12) under a total power constraint.

Bergmans (1974) proved that Gaussian coding is optimal for the scalar GBC. Let $\mathcal{R}^{\mathrm{G}}(\mathbf{S}, \mathbf{W}_{1\ldots m})$ denote the achievable rate region due to Gaussian coding and let $\overline{R}^o = (R_1^o, \ldots, R_m^o) \in \mathcal{R}^{\mathrm{G}}(\mathbf{S}, \mathbf{W}_{1\ldots m})$ be a rate vector that lies on the boundary of $\mathcal{R}^{\mathrm{G}}(\mathbf{S}, \mathbf{W}_{1\ldots m})$. Bergmans used the entropy power inequality (EPI) to show that a rate vector $\overline{R}$ is not achievable if it is element-wise larger than $\overline{R}^o$ $(\overline{R} > \overline{R}^o)$ such that $R_i \geq R_i^o, \forall i$, with strict inequality for some $i \in \{1, \ldots, m\}$. As all rate vectors $\overline{R} \notin \mathcal{R}^G(\mathbf{S}, \mathbf{W}_{1\ldots m})$ are element-wise larger than some boundary point of $\mathcal{R}^G(\mathbf{S}, \mathbf{W}_{1\ldots m})$, it is clear that $\mathcal{R}^G(\mathbf{S}, \mathbf{W}_{1\ldots m})$ must be the capacity region.

Yet this proof does not directly extend to the vector case of expression (19.14). Let $\mathbf{B}_{1\ldots m}^o$ achieve a boundary point of $\mathcal{R}^G(\mathbf{S}, \mathbf{W}_{1\ldots m})$ such that $R_i^o = R_i^G(\mathbf{B}_{1\ldots m}, \mathbf{H}, \tilde{\mathbf{N}}_{1\ldots m}), \forall i$, where $\mathbf{H} = \mathbf{I}$ and $\tilde{\mathbf{N}}_i = \mathbf{W}_i - \mathbf{W}_{i-1}$. An examination of Bergmans' proof reveals that the use of the EPI requires that Minkowski's inequality is satisfied with equality, namely, that the matrix $\left(\mathbf{W}_k + \sum_{i=1}^{k} \mathbf{B}_j\right)$ be proportional to $\mathbf{W}_{k+1} - \mathbf{W}_k$ for all k. In general, this condition does not hold for the vector GBC. To circumvent this problem, we introduce a new notion of an *enhanced channel*. Consider two aligned and degraded GBCs as given in (19.14) with noise covariance matrices $\mathbf{W}_{1\ldots m}'$ and $\mathbf{W}_{1\ldots m}$. We say that one is an enhanced version of the other if $\mathbf{W}_i' \preceq \mathbf{W}_i \ \forall i$. Clearly, the capacity region of an enhanced channel contains that of the other channel. The following theorem shows that we can find unique enhanced channels that allow us to use Bergmans' proof on an enhanced channel.

Lemma 19.1 (Weingarten et al., 2004) *For a set of Gaussian coding covariance matrices $\mathbf{B}_{1\ldots m}^o$ that achieve a boundary point on the Gaussian region of (19.14), there exists an enhanced channel with noise covariances $\mathbf{W}_{1\ldots m}'$ such that the following properties hold:*

(i) *Proportionality: there exist $\alpha_k \geq 0$, $k = 1, \ldots, m-1$, such that*

$$\alpha_k \left(\mathbf{W}_k' + \sum_{i=1}^{k} \mathbf{B}_i^o\right) = (\mathbf{W}_{k+1}' - \mathbf{W}_k'), \quad \forall k = 1, \ldots, m-1. \quad (19.15)$$

(ii) *Rate preservation and optimality preservation: the Gaussian coding covariance matrices $\mathbf{B}_1, \ldots, \mathbf{B}_m$ achieve the exact same rates in the enhanced channel as in the original channel. Furthermore, these rates lie on the boundary of the Gaussian rate region of the enhanced chan-nel as well as that of the original channel.*

Fig. 19.1(a) demonstrates this result and shows the Gaussian region of the channel in expression (19.14) and two Gaussian regions of enhanced channels that correspond to two different points on the boundary of the first region.

Therefore, we conclude that for every rate vector $\overline{R}^o$ that lies on the bound-ary of the Gaussian region of the BC in (19.14), we can find an enhanced channel, a different one for every boundary point, such that $\overline{R}^o$ also lies on

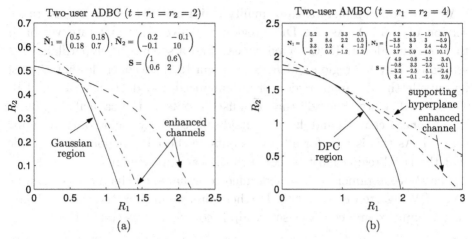

Fig. 19.1. (a) Capacity region of an aligned and degraded MIMO GBC (ADBC). (b) Capacity region of an aligned and nondegraded MIMO GBC (AMBC).

the boundary of its Gaussian region. Yet, by Lemma 19.1, the proportionality condition holds for the enhanced channel (19.15); therefore, we can use Bergmans' proof to show that all rate vectors $\overline{R} > \overline{R}^o$ are not achievable in the enhanced channel and, hence, also not in our original channel.

As the proof of Theorem 19.3 relies on uplink-downlink duality (Theorem 19.1), it only holds under a total power constraint. To extend the DPC optimality claim to a more general constraint settings (e.g., per antenna power constraint), we present in the following lemma an extension of Lemma 19.1 to the case where the MIMO GBC is not degraded. This will allow us to state results regarding the capacity region of the nondegraded MIMO BC (19.1) without using the DSM bound and the uplink-downlink duality. We still focus on the aligned channel as in (19.14), however, we no longer require that this channel be degraded.

Lemma 19.2 (Weingarten *et al.*, 2006) *Consider the DPC region of an aligned and not necessarily degraded MIMO GBC under a covariance input constraint. In addition, let $\{(R_1, \ldots, R_m) \mid \sum_{i=1}^m \gamma_i R_i = b\}$ be a supporting hyperplane of this region ($\gamma_i \geq 0$, $\forall i$). Then, there exists an enhanced, aligned, and degraded MIMO GBC with noise covariances $\mathbf{W}'_i \preceq \mathbf{W}_i, \forall i$ whose Gaussian rate region is supported by the same hyperplane.*

Fig. 19.1(b) illustrates the above lemma and shows a DPC region of an aligned and nondegraded BC, a supporting hyperplane, and a Gaussian region of an enhanced, aligned, and degraded MIMO GBC that is supported by the same hyperplane and lies between the two curves.

We can now complete the optimality claim of DPC for the nondegraded version of (19.14). As the DPC region is convex, for any rate vector $\overline{R}$ that lies outside the DPC region we can find a separating hyperplane that separates the DPC region and $\overline{R}$. By Lemma 19.2, we conclude that $\overline{R}$ must lie outside the Gaussian region of an enhanced and degraded BC, and by Theorem 19.4, we conclude that $\overline{R}$ must lie outside the capacity region of the enhanced channel and, hence, outside the capacity region of the original channel. As this is true for all points outside the DPC region, we conclude that the DPC region must be equal to the capacity region.

The above argument can be extended to nonaligned channels as given by (19.1) (Weingarten *et al.*, 2006). Furthermore, we may extend the constraint on the input to any compact set of input covariances as stated below.

Theorem 19.5 (Weingarten *et al.*, 2006)   *Let S be a compact set of positive semidefinite matrices and consider the MIMO BC in (19.1) with an input constraint $\mathbb{E}[\mathbf{x}\mathbf{x}^{\mathsf{H}}] \preceq \mathbf{S}$ for some $\mathbf{S} \in S$. Then, the capacity region of this channel is given by*

$$\mathcal{C}^{\mathrm{BC}}(S, \mathbf{H}_{1\dots m}) = \bigcup_{\mathbf{S} \in S} \mathcal{R}^{\mathrm{DPC}}(\mathbf{S}, \mathbf{H}_{1\dots m}).$$

19.6 Comparisons

In this section, we compare the achievable rate region of the MISO GBC under different strategies in the simple case of two users. Albeit a toy example, this case is nevertheless useful as an exercise to clarify ideas.

Assume that the rows of $\mathbf{H} \in \mathbb{C}^{2\times 2}$ ($\mathbf{h}_1$ and $\mathbf{h}_2$) are linearly independent.[†] With channel inversion, the transmit matrix is given by $\mathbf{G} = \mathbf{H}^{-1}$. The achievable rate region is the closure of the convex hull (justified by a standard time-sharing argument) of the points (R_1, R_2) such that

$$R_1 \leq \log\left(1 + \frac{|\det(\mathbf{H})|^2}{|\mathbf{h}_2|^2} q_1\right), \quad R_2 \leq \log\left(1 + \frac{|\det(\mathbf{H})|^2}{|\mathbf{h}_1|^2} q_2\right) \quad (19.16)$$

for all $q_1 + q_2 \leq P$.

In the case of linear MMSE pre-equalization, it is simpler to work with the dual uplink channel. The achievable rate region is the closure of the convex hull of the points (R_1, R_2) such that

$$R_1 \leq \log(1 + \gamma_1), \quad R_2 \leq \log(1 + \gamma_2), \quad (19.17)$$

[†]If the two rows are linearly dependent, then the channel reduces trivially to a single-input degraded GBC, for which the capacity region is well known and easily computed.

where γ_1 and γ_2 are the SINRs of the dual uplink channel under linear MMSE detection, with individual user powers p_1, p_2 such that $p_1 + p_2 \leq P$. Explicitly, we have

$$\gamma_1 = \frac{p_1|\mathbf{h}_1|^2 + p_1 p_2 |\det \mathbf{H}|^2}{1 + p_2 |\mathbf{h}_2|^2} \tag{19.18}$$

(for γ_2 just exchange indices 1 and 2).

Next, we consider LQ decomposition and successive DPC encoding (RKI). In this case, the achievable rate region is the closure of the convex hull of the rate points

$$R_1 \leq \log\left(1 + |\mathbf{h}_1|^2 q_1\right), \qquad R_2 \leq \log\left(1 + \frac{|\det \mathbf{H}|^2}{|\mathbf{h}_1|^2} q_2\right),$$

$$R_1 \leq \log\left(1 + \frac{|\det \mathbf{H}|^2}{|\mathbf{h}_2|^2} q_1\right), \quad R_2 \leq \log\left(1 + |\mathbf{h}_2|^2 q_2\right) \tag{19.19}$$

for all $q_1 + q_2 \leq P$.

By comparing (19.19) and (19.16), we notice that for any given q_1, q_2, the rate point achieved by successive encoding dominates component-wise the rates achieved by channel inversion. In fact, by the Hadamard inequality, $|\mathbf{h}_1|^2 |\mathbf{h}_2|^2 \geq |\det \mathbf{H}|^2$ with equality if and only if $\mathbf{h}_1$ and $\mathbf{h}_2$ are mutually orthogonal.

From the previous sections, we know that the full capacity region is achieved by dirty-paper successive encoding jointly with the optimal MMSE-DFE beamforming matrix. Again, in this case it is easier to work with the dual uplink channel, considering successive decoding order 1,2 and 2,1, and MMSE-DFE detection. The capacity region is given by the closure of the convex hull of the points (19.17), where now the uplink SINRs are given by

$$\gamma_1 = \frac{p_1|\mathbf{h}_1|^2 + p_1 p_2 |\det \mathbf{H}|^2}{1 + p_2 |\mathbf{h}_2|^2}, \quad \gamma_2 = |\mathbf{h}_2|^2 p_2 \tag{19.20}$$

for decoding order 1,2, and by

$$\gamma_1 = |\mathbf{h}_1|^2 p_1, \quad \gamma_2 = \frac{p_2|\mathbf{h}_2|^2 + p_1 p_2 |\det \mathbf{H}|^2}{1 + p_1 |\mathbf{h}_1|^2} \tag{19.21}$$

for decoding order 2,1. In general, the linear MMSE achievable region (optimal linear beamforming region) touches the capacity region only in the trivial points where $p_1 = P$ or $p_2 = P$.

Assuming without loss of generality that $|\mathbf{h}_1| \geq |\mathbf{h}_2|$, the maximum sum rate (throughput) can be easily obtained from expressions (19.21), and it is

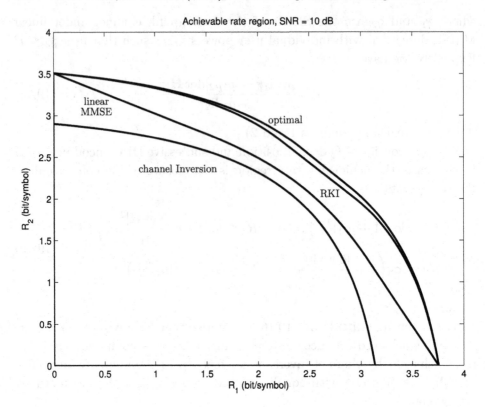

Fig. 19.2. Achievable rate regions under different coding strategies.

explicitly given by Caire and Shamai (Shitz) (2003) as

$$R = \begin{cases} \log(1 + |\mathbf{h}_1|^2 P), & P \leq P^* \\ \log \dfrac{(P|\det \mathbf{H}|^2 + |\mathbf{h}_1|^2 + |\mathbf{h}_2|^2)^2 - 4|\mathbf{h}_2 \mathbf{h}_1^{\mathsf{H}}|^2}{4|\det \mathbf{H}|^2}, & P > P^*, \end{cases} \quad (19.22)$$

where $P^* = (|\mathbf{h}_1|^2 - |\mathbf{h}_2|^2)/|\det \mathbf{H}|^2$. For $P \leq P^*$, beamforming to the best user is, throughput-wise, optimal. Therefore, as expected, the multiplexing gain of multiple antennas in the downlink manifests itself only for sufficiently large power. Generally speaking, in all cases considered here and for $\mathbf{H}$ of rank m, the multiplexing gain is m for all the mentioned strategies.

Fig. 19.2 shows the achievable rate regions in all cases discussed above for the channel matrix

$$\mathbf{H} = \begin{bmatrix} 1.0 & 0.5 \\ 0.2 & 1.0 \end{bmatrix}$$

and for $P = 10$.

19.7 Conclusions

In this succinct overview of the information-theoretic aspects of the MIMO GBC, we highlighted the central and essentially recent impact of information-theoretic techniques to the end of determining the full capacity region of this nondegraded broadcast channel model. While pointing out the basic notions that led to the progress and satisfactory solution of this problem—primarily, Marton region; Sato bounds; dirty paper coding; duality principles; degraded same marginal bounds; and the enhanced channel concept—we were able, within the space limitations, to provide very little on the wide-scope perspective of these results. Neither have we discussed a variety of applications and implications at the system and network levels such as scheduling, opportunistic approaches, feedback information, cooperation, scaling laws, queuing aspects. These and many other information-theoretic, systems, and network issues are discussed in a wide-scope tutorial (Caire *et al.*, 2006), which also provides an extensive reference list.

References

Bennatan, A., Burstein, D., Caire, G., and Shamai (Shitz), S. (2004). Superposition coding for side information channels. In *Proc. Int. Symp. Inf. Theory and its Applications.*

—— (2006). Superposition coding for side information channels. *IEEE Trans. Inf. Theory.* In press.

Bergmans, P. (1974). A simple converse for broadcast channels with additive white Gaussian noise. *IEEE Trans. Inf. Theory*, **20** (2), 279–280.

Boyd, S. and Vandenberghe, L. (2004). *Convex Optimization* (Cambridge Univ. Press, Cambridge, U.K.).

Caire, G. and Shamai (Shitz), S. (2001). On the multiple antenna broadcast channel (*invited paper*). In *Proc. 35th Asilomar Conf. Signals, Syst., Comput.*, vol. 2, 1188–1193.

—— (2003). On the achievable throughput of a multiantenna Gaussian broadcast channel. *IEEE Trans. Inf. Theory*, **49** (7), 1691–1706.

Caire, G., Shamai (Shitz), S., Steinberg, Y., and Weingarten, H. (2006). MIMO broadcast channels: Theoretical and system aspects. *to be submitted to Foundations and Trends in Commun. and Inf. Theory (FnT).*

Costa, M. H. M. (1983). Writing on dirty paper. *IEEE Trans. Inf. Theory*, **29** (3), 439–441.

Cover, T. M. (1972). Broadcast channels. *IEEE Trans. Inf. Theory*, **18** (1), 2–14.

Cover, T. M. and Thomas, J. A. (1991). *Elements of Information Theory* (John Wiley & Sons Ltd., New York).

Dimic, G. and Sidiropoulos, N. (2004). Low-complexity downlink beamforming for maximum sum capacity. In *Proc. IEEE Int. Conf. Acoust., Speech, Signal Process.*, vol. 4, 701–704.

Gelfand, S. and Pinsker, M. (1980). Coding for channel with random parameters. *Probl. Contr. and Inf. Theory*, **9** (1), 19–31.

Goldsmith, A., Jafar, S. A., Jindal, N., and Vishwanath, S. (2003). Capacity limits of MIMO channels. *IEEE J. Sel. Areas Commun.*, **21** (6), 684–702.

Jindal, N., Jafar, S. A., Vishwanath, S., and Goldsmith, A. (2002). Sum power iterative water-filling for multi-antenna Gaussian broadcast channels. In *36th Asilomar Conf. Signals, Syst., Comput.*, vol. 2, 1518–1522.

Jindal, N., Vishwanath, S., and Goldsmith, A. (2004). On the duality of Gaussian multiple-access and broadcast channels. *IEEE Trans. Inf. Theory*, **50** (5), 768–783.

Marton, K. (1979). A coding theorem for the discrete memoryless broadcast channel. *IEEE Trans. Inf. Theory*, **25** (3), 306–311.

Sato, H. (1978). An outer bound on the capacity region of broadcast channels. *IEEE Trans. Inf. Theory*, **24** (3), 374–377.

Schubert, M. and Boche, H. (2002). Joint 'dirty paper' pre-coding and downlink beamforming. In *Proc. IEEE Int. Symp. Spread Spectrum Tech. and their Applications*, vol. 2, 536–540.

—— (2004). Solution of the multiuser downlink beamforming problem with individual SINR constraints. *IEEE Trans. Veh. Technol.*, **53** (1), 18–28.

Tse, D. and Viswanath, P. (2002). On the capacity of the multiple antenna broadcast channel. In Foschini, G. J. and Verdú, S. (eds.), *DIMACS Series in Discrete Math. and Theoretical Comput. Science*, vol. 62, 67–105.

—— (2005). *Fundamentals of Wireless Communication* (Cambridge Univ. Press, Cambridge, U.K.). URL http://degas.eecs.berkeleyedu/ ~dtse/book html.

Ulukus, S. and Yates, R. (1998). Adaptive power control and MMSE interference suppression. *Baltzer/ACM Wireless Netw.*, **4** (6), 489–496.

Vishwanath, S., Jindal, N., and Goldsmith, A. (2003). Duality, achievable rates, and sum-rate capacity of Gaussian MIMO broadcast channels. *IEEE Trans. Inf. Theory*, **49** (10), 2658–2668.

Vishwanath, S., Kramer, G., Shamai (Shitz), S., Jafar, S., and Goldsmith, A. (2002). Capacity bounds for Gaussian vector broadcast channels. In Foschini, G. J. and Verdú, S. (eds.), *DIMACS Series in Discrete Math. and Theoretical Comput. Science*, vol. 62, 107–122.

Vishwanath, S., Rhee, W., Jindal, N., Jafar, S., and Goldsmith, A. (2004). Sum power iterative waterfilling for Gaussian vector broadcast channels. In *Proc. IEEE Int. Symp. Inf. Theory*, 467.

Viswanath, P. and Tse, D. (2003). Sum capacity of the vector Gaussian channel and uplink-downlink duality. *IEEE Trans. Inf. Theory*, **49** (8), 1912–1921.

Weingarten, H., Steinberg, Y., and Shamai (Shitz), S. (2004). The capacity region of the Gaussian MIMO broadcast channel. In *Proc. Conf. Inf. Sciences and Syst.*, 17–19.

—— (2006). The capacity region of the Gaussian MIMO broadcast channel. *IEEE Trans. Inf. Theory*. In press.

Wiesel, A., Eldar, Y. C., and Shamai (Shitz), S. (2006). Linear precoding via conic optimization for fixed MIMO receivers. *IEEE Trans. Signal Process.*, **54** (1), 161–176.

Windpassinger, C., Fischer, R. F. H., and Huber, J. B. (2004). Lattice-reduction-aided broadcast precoding. *IEEE Trans. Commun.*, **52** (12), 2057–2060.

Yu, W. (2006). Uplink-downlink duality via minimax duality. *IEEE Trans. Inf. Theory*, **52** (2), 361–374.

Yu, W. and Cioffi, J. (2004). The sum capacity of a Gaussian vector broadcast channel. *IEEE Trans. Inf. Theory*, **50** (9), 1875–1892.

Yu, W. and Lan, T. (2004). Minimax duality of Gaussian vector broadcast channels. In *Proc. IEEE Int. Symp. Inf. Theory*, 177.

Yu, W., Sutivong, A., Julian, D., Cover, T. M., and Chiang, M. (2001). Writing on colored paper. In *Proc. IEEE Int. Symp. Inf. Theory*, 302.

Zamir, R., Shamai (Shitz), S., and Erez, U. (2002). Nested linear/lattice codes for structured multiterminal binning. *IEEE Trans. Inf. Theory*, **48** (6), 1250–1276.

20

Multiuser MIMO systems

H. Vincent Poor
Princeton University

Daryl Reynolds
West Virginia University

Xiaodong Wang
Columbia University

20.1 Receiver design for multiuser MIMO systems

20.1.1 Introduction

The seemingly insatiable demand for performance and capacity in cellular wireless systems has prompted the development of myriad receiver-based signal processing techniques for using system resources more efficiently. One of the most powerful coding and signal processing paradigms for this purpose is space-time processing with multiple transmit and/or receive antennas. As discussed in previous chapters, information theoretic-results show that multiple antennas can increase rate (multiplexing gain), reliability (diversity gain), or both. In this section, we propose specific coding and signal processing schemes for blind adaptive space-time processing of MIMO multiuser systems. We make use of the Alamouti space-time block code (STBC) (Alamouti, 1998; Tarokh *et al.*, 1999) for uplink transmission, which has been adopted in a number of 3G WCDMA standards (Lucent Technologies, 1999), and blind adaptive MMSE joint multiuser detection (Reynolds and Wang, 2001; Wang and Poor, 2004; Verdú, 1998; Honig *et al.*, 1995), and space-time decoding for base station reception. Related work includes the results by Hochwald *et al.* (2001), Papadias and Huang (2001).

Some of the results in this section have been published previously in Reynolds *et al.* (2002b).

20.1.2 Blind space-time multiuser MIMO reception

We consider a K-user code division multiple access (CDMA) wireless cellular system with processing gain N operating in flat block fading[†] with M_B

[†]Extensions to frequency-selective intersymbol interference channels appear in Reynolds *et al.* (2002b).

base station antennas and M_U antennas at each mobile unit. For simplicity of exposition, we will consider only $M_U = M_B = 2$ and BPSK modulation in this section. Extensions to other antenna configurations and modulation techniques are straightforward. The Alamouti STBC specifies full-rate, two-antenna transmission over two-symbol blocks. Let $b_{1,k}[i], b_{2,k}[i]$ denote the BPSK symbols transmitted from user k during the ith block. During the first symbol interval of block i, user k transmits $(b_{1,k}[i], b_{2,k}[i])$ from the two transmit antennas. During the second symbol interval, user k transmits $(-b_{2,k}[i], b_{1,k}[i])$. The discrete-time received N-vector at base station antenna 1 during the two symbol periods for block i is

$$r_1^{(1)}[i] = \sum_{k=1}^{K} \left(h_k^{(1,1)} b_{1,k}[i] s_{1,k} + h_k^{(2,1)} b_{2,k}[i] s_{2,k} \right) + n_1^{(1)}[i], \qquad (20.1)$$

$$r_2^{(1)}[i] = \sum_{k=1}^{K} \left(-h_k^{(1,1)} b_{2,k}[i] s_{2,k} + h_k^{(2,1)} b_{1,k}[i] s_{1,k} \right) + n_2^{(1)}[i], \qquad (20.2)$$

and the corresponding signals received at antenna 2 are

$$r_1^{(2)}[i] = \sum_{k=1}^{K} \left(h_k^{(1,2)} b_{1,k}[i] s_{1,k} + h_k^{(2,2)} b_{2,k}[i] s_{2,k} \right) + n_1^{(2)}[i], \qquad (20.3)$$

$$r_2^{(2)}[i] = \sum_{k=1}^{K} \left(-h_k^{(1,2)} b_{2,k}[i] s_{2,k} + h_k^{(2,2)} b_{1,k}[i] s_{1,k} \right) + n_2^{(2)}[i], \qquad (20.4)$$

where $h_k^{(a,b)}$ denotes the complex Gaussian channel gain between transmit antenna a of user k and receive antenna b, $n_a^{(b)}[i]$, $a, b \in \{1, 2\}$, are independent and identically distributed complex Gaussian noise vectors, and $s_{1,k}, s_{2,k}$ are the two spreading codes assigned to user k to facilitate blind channel identification. We stack the received signal vectors and denote

$$\tilde{r}[i] \triangleq \begin{bmatrix} r_1^{(1)}[i] \\ \left(r_2^{(1)}[i] \right)^* \\ r_1^{(2)}[i] \\ \left(r_2^{(2)}[i] \right)^* \end{bmatrix}, \qquad \tilde{n}[i] \triangleq \begin{bmatrix} n_1^{(1)}[i] \\ \left(n_2^{(1)}[i] \right)^* \\ n_1^{(2)}[i] \\ \left(n_2^{(2)}[i] \right)^* \end{bmatrix},$$

$$h_k \triangleq \begin{bmatrix} h_k^{(1,1)} \\ \left(h_k^{(2,1)} \right)^* \\ h_k^{(1,2)} \\ \left(h_k^{(2,2)} \right)^* \end{bmatrix}, \qquad \bar{h}_k \triangleq \begin{bmatrix} h_k^{(2,1)} \\ \left(-h_k^{(1,1)} \right)^* \\ h_k^{(2,2)} \\ \left(-h_k^{(1,2)} \right)^* \end{bmatrix}.$$

Then we have

$$\tilde{r}[i] = \sum_{k=1}^{K} \left(b_{1,k}[i] h_k \otimes s_{1,k} + b_{2,k}[i] \bar{h}_k \otimes s_{2,k} \right) + \tilde{n}[i]$$

$$= \tilde{S} b[i] + \tilde{n}[i], \tag{20.5}$$

where

$$\tilde{S} \triangleq \left[h_1 \otimes s_{1,1} \quad \bar{h}_1 \otimes s_{2,1} \quad \cdots \quad h_K \otimes s_{1,K} \quad \bar{h}_K \otimes s_{2,K} \right]_{4N \times 2K},$$

$$b[i] \triangleq \left[b_{1,1}[i] \quad b_{2,1}[i] \quad b_{1,2}[i] \quad b_{2,2}[i] \quad \cdots \quad b_{1,K}[i] \quad b_{2,K}[i] \right]^T_{2K \times 1},$$

and where $\otimes$ denotes the Kronecker product. The autocorrelation matrix of the stacked signal $\tilde{r}[i]$, C, and its eigendecomposition are given by

$$C = \mathsf{E}\left\{ \tilde{r}[i] \tilde{r}[i]^H \right\} = \tilde{S}\tilde{S}^H + \sigma^2 I_{4N}$$

$$= U_s \Lambda_s U_s^H + \sigma^2 U_n U_n^H,$$

where $\Lambda_s = \mathrm{diag}\{\lambda_1, \lambda_2, \ldots, \lambda_{2K}\}$ contains the largest $2K$ eigenvalues of C, the columns of U_s are the corresponding eigenvectors; and the columns of U_n are the $4N - 2K$ eigenvectors corresponding to the smallest eigenvalue σ^2.

The blind linear space-time MMSE filter for joint suppression of multiple access interference (MAI) and space-time decoding for symbol $[b[i]]_1 = b_{1,1}[i]$ is given by the solution to the optimization problem

$$w_{1,1} \triangleq \arg \min_{w \in \mathbb{C}^{4N}} \mathsf{E}\left\{ \left| b_{1,1}[i] - w^H \tilde{r}[i] \right|^2 \right\}. \tag{20.6}$$

It has been shown by Wang and Poor (2004) that a scaled version of the solution can be written in terms of the signal subspace components as

$$w_{1,1} = U_s \Lambda_s^{-1} U_s^H \left(h_1 \otimes s_{1,1} \right), \tag{20.7}$$

and the decision is made according to

$$z_{1,1}[i] = w_{1,1}^H \tilde{r}[i],$$

$$\hat{b}_{1,1}[i] = \mathrm{sign}\left[\Re\left(z_{1,1}[i] \right) \right] \quad \text{(coherent detection)},$$

$$\hat{\beta}_{1,1}[i] = \mathrm{sign}\left[\Re\left(z_{1,1}[i-1]^* z_{1,1}[i] \right) \right] \quad \text{(differential detection)}.$$

Before we address specific batch and sequential adaptive algorithms, we note that these algorithms can also be implemented using linear *group-blind* multiuser detectors, which, in contrast to their blind counterparts, are constructed with knowledge of the spreading codes of a subset of the active users. They would be appropriate, for example, in uplink environments in which the base station knows the signature waveforms of all of the users in the cell, but not those of users outside the cell. Specifically, we may rewrite (20.5) as

$$\tilde{r}[i] = \check{S}\check{b}[i] + \bar{S}\bar{b}[i] + \tilde{n}[i],$$

where we have separated the users into two groups. The signature sequences of the known users are the columns of $\check{S}$. The unknown users' sequences are the columns of $\bar{S}$. Then the group-blind linear hybrid detector for symbol $b_{1,1}[i]$ is given by (Wang and Høst-Madsen, 1999)

$$w_{1,1}^{GB} = U_s \Lambda_s^{-1} U_s^H \check{S} \left[\check{S}^H U_s \Lambda_s^{-1} U_s^H \check{S} \right]^{-1} (h_1 \otimes s_{1,1}).$$

This detector offers a significant performance improvement over blind implementations of (20.7) for environments in which the signature sequences of some of the interfering users are known.

Batch blind linear space-time multiuser detection

Implementation of (20.7) requires knowledge of the signal subspace components and the channel. The subspace components can be estimated blindly from the received signal using the sample autocorrelation matrix of the received signal. To obtain an estimate of h_1, we make use of the orthogonality between the signal and noise subspaces, i.e., the fact that $U_n^H (h_1 \otimes s_{1,1}) = 0$. In particular, we have

$$\begin{aligned}
\hat{h}_1 &= \arg \min_{h \in \mathbb{C}^4} \left\| U_n^H (h \otimes s_{1,1}) \right\|^2 \\
&= \arg \max_{h \in \mathbb{C}^4} \left\| U_s^H (h \otimes s_{1,1}) \right\|^2 \\
&= \arg \max_{h \in \mathbb{C}^4} \left(h^H \otimes s_{1,1}^T \right) U_s U_s^H \left(h \otimes s_{1,1} \right) \\
&= \arg \max_{h \in \mathbb{C}^4} h^H \underbrace{\left[\left(I_4 \otimes s_{1,1}^T \right) U_s U_s^H \left(I_4 \otimes s_{1,1} \right) \right]}_{Q} h \quad (20.8) \\
&= \text{principal eigenvector of } Q. \quad (20.9)
\end{aligned}$$

In (20.9), $\hat{h}_1$ specifies h_1 up to an arbitrary complex scale factor α, i.e., $\hat{h}_1 = \alpha h_1$, but this ambiguity can be circumvented using differential modulation and detection. The following is the summary of a batch blind space-time multiuser detection algorithm for the two transmitter antenna/two receiver antenna configuration. The channel is assumed to be constant for at least the duration of the batch size M.

Algorithm 20.1 [Batch blind linear space-time multiuser detector—synchronous CDMA, two transmitter antennas and two receiver antennas]

- *Estimate the signal subspace:*

$$\hat{C} = \frac{1}{M} \sum_{i=0}^{M-1} \tilde{r}[i] \tilde{r}[i]^{\mathsf{H}}$$
$$= \hat{U}_s \hat{\Lambda}_s \hat{U}_s^{\mathsf{H}} + \hat{U}_n \hat{\Lambda}_n \hat{U}_n^{\mathsf{H}}.$$

- *Estimate the channels:*

$$\hat{Q}_1 = \left(I_4 \otimes s_{1,1}^{\mathsf{T}} \right) \hat{U}_s \hat{U}_s^{\mathsf{H}} \left(I_4 \otimes s_{1,1} \right),$$
$$\hat{Q}_2 = \left(I_4 \otimes s_{2,1}^{\mathsf{T}} \right) \hat{U}_s \hat{U}_s^{\mathsf{H}} \left(I_4 \otimes s_{2,1} \right),$$
$$\hat{h}_1 = \text{principal eigenvector of } \hat{Q}_1,$$
$$\hat{\tilde{h}}_1 = \text{principal eigenvector of } \hat{Q}_2.$$

- *Form the detectors:*

$$\hat{w}_{1,1} = \hat{U}_s \hat{\Lambda}_s^{-1} \hat{U}_s^{\mathsf{H}} \left(\hat{h}_1 \otimes s_{1,1} \right), \qquad (20.10)$$
$$\hat{w}_{2,1} = \hat{U}_s \hat{\Lambda}_s^{-1} \hat{U}_s^{\mathsf{H}} \left(\hat{\tilde{h}}_1 \otimes s_{2,1} \right). \qquad (20.11)$$

- *Perform differential detection: for $i = 0, \ldots, M-1$,*

$$z_{1,1}[i] = \hat{w}_{1,1}^{\mathsf{H}} \tilde{r}[i],$$
$$z_{2,1}[i] = \hat{w}_{2,1}^{\mathsf{H}} \tilde{r}[i],$$
$$\hat{\beta}_{1,1}[i] = \text{sign}\left(\Re\left\{ z_{1,1}[i] z_{1,1}[i-1]^* \right\} \right),$$
$$\hat{\beta}_{2,1}[i] = \text{sign}\left(\Re\left\{ z_{2,1}[i] z_{2,1}[i-1]^* \right\} \right).$$

A batch group-blind space-time multiuser detector algorithm can be implemented with simple modifications to (20.10) and (20.11).

Adaptive blind linear space-time multiuser detection

To form a sequential blind adaptive receiver, we need adaptive algorithms for sequentially estimating the channel and the signal subspace components U_s and Λ_s. First, we address sequential adaptive channel estimation. Denote by $z[i]$ the projection of the stacked signal $\tilde{r}[i]$ onto the noise subspace, that is,

$$z[i] = \tilde{r}[i] - U_s U_s^H \tilde{r}[i]$$
$$= U_n U_n^H \tilde{r}[i].$$

Since $z[i]$ lies in the noise subspace, it is orthogonal to any signal in the signal subspace, and in particular, it is orthogonal to $(h_1 \otimes s_{1,1})$. Hence, h_1 is the solution to the following constrained optimization problem:

$$\hat{h}_1 = \min_{h_1 \in \mathbb{C}^4} \; \mathsf{E}\left\{ \left\| z[i]^H (h_1 \otimes s_{1,1}) \right\|^2 \right\}$$

$$= \min_{h_1 \in \mathbb{C}^4} \; \mathsf{E}\left\{ \left\| z[i]^H (I_4 \otimes s_{1,1}) h_1 \right\|^2 \right\}$$

$$= \min_{h_1 \in \mathbb{C}^4} \; \mathsf{E}\left\{ \left\| \left[\left(I_4 \otimes s_{1,1}^T \right) z[i] \right]^H h_1 \right\|^2 \right\} \qquad \text{s.t. } \|h_1\| = 1. \qquad (20.12)$$

To obtain a sequential algorithm to solve the above optimization problem, we write it in the following (trivial) state space form

$$h_1[i+1] = h_1[i], \qquad\qquad \text{state equation}$$

$$0 = \left[\left(I_4 \otimes s_{1,1}^T \right) z[i] \right]^H h_1[i], \qquad \text{observation equation.}$$

Defining $x[i] \triangleq \left(I_4 \otimes s_{1,1}^T \right) z[i]$, the standard Kalman filter can then be applied to the above system as follows:

$$k[i] = \Sigma[i-1] x[i] \left(x[i]^H \Sigma[i-1] x[i] \right)^{-1},$$

$$h_1[i] = \frac{h_1[i-1] - k[i] \left(x[i]^H h_1[i-1] \right)}{\left\| h_1[i-1] - k[i] \left(x[i]^H h_1[i-1] \right) \right\|},$$

$$\Sigma[i] = \Sigma[i-1] - k[i] x[i]^H \Sigma[i-1].$$

Once we have obtained channel estimates at block i, we can combine them with estimates of the signal subspace components to form the detector given in (20.7). Subspace tracking algorithms of various complexities exist in the literature. Since we are stacking received signal vectors and because subspace tracking complexity increases at least linearly with signal subspace dimension, it is imperative that we choose an algorithm with minimal complexity. The best existing low-complexity algorithm for this purpose appears to be noise-averaged Hermitian–Jacobi fast subspace tracking (NAHJ-FST)

(Reynolds and Wang, 2001). This algorithm has the lowest complexity of any algorithm used for similar purposes and has performed well when used for signal subspace tracking in multipath fading environments. Since the size of U_s is $4N \times 2K$, the complexity is $40 \cdot 4N \cdot 2K + 3 \cdot 4N + 7.5(2K)^2 + 7 \cdot 2K$ floating operations per iteration. The algorithm and a multiuser detection application are presented by Reynolds and Wang (2001). The application to the current tracking problem is straightforward and will not be discussed in detail.

Algorithm 20.2 [Blind adaptive linear space-time multiuser detector—synchronous CDMA, two transmitter antennas and two receiver antennas]

- *Using a suitable signal subspace tracking algorithm, e.g., NAHJ-FST,*
  update the signal subspace components $U_s[i]$ and $\Lambda_s[i]$ at each block i.
- *Track the channel $h_1[i]$ and $\bar{h}_1[i]$ according to the following:*

$$z[i] = \tilde{r}[i] - U_s[i]U_s[i]^{\mathsf{H}}\tilde{r}[i],$$

$$x[i] = \left(I_4 \otimes s_{1,1}^{\mathsf{T}}\right) z[i],$$

$$\bar{x}[i] = \left(I_4 \otimes s_{2,1}^{\mathsf{T}}\right) z[i],$$

$$k[i] = \Sigma[i-1]\, x[i] \left(x[i]^{\mathsf{H}}\Sigma[i-1]x[i]\right)^{-1},$$

$$\bar{k}[i] = \bar{\Sigma}[i-1]\, \bar{x}[i] \left(\bar{x}[i]^{\mathsf{H}}\bar{\Sigma}[i-1]\bar{x}[i]\right)^{-1},$$

$$h_1[i] = \frac{h_1[i-1] - k[i]\left(x[i]^{\mathsf{H}}h_1[i-1]\right)}{\|h_1[i-1] - k[i]\left(x[i]^{\mathsf{H}}h_1[i-1]\right)\|},$$

$$\bar{h}_1[i] = \frac{\bar{h}_1[i-1] - \bar{k}[i]\left(\bar{x}[i]^{\mathsf{H}}\bar{h}_1[i-1]\right)}{\|\bar{h}_1[i-1] - \bar{k}[i]\left(\bar{x}[i]^{\mathsf{H}}\bar{h}_1[i-1]\right)\|},$$

$$\Sigma[i] = \Sigma[i-1] - k[i]\, x[i]^{\mathsf{H}}\Sigma[i-1],$$

$$\bar{\Sigma}[i] = \bar{\Sigma}[i-1] - \bar{k}[i]\, \bar{x}[i]^{\mathsf{H}}\bar{\Sigma}[i-1].$$

- *Form the detectors:*

$$\hat{w}_{1,1}[i] = U_s[i]\Lambda_s^{-1}[i]U_s[i]^{\mathsf{H}}\left(h_1[i] \otimes s_{1,1}\right),$$

$$\hat{w}_{2,1}[i] = U_s[i]\Lambda_s^{-1}[i]U_s[i]^{\mathsf{H}}\left(\bar{h}_1[i] \otimes s_{2,1}\right).$$

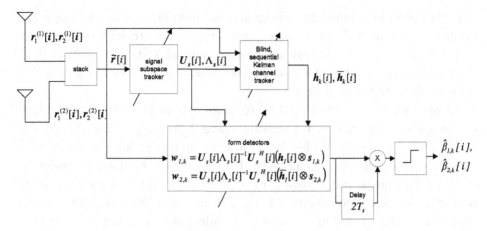

Fig. 20.1. Adaptive receiver structure for linear space-time multiuser detectors.

- *Perform differential detection:*

$$z_{1,1}[i] = \hat{\boldsymbol{w}}_{1,1}[i]^{\mathsf{H}} \tilde{\boldsymbol{r}}[i],$$

$$z_{2,1}[i] = \hat{\boldsymbol{w}}_{2,1}[i]^{\mathsf{H}} \tilde{\boldsymbol{r}}[i],$$

$$\hat{\beta}_{1,1}[i] = \mathrm{sign}\left(\Re\left\{z_{1,1}[i]\, z_{1,1}[i-1]^*\right\}\right),$$

$$\hat{\beta}_{2,1}[i] = \mathrm{sign}\left(\Re\left\{z_{2,1}[i]\, z_{2,1}[i-1]^*\right\}\right).$$

A group-blind sequential adaptive space-time multiuser detector can be implemented in a similar fashion. The adaptive receiver structure is illustrated in Fig. 20.1.

20.2 Transmitter precoding for ultralow complexity multiuser MIMO reception

20.2.1 Introduction

As signal processing techniques for communication become more sophisticated, they place an ever increasing computational burden (cost) on detectors, demodulators, and decoders. In many applications, however, it is useful to have the option of moving complexity away from the receiver to the transmitter. Cellular service providers, for example, would prefer to keep mobile unit costs to a minimum so they can continue to entice customers with free phones. With this in mind, we propose, in this section, downlink *transmitter*-based techniques for joint suppression of multiple-access interference and diversity exploitation for multiuser MIMO systems. As before, our focus is on low-complexity designs, so we restrict ourselves to linear processing, which we call transmitter precoding, and matched-filter detection at the mobile units that does not require channel state information (CSI) (Reynolds *et al.*, 2002a). We consider scenarios where perfect or partial CSI is available at the transmitter.

As previously developed in the literature, linear precoding for multiple access systems focused on transmitter-based MAI suppression. Vojčić and Jang (1998), for example, developed MMSE precoders for CDMA in additive white Gaussian noise channels. They also presented an extension to multipath channels, but it requires a rake receiver and that the channel be perfectly known at the receiver. Brandt-Pearce and Dharap (2000) considered transmitter precoding for multipath fading channels, but in contrast to the present work, their prefilter is applied to the output of the spread spectrum encoder, rather than applying the filter first, followed by spreading. It was shown that this approach has inferior average performance unless the spreading codes themselves are allowed to be adaptive.

Some of the results in this section have been published previously in Reynolds *et al.* (2005).

20.2.2 System model

Without receiver CSI, it is difficult to fully exploit receive antenna diversity because the only diversity combining available at the receiver is (noncoherent) addition of the antenna outputs. This provides no diversity in fading environments (Hochwald *et al.*, 2001). In a block fading environment with transmitter CSI, however, we can employ selection diversity with multiple receive antennas simply by adding a few bits to each frame to instruct the receiver to use the "best" antenna. This possibility notwithstanding, we consider a K-user downlink BPSK-modulated CDMA system in flat block-fading[†] with two transmit antennas and a single receive antenna for each user. Extensions to more than two transmit antennas and other modulation techniques are straightforward. The discrete-time, BPSK-modulated signal transmitted from antenna $a \in \{1, 2\}$ is

$$\boldsymbol{x}^{(a)} = \alpha \boldsymbol{S} \boldsymbol{M}^{(a)} \boldsymbol{b}, \tag{20.13}$$

where the columns of $\boldsymbol{S} \in \mathbb{C}^{N \times K}$ are the normalized spreading codes of the K users, $\boldsymbol{b} \in \{\pm 1\}^K$ contains the downlink symbols corresponding to the

---

[†]Extensions to frequency selective and intersymbol interference channels are given by Reynolds *et al.* (2005).

K users, N is the processing gain, and $\boldsymbol{M}^{(a)} \in \mathbb{C}^{K \times K}$ is a complex precoding matrix used for multiple-access interference suppression and transmitter antenna diversity exploitation. It is optimized in later sections. The scalar α is a transmit power factor that will be addressed in a later section. For now, we assume $\alpha = 1$.

The goal is to choose $\boldsymbol{M}^{(1)}$ and $\boldsymbol{M}^{(2)}$ to optimize downlink performance when no receiver CSI is available and the receiver is constrained to matched-filter detection. The precoders must not only suppress interference, but they must also exploit available diversity. We are interested in situations in which we have either perfect or partial CSI available at the transmitter.

20.2.3 Orthogonal spreading codes and perfect transmitter CSI

After chip-matched filtering, the noise free received signal at user 1's mobile unit is

$$\begin{aligned}
\boldsymbol{r}_1 &= h_1^{(1)} \boldsymbol{x}^{(1)} + h_1^{(2)} \boldsymbol{x}^{(2)} \\
&= h_1^{(1)} \boldsymbol{SM}^{(1)} \boldsymbol{b} + h_1^{(2)} \boldsymbol{SM}^{(2)} \boldsymbol{b},
\end{aligned}$$

where $h_b^{(a)}$ is the complex channel gain between transmit antenna a and user b's mobile unit and where we have set $\alpha = 1$. The channel gains are assumed to be mutually independent. The mobile units are restricted to (spreading-code) matched-filter detection. If $\boldsymbol{s}_1 \triangleq [\boldsymbol{S}]_{:,1}{}^\dagger$ and we have orthogonal spreading codes, the decision statistic for user 1 is

$$\begin{aligned}
d_1 &= \boldsymbol{s}_1^{\mathsf{H}} \boldsymbol{r}_1 + \sigma n_1 \\
&= h_1^{(1)} \left[\boldsymbol{M}^{(1)} \boldsymbol{b} \right]_1 + h_1^{(2)} \left[\boldsymbol{M}^{(2)} \boldsymbol{b} \right]_1 + \sigma n_1,
\end{aligned} \tag{20.14}$$

where $n_1 \sim \mathcal{N}_c(0, 1)$ is independent of $\boldsymbol{b}$ and the channel and σ^2 is the noise power. For now, define $\boldsymbol{M}^{(a)}$, $a \in \{1, 2\}$, to be diagonal matrices whose elements are given by

$$\left[\boldsymbol{M}^{(a)} \right]_{i,i} = \frac{h_i^{(a)*}}{\sqrt{\left| h_i^{(1)} \right|^2 + \left| h_i^{(2)} \right|^2}}. \tag{20.15}$$

Then, we have

$$d_1 = \sqrt{\left| h_1^{(1)} \right|^2 + \left| h_1^{(2)} \right|^2}\, b_1 + \sigma n_1,$$

$\dagger$The notation $[\boldsymbol{S}]_{:,i}$ indicates the ith column of the matrix $\boldsymbol{S}$.

and the corresponding symbol estimate is

$$\hat{b}_1 = \text{sign}\{\Re[d_1]\}.$$

This achieves an instantaneous SNR of

$$\text{SNR}_1 = \frac{|h_1^{(1)}|^2 + |h_1^{(2)}|^2}{\sigma^2},$$

which provides full two-branch diversity for every user and has a χ_4^2 distribution when the channel gains are independent complex Gaussian random variables. Precoding for this scenario reduces to maximal ratio weighting (Hochwald *et al.*, 2001), which has the same performance as beamforming to a single receive antenna. Note that $M^{(1)}, M^{(2)}$ in (20.15) are normalized in the sense that the sum of the average (with respect to b) transmit power from both antennas is K. Specifically, we have that

$$E_b\left\{\left\|SM^{(1)}b\right\|^2\right\} + E_b\left\{\left\|SM^{(2)}b\right\|^2\right\}$$

$$= \text{tr}\left(M^{(1)\mathsf{H}}M^{(1)}\right) + \text{tr}\left(M^{(2)\mathsf{H}}M^{(2)}\right) = K$$

for every channel realization. Therefore, we can set $\alpha = 1$ in (20.13).

20.2.4 Nonorthogonal spreading codes and perfect transmitter CSI

Here we assume nonorthogonal spreading codes and allow the precoding matrix to be nondiagonal. Let $\rho_1^{\mathsf{T}} \triangleq s_1^{\mathsf{H}} S$. The decision statistic for user 1, assuming for the moment that $\alpha = 1$, is

$$d_1 = \underbrace{h_1^{(1)} \rho_1^{\mathsf{T}} M^{(1)} b}_{d_1^{(1)}} + \underbrace{h_1^{(2)} \rho_1^{\mathsf{T}} M^{(2)} b}_{d_1^{(2)}} + \sigma n_1.$$

Our goal is to choose $M^{(1)}, M^{(2)}$ to maximize the collective performance of all users in some sense, assuming no receiver CSI and matched-filter detection. We form MMSE cost functions for the optimization of $M^{(1)}, M^{(2)}$ by stacking $d_k^{(1)}$ ($1 \leq k \leq K$) and $d_k^{(2)}$ ($1 \leq k \leq K$), respectively. The result for transmit antenna $a \in \{1, 2\}$ is

$$
J^{(a)} = \mathsf{E} \left\{ \left\| \begin{bmatrix} |h_1^{(a)}|^2 \left(|h_1^{(1)}|^2 + |h_1^{(2)}|^2 \right)^{-\frac{1}{2}} b_1 \\ |h_2^{(a)}|^2 \left(|h_2^{(1)}|^2 + |h_2^{(2)}|^2 \right)^{-\frac{1}{2}} b_2 \\ \vdots \\ |h_K^{(a)}|^2 \left(|h_K^{(1)}|^2 + |h_K^{(2)}|^2 \right)^{-\frac{1}{2}} b_K \end{bmatrix} \right. \right.
$$

$$
\left. \left. - \begin{bmatrix} h_1^{(a)} \boldsymbol{\rho}_1^{\mathsf{T}} \\ h_2^{(a)} \boldsymbol{\rho}_2^{\mathsf{T}} \\ \vdots \\ h_K^{(a)} \boldsymbol{\rho}_K^{\mathsf{T}} \end{bmatrix} \boldsymbol{M}^{(a)} \boldsymbol{b} - \begin{bmatrix} \sigma n_1 \\ \sigma n_2 \\ \vdots \\ \sigma n_K \end{bmatrix} \right\|^2 \right\} \tag{20.16}
$$

$$
= \mathsf{E} \left\{ \left\| \boldsymbol{D}^{(a)} \boldsymbol{b} - \boldsymbol{H}^{(a)} \boldsymbol{R} \boldsymbol{M}^{(a)} \boldsymbol{b} - \boldsymbol{n} \right\|^2 \right\}, \tag{20.17}
$$

where

$$
\boldsymbol{D}^{(a)} \triangleq \mathrm{diag}\left(|h_1^{(a)}|^2 \left(|h_1^{(1)}|^2 + |h_1^{(2)}|^2 \right)^{-\frac{1}{2}}, |h_2^{(a)}|^2 \left(|h_2^{(1)}|^2 + |h_2^{(2)}|^2 \right)^{-\frac{1}{2}}, \ldots, \right.
$$

$$
\left. |h_K^{(a)}|^2 \left(|h_K^{(1)}|^2 + |h_K^{(2)}|^2 \right)^{-\frac{1}{2}} \right), \tag{20.18}
$$

$$
\boldsymbol{H}^{(a)} \triangleq \mathrm{diag}\left(h_1^{(a)}, h_2^{(a)}, \ldots, h_K^{(a)} \right), \quad a \in \{1, 2\}, \ \boldsymbol{R} \triangleq \boldsymbol{S}^{\mathsf{H}} \boldsymbol{S},
$$

$$
\boldsymbol{n} \triangleq \begin{bmatrix} \sigma n_1 & \sigma n_2 & \cdots & \sigma n_K \end{bmatrix}^{\mathsf{T}},
$$

and where the expectations are with respect to $\boldsymbol{n}$ and $\boldsymbol{b}$. Through the construction of these cost functions, we assume for the moment that the channel gains are deterministic and known at the transmitter.

The motivation behind the construction of the cost functions is self evident except, perhaps, for the presence of $\boldsymbol{D}^{(1)}$ and $\boldsymbol{D}^{(2)}$. This is related to the transmit power constraint and power loading. If we allow for an infinite peak-to-average power ratio at the transmitter, we can replace $\boldsymbol{D}^{(1)}$ and $\boldsymbol{D}^{(2)}$ in (20.17) with $\boldsymbol{I}_K$, and the resulting optimal precoding matrix will completely eliminate the detrimental effects of fading.[†] Because real transmitters cannot operate with an infinite dynamic range, this is not a

---

[†]The precoding matrix for this situation can be found using (20.19)-(20.20) and by solving for α using the procedure in Section 20.2.6. Essentially, the transmitter will increase power (perhaps without bound) during fades and decrease power during channel peaks, resulting in infinite peak-to-average transmit power.

reasonable assumption. Therefore, we will insist that the sum of the average (with respect to b) transmit power from all antennas be equal to the number of users. For diversity transmission (instead of multiplexing) with this power constraint and orthogonal codes, the best precoding scheme is maximum ratio weighting as in (20.15). It is therefore important that the precoding matrices that minimize the cost functions $J^{(1)}$, $J^{(2)}$ reduce to (20.15) when spreading codes are orthogonal. It is easy to verify using the results in Section 20.2.5 that this is true when $D^{(a)}$ satisfies (20.18).

20.2.5 Optimizing M

Proposition 20.1 *The choice of $M^{(1)}$ that minimizes $J^{(1)}$ and the choice of $M^{(2)}$ that minimizes $J^{(2)}$ are given by*

$$M^{(1)} = R^{-1} \left[H^{(1)} \right]^{-1} D^{(1)}, \tag{20.19}$$

$$M^{(2)} = R^{-1} \left[H^{(2)} \right]^{-1} D^{(2)}. \tag{20.20}$$

The proof is given by Reynolds *et al.* (2005).

These results show that optimal precoding with perfect transmitter CSI and nonorthogonal codes is maximum ratio weighting followed by transmitter-based decorrelation. It is easy to see that for orthogonal spreading code sets ($R = I_K$), equations (20.19) and (20.20) reduce to (20.15), i.e., the results for the optimization of $M^{(1)}$, $M^{(2)}$ agree with our intuition for orthogonal codes.

20.2.6 Performance and achievable diversity

We have seen that with perfect channel knowledge at the transmitter and orthogonal spreading codes, we can achieve full transmit diversity with precoding. We will see here that nonzero cross-correlations among spreading codes lead to an SNR loss, but full diversity is still achievable.

The general case

Stacking decision statistics from all users obtained using the optimal $M^{(1)}$, $M^{(2)}$, we define the composite received signal as

$$r \triangleq \begin{bmatrix} d_1 & d_2 & \cdots & d_K \end{bmatrix}^{\mathsf{T}}$$

$$= \alpha \left[H^{(1)} R M^{(1)} + H^{(2)} R M^{(2)} \right] b + \sigma n$$

$$= \alpha E b + \sigma n,$$

where

$$E \triangleq \mathrm{diag}\left(\sqrt{|h_1^{(1)}|^2 + |h_1^{(2)}|^2}, \sqrt{|h_2^{(1)}|^2 + |h_2^{(2)}|^2}, \ldots, \sqrt{|h_K^{(1)}|^2 + |h_K^{(2)}|^2} \right).$$

Because E is diagonal, multiple access interference is completely eliminated. Furthermore, we have seen in Section 20.2.3 that with orthogonal codes, we can set $\alpha = 1$ (i.e., no transmit power adjustment is necessary) and achieve full transmit diversity. In general, however, we must set $\alpha \leq 1$ to constrain average transmit power. For our purposes, average transmit power normalization requires

$$\alpha^2 \mathsf{E}_b \left\{ \left\| SM^{(1)}b \right\|^2 \right\} + \alpha^2 \mathsf{E}_b \left\{ \left\| SM^{(2)}b \right\|^2 \right\} = K \qquad (20.21)$$

for every channel realization. That is, the sum of the average transmit power from the two antennas is equal to the number of users. Dropping the antenna superscripts for notational convenience, we have

$$\mathsf{E}_b\left\{ \|SMb\|^2 \right\} = \mathrm{tr}\left(M^H RM \right)$$
$$= \mathrm{tr}\left(\left(H^{-1}D\right)^H R^{-1} \left(H^{-1}D\right) \right).$$

For transmit antenna $a \in \{1,2\}$, the diagonal structures of D and H yield

$$\mathrm{tr}\left(M^{(a)H} RM^{(a)} \right) = \sum_{i=1}^{K} [R^{-1}]_{i,i} \frac{|h_i^{(a)}|^2}{|h_i^{(1)}|^2 + |h_i^{(2)}|^2}.$$

Summing the average transmit power contributions from each antenna, we have

$$\mathrm{tr}\left(M^{(1)H} RM^{(1)} \right) + \mathrm{tr}\left(M^{(2)H} RM^{(2)} \right) = \sum_{i=1}^{K} [R^{-1}]_{i,i}$$
$$= \sum_{i=1}^{K} \frac{1}{\lambda_i}, \qquad (20.22)$$

where $\{\lambda_i\}_{i=1}^{K}$ are the eigenvalues of R. Therefore, by (20.21), the power scaling factor α must satisfy

$$\alpha = \sqrt{\frac{1}{\dfrac{1}{K}\displaystyle\sum_{i=1}^{K}\dfrac{1}{\lambda_i}}}.$$

Notice that α^2 is simply the inverse of the average of the diagonal elements of $\boldsymbol{R}^{-1}$. It is interesting to relate this result to the performance of receiver-based decorrelating multiuser detection (MUD), in which the performance of user k is dependent upon the inverse of $\left[\boldsymbol{R}^{-1}\right]_{k,k}$, but is not dependent upon the other diagonal elements of $\boldsymbol{R}^{-1}$. In this sense, we can think of the performance of precoding, which is the same for every user, as the performance of decorrelating MUD "averaged" over every user. This interpretation is supported by the simulation results reported by Vojčić and Jang (1998).

Assuming all channel gains are independent and have the same statistics, the average bit-error probability (BEP) of every user will be the same and is given by

$$\overline{\mathrm{Pr}}(\epsilon) = \mathsf{E}\left\{Q\left(\frac{\alpha}{\sigma}\sqrt{|h_1^{(1)}|^2 + |h_1^{(2)}|^2}\right)\right\}$$

$$= \frac{1}{4}(\mu^3 - 3\mu + 2), \tag{20.23}$$

where

$$\mu \triangleq \sqrt{\frac{\gamma}{1+\gamma}}, \quad \gamma \triangleq \frac{E_h\alpha^2}{2\sigma^2}, \quad E_h \triangleq \mathsf{E}\left\{|h_i^{(a)}|^2\right\}, \quad a = 1,2; \quad i = 1,\ldots,K.$$

This performance is the same as two-branch maximum ratio combining with an SNR penalty of $10\log_{10}\alpha^2$ dB. Therefore, diversity, defined here as the slope of the BEP curve, is unaffected by signature waveform cross-correlations, but we do suffer SNR loss.

Equicorrelated spreading codes

As an important special case, we consider the scenario in which the normalized spreading code cross-correlations satisfy

$$\boldsymbol{s}_k^{\mathsf{H}}\boldsymbol{s}_l = \begin{cases} 1, & k = l \\ \rho, & k \neq l \end{cases} \tag{20.24}$$

for some $\rho \in [0,1)$. It is easy to show using the matrix inversion lemma that

$$\boldsymbol{R}^{-1} = \frac{1}{1-\rho}\boldsymbol{I}_K - \underbrace{\frac{\rho}{(1-K)\rho^2 + (K-2)\rho + 1}}_{\check{\rho}}\boldsymbol{1}_{K,K}, \tag{20.25}$$

which yields

$$\mathrm{tr}\left(\boldsymbol{M}^{(1)\mathsf{H}}\boldsymbol{R}\boldsymbol{M}^{(1)}\right) + \mathrm{tr}\left(\boldsymbol{M}^{(2)\mathsf{H}}\boldsymbol{R}\boldsymbol{M}^{(2)}\right) = K\left(\frac{1}{1-\rho} - \check{\rho}\right)$$

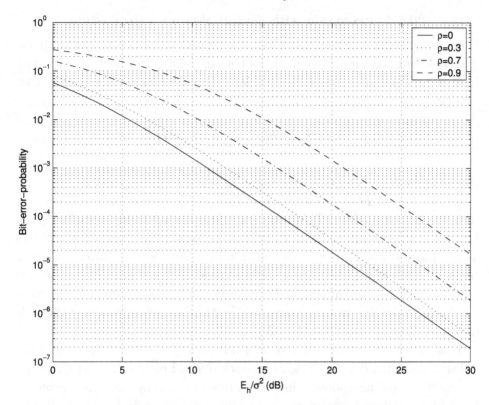

Fig. 20.2. The bit-error probability for the equicorrelated spreading code case, averaged over all twenty users and their channel gains, versus E_h/σ^2 for precoding with two transmit antennas and one receiver antenna and for cross-correlation values of $\rho = 0, 0.3, 0.7, 0.9$. The transmit energy per user per bit is 1.

and

$$\alpha(\rho, K) = \left[\frac{1}{1-\rho} - \frac{\rho}{(1-K)\rho^2 + (K-2)\rho + 1} \right]^{-\frac{1}{2}}.$$

Clearly,

$$\lim_{K \to \infty} \alpha(\rho, K) = \sqrt{1 - \rho}.$$

In fact, $\alpha(\rho, K)$ tends to its limit rather quickly. Therefore, for a moderate or large number of users, the performance is nearly equivalent to two-branch maximum ratio combining with an SNR penalty of $10 \log_{10}(1-\rho)$ dB. Fig. 20.2 illustrates the BEP performance, calculated using (20.23), for twenty users and various values of the cross-correlation parameter ρ. The $10 \log_{10}(1 - \rho)$ dB SNR penalty is clearly visible.

20.2.7 Precoding with partial channel information

Here, we consider a scenario in which the transmitter has only partial information about the current channel. As before, the receiver has no CSI of any kind and is limited to matched-filter detection. We create the following cost function as a modification of $J^{(1)}, J^{(2)}$:

$$J_p^{(a)} = \mathsf{E}\left\{\left\|\boldsymbol{D}^{(a)}\boldsymbol{b} - \boldsymbol{H}^{(a)}\boldsymbol{R}\boldsymbol{M}^{(a)}\boldsymbol{b} - \boldsymbol{n}\right\|^2 \,\middle|\, \hat{\boldsymbol{H}}^{(1)}, \hat{\boldsymbol{H}}^{(2)}\right\}, \qquad a \in \{1, 2\},$$

where $\hat{\boldsymbol{H}}^{(1)}, \hat{\boldsymbol{H}}^{(2)}$ are quantities that are statistically dependent upon the channel matrices $\boldsymbol{H}^{(1)}$ and $\boldsymbol{H}^{(2)}$, and where the expectation is with respect to $\boldsymbol{b}$, $\boldsymbol{n}$, $\boldsymbol{H}^{(a)}$, and $\boldsymbol{D}^{(a)}$. This practical CSI approach is motivated, in part, by Jakes' model (Jakes, 1974, p. 69), which treats the channel coefficients as samples of a stationary Gaussian random process with autocorrelation function $J_0(2\pi f_m \tau)$, where $J_0(\cdot)$ is the zero-order Bessel function of the first kind, f_m denotes the maximum Doppler frequency, and τ denotes the time lag. In this context, we can think of $\hat{\boldsymbol{H}}^{(1)}, \hat{\boldsymbol{H}}^{(2)}$ as *old* estimates of the true downlink channel matrices $\boldsymbol{H}^{(1)}, \boldsymbol{H}^{(2)}$.

This partial CSI scenario will require differential encoding/decoding of the BPSK modulated data because of the lack of good phase information at the transmitter. To incorporate this constraint into the optimization problem (and to avoid trivial "all zero" precoding solutions), $\hat{\boldsymbol{H}}^{(1)}, \hat{\boldsymbol{H}}^{(2)}$ will also represent perfect knowledge of the channel phase.

Optimizing $\boldsymbol{M}$, revisited

Proposition 20.2 *Define the following correlation matrices:*

$$\boldsymbol{C}_{HH}^{(a)} \triangleq \mathsf{E}\left\{\boldsymbol{H}^{(a)\mathsf{H}}\boldsymbol{H}^{(a)} \,\middle|\, \hat{\boldsymbol{H}}^{(1)}, \hat{\boldsymbol{H}}^{(2)}\right\}, \qquad a \in \{1, 2\}, \tag{20.26}$$

$$\boldsymbol{C}_{HD}^{(a)} \triangleq \mathsf{E}\left\{\boldsymbol{H}^{(a)\mathsf{H}}\boldsymbol{D}^{(a)} \,\middle|\, \hat{\boldsymbol{H}}^{(1)}, \hat{\boldsymbol{H}}^{(2)}\right\}, \qquad a \in \{1, 2\}. \tag{20.27}$$

The choice of $\boldsymbol{M}^{(1)}$ that minimizes $J_p^{(1)}$ and the choice of $\boldsymbol{M}^{(2)}$ that minimizes $J_p^{(2)}$ are given by

$$\boldsymbol{M}^{(1)} = \boldsymbol{R}^{-1}\left[\boldsymbol{C}_{HH}^{(1)}\right]^{-1}\boldsymbol{C}_{HD}^{(1)}, \tag{20.28}$$

$$\boldsymbol{M}^{(2)} = \boldsymbol{R}^{-1}\left[\boldsymbol{C}_{HH}^{(2)}\right]^{-1}\boldsymbol{C}_{HD}^{(2)}. \tag{20.29}$$

The proof is given by Reynolds *et al.* (2005).

An application: precoding with old channel estimates

We characterize the channel as Rayleigh fading, following Jakes' model (Jakes, 1974, p. 69). That is, we assume the channel coefficients are samples of a stationary Gaussian random process with an autocorrelation function proportional to $J_0(2\pi f_m \tau)$. Specifically, the time-varying channel between transmit antenna a and user k is governed by

$$\frac{1}{E_h} \mathsf{E}\left\{ h_k^{(a)*}(t) \cdot h_k^{(a)}(t-\tau) \right\} = J_0(2\pi f_m \tau),$$

where, as before, $E_h \triangleq \mathsf{E}\{|h_k^{(a)}(t)|^2\}$. This approach is useful for developing linear precoders for a system that produces good but delayed channel estimates at the transmitter. Because the focus here is on characterizing the effects of the channel variation, independent of the channel estimation algorithm employed, we assume that the outdated fading estimates are made perfectly.

For notational convenience, we write

$$h_k^{(a)}(t) = h_k^{(a)} = h_r^{(a)} + j h_i^{(a)},$$
$$h_k^{(a)}(t-\tau) = \check{h}_k^{(a)} = \check{h}_r^{(a)} + j \check{h}_i^{(a)}.$$

Then $h_r^{(a)}$ and $h_i^{(a)}$ are mutually independent, as are $\check{h}_r^{(a)}$ and $\check{h}_i^{(a)}$. Channel coefficients are also independent across antennas. Dropping the antenna superscript for the moment, the conditional distributions of $h_r, \check{h}_r$ and $h_i, \check{h}_i$ are given by

$$p(h|\check{h}) = \left(2\pi \sigma_{h|\check{h}}^2 \right)^{-\frac{1}{2}} \cdot \exp\left\{ -\frac{1}{2\sigma_{h|\check{h}}^2}(h - \mu_{h|\check{h}})^2 \right\},$$

where we have also dropped the r or i subscript for notational convenience. If the channels have zero mean, the conditional mean and variance are given by

$$\mu_{h|\check{h}} = \check{h}\rho_{h|\check{h}}, \qquad\qquad \sigma_{h|\check{h}}^2 = \frac{E_h}{2}(1 - \rho_{h|\check{h}}^2),$$

where

$$\rho_{h|\check{h}} = \frac{\mathsf{E}\{\check{h}h\}}{E_h/2} = J_0(2\pi f_m \tau).$$

For this scenario, the correlation matrices $\boldsymbol{C}_{HH}^{(a)}$ and $\boldsymbol{C}_{HD}^{(a)}$ are diagonal with entries given by

$$\left[C_{HH}^{(a)}\right]_{i,i} = \mathsf{E}\left\{|h_i^{(a)}|^2 \;\Big|\; \check{h}_i^{(a)}\right\},$$

$$\left[C_{HD}^{(a)}\right]_{i,i} = \mathsf{E}\left\{\frac{h_i^{(a)*}|h_i^{(a)}|^2}{\sqrt{|h_i^{(1)}|^2 + |h_i^{(2)}|^2}} \;\Bigg|\; \check{h}_i^{(1)}, \check{h}_i^{(2)}\right\}.$$

The former expectation can be evaluated analytically as

$$\left[C_{HH}^{(a)}\right]_{i,i} = J_0^2(2\pi f_m \tau)|\check{h}_i^{(a)}|^2 + E_h[1 - J_0^2(2\pi f_m \tau)],$$

and the latter can be evaluated via Monte-Carlo simulation for every realization of the past channel.

20.3 Conclusions

In this chapter, we have investigated transmitter and receiver based techniques for suppressing interference and exploiting diversity in multiuser MIMO systems. Our focus has been on linear processing, which produces low-complexity designs that provide high performance and the flexibility to operate in adaptive mode with perfect or partial channel state information.

In some situations, additional performance gains can be obtained by embedding these linear techniques within iterative algorithms in turbo or interference canceling receiver structures. Space does not permit a review of these methods here, but further details can be found in the works of Poor (2004), Wang and Poor (2004) and in the references therein.

References

Alamouti, S. (1998). A simple transmit diversity technique for wireless communications. *IEEE J. Sel. Areas Commun.*, **16** (8), 1451–1458.

Brandt-Pearce, M. and Dharap, A. (2000). Transmitter-based multiuser interference rejection for the down-link of a wireless CDMA system in a multipath environment. *IEEE J. Sel. Areas Commun.*, **18** (3), 407–417.

Hochwald, B., Marzetta, T. L., and Papadias, C. B. (2001). A transmitter diversity scheme for wideband CDMA systems based on space-time spreading. *IEEE J. Sel. Areas Commun.*, **19** (1), 48–60.

Honig, M., Madhow, U., and Verdú, S. (1995). Blind adaptive multiuser detection. *IEEE Trans. Inf. Theory*, **41** (4), 944–960.

Jakes, W. C. (1974). *Microwave Mobile Communications* (Wiley).

Lucent Technologies (1999). Downlink diversity improvements through space-time spreading. proposed 3GPP2-C30-19990817-014 to the CDMA-2000 standard, Aug. 1999.

Papadias, C. B. and Huang, H. (2001). Linear space-time multiuser detection for multipath CDMA channels. *IEEE J. Sel. Areas Commun.*, **19** (2), 254–265.

Poor, H. V. (2004). Iterative multiuser detection. *IEEE Signal Process. Mag.*, **21** (1), 81–88.

Reynolds, D., Høst-Madsen, A., and Wang, X. (2002a). Adaptive transmitter precoding for time division duplex CDMA in fading multipath channels: Strategy and analysis. *EURASIP J. Appl. Signal Process.*, **2002** (12), 1325–1334.

Reynolds, D. and Wang, X. (2001). Adaptive group-blind multiuser detection based on a new subspace tracking algorithm. *IEEE Trans. Commun.*, **49** (7), 1135–1141.

Reynolds, D., Wang, X., and Modi, K. N. (2005). Interference suppression and diversity exploitation for multiantenna CDMA with ultra-low complexity receivers. *IEEE Trans. Signal Process.*, **53** (8), 3226–3237.

Reynolds, D., Wang, X., and Poor, H. V. (2002b). Blind adaptive space-time multiuser detection with multiple transmitter and receiver antennas. *IEEE Trans. Signal Process.*, **50** (6), 1261–1276.

Tarokh, V., Jafarkhani, H., and Calderbank, A. R. (1999). Space-time block codes from orthogonal designs. *IEEE Trans. Inf. Theory*, **45** (5), 1456–1467.

Verdú, S. (1998). *Multiuser Detection* (Cambridge Univ. Press, Cambridge, U.K.).

Vojčić, B. R. and Jang, W. M. (1998). Transmitter precoding in synchronous multiuser communications. *IEEE Trans. Commun.*, **46** (10), 1346–1355.

Wang, X. and Høst-Madsen, A. (1999). Group-blind multiuser detection for uplink CDMA. *IEEE J. Sel. Areas Commun.*, **17** (11), 1971–1984.

Wang, X. and Poor, H. V. (2004). *Wireless Communication Systems: Advanced Techniques for Signal Reception* (Prentice-Hall, Upper Saddle River, NJ).

21

Opportunistic communication: a system view

Pramod Viswanath

University of Illinois, Urbana-Champaign

The wireless medium is often called a *fading* channel: the pejorative adjective suggests that the intrinsic temporal and frequency variations are an impediment to reliable communication. While not untrue, the channel fluctuations are turned from *foe to friend* in some scenarios. A concrete situation is when the time scale of communication is much larger than that of the channel fluctuations: the so-called ergodic fading channel. In a point-to-point ergodic fading channel, the transmitter can make good use of the channel state information (CSI): by devoting more power to when the channel is good and less (or even none) when the channel is bad, the rate of reliable communication is improved. The improvement is significant when the operating signal-to-noise ratio (SNR) is small; this is a simple instance of opportunistic communication. Another important instance is in cellular communication: by scheduling user transmissions when their channel conditions are good, the system throughput is improved. This effect is called *multiuser diversity* and its role in cellular communication is the focus of this chapter.

21.1 Multiuser diversity gain

Consider the single antenna downlink flat fading channel with K users:

$$y_k[m] = h_k[m]x[m] + w_k[m], \qquad k = 1, \ldots, K, \tag{21.1}$$

where $\{h_k[m]\}_m$ is the channel fading process of user k. There is an average power constraint of P on the transmit signal and $w_k[m] \sim \mathcal{CN}(0, N_0)$ are i.i.d. in time m (for each user $k = 1, \ldots, K$). For concreteness, consider the symmetric case: $\{h_k[m]\}_m$ are identically distributed processes for $k = 1, \ldots, K$. Further, let us also suppose that the processes $\{h_k[m]\}_m$

426

are ergodic (i.e., the time average of every realization equals the statistical average).

When the single transmitter tracks the channel fluctuations of all the users (with each user tracking its own channel fluctuation), the sum capacity of the downlink is achieved by transmitting only to the best user (Tse, 1997); as the channels vary, we can pick the best user at each time and further allocate it an appropriate waterfilling power subject to the average power constraint. The corresponding sum rate can be interpreted as the waterfilling capacity of a point-to-point link with a power constraint equal to the transmit power, and a fading process whose magnitude varies as $\{\max_k |h_k[m]|\}$. Compared with a system with a single transmitting user, the multiuser gain comes from the fact that the effective channel gain at time m is improved from $|h_1[m]|^2$ to $\max_{1 \leq k \leq K} |h_k[m]|^2$. This effect is entirely due to the ability to dynamically schedule resources among the users as a function of the channel state.

The corresponding full CSI (both at the transmitter and the receivers) sum capacity is increased due to the *multiuser diversity* effect: when there are many users that fade *independently*, at any one time there is a high probability that one of the users will have a strong channel. By allowing only that user to transmit, the shared channel resource is used in the most efficient manner and the total system throughput is maximized. The larger the number of users, the stronger tends to be the strongest channel, and the more the multiuser diversity gain.

The amount of multiuser diversity gain depends crucially on the *tail* of the fading distribution $|h_k|^2$: the heavier the tail, the more likely there is a user with a very strong channel, and the larger the multiuser diversity gain. This is shown in Fig. 21.1, where the sum capacity is plotted as a function of the number of users for both Rayleigh and Ricean fading with κ-factor equaling to 5, with the SNR fixed at 0 dB. Ricean fading models the situation when there is a strong specular line-of-sight path plus many small reflected paths. The parameter κ is defined as the ratio of the energy in the specular line-of-sight path to the energy in the diffused components. Because of the line-of-sight component, the Ricean fading distribution is less "random" and has a lighter tail than the Rayleigh distribution with the same average channel gain. As a consequence, it can be seen from Fig. 21.1 that the multiuser diversity gain is significantly smaller in the Ricean case compared with the Rayleigh case.

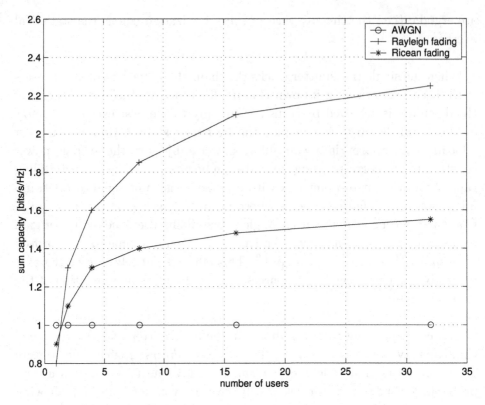

Fig. 21.1. Multiuser diversity gain for Rayleigh and Ricean fading channels ($\kappa = 5$); SNR $= P/N_0 = 0$ dB.

21.2 Multiuser versus classical diversity

We have called the above explained phenomenon multiuser diversity. Like the classical diversity techniques such as temporal, spatial and frequency diversity, multiuser diversity also arises from the existence of independently faded signal paths, in this case from the multiple users in the network. However, there are several important differences. First, the main objective of the classical diversity techniques is to improve the *reliability* of point-to-point communication in slowly fading channels; in contrast, the role of multiuser diversity is to increase the total throughput of the downlink over fast fading channels. Under the sum capacity-achieving strategy, a user has no guarantee of a high rate in any particular slow-fading state; only by averaging over the variations of the channel a high long-term average throughput is attained. Second, while the classical diversity techniques are designed to *counteract* the adverse effect of fading, multiuser diversity improves system performance by *exploiting* channel fading: channel fluctuations due to fading ensure that with high probability there is a user with a channel strength much larger than the mean level; by allocating all the system resources to that user, the benefit of this strong channel is fully capitalized. Third, while the classical diversity techniques pertain to a point-to-point link, the benefit of multiuser diversity is *systemwide*, across the users in the network. This aspect of multiuser diversity has ramifications on the implementation of multiuser diversity in a cellular system. We discuss this next.

21.3 Multiuser diversity: system aspects

The cellular system requirements to extract the multiuser diversity benefits are:

- the base station has access to channel quality measurements: in the downlink, we need each receiver to track its own channel SNR, through say a common downlink pilot, and feed back the instantaneous channel quality to the base station (assuming a frequency division duplex (FDD) system).
- the ability of the base station to schedule transmissions among the users as well as to adapt the data rate as a function of the instantaneous channel quality.

These features are already present in the designs of many third-generation systems. Nevertheless, in practice there are several considerations to take into account before realizing such gains. In particular, there are three main hurdles toward a system implementation of the multiuser diversity idea.

(i) **Fairness and delay**: to implement the idea of multiuser diversity in a real system, one is immediately confronted with two issues: fairness and delay. In the ideal situation when users' fading *statistics* are the same, the strategy of communicating with the user having the best channel, maximizes not only the total capacity of the system but also the throughput of individual users (Tse and Viswanath, 2005, Section 6.4.2). In reality, the statistics are not symmetric; there are users who are closer to the base station with a better average SNR; there are users who are stationary and some that are moving; there are users who are in a rich scattering environment and some with no scatterers around them. Moreover, the strategy is only concerned with maximizing long-term average throughputs; in practice, there are latency requirements, in which case the average throughputs over the delay time-scale is the performance metric of interest. The challenge is to address these issues while at the same time exploiting the multiuser diversity gain inherent in a system with users having independent, fluctuating channel conditions. There are many schedulers that harness multiuser diversity while addressing the real-world fairness issues; Tse and Viswanath (2005, Chapter 6) has a detailed study of one of them: the *proportional fair* scheduler.

(ii) **Channel measurement and feedback**: one of the key system requirements to harness multiuser diversity is to have scheduling decisions by the base station be made as a function of the channel states of the users. In the downlink, the users have access to their channel states but need to feedback these values to the base station. Both the error in channel state measurement and the delay in feeding it back constitute a significant bottleneck in extracting the multiuser diversity gains.

The prediction error is due to two effects: the error in measuring the channel from the pilot and the delay in feeding back the information to the base station. In the downlink, the pilot is shared between many users and is strong; so, the measurement error is quite small and the prediction error is mainly due to the feedback delay.

One remedy to reduce the feedback delay is to shrink the size of the scheduling time slot. However, this increases the feedback frequency in the uplink and thus, increases the system overhead. There are ways to reduce this feedback and Tse and Viswanath (2005, Chapter 6) studies some of these techniques.

(iii) **Slow and limited fluctuations**: we have observed that the multiuser diversity gains depend on the distribution of channel fluctuations. In particular, larger and faster variations in a channel are preferred over slow ones. However, there may be a line-of-sight path and little scattering in the environment, and hence the dynamic range of channel fluctuations may be small. Further, the channel may fade very slowly compared with the delay constraints of the application so that transmissions cannot wait until the channel reaches its peak. Effectively, the dynamic range of channel fluctuations is small within the time scale of interest. Both are important sources of hindrance to implementing multiuser diversity in a real system. We will next see a simple and practical scheme using an antenna array at the base station that creates fast and large channel fluctuations even when the channel is originally slowly fading with a small range of fluctuation.

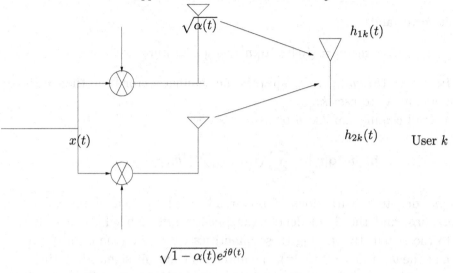

Fig. 21.2. The same signal is transmitted over the two antennas with time-varying phase and powers.

21.4 Opportunistic beamforming using dumb antennas

The amount of multiuser diversity depends on the rate and dynamic range of channel fluctuations. In environments where the channel fluctuations are small, a natural idea comes to mind: why not amplify the multiuser diversity gain by *inducing* faster and larger fluctuations? Focusing on the downlink, we describe a technique that does this using multiple transmit antennas at the base station as illustrated in Fig. 21.2.

Consider a system with n_t transmit antennas at the base station. Let $h_{lk}[m]$ be the complex channel gain from antenna l to the kth user at time m. At time m, the same symbol $x[m]$ is transmitted from all of the antennas except that it is multiplied by a complex number $\sqrt{\alpha_l[m]}e^{j\theta_l[m]}$ at antenna l, for $l = 1, \ldots, n_t$, such that $\sum_{l=1}^{n_t} \alpha_l[m] = 1$, preserving the total transmit power. The received signal at user k [see the basic downlink fading channel model in (21.1) for comparison] is given by:

$$y_k[m] = \left(\sum_{l=1}^{n_t} \sqrt{\alpha_l[m]}e^{j\theta_l[m]}h_{lk}[m] \right) x[m] + w_k[m]. \qquad (21.2)$$

In vector form, the scheme transmits $\mathbf{q}[m]x[m]$ at time m, where

$$\mathbf{q}[m] := \begin{bmatrix} \sqrt{\alpha_1[m]}e^{j\theta_1[m]} \\ \vdots \\ \sqrt{\alpha_{n_t}[m]}e^{j\theta_{n_t}[m]} \end{bmatrix}$$

is a unit vector and

$$y_k[m] = (\mathbf{h}_k[m]^* \mathbf{q}[m])\, x[m] + w_k[m]$$

where $\mathbf{h}_k[m]^* := (h_{1k}[m], \ldots, h_{n_t,k}[m])$ is the channel vector from the transmit antenna array to user k.

The overall channel gain seen by user k is now

$$\mathbf{h}_k[m]^* \mathbf{q}[m] = \sum_{l=1}^{n_t} \sqrt{\alpha_l[m]}\, e^{j\theta_l[m]} h_{lk}[m].$$

The $\alpha_l[m]$ denote the fractions of power allocated to each of the transmit antennas, and the $\theta_l[m]$ denote the phase shifts applied at each antenna to the signal. By varying these quantities over time (the $\alpha_l[m]$ from 0 to 1 and the $\theta_l[m]$ from 0 to 2π), the antennas transmit signals in a time-varying direction, and fluctuations in the overall channel can be induced even if the physical channel gains $\{h_{lk}[m]\}$ have very little fluctuations (see Fig. 21.3).

As in the single transmit antenna system, each user k feeds back the overall received SNR of its own channel, $|\mathbf{h}_k[m]^* \mathbf{q}[m]|^2/N_0$, to the base station (or, equivalently, the data rate that the channel can currently support) and the base station schedules transmissions to users accordingly. There is no need to measure the individual channel gains $h_{lk}[m]$ (phase or magnitude); in fact, the existence of multiple transmit antennas is completely transparent to the users. Thus, only a single pilot signal is needed for channel measurement (as opposed to a pilot to measure each antenna gain). The pilot symbols are repeated at each transmit antenna, exactly like the data symbols.

The rate of variation of $\{\alpha_l[m]\}$ and $\{\theta_l[m]\}$ in time (or, equivalently, of the transmit direction $\mathbf{q}[m]$) is a design parameter of the system. We would like it to be as fast as possible to provide full channel fluctuations within the latency time scale of interest. On the other hand, there is a practical limitation as to how fast this can be. The variation should be slow enough and should happen at a time scale that allows the channel to be reliably estimated by the users and the SNR fed back. Further, the variation should be slow enough to ensure that the channel seen by a user does not change abruptly and thus maintains stability of the channel tracking loop.

21.4.1 Slow fading: opportunistic beamforming

To get some insight into the performance of this scheme, consider the case of slow fading where the channel gain vector of each user k remains constant,

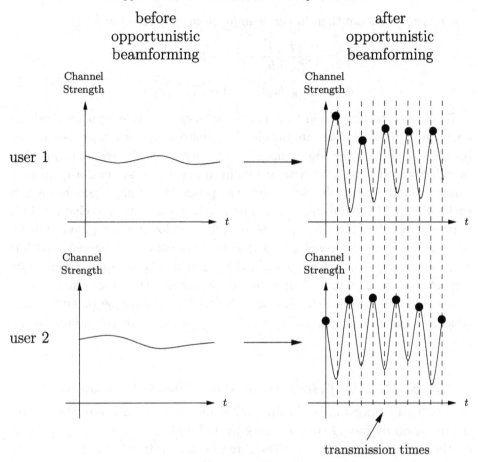

Fig. 21.3. Pictorial representation of the slowly fading channels of two users before (above) and after (below) applying opportunistic beamforming.

i.e., $\mathbf{h}_k[m] = \mathbf{h}_k$, for all m (in practice, this means: for all m over the latency time scale of interest). The received SNR for this user would have remained constant if only one antenna were used. If all users in the system experience such slow fading, no multiuser diversity gain can be exploited. Under the proposed scheme, on the other hand, the overall channel gain $\mathbf{h}_k[m]^*\mathbf{q}[m]$ for each user k varies in time and provides an opportunity to exploit multiuser diversity.

Let us focus on a particular user k. Now if $\mathbf{q}[m]$ varies across all directions, the amplitude squared of the channel $|\mathbf{h}_k[m]^*\mathbf{q}[m]|^2$ seen by user k varies from 0 to $\sum_{l=1}^{n_t} |h_{lk}|^2$. The peak value occurs when the transmission is aligned along the direction of the channel of user k, i.e., $\mathbf{q}[m] = \mathbf{h}_k/\|\mathbf{h}_k\|$. The power and phase values are then in the *beamforming configuration* :

$$\alpha_l = \frac{\mid h_{lk} \mid^2}{\sum_{j=1}^{n_t} \mid h_{jk} \mid^2}, \qquad l = 1, \ldots, n_t,$$

$$\theta_l = -\arg(h_{lk}), \qquad l = 1, \ldots, n_t.$$

To be able to beamform to a particular user, the base station needs to know individual channel amplitude and phase responses from all the antennas, which requires much more information to feed back than just the overall SNR. However, if there are many users in the system, a good multiuser diversity harnessing scheduler (the proportional fair algorithm is one such) will schedule transmission to a user only when its overall channel SNR is near its peak. Thus, it is plausible that in a slowly fading environment, the technique can approach the performance of coherent beamforming but with only overall SNR feedback. In this context, the technique can be interpreted as *opportunistic beamforming*: by varying the phases and powers allocated to the transmit antennas, a beam is randomly swept and at any time transmission is scheduled to the user that is currently closest to the beam.

21.4.2 Fast fading: increasing channel fluctuations

We see that opportunistic beamforming can significantly improve performance in slowly fading environments by adding fast time-scale fluctuations on the overall channel quality. The rate of channel fluctuation is artificially sped up. Can opportunistic beamforming help if the underlying channel variations are already fast (fast compared with the latency time-scale)?

The long term throughput under fast fading depends only on the stationary distribution of the channel gains. The impact of opportunistic beamforming in the fast fading scenario then depends on how the stationary distributions of the overall channel gains can be modified by power and phase randomization. Intuitively, better multiuser diversity gain can be exploited if the dynamic range of the distribution of h_k can be increased, so that the maximum SNRs can be larger. We consider two examples of common fading models.

Independent Rayleigh fading In this model, appropriate for an environment where there is full scattering and the transmit antennas are spaced sufficiently apart, the channel gains $h_{1k}[m], \ldots, h_{n_tk}[m]$ are i.i.d. complex Gaussian random variables. In this case, the channel vector $\mathbf{h}_k[m]$ is isotropically distributed, and $\mathbf{h}_k[m]^*\mathbf{q}[m]$ is circularly symmetric Gaussian for any choice of $\mathbf{q}[m]$; moreover, the overall gains are independent across the users. Hence, the stationary statistics of the channel are *identical* to the original situation with one transmit antenna. Thus, in an independent fast Rayleigh fading environment, the opportunistic beamforming technique does not provide any performance gain.

Independent Ricean fading In contrast to the Rayleigh fading case, opportunistic beamforming has a significant impact in a Ricean environment, particularly when the κ-factor is large. In this case, the scheme can significantly increase the dynamic range of the fluctuations. This is because the fluctuations in the underlying Ricean fading process come from the diffused component, while with randomization of phase and powers, the fluctuations are from the coherent addition and cancellation of the direct path components in the signals from the different transmit antennas, in addition to the fluctuation of the diffused components. If the direct path is much stronger than the diffused part (i.e., if κ is large), then much larger fluctuations can be created with this technique.

This intuition is substantiated in Fig. 21.4, which plots the total throughput with the proportional fair algorithm (with large latency time scale) for Ricean fading with $\kappa = 10$. We see that there is a considerable improvement in performance going from the single transmit antenna case to dual transmit antennas with opportunistic beamforming. For comparison, we also plot the analogous curves for pure Rayleigh fading; as expected, there is no improvement in performance in this case. Fig. 21.5 compares the stationary distributions of the overall channel gain $|\mathbf{h}_k[m]^*\mathbf{q}[m]|$ in the single-antenna and dual-antenna cases; one can see the increase in dynamic range due to opportunistic beamforming.

21.5 Antennas: dumb, smart and smarter

It is insightful to compare the opportunistic beamforming technique with the two other important point-to-point transmit antenna techniques:

- *space-time codes* like the Alamouti scheme. They are primarily used to increase the diversity in slowly fading point-to-point links.
- *transmit beamforming*. In addition to providing diversity, a power gain is also obtained through the coherent addition of signals at the users.

The three techniques have different system requirements. Coherent space-time codes like the Alamouti scheme require the users to track all the *individual* channel gains (amplitude and phase) from the transmit antennas. This

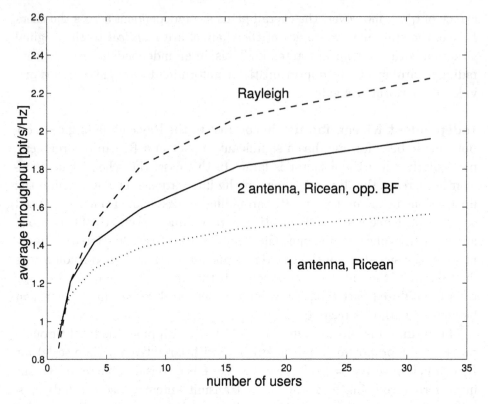

Fig. 21.4. Total throughput as a function of the number of users under Ricean fast fading, with and without opportunistic beamforming. The power allocations $\alpha_l[m]$ are uniformly distributed in $[0, 1]$ and the phases $\theta_l[m]$ uniform in $[0, 2\pi]$. The total throughput under Rayleigh fast fading is also shown for comparison.

requires separate pilot symbols on each of the transmit antennas. Transmit beamforming has an even stronger requirement that the channel should be known at the transmitter. In an FDD system, this means feedback of the individual channel gains (amplitude and phase). In contrast to these two techniques, the opportunistic beamforming scheme requires no knowledge of the individual channel gains, neither at the users nor at the transmitter. In fact, the users are *completely ignorant* of the fact that there are multiple transmit antennas and the receiver is identical to that in the single transmit antenna case. Thus, they can be termed *dumb antennas*. Opportunistic beamforming does rely on multiuser diversity scheduling, which requires the feedback of the overall SNR of each user. However, this only needs a *single* pilot to measure the overall channel.

What is the performance of these techniques when used in the downlink? In a slowly fading environment, we have already remarked that opportunis-

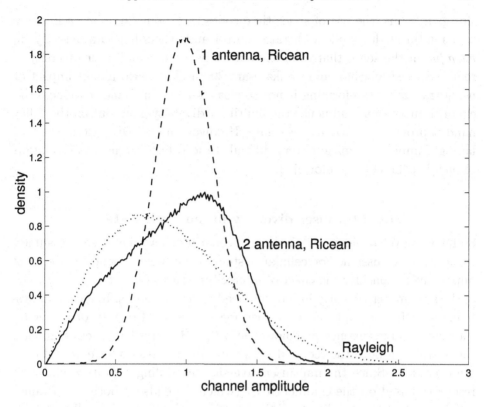

Fig. 21.5. Comparison of the distribution of the overall channel gain with and without opportunistic beamforming using two transmit antennas, Ricean fading. The Rayleigh faded channel is unaffected by opportunistic beamforming.

tic beamforming approaches the performance of transmit beamforming when there are many users in the system. On the other hand, space-time codes do not perform as well as transmit beamforming since they do not capture the array power gain. This means, for example, using the Alamouti scheme on dual transmit antennas in the downlink is 3 dB worse than using opportunistic beamforming combined with multiuser diversity scheduling when there are many users in the system. Thus, dumb antennas together with smart scheduling can surpass the performance of smart space-time codes and approach that of the even smarter transmit beamforming.

How about in a fast Rayleigh fading environment? In this case, we have observed that dumb antennas have no effect on the overall channel as the full multiuser diversity gain has already been realized. Space-time codes, on the other hand, *increase* the diversity of the point-to-point links and consequently *decrease* the channel fluctuations and hence the multiuser diversity gain. Thus, the use of space-time codes as a point-to-point technology in a multiuser downlink with rate control and scheduling can actually be *harmful,* in the sense that even the naturally present multiuser diversity is removed (Viswanathan and Venkatesan, 2004). The performance impact of using transmit beamforming is not so clear: on the one hand it reduces the channel fluctuation and hence the multiuser diversity gain, but on the other hand it provides an array power gain. However, in an FDD system the fast fading channel may make it very difficult to feed back so much information to enable coherent beamforming.

21.6 Multiuser diversity in multicell systems

So far we have considered a single-cell scenario, where the noise is assumed to be white Gaussian. For cellular systems with full frequency reuse, it is important to consider the effect of intercell interference on the performance of the system, particularly in interference-limited scenarios. In a cellular system, this effect is captured by measuring the channel quality of a user by the signal-to-interference-plus-noise ratio (SINR). In a fading environment, the energies in both the received signal and the received interference fluctuate over time. Since the multiuser diversity scheduling algorithm allocates resources based on the channel SINR (which depends on both the channel amplitude and the amplitude of the interference), it automatically exploits both the fluctuations in the energy of the received signal as well as that of the interference: the algorithm tries to schedule resource to a user whose instantaneous channel is good *and* the interference is weak. Thus, multiuser diversity naturally takes advantage of the time-varying interference to increase the spatial reuse of the network.

From this point of view, amplitude and phase randomization at the base station transmit antennas plays an additional role: it increases not only the amount of fluctuations of the received signal to the intended users *within* the cells, it also increases the fluctuations of the interference the base station causes in *adjacent* cells. Hence, opportunistic beamforming has a dual benefit in an interference-limited cellular system. In fact, opportunistic beamforming performs *opportunistic nulling* simultaneously: while randomization of amplitude and phase in the transmitted signals from the antennas allows near coherent beamforming to some user within the cell, it will create near nulls at some other user in adjacent cells. This in effect allows *interference avoidance* for that user if it is currently being scheduled.

Let us focus on the downlink in a slowly flat-fading scenario to get some insight into the performance gain from opportunistic beamforming and nulling.

Under amplitude and phase randomization at all base stations, the received signal of a typical user that is interfered by J adjacent base stations is given by

$$y[m] = (\mathbf{h}^*\mathbf{q}[m])\, x[m] + \sum_{j=1}^{J} (\mathbf{g}_j^*\mathbf{q}_j[m])\, u_j[m] + z[m].$$

Here, $x[m], \mathbf{h}, \mathbf{q}[m]$ are respectively the signal, channel vector and random transmit direction from the base station of interest; $u_j[m], \mathbf{g}_j, \mathbf{q}_j[m]$ are respectively the interfering signal, channel vector and random transmit direction from the jth base station. All base stations have the same transmit power P and n_t transmit antennas, and perform amplitude and phase randomization independently.

By averaging over the signal $x[m]$ and the interference terms $u_j[m]$, the (time-varying) SINR of user k can be computed to be

$$\mathsf{SINR}_k[m] = \frac{P|\mathbf{h}^*\mathbf{q}[m]|^2}{P\sum_{j=1}^{J}|\mathbf{g}_j^*\mathbf{q}_j[m]|^2 + N_0}. \tag{21.3}$$

As the random transmit directions $\mathbf{q}[m], \mathbf{q}_j[m]$ vary, the overall SINR changes over time. This is due to the variations of the overall gain from the base station of interest as well as those from the interfering base stations. The SINR is high when $\mathbf{q}[m]$ is closely aligned to the channel vector $\mathbf{h}$, and/or for many j, $\mathbf{q}_j[m]$ is nearly orthogonal to $\mathbf{g}_j$, i.e., the user is near a null of the interference pattern from the jth base station. In a system with many other users, the proportional fair scheduler will serve this user while its SINR is at its peak $P\|\mathbf{h}\|^2/N_0$, i.e., when the received signal is the strongest and the interference is completely nulled out. Thus, the opportunistic nulling and beamforming technique has the potential of shifting a user from a low SINR, interference-limited regime to a high SINR, noise-limited regime.

21.7 A concluding system view

A new design principle for wireless systems can now be seen through the lens of multiuser diversity. In the classical viewpoint, the design techniques center on making the *individual* point-to-point links as close to AWGN channels as possible, with a reliable channel quality that is constant over time. This is accomplished by *channel averaging*, and includes the use of diversity techniques such as multipath combining, time interleaving and antenna diversity that attempt to keep the channel fading constant in time, as well as interference management techniques such as interference averaging by means of spreading.

However, if one shifts from the view of the wireless system as a set of point-to-point links to the view of a system with multiple users sharing the same resources (spectrum and time), then quite a different design objective suggests itself. Indeed, the results in this chapter suggest that one should instead try to *exploit* the channel fluctuations. This is done through an appropriate scheduling algorithm that "rides the peaks", i.e., each user is scheduled when it has a very strong channel, while taking into account real world traffic constraints such as delay and fairness. The technique of dumb antennas goes one step further by *creating* variations when there are none. This is accomplished by varying the strengths of *both* the signal and the interference that a user receives through opportunistic beamforming and nulling.

The viability of the opportunistic communication scheme depends on traffic that has some tolerance to scheduling delays. On the other hand, there are some forms of traffic that are not so flexible. The functioning of the wireless systems is supported by the overhead control channels, which are "circuit-switched" and hence have very tight latency requirements, unlike data channels, which have the flexibility to allow dynamic scheduling. From the perspective of these signals, it is preferable that the channel remained unfaded; a requirement that is contradictory to our scheduler-oriented observation that we would prefer the channel to have fast and large variations.

This issue suggests the following design perspective: separate very-low latency signals (such as control signals) from flexible latency data. One way to achieve this separation is to split the bandwidth into two parts. One part is made as flat as possible (by spreading over this part of the bandwidth, say) and is used to transmit flows with very low latency requirements. The performance metric here is to make the channel as reliable as possible (equivalently keeping the probability of outage low) for some fixed data rate. The second part uses opportunistic beamforming to induce large and fast channel fluctuations and a scheduler to harness the multiuser diversity gains. The performance metric on this part is to maximize the multiuser diversity gain.

The gains of the opportunistic beamforming and nulling depend on the probability that the received signal is nearly beamformed *and* all the interference is nearly null. In the interference-limited regime and when $P/N_0 \gg 1$, the performance depends mainly on the probability of the latter event. In the downlink, this probability is large since there are only one or two base stations contributing most of the interference. The uplink poses a contrasting picture: there is interference from many mobiles allowing interference averaging. Now the probability that the *total* interference is near null is much smaller. Interference averaging, which is one of the principle design features of wideband full reuse systems, is actually unfavorable for the opportunistic scheme described here, since it reduces the likelihood of the nulling of the interference and hence the likelihood of the peaks of the SINR.

In a typical cell, there will be a distribution of users, some closer to the base station and some closer to the cell boundaries. Users close to the base station are at high SINR and are noise-limited; the contribution of the inter-cell interference is relatively small. These users benefit mainly from opportunistic beamforming. Users close to the cell boundaries, on the other hand, are at low SINR and are interference-limited; the average interference power can be much larger than the background noise. These users benefit both from opportunistic beamforming and from opportunistic nulling of intercell interference. Thus, the cell-edge users benefit more in this system than users in the interior. This is rather desirable from a system fairness point-of-view, as the cell-edge users tend to have poorer service. This feature is particularly important for a system without soft handoff (which is difficult to implement in a packet data scheduling system). To maximize the opportunistic nulling benefits, the transmit power at the base station should be set as large as possible, subject to regulatory and hardware constraints.

We have seen the multiuser diversity as primarily a form of power gain. The opportunistic beamforming technique of using an array of multiple transmit antennas has approximately an n_t-fold improvement in received SNR to a user in a slowly fading environment (Viswanath *et al.*, 2002, Theorem 1), as compared with the single-antenna case. With an array of n_r *receive* antennas at each mobile (and say a single transmit antenna at the base station), the received SNR of any user gets an n_r fold improvement as compared with a single receive antenna; this gain is realized by *receiver beamforming*. This operation is easy to accomplish since the mobile has full channel information at each of the antenna elements. Hence the gains of opportunistic beamforming are about the same order as that of installing sets of receive antenna arrays at *each* of the mobiles.

Thus, for a system designer, the opportunistic beamforming technique provides a compelling case for implementation, particularly in view of the constraints of space and cost of installing multiple antennas on *each* mobile device. Further, this technique neither needs any extra processing on part of any user, nor any updates to an existing air-link interface standard. In other words, the mobile receiver can be completely ignorant to the use or nonuse of this technique. This means that it does not have to be "designed in" (by appropriate inclusions in the air interface standard and the receiver design) and can be added/removed at any time. This is one of the important benefits of this technique from an overall system design point of view.

In traditional cellular wireless systems, the cell is sectorized to allow better focusing of the power transmitted from the antennas and also to reduce the interference seen by mobile users from transmissions of the same base station but intended for users in different sectors. This technique is particularly gainful in scenarios when the base station is located at a fairly large height and thus there is limited scattering around the base station. In contrast, in systems with far denser deployment of base stations (a strategy that can be expected to be a good one for wireless systems aiming to provide mobile broadband data services), it is unreasonable to stipulate that the base stations be located high above the ground so that the local scattering (around the base station) is minimal. In an urban environment, there is substantial local scattering around a base station and the gains of sectorization are minimal; users in a sector also see interference from the same base station (due to the local scattering) intended for another sector. The opportunistic beamforming scheme can be thought of as sweeping a random beam and scheduling transmissions to users when they are beamformed. Thus, the gains of sectorization are automatically realized. We conclude that the opportunistic beamforming technique is particularly suited to harness sectorization gains even in low height base stations with plenty of local scattering. In a cellular system, the opportunistic beamforming scheme also obtains the gains of nulling, a gain traditionally obtained by coordinated transmissions from neighboring base stations in a full frequency reuse system or by appropriately designing the frequency reuse pattern.

Acknowledgments

The research that led to the material in this chapter was supported in part by the National Science Foundation under grant CCR 0312413 and by the Motorola Center for Communication. The material in this chapter is adapted from Viswanath *et al.* (2002) and Tse and Viswanath (2005, Chapter 6).

References

Tse, D. (1997). Optimal power allocation over parallel Gaussian broadcast channels. In *Proc. IEEE Int. Symp. Inf. Theory*, 27.

Tse, D. and Viswanath, P. (2005). *Fundamentals of Wireless Communications* (Cambridge Univ. Press).

Viswanath, P., Tse, D., and Laroia, R. (2002). Opportunistic beamforming using dumb antennas. *IEEE Trans. Inf. Theory*, **48** (6), 1277–1294.

Viswanathan, H. and Venkatesan, S. (2004). The impact of antenna diversity in packet data systems with scheduling. *IEEE Trans. Commun.*, **52** (4), 546–549.

22

System level performance of MIMO systems

Ninoslav Marina

École Polytechnique Fédérale de Lausanne (EPFL)

Olav Tirkkonen and Pirjo Pasanen

Nokia Research Center

22.1 Introduction

The fundamental limitation of communication over the radio interface is the scarcity of bandwidth. To support higher data rates and increasing numbers of subscribers, a substantial increase in spectral efficiency has to be achieved. Theoretically, MIMO technologies promise a significant increase in capacity. However, due to size limitations, the number of antennas is strictly limited, especially at the mobile transceiver. Taking different polarizations into account, four antennas at the mobile terminal may be considered an upper limit. In the base station the number of antennas can be larger, but in a sectorized cell, four antennas per sector may be a reasonable estimate.

When designing a cellular system, multiuser and multicellular capacity are of primary importance. For single-input single-output (SISO) channels, cellular CDMA and/or TDMA capacities with different reuse patterns have been widely investigated, e.g., by Viterbi (1995), Stüber (1996), Lee and Steele (1995), Caire *et al.* (1998), and Shamai and Verdú (2001).

In this chapter, we address the problem of cellular MIMO system capacity using information-theoretic tools. We expand the existing MIMO work (Catreux *et al.*, 2001; Farrokhi *et al.*, 2002; Lozano and Tulino, 2002; Gray *et al.*, 2003) by providing a semianalytic treatment of the out-of-cell interference for any reuse pattern. To this end, we use tools introduced by Viterbi (1995), Stüber (1996), Lee and Steele (1995), and Caire *et al.* (1998).

We consider flat fading channels, and use the acronym TDMA to stand for all orthogonal multiple-access strategies where one user is accessing the channel at a given time, and the acronym CDMA to stand for all nonorthogonal multiple-access strategies where multiple users access the channel at a given time. The results discussed here are directly applicable to a multicarrier-based wideband system with cyclic prefix: orthogonal frequency-division multiple access (OFDMA) and TDMA-OFDM are examples of orthogonal strategies. For single-carrier wideband systems and multicarrier CDMA (MC-CDMA), our results are not valid. To assess such systems, interpath interference and equalizer performance need to be properly modeled.

The analysis presented here is for continuous coverage cellular systems. System capacity is measured as spectral efficiency per unit area, where the unit area is the area of a cell. For noncellular systems, such as WLAN hot spots, the methods discussed here are not needed.

22.2 Multiple-access channel model

Single-user channel We assume flat fading, memoryless, multiple-input multiple-output channels, with N_t transmit antennas and N_r receive antennas. The baseband signal model for a single user is $\mathbf{y} = \mathbf{F}\mathbf{x} + \mathbf{z}$, where $\mathbf{F}$ is the $N_r \times N_t$ fading matrix, $\mathbf{x}$ is the vector of transmitted signals from the N_t antennas during one channel use, and $\mathbf{z}$ is the vector of additive noise at each receive antenna. Flat fading is the most important simplification used in this chapter, restricting the immediate applicability of the results to narrowband systems.

Multiple-access channel Multiple users may be accessing the channel within a given geographic area, known as a cell. Assuming a uniform user density ρ, the number of active users per cell is

$$N_u = \rho\, S_c, \tag{22.1}$$

where S_c is the area of the cell. In this chapter, we concentrate on multiple-access methods, so the most straightforward interpretation is in terms of the uplink (reverse) direction of cellular systems. However, also the downlink can be modeled with the tools discussed here, if we ignore the broadcast channel aspects of a single-cell downlink (forward link), and interpret the multiple-access channel in terms of multiple geographically distributed base stations accessing the channel. When not explicitly stated, however, the terminology will be that of the uplink. For simplicity, we assume all users to have the same N_t, and all cells to have the same N_u. All receivers (base stations) are equipped with N_r antennas.

Different multiple-access schemes translate into different numbers of active users in a cell. In a fully nonorthogonal multiple-access scheme, like CDMA, all users are active all the time, and the channel model reads

$$\mathbf{y} = \sum_{u=1}^{N_u} \mathbf{F}_u \mathbf{x}_u + \mathbf{z}. \tag{22.2}$$

Here, $\mathbf{F}_u$ are the fading matrices for the different users, and $\mathbf{z}$ describes, apart from additive noise, the out-of-cell interference. The interference originating from within the same cell is explicitly accounted for by the system model.

Large-scale effects and fast fading As we are interested in capacities of cellular systems, large-scale effects have to be taken into account when discussing the realizations of the fading matrices. We assume that the large-scale effects are the same for all antenna pairs, so that

$$\mathbf{F}_u = \sqrt{\Theta_u} \, \mathbf{H}_u, \tag{22.3}$$

where the scalar factor Θ_u describes the large-scale effects, and the matrix $\mathbf{H}_u$ describes fast fading, which is different for different antenna pairs. For simplicity, we assume that the fast fading matrices $\mathbf{H}_u$ have i.i.d. zero-mean complex Gaussian entries h_{ij} with unit variance, so that $E[|h_{ij}|^2] = 1$.

Path loss and shadowing When path loss and shadowing effects are taken into account, the large scale effects may be modeled as (Friis, 1946)

$$\Theta(r, \xi) = A \, r^{-\beta} \cdot 10^{\frac{\xi}{10}}. \tag{22.4}$$

Here, r is the distance between the user and the base station, and A is an attenuation factor that depends on filter, antenna, and line losses, the antenna gains, etc. The path loss exponent β can take values from 1.6 to 6, with a typical value for urban cells being 3.5. The geographical design of a cellular network is largely a trade-off between coverage and base station density, which is governed by β.

Log-normal shadowing models the large-scale effects of the channel that are not directly dependent on the distance r, e.g., effects of buildings, foliage, and so on It is described by the zero-mean normal random variable ξ with standard deviation σ_s varying between 5 dB and 12 dB, with 8 dB being a typical value. The average effect of the shadowing is

$$O(\sigma_s) = E[10^{\xi/10}] = \int 10^{\xi/10} \frac{e^{-\frac{\xi^2}{2\sigma_s^2}}}{\sqrt{2\pi}\sigma_s} d\xi = e^{\frac{1}{2}\left(\frac{\sigma_s \ln 10}{10}\right)^2}. \tag{22.5}$$

Note that with increasing shadowing standard deviation, the average channel gain becomes slightly *bigger*. This is a consequence of the log-normal shadowing model.

Power equalization and power control Side information on the large-scale effects of the channel can be used at the transmitter to invoke *Power Equalization* (PE), a.k.a. slow power control. The power of mobile units (in the uplink) is adjusted so that the mean total received power is the same for all units. In a multiuser CDMA system with matched filter (rake) receiver, power equalization is necessary to mitigate the near-far effect.

Similarly, side information on instantaneous fading realizations can be used to invoke *Power Control* (PC), which keeps the instantaneous total received power of a user approximately constant. Here, we shall not consider PC, nor other methods requiring extensive side information.

Power constraint Multiaccess capacity strongly depends on the power constraint used. Here, we consider mainly

$$\sum_u E_t\left[P_u(t)\right] \le P_{\text{tot}}, \qquad (22.6)$$

i.e., we constrain the *total average power radiated into a cell by all active users*. Alternatively, the instantaneous transmit power of each user may be constrained as

$$P_u(t) \le P_{\text{inst}} \quad \forall\, u, t. \qquad (22.7)$$

By definition, P_{tot} is more closely related to E_b/N_0 than P_{inst}. Applying the latter allows a higher P_{tot} for a CDMA (or FDMA) system, where multiple users are active simultaneously, than for a TDMA system with a single active user.

The MIMO received power per receive antenna observed at the base station from a user at distance r, *averaged over fast fading*, is

$$P_R\left(r, \xi, P_t\right) = \Theta(r, \xi)\, E_{\mathbf{H}}\left[\text{Tr}\left(\mathbf{H}^\dagger\mathbf{H}\right)/N_r\right] P_t/N_t = \Theta(r, \xi)\, P_t\,. \qquad (22.8)$$

Note that P_t is the sum of the transmitted power from all antennas. Also note that (22.8) does not assume anything about the distribution of the transmit power across antennas.

With perfect power equalization, the expected received power per antenna is a constant P_R, and the transmitted power is a function of the distance and the shadowing, $P_t(r, \xi) = P_R/\Theta(r, \xi)$. Assuming N_u uniformly distributed users in a circular cell with radius c, the density of users is

$$\rho = N_u/(\pi c^2). \qquad (22.9)$$

Using (22.4), (22.5), and (22.8), the average total transmit power is

$$P_t^{(\text{tot})} = N_u \, E_\xi \left[\frac{2}{c^2} \int_0^c r \, P_t(r, \xi) \, dr \right] = N_u \, O(\sigma_s) \, \frac{2c^\beta P_R}{A(\beta + 2)} \equiv N_u \, L \, P_R.$$

(22.10)

The last equality defines an average attenuation factor L. Note that the area of an infinitesimal ring with radius r is $2\pi r dr$. Combined with the density of users, this gives the probability density $2r/c^2$ for users at distance r.

22.3 Multicellular network model

In a *cellular* network, there are multiple base stations distributed in a geographical region, and the region is divided into cells, with one base station per cell. The cell belonging to a base station is the area where the signal to/from this station is strongest. Every point in the region belongs to a cell, so that (almost) 100% coverage is guaranteed; wherever a user is, it will be able to connect to the network. Often, individual cells are further divided into sectors, using, e.g., directive antennas. For example, one cell may be divided into three sectors of 120°.

In a *noncellular* network, continuous coverage is not guaranteed. This is the case for, e.g., typical WLAN systems, where service is concentrated to isolated hot spots.

Reuse In a cellular network, interference becomes a problem. On any line connecting two neighboring base stations, at least one point exists where the signal to or from both stations is exactly the same. The resulting cochannel interference problem may be solved by frequency planning. The total system bandwidth is divided into bands, and the same band is not used in neighboring cells. The distribution of frequency bands among cells is called a *reuse pattern*.

In TDMA systems, such as GSM, frequency planning has been the primary way of solving the out-of-cell interference problem. In contrast, CDMA systems may apply reuse 1—the full system bandwidth may be used in every cell. This is possible due to the spreading nature of CDMA. At a given time, the interference from the users in a neighboring cell becomes averaged over an almost uniform distribution of locations. In a TDMA system, at a given time, there is only one active user per cell, and interference is not averaged. Modern developments of CDMA systems aim at higher single-user data rates, which leads to smaller interference averaging and increasing interference problems.

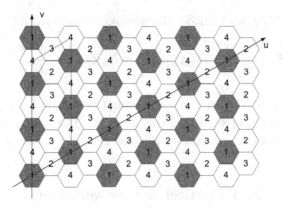

Fig. 22.1. Grid of hexagons with hexagonal coordinate axes and reuse pattern 4.

Hexagonal cellular network To analyze a cellular network, the most convenient model is a grid of regular hexagons of radius a (MacDonald, 1979). The radius of a regular hexagon is the distance from the center to one of the vertices, which is equal to the length of an edge. The area of a hexagonal cell is $3\sqrt{3}a^2/2$, and the density of users would be

$$\rho = 2N_u/(3\sqrt{3}a^2), \tag{22.11}$$

instead of (22.9). In a hexagonal cellular network, the possible values of the reuse factor are

$$q = i^2 + ij + j^2, \tag{22.12}$$

described by two nonnegative integers i and j with $j \le i$, see Stüber (1996), Parsons (2001). The smallest possible values are 3,4,7,9,12,13,16,19,21,25. In Fig. 22.1, we see the reuse pattern for $q = 4$. We assume that the base station is always in the middle of the cell, and that the mobile stations (MS) are uniformly distributed over the cell.

Note that in a realistic network, base stations are not evenly distributed on a regular lattice, and the medium is not homogeneous, so that boundaries of cells will not be straight lines.

Hexagonal coordinate system To describe the geometry of a hexagonal network, we use hexagonal coordinates (MacDonald, 1979; Parsons, 2001), shown in Fig. 22.1. With coordinates normalized to cell radius a, the distance between two points (u_1, v_1) and (u_2, v_2) is

$$d_{12} = a\sqrt{3}\sqrt{(u_2 - u_1)^2 + (u_2 - u_1)(v_2 - v_1) + (v_2 - v_1)^2}. \tag{22.13}$$

With reuse factor q, the hexagonal coordinates of the positions of all base stations using the same frequency as the one at the origin are given by

$$u(t, n_u) = (t \cdot i) \mod q + n_u \cdot q, \qquad n_u \in \mathbb{N}$$
$$v(t, n_v) = (t \cdot j) \mod q + \frac{n_v \cdot q}{\gcd(i,j)}, \qquad n_l \in \mathbb{N}, \qquad (22.14)$$

where $t \in \{0, \dots, q/\gcd(i,j) - 1\}$, and $\gcd(i,j)$ is the greatest common divisor of i and j. This parameterization finds all the cells with centers in the sixth of the plane with polar angles $\in [\pi/6, \pi/2)$. The distances of the centers of these cells from the receiving base station at the origin may be calculated from (22.13), and are denoted by b_{t,n_u,n_v}.

Handover In a typical cellular system users are not aware of their position relative to the serving base station. They measure pilot signals from several base stations and select the serving station accordingly, in a process known as handover, see, e.g., Viterbi (1995). In *soft handover*, users may be connected to multiple base stations simultaneously. In a simpler *hard handover* scenario, the user selects the base station that is most likely to have the strongest channel, based on long term measurements. The simplest way to model hard handover is to consider a user to be connected to the nearest cell, irrespectively of the fast fading and shadowing realization.

22.4 Interference modeling

The N_u users in the multiaccess MIMO channel (22.2) obey the same statistics, and are thus statistically indistinguishable. Then, the throughput capacity is N_u times the capacity of the first user. Therefore, we analyze what happens with the capacity of the first user only. With *perfect power equalization*, we have the signal model

$$\mathbf{y} = \mathbf{H}_1 \tilde{\mathbf{x}}_1 + \mathbf{z}_{\text{in}} + \mathbf{z}_{\text{out}} + \mathbf{z}_n, \qquad (22.15)$$

where $\mathbf{z}_n$ is additive white zero-mean Gaussian noise with power spectral density N_0, $\mathbf{z}_{\text{out}}$ is the interference caused by out-of-cell users, and the cochannel interference caused by the N_u in-cell users is

$$\mathbf{z}_{\text{in}} = \sum_{u=2}^{N_u} \mathbf{H}_u \tilde{\mathbf{x}}_u . \qquad (22.16)$$

The transmitted signals $\tilde{\mathbf{x}}_u = \sqrt{\Theta(r, \xi)} \mathbf{x}_u$ have been scaled to take into account the path loss and shadowing. For user u, the covariance of the transmitted symbols is

$$\mathbf{Q}_u = E[\tilde{\mathbf{x}}_u \tilde{\mathbf{x}}_u^{\dagger}] = \mathrm{diag}\left[P_1^{(u)} \; P_2^{(u)} \; \cdots \; P_{N_t}^{(u)}\right]. \qquad (22.17)$$

With perfect power equalization, the input power constraint is

$$\mathrm{tr}(\mathbf{Q}_u) = \sum_{m=1}^{N_t} P_m^{(u)} = P_R, \qquad (22.18)$$

where P_R is the expected received power per antenna, as discussed in Section 22.2. The power constraint of the true transmitted symbols is given by $\mathrm{tr}(Ex_u x_u^{\dagger}) = P_R/\Theta(r_u, \xi_u)$.

If N_u is big enough, the central limit theorem may be invoked and $\mathbf{z}_{\mathrm{in}}$ also modeled as zero-mean Gaussian, with covariance matrix $\mathbf{Q}_I$. We made a similar assumption before also for the out-of-cell interference. In this section, we consider these interference factors and their covariances, along with the accuracy of the Gaussian approximation.

22.4.1 In-cell interference

Assuming independent users, the covariance matrix of $\mathbf{z}_{\mathrm{in}}$ of (22.16) is

$$\mathbf{Q}_{\mathrm{in}} = E\left[\sum_{u=2}^{N_u} \mathbf{H}_u \mathbf{x}_u \sum_{v=2}^{N_u} \mathbf{x}_v^{\dagger} \mathbf{H}_v^{\dagger}\right] = \sum_{u=2}^{N_u} E\left[\mathbf{H}_u \mathbf{x}_u \mathbf{x}_u^{\dagger} \mathbf{H}_u^{\dagger}\right].$$

Furthermore, when there is no side information at the transmitter, the symbols and fading matrices are independent. Using the covariance (22.17) and the i.i.d. property of fading, we find that $\mathbf{Q}_{\mathrm{in}}$ is a diagonal matrix with entries

$$I_{\mathrm{in}} = \sum_{u=2}^{N_u} \sum_{m=1}^{N_t} E\left[\left|h_{nm}^{(u)}\right|^2\right] P_{u,m} = (N_u - 1)P_R . \qquad (22.19)$$

The last equality follows from (22.18). Thus the expected interference is constant across receive antennas, and

$$\mathbf{Q}_{\mathrm{in}} = (N_u - 1)P_R \, \mathbf{I} \equiv I_{\mathrm{in}} \, \mathbf{I}. \qquad (22.20)$$

This result applies to the CDMA uplink, where the users simultaneously transmit independent signals. In TDMA and downlink CDMA, the users are not independent, but orthogonal. Then I_{in} is solely due to the multipath nature of the channel. To calculate the interference in this case, an appropriate model needs to be developed. This is beyond the scope of this chapter.

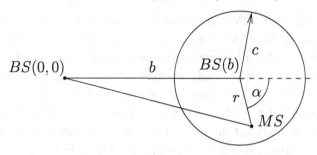

Fig. 22.2. Geometry of interference from an out-of-cell mobile station.

22.4.2 Out-of-cell interference

As above, the covariance of the out-of-cell interference $\mathbf{z}_{\text{out}}$ is diagonal, $\mathbf{Q}_{\text{out}} = I_{\text{out}} \, \mathbf{I}$. The out-of-cell interference for CDMA was determined in Viterbi (1995) and Stüber (1996). In the former, perfect power equalization was assumed, in the latter, perfect power control. For simplicity, we follow Viterbi (1995) here.

In Fig. 22.2, the base station in the cell of interest is denoted by $BS(0,0)$, a base station at distance b is denoted by $BS(b)$, and a mobile station power equalized by $BS(b)$ by MS. Accordingly, the distance r of MS from $BS(b)$ and the corresponding shadowing realization ξ_r determine the transmit power used by MS. The distance from MS to $BS(0,0)$ is y. In addition, a circular approximation of the cell surrounding $BS(b)$ is depicted.

Using (22.4) and (22.8), it is easy to see that the interference at $BS(0,0)$ from the power equalized MS is

$$I = P_t \, \Theta(y, \xi_y) = P_t \, A \, y^{-\beta} \, 10^{\xi_y/10} = P_R \left(\frac{r}{y}\right)^{\beta} 10^{(\xi_y - \xi_r)/10},$$

where ξ_y is the shadowing realization from MS to $BS(0,0)$. The expected interference experienced at $BS(0,0)$ from the users in an out-of-cell region S becomes

$$E\left[I_S\right] = P_R \, O(\tilde{\sigma}_s) \int_S dA \, \rho \left(\frac{r}{y}\right)^{\beta}. \tag{22.21}$$

The integral is over the area S, and the integration measure is an area element. Note that the density of users ρ is incorporated into the model. The distance y is always measured with respect to $BS(0,0)$, whereas the distance r is with respect to the base station in the multicellular network serving MS according to the handover scenario used. The effect of shadowing is denoted by $O(\tilde{\sigma}_s) = E\left[10^{(\xi_y - \xi_r)/10}\right]$, which is the expected value of the log-normal fading random variable $\xi_y - \xi_r$.

Shadowing correlation Before proceeding, we address the problem of shadowing correlation. If variables ξ_r and ξ_y are completely correlated, the effect of shadowing on the interference vanishes completely, and $O(\tilde{\sigma}_s) = 1$. If the shadowing realizations are i.i.d., we have $O(\tilde{\sigma}_s) = O(\sigma_s)^2$, where $O(\sigma_s)$ is the expected log-normal shadowing with standard deviation σ_s, expressed in (22.5). As shadowing models the effect of large-scale objects, it makes sense to assume a correlation that lies in between these two extremes. Often a shadowing correlation of 0.5 is assumed, giving $O(\tilde{\sigma}_s) = O(\sigma_s)$.

Approximate interference from a cell The average interference from a cell can be obtained by integrating (22.21) over the area of a cell. An approximation can be found by considering the circular cell depicted in Fig. 22.2. From the figure, we have $y^2 = b^2 + r^2 + 2br \cos \alpha$. The average interference coming from a cell of radius c at distance b is

$$I(b) = \frac{N_u}{\pi c^2} \, P_R \, O(\tilde{\sigma}_s) \int_0^c dr \; r^{1+\beta} \int_0^{2\pi} (b^2 + r^2 + 2br \cos \alpha)^{-\beta/2} d\alpha, \quad (22.22)$$

and can be solved in closed form if β is an even integer. The user density (22.9) was used in (22.22). Numerical results show that (22.22) is a good approximation for hexagonal cells. In Marina (2004), a doughnut-sector approximation was made for the interference calculation from a cell, resulting in a one-dimensional integral replacing (22.22) for generic β. It was observed that the two-dimensional approximation (22.22) was more exact. As numerical methods are needed, (22.22) may be used instead of a one-dimensional approximation.

Note that the average interference is scale invariant, i.e., it does not depend on the size of the cell. Thus, we can put $c = 1$ without losing any information. An absolute length scale appears if an absolute power constraint is assumed.

22.4.3 CDMA out/in-cell interference ratio

With shadowing correlation 0.5, the ratio between the out-of-cell and the in-cell interference for a CDMA system becomes

$$f = \frac{E[I_{\text{out}}]}{E[I_{\text{in}}]} = \frac{2 \, O(\sigma_s)}{3\sqrt{3}} \int_{S_0} dA \left(\frac{r}{y}\right)^\beta, \quad (22.23)$$

where the integration is over the complement S_0 of the hexagonal cell of $BS(0,0)$ located at the origin. The in-cell interference (22.20) and the user density (22.11) were used. The interference factor is scale invariant, and the size of the cell is omitted from the expression for the user density. In the integration, the omitted cell at the origin has radius $a = 1$.

The integral in (22.23) was numerically solved in Viterbi (1995). With standard deviation $\sigma_s = 8$ dB and path-loss exponent $\beta = 4$, f was found to be approximately 0.6. For lower β, f is higher. When changing the shadowing standard deviation from 8 dB to 0 dB, f drops from 0.6 to 0.52.

22.4.4 Out-of-cell interference with reuse

In this section, we compute the out-of-cell interference for a system with reuse. That is, in (22.23), the integration area S_0 is restricted to a subset of cochannel cells, which depends on the reuse pattern. The problem of signal-to-interference ratio calculation for a TDMA system with power equalization and reuse factor 3 was treated in Lee and Steele (1995) using Monte Carlo simulation. In Caire *et al.* (1998), a few reuse factors were similarly treated with Monte Carlo simulations, and SISO system capacities in terms of the number of users with given outage probability were reported. In Marina (2004), a semianalytic treatment was used, where the interference is approximated with one-dimensional integrals.

When the reuse factor q is greater than one, we need to sum interference that comes from all interfering cells that are sharing the same frequency, according to the reuse pattern. For this, we may use the hexagonal coordinate system discussed in Section 22.3. Since the system is hexagonally symmetric, the summation over the reuse pattern needs to be done only in one of the six hexagonal directions. If omnidirectional antennas are assumed, the interference coming from the whole plane is six times the interference coming from one hexagonal direction.

Here, however, we assume perfectly sectored antennas, so that there are three 120° sectors per cell, and there is no interference between sectors. Then the interference experienced by a sector is twice the interference coming from one hexagonal direction, as a hexagonal direction corresponds to a 60° angle at the base station.

Knowing the average interference $I(b)$ of (22.22) that comes from a cell at distance b, we can calculate the total mean interference factor as

$$I_q = 2 \sum_{t=0}^{(q/\gcd(i,j))-1} \sum_{\substack{n_u=0 \\ n_u+t\neq 0}}^{\infty} \sum_{n_v=0}^{\infty} I(b_{t,n_u,n_v})/P_R. \qquad (22.24)$$

Here, the base station distances b are parameterized as in (22.14). Note that I_q is normalized through division by P_R. The full sector is recovered

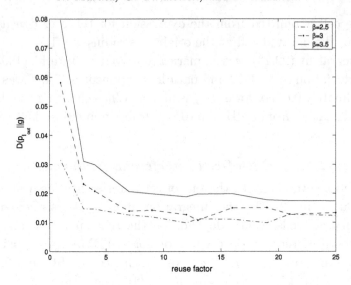

Fig. 22.3. Kullback–Leibler distance as a function of the reuse factor for different β.

multiplying by two, and the restriction on the sum over n_u removes the cell at the origin.

22.4.5 Validity of the Gaussian approximation

To measure the validity of applying the central limit theorem to the out-of-cell interference, we observe the relative entropy (Kullback–Leibler distance) between two probability density functions $p(x)$ and $g(x)$, defined as

$$D(p\|g) = \int p(x) \log \frac{p(x)}{g(x)} dx.$$

In Fig. 22.3, we plot the relative entropy between the real probability density function (pdf) of the out-of-cell interference $p_{I_{out}}(x)$ and the Gaussian pdf $g(x)$ with the same mean and variance, as a function of the reuse factor for different β.

For reuse factor $q = 1$, the relative entropy distance is significantly bigger than for a higher q. Thus for TDMA with reuse factor at least 3, the Gaussian approximation makes sense. For TDMA with reuse one, however, the out-of-cell interference is far from Gaussian, because of the much stronger contribution of the adjacent cells. Similarly, it may be observed that for CDMA with in-cell interference from at least fifteen users, the Gaussian approximation works well.

22.5 Multicellular capacity

Before addressing the system capacity, the link capacity of users with the signal model 22.15, operating on a bandwidth W should be considered. When the central limit theorem applies, the covariance of the interference and noise is $\mathbf{Q_z} = (I_{\text{in}} + I_{\text{out}} + N_0 W)\,\mathbf{I}$. Users' symbols have diagonal covariance (22.17) with power constraint (22.18). The results of Telatar (1999) may be directly applied: capacity is achieved by a Gaussian input at each antenna, with the covariance (22.17) proportional to identity, $\mathbf{Q_u} = P_R/N_t\,\mathbf{I}$. The link capacity (in bits/s) is

$$
C = W \sum_{i=1}^{\min(N_t,N_r)} E\left[\log_2\left(1 + \frac{\lambda_i P_R/N_t}{I_{\text{out}} + I_{\text{in}} + N_0 W}\right)\right],
\tag{22.25}
$$

where λ_i are the $\min(N_t, N_r)$ nonzero eigenvalues of $\mathbf{H}^\dagger\mathbf{H}$, and the expectation is over the distribution of the eigenvalues.

22.5.1 Processing gain from direct sequence spreading

It is straightforward to understand (22.25) with I_{in} of equation (22.19) in terms of properties of pseudorandom spreading sequences.

With a Fourier bandwidth W and a time T, the number of real dimensions in the signal space is $2TW$. With spreading, the signal waveform covers only part of the dimensionality of the signal space. Consider direct sequence spreading with a spreading factor N_s, realized by a *pseudorandom spreading sequence* $\mathbf{s} = [s_1, \ldots, s_{N_s}]^T/\sqrt{N_s}$, $s_i \in \{-1, 1\}$. The chip duration is $1/W$, with W the Fourier bandwidth used in the cell. Each symbol is transmitted over a sequence of N_s chips, so the symbol period is $T = N_s/W$. Out of the $2N_s$ possible waveforms, only two (the I- and Q-branch) are used by the user, defining the *Shannon bandwidth* $W_s = W/N_s$, see Massey (1994).

Omitting fading and MIMO for the time being, the channel model during one symbol period is $\mathbf{y} = x\mathbf{s} + \mathbf{z}$, where $\mathbf{z}$ describes the interference (both in- and out-of-cell) and noise. The vectors are N_s-vectors in *discrete* time, and accordingly, the covariance of $\mathbf{z}$ is $\mathbf{Q_z} = (I/W + N_0)\,\mathbf{I}$, where $I/W + N_0$ is the interference plus noise *energy* during a chip interval. The expected symbol energy is $E\left[|x|^2\right] = P_R T = N_s P_R/W$. An equivalent channel model can be constructed by matched filtering with the spreading sequence, resulting in $\tilde{y} = x + \mathbf{s}^\dagger\mathbf{z}$. The noise covariance in this equivalent signal model is $E\left[\mathbf{s}^\dagger \mathbf{z}\mathbf{z}^\dagger \mathbf{s}\right] = I/W + N_0$. Here we can see the *processing gain* of the direct sequence spreading. The noise energy in the equivalent channel is not amplified by N_s, whereas the symbol energy is. With Gaussian x, the capacity of this *discrete time* system is $\log_2(1 + (N_s\, P_R/W)/(I/W + N_0))$ bits per N_s/W seconds, i.e.,

$$C = \frac{W}{N_s} \log_2 \left(1 + \frac{N_s\, P_R}{I + N_0 W} \right) \quad [\text{bit/s}]$$

Here, we see that the processing gain N_s multiplies the power within the logarithm. Also, the bandwidth multiplying the logarithm is the Shannon bandwidth, taking into account the dimensionality of the used part of the signal space. For more discussion on this, see Massey (1994).

In an asynchronous direct sequence spread spectrum (DSSS) CDMA uplink, different users have different pseudorandom spreading sequences. Then, the chips of different users are independent, and the input covariance between users vanishes. Then, the assumptions of Section 22.4.1 are fulfilled, and the interference power I_{in} is as calculated there. This can be directly verified by calculating the expected interference between two pseudorandom spreading sequences. Standard random walk analysis (see, e.g., Verdú (2002)) results in

$$E\left[\left(\mathbf{s}_u^\dagger \mathbf{s}_v\right)^2\right] = \begin{cases} 1 & u = v, \\ \frac{1}{N_s} & u \neq v. \end{cases}$$

The processing gain is visible as the $1/N_s$ damping for the interference from another spreading code.

22.5.2 CDMA and TDMA capacity expressions

We can now write exact expressions for CDMA and TDMA with out-of-cell interference. The results pertain both to uplink and downlink of a MIMO system, depending on the interpretation.

For a TDMA system with N_u users at each cell, we have only one active user i during the time interval of T_i, where $\sum_i T_i = T$. The total bandwidth available for the multicellular system is W. If the reuse factor is q, this is divided into q bands, so that in each cell (sector) we use the bandwidth W/q. Thus, the system capacity is

$$\mathcal{C}_{\text{TDMA}} = \frac{W}{q} \sum_{i=1}^{\min(N_t, N_r)} E\left[\log_2 \left(1 + \frac{\lambda_i P_R/N_t}{I_{\text{out}}(q) + N_0 W/q} \right) \right].$$

Note that $I_{\text{out}}(q)$ decreases in the reuse factor q, see (22.24).

For DSSS-CDMA with spreading factor N_s and N_u users per cell, we have

$$C_{\text{CDMA}} = N_u \frac{W}{N_s} \sum_{i=1}^{\min(N_t, N_r)} E\left[\log_2\left(1 + \frac{\lambda_i N_s P_R / N_t}{(N_u - 1)(1 + f)P_R + N_0 W}\right)\right].$$

$$(22.26)$$

Here, we see the processing gain, the Shannon bandwidth, and the factor N_u multiplying the single-user capacity.

In Equation (22.24), the interference factor I_q was defined for a system with reuse greater than one. Generalizing to reuse 1, I_1 is defined as the ratio of the total interference and P_R, damped by the processing gain, i.e., $I_1 = (1 + f)(N_u - 1)/N_s$. The capacity expressions above combine to

$$C = g\,W \sum_{i=1}^{\min(N_t, N_r)} E\left[\log_2\left(1 + \frac{\text{SNR} \cdot \lambda_i / N_t}{I_q \cdot \text{SNR} + g}\right)\right]. \qquad (22.27)$$

Here g is an inverse effective reuse factor. For CDMA, $g = N_u/N_s$, and for TDMA we have $g = 1/q$.

It should be stressed that here the SNR is the *system signal-to-noise ratio*. It is related to the power constraint (22.6), and defined in terms of the total power (22.10) radiated into the cell, the noise power spectral density, and the total Fourier bandwidth occupied by the system as

$$\text{SNR} \equiv \frac{P_t^{(\text{tot})}}{N_0 W L} = \frac{N_u P_R}{N_0 W}.$$

Note that the same system SNR is used both for TDMA and CDMA, even though in the former case, only a fraction of the full bandwidth is used in a given cell. The average attenuation L defined in (22.10) has been divided away for convenience. The SNR related to an individual user link is $\text{SNR}_{\text{link}} = \text{SNR}/N_u$.

22.5.3 High-SNR capacity

The interference factor I_q determines the asymptotic behavior of the capacity at high SNR,

$$\lim_{\text{SNR}\to\infty} C = g\,W \sum_{i=1}^{\min(N_t, N_r)} E\left[\log_2\left(1 + \frac{\lambda_i}{I_q \cdot N_t}\right)\right].$$

As soon as we have nonvanishing interference, the capacity is interference limited. If $I_q = 0$, we get the expression for TDMA capacity without out-of-cell interference, which increases without limit as SNR increases.

Note that we have considered CDMA capacity for a fixed number of users. Theoretically, N_u can be higher than N_s, so that $g > 1$, and the effective reuse would be < 1. As shown in Verdú (2002), for random spreading CDMA in an isolated cell, $N_u \to \infty$ gives the highest throughput $(1/\ln 2)$ that depends neither on the SNR nor on the interference. For a multicell MIMO CDMA system (22.26) with interference, as calculated above, the limiting capacity for an infinite number of users can be obtained by using the Taylor expansion of $\log(1 + x)$,

$$\lim_{N_u/N_s \to \infty} \lim_{\text{SNR} \to \infty} C = \lim_{\text{SNR} \to \infty} \frac{W}{\ln 2} \cdot \frac{\text{SNR } E\left[\sum_i \lambda_i\right]/N_t}{(1+f)\text{SNR} + 1} = \frac{W \, N_r}{\ln 2 \, (1+f)},$$

where we used the fact that for a channel matrix with zero-mean and unit-variance entries $\sum_{i=1}^{\min(N_t,N_r)} E\left[\lambda_i\right] = E\left[\text{tr}(\mathbf{H}\mathbf{H}^\dagger)\right] = N_t N_r$.

In practical CDMA systems, the highest feasible N_u does not exceed N_s. In multipath propagation, each user gives rise to multiple fingers to be detected, which fills dimensions in the signal space so that a feasible N_u is less than N_s. Here, we are not discussing multipath effects, and consider $N_u = N_s$.

22.5.4 Capacity as a function of E_b/N_0 and low-SNR capacity

Above we have expressed the capacity as a function of SNR, but it is useful to consider it as a function of the transmitted energy per bit E_b/N_0. If we express E_b/N_0 as a function of the spectral efficiency $C = \mathcal{C}/W$, we get

$$\left(\frac{E_b}{N_0}\right)(C) = \frac{\text{SNR}}{C(\text{SNR})} = \frac{\text{SNR}_{\text{link}}}{C_{\text{link}}(\text{SNR}_{\text{link}})}.$$

There is always a nonzero minimum E_b/N_0 below which no reliable communication is possible. It can be calculated explicitly as (Verdú, 2002)

$$(E_b/N_0)_{\min} = \frac{1}{\dot{C}(0)} = \frac{N_t \ln 2}{E[\sum_{i=1}^{\min(N_t,N_r)} \lambda_i]} = \frac{\ln 2}{N_r}, \qquad (22.28)$$

where it was assumed that the fading matrix $\mathbf{H}$ has zero-mean entries with unit variance. Thus, there is a 3 dB gain in $(E_b/N_0)_{\min}$ for each doubling of the number of receive antennas. This is strictly due to the increased total received power when increasing the number of antennas.

Close to $(E_b/N_0)_{\min}$, the spectral efficiency is dominated by its first expansion coefficient. Thus, it is of interest to find the bandwidth slope, or the slope of the spectral efficiency at $(E_b/N_0)_{\min}$ (Verdú, 2002). It can be calculated as $S_0 = -2 \ln 2 \cdot (\dot{C}(0))^2/\ddot{C}(0)$. The second derivative of (22.27) is

$$\ddot{C}(0) = - \sum_{i=1}^{\min(N_t,N_r)} \frac{E\left[\lambda_i^2\right] + 2E\left[\lambda_i\right]\,N_t\,I_q}{g\,\ln 2\,N_t^2}.$$

If $\mathbf{H}$ has zero-mean, unit-variance entries, this may be evaluated in closed form using $\sum_{i=1}^{\min(N_t,N_r)} E\left[\lambda_i^2\right] = E\left[\mathrm{tr}\left(\mathbf{H}^\dagger\mathbf{H}\mathbf{H}^\dagger\mathbf{H}\right)\right] = N_r N_t^2 + N_t N_r^2$. The bandwidth slope is thus (Marina, 2004)

$$S_0 = \frac{2\,g\,N_r}{1 + 2I_q + N_r/N_t}. \tag{22.29}$$

The bandwidth slope has the intuitively appealing behavior that the larger N_r/N_t is, the less impact the interference, characterized by I_q, has on the slope and on the low-SNR capacity. That is, increasing N_r improves tolerance to interference. In contrast, increasing the effective reuse factor (decreasing g) decreases S_0 linearly. For SISO interference channels, the bandwidth slope was analyzed in Caire *et al.* (2004).

22.6 Numerical capacity results

In this section, we show numerical system capacity results for cellular CDMA and TDMA systems. In the simulations, the path loss exponent is $\beta = 3.5$, the shadowing standard deviation $\sigma_s = 8\mathrm{dB}$, the shadowing correlation is 0.5, fast fading is i.i.d. Rayleigh, power equalization is used, but not power control; handovers are hard. The cellular layout is hexagonal with three sectors per cell. For CDMA, there are $N_u = 16$ spreading factor $N_s = 16$ users per sector. For TDMA, $N_u = N_s = 1$, and reuse factors $q = 3, 4, 7$ are applied. We consider three kinds of receivers for CDMA. In addition to single-user RAKE receivers, we consider multiuser detectors (Verdú, 1998) with a hypothetical 100% efficiency (i.e., 100% of the in-cell interference is canceled), and a more realistic 50% efficiency.

In a cellular system, spectral efficiency has to be measured in units of bits/s/Hz per unit area. In the figures, this unit area is selected as a sector, which is natural in the scenario discussed here. To translate these results into spectral efficiencies per m^2, absolute values of power constraints and cellular radii should be considered.

Spectral efficiency is plotted in terms of E_b/N_0, as discussed in the previous section, in contrast to, e.g., Gray *et al.* (2003), where results were presented in terms of $\mathrm{SNR}_{\mathrm{link}}$.

In the numerical experiments, we assumed i.i.d. Rayleigh fading for the MIMO channels. With more realistic channel models, the capacities would be smaller. In Farrokhi *et al.* (2002) and Gray *et al.* (2003), the effect of antenna correlation on MIMO system capacities was discussed for the downlink and the uplink, respectively.

Uplink interpretation When presenting uplink results in terms of E_b/N_0, the reached spectral efficiency is presented as a function of the total power drawn from the batteries of the mobile stations. Battery lifetime is an essential part of the user experience, and when plotting in E_b/N_0, different multiple-access schemes are compared at constant battery lifetime. If plotting in $\mathrm{SNR_{link}}$, the potential of nonorthogonal schemes (CDMA) to increase system capacity at the price of increased power consumption would become visible.

Downlink interpretation If no side information is used to tune the transmission, there is, in principle, no difference between downlink CDM and TDM in a frequency-flat channel. From the point of view of the analysis presented here, there is only one active user per sector (the transmitting base station), and there is no difference between SNR and $\mathrm{SNR_{link}}$. The TDMA plots for reuse 3,4,7 are directly applicable, whereas the CDMA results with 100% MUD efficiency give an upper limit to the downlink capacity with reuse 1. The fact that this is an upper limit follows from the Gaussian approximation, see Section 22.4.5.

Simulation results SISO and 4×4 MIMO results are shown in Fig. 22.4. Both CDMA and TDMA capacities are interference limited. In Gray *et al.* (2003), it was observed that there is always a crossover between the capacity of a CDMA system and that of a TDMA system with reuse and no out-of-cell interference, the former giving more capacity at low SNR. This is due to the fact that the interference that diminishes CDMA capacity becomes negligible at low SNR, whereas the reuse factor, which diminishes TDMA capacity, does not depend on SNR. Here, we observe that the crossover persists even when the out-of-cell interference is taken into account for TDMA, and when we are comparing with the same E_b/N_0. In the crossover regime, CDMA is interference limited, whereas TDMA is not yet interference limited. Note that even with a 4×4 MIMO system, only a system spectral efficiency of 4.7 bps/Hz/sector can be reached.

Fig. 22.5 shows the dramatic effect of out-of-cell interference on TDMA capacity. Comparing the plots for different antenna configurations, we can see

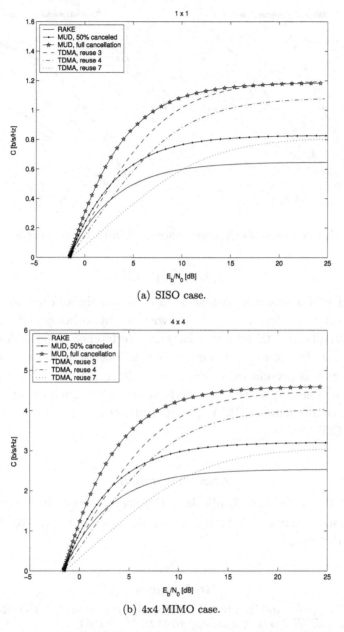

(a) SISO case.

(b) 4x4 MIMO case.

Fig. 22.4. System capacities of CDMA with different MUD efficiencies and TDMA with different reuse factors.

the 3 dB shifts in $(E_b/N_0)_{\min}$ predicted by (22.28). Also, the effect observed in equation (22.29) that the bandwidth slope decreases with increasing interference and decreasing number of transmit antennas is clearly visible.

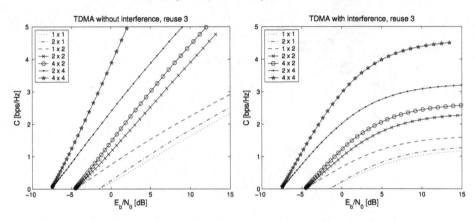

Fig. 22.5. Effect of out-of-cell interference on TDMA capacity, reuse factor 3.

22.7 Conclusions

The capacity of a practical system is interference limited already at rather modest spectral efficiencies, compared with the hypothetical efficiencies promised by single-cell MIMO analysis. The interference in CDMA-based systems with nonideal receivers is higher than in TDMA, and due to out-of-cell interference, it is present even if multiuser detection is used against in-cell interference. There is a crossover SNR value, beyond which an orthogonal multiple-access scheme (TDMA) provides higher capacity than a nonorthogonal one (CDMA).

Acknowledgment

We wish to thank Dr. Kari Kalliojärvi, Dr. Steven Gray, Dr. Rinat Kashaev, and Prof. Bixio Rimoldi for their insightful comments related to the subject matter of this chapter.

References

Caire, G., Knopp, R., and Humblet, P. (1998). System capacity of F-TDMA cellular systems. *IEEE Trans. Commun.*, **46** (12), 1649–1661.

Caire, G., Tuninetti, D., and Verdú, S. (2004). Suboptimality of TDMA in the low-power regime. *IEEE Trans. Inf. Theory*, **50** (4), 608–620.

Catreux, S., Driessen, P. F., and Greenstein, L. J. (2001). Attainable throughput of an interference-limited multiple-input multiple-output (MIMO) cellular system. *IEEE Trans. Commun.*, **49** (8), 1307–1311.

Farrokhi, F. R., Lozano, A., Foschini, G. J., and Valenzuela, R. A. (2002). Spectral efficiency of FDMA/TDMA wireless systems with transmit and receive arrays. *IEEE Trans. Wireless Commun.*, **1** (4), 591–599.

Friis, H. T. (1946). A note on a simple transmission formula. *Proc. IRE*, **34** (5), 254–256.

Gray, S., Pasanen, P., and Tirkkonen, O. (2003). Application of information theory to the design of 4th generation cellular communication systems. *Wireless Pers. Commun.*, **26** (2-3), 203–216.

Lee, C.-C. and Steele, R. (1995). Signal-to-interference calculations for modern TDMA cellular communication systems. *IEE Proc. Commun.*, **142** (1), 21–30.

Lozano, A. and Tulino, A. M. (2002). Capacity of multiple-transmit multiple-receive antenna architectures. *IEEE Trans. Inf. Theory*, **48** (12), 3117–3128.

MacDonald, V. H. (1979). The cellular concept. *Bell Syst. Tech. J.*, **58** (1), 15–41.

Marina, N. (2004). *Successive Decoding*. Ph.D. thesis, School of Computer and Communication Sciences, École Polytechnique Fédérale de Lausanne (EPFL), Lausanne.

Massey, J. L. (1994). Information theory aspects of spread-spectrum communications. In *Proc. IEEE Int. Symp. Spread Spectrum Techn. Appl.*, 16–21.

Parsons, J. D. (2001). *The Mobile Radio Propagation Channel* (John Wiley & Sons, Chichester, U.K.), second edn.

Shamai, S. and Verdú, S. (2001). The impact of frequency-flat fading on the spectral efficiency of CDMA. *IEEE Trans. Inf. Theory*, **47** (4), 1302–1327.

Stüber, G. L. (1996). *Principles of Mobile Communications* (Kluwer Academic Publishers, Norwell, MA, U.S.A.).

Telatar, E. (1999). Capacity of multi-antenna Gaussian channels. *Eur. Trans. Telecommun.*, **10** (6), 585–595.

Verdú, S. (1998). *Multiuser Detection* (Cambridge Univ. Press, New York, NY, U.S.A.).

—— (2002). Spectral efficiency in the wideband regime. *IEEE Trans. Inf. Theory*, **48** (2), 1319–1343.

Viterbi, A. J. (1995). *CDMA: Principles of Spread Spectrum Communications* (Addison-Wesley, Reading, MA, U.S.A.).

Part V

Implementations, measurements, prototypes, and standards

23

What we can learn from multiantenna measurements

Ernst Bonek

Technische Universität Wien

Werner Weichselberger

woolf solutions IT-consulting and development

23.1 Why multiantenna measurements?

"The propagation channel is at the heart of any wireless system; it sets the ultimate limits for other fields of wireless communications engineering."

Nowhere is this statement more true than in MIMO systems. The MIMO radio channel crucially determines the characteristics of the entire system. It forms the basis upon which system engineers and signal processing specialists can build MIMO systems. Multiantenna measurements serve to characterize the MIMO propagation channel. We need them to construct models of MIMO transmission, to validate our models, but also in operation to obtain information at the transmitter (Tx) and receiver (Rx) sides about the current channel (channel state information, see Chapter 6). Detailed knowledge about wave propagation has clarified information-theoretic viewpoints and inspired new MIMO models.

We will touch on pros and cons of some measurement methods, and point out some common pitfalls when interpreting measurements. We will highlight calibration, mutual coupling, normalization, and the question of what to consider as an appropriate ensemble of data points. How do the antenna arrays compound the environment to the MIMO channel, and which limits does the radio channel impose on the MIMO communication system? Precise information about the spatio-temporal characteristics of the environment ("clustering") bears a wealth of information, some of which is still untapped. Spatial correlation and spatial stationarity are important characteristics that multiantenna measurements provide us with. An example of validation of MIMO channel models through experiments will reveal the benefits and shortcomings of some common performance figures.

23.1.1 Setting the scene

As an introduction, let us have a look at how the channel characteristics determine in detail which of the well-known benefits of MIMO can be exploited, viz. beamforming gain, spatial diversity, and spatial multiplexing.

Beamforming gain By beamforming, the transmit and receive antenna patterns can be focused into a specific angular direction by the appropriate choice of complex baseband antenna weights. When no instantaneous channel information is available, the resulting increase in antenna gain is limited by the channel properties and the number of antenna elements, n_R for the receiver and n_T for the transmitter. The more directive the channel, the better for beamforming. Under ideal channel conditions for beamforming, e.g., line-of-sight, the Rx and Tx gains add up, leading to an upper limit of $n_R n_T$ for the beamforming gain of a MIMO system. Interest in beamforming is increasing again as a means to combat interference.

Spatial diversity Multiple replicas of the radio signal from different points in space give rise to spatial diversity, which increases the reliability of the fading radio link. Under ideal conditions for diversity, e.g., a spatially white MIMO channel, the diversity order is limited to $n_R n_T$. Spatial correlation of the antenna signals reduces the diversity order. Spatial diversity can be exploited either at the receiver or the transmitter, or both.

Spatial multiplexing As explained in Part II, MIMO channels can support spatially parallel data streams by transmitting and receiving on orthogonal spatial filters. The number of multiplexed streams depends on the rank of the instantaneous channel matrix $\boldsymbol{H}$, which, in turn, depends on the spatial properties of the radio environment. The spatial multiplexing gain varies from one, e.g., for line of sight, to the minimum of n_R and n_T in a sufficiently rich scattering environment.

Beamforming, diversity, and multiplexing are rivaling techniques. Full beamforming excludes diversity and multiplexing; full diversity excludes beamforming and multiplexing; finally, full multiplexing excludes beamforming and—note the exception—*reduces* diversity. Even in the case of full multiplexing, diversity on the signal streams can be achieved (Hottinen *et al.*, 2003).

To illustrate the role of the propagation channel, the threefold trade-off between beamforming, diversity, and multiplexing can be broken down into various dichotomous trade-offs (Weichselberger, 2003), see Fig. 23.1. Many

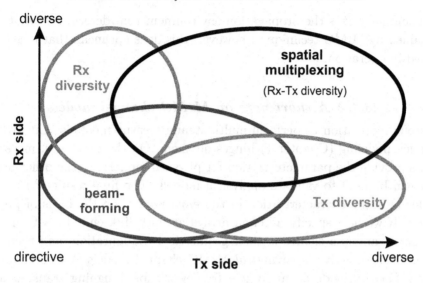

Fig. 23.1. The extent to which the channel supports beamforming, diversity, or multiplexing depends on the directivity and diversity of both link ends.

scattering objects in a broad angular region around the Tx array make the Tx side "diverse" and allow for diversity or spatial multiplexing. In contrast, a "directive" channel, i.e., a single strong scatterer or a group of closely-spaced scatterers, may support beamforming based on long-term channel knowledge. In a directive environment, signals between the Tx array and the Tx-relevant scatterers propagate more or less coherently. The same holds true for the Rx link end.

The partial overlap of the ellipses in Fig. 23.1 indicates that there is a gradual transition between the pure realizations of any MIMO benefit. If one side of the MIMO link is purely directive, no multiplexing is possible, but diversity can be exploited at the uncorrelated link end. If, however, both link ends of the MIMO channel are (partly) decorrelated, both diversity and multiplexing are possible.[†] In this case, the optimal trade-off between diversity and multiplexing will be determined by system requirements, e.g., desired data rate and reliability of transmission. In channels that are not fully diverse, partial beamforming gain can be achieved by restricting the Tx and/or Rx filters to the subspace spanned by the spatial correlation matrices.

[†]Even in the case of completely uncorrelated antennas at both link ends, the radio channel is not guaranteed to support spatial multiplexing ("pinhole" or "keyhole" channel). Given the enormous effort necessary to actually find and measure such a rank-one keyhole channel (Almers *et al.*, 2003), its occurrence in practical situations is questionable, though. The rank-*reduced* "pinhole" (Gesbert *et al.*, 2000), irrespective of spatial correlation at Rx or Tx, is more likely.

In summary, it is the propagation environment that determines what *can* be gained by MIMO techniques; however, this does not mean that it *will* be gained in operation.

23.1.2 A short note on MIMO channel models

A strong motivation to perform multiantenna measurements is that MIMO channel modeling (Chapter 1) hinges on them. On the one hand, measurements provide the parameterization for physical models; on the other hand, they can be used to validate analytical models. The final result of *physical* modeling is the characterization of the *environment* on the basis of propagation. If we now specify antenna arrays at both link ends by setting the number of antenna elements, their geometrical configuration, and their polarizations, we arrive at *analytical* MIMO channel models (Weichselberger, 2003). These provide an analytical framework for designing transmit and receive signaling techniques, e.g., space-time codes.

23.2 About performing multiantenna measurements

23.2.1 Two different views

Before we describe multiantenna measurements in detail, a word of caution is advised. There are two opposing views on how to perform and interpret channel measurements: should one *include* the effects of the antenna arrays in the characterization of the channel, or *separate* the propagation environment from the arrays? One can argue that the antennas are an integral part of the radio channel. First, antennas in close proximity to each other interact electromagnetically (mutual coupling); second, by their sheer existence, they modify the electromagnetic field they sample. So, multiantenna measurements include the influence of the antenna arrays used. But this view necessitates innumerable multiantenna measurement campaigns, given the infinite number of possibilities of arranging a variable number of antenna elements with arbitrary polarizations into an array. If you are already sure of the optimality of the arrays you have chosen for your MIMO system, then go ahead. However, the array configuration that will be best in exploiting the propagation channel's properties is usually not known a priori.

For the purpose of optimizing arrays independently of the radio propagation between them, we therefore recommend that the electromagnetic characteristics of the environment and the antenna arrays are studied *separately*. Even if this is difficult to achieve in practice, we should strive for this goal as it prevents us—in good reductionist practice—from mixing up different

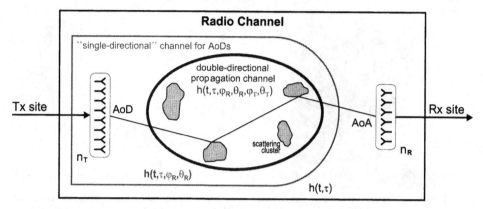

Fig. 23.2. The double-directional impulse response, $h(t, \tau, \varphi_R, \theta_R, \varphi_T, \theta_T)$ character-izes the MIMO propagation environment. The symbols t, τ, φ, and θ denote time, delay, azimuth, and elevation.

problems. How antennas distort the field they are supposed to measure and interact with each other in close proximity is one issue; how electromagnetic waves, once radiated, propagate between antennas, and how the environment delivers these waves to the receiving array is a second, independent issue.

Once we know how a certain environment propagates the waves and, sec-ondly, how antennas interact with each other on short range, we can combine the results to get the full picture of the MIMO channel. This will enable us to find optimal antenna arrays at either link end of any MIMO system, and to exploit the actual channel in the best possible way.

23.2.2 Antenna arrays for MIMO measurements

Since we are interested in the spatial characterization of the radio chan-nel, we need *spatially resolved* measurements of the channel. The double-directional viewpoint (Steinbauer *et al.*, 2001) of the radio channel describes angles-of-departure (AoD) and angles-of-arrival (AoA) of electromagnetic waves (see Fig. 23.2). A full specification of a direction requires two angles, azimuth and elevation.

Array shapes for MIMO measurements

Antenna arrays for measurements are designed to provide a high and unam-biguous spatial resolution over a preferably large angular view. Unambiguity is achieved by limiting the maximum element distance, e.g., half the wave-length for uniform linear arrays (ULAs), and by shielding ambiguous angular regions by means of a reflecting panel behind the array. Dummy elements,

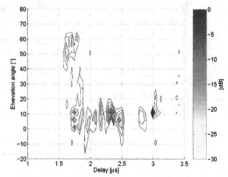

(a) 3-D array of 32 dual-polarized patch antennas, center frequency 2 GHz.

(b) Elevation-delay power spectrum measured with the array of (a).

Fig. 23.3. Courtesy: K. Kalliola, Helsinki University of Technology, now with Nokia Research Center, Helsinki.

which are not connected to a receiver chain, are used at the edges of an array to prevent edge effects and to facilitate the correction of mutual coupling.

The number of antenna elements is limited by practical reasons, both for measurements and communication applications.[†] The larger the array, the better the angular resolution. This well-known restriction of antenna arrays has been lifted, in principle, by subspace estimation techniques of directions (see Chapter 13), but it re-enters through the back door of the SNR achievable by the measurement.

Today's MIMO measurements apply linear, circular, planar, and sophisticated 3-D arrays with single or dual polarized antenna elements. While linear arrays allow for the measurement of a single angular domain only, the other array geometries cover a 2-D angular view. The array geometries also differ in the ease of postprocessing applications.

Ingenious shapes have been devised and built to provide a full, all-around 3-D view of the environment. Kimmo Kalliola's "football" (see Fig. 23.3(a)) with 32 dual-polarized patch antennas (Kalliola *et al.*, 2000) pioneered the field and delivered surprising results (Kalliola *et al.*, 2003), e.g., strong high-elevation multipath in urban environments. Cylindrical shapes (Ylitalo *et al.*, 2004; Thomä *et al.*, 2005) are more conducive to postprocessing by super-resolution techniques, but a slight dependence on the elements' elevation pattern remains.

Rotating directive antennas have been used in the past (Gans *et al.*, 2002).

[†]Note that previous applications of antenna arrays had different requirements and limitations: radar requires a resolution that is higher by orders of magnitude; radio astronomy handles antenna spacings of thousands of kilometers.

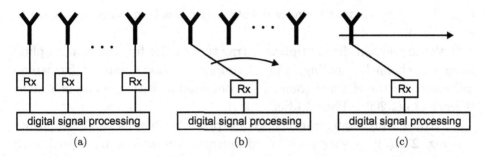

Fig. 23.4. (a) Real, (b) multiplexed, and (c) virtual antenna arrays.

Besides requiring a static environment and postprocessing deconvolution of antenna pattern and measured angular power spectrum (APS), they suffer from artifact responses due to pattern side lobes and the continuous change of the antenna orientation.

Polarization is an attractive property of the electromagnetic field to exploit in MIMO systems. It offers a dramatic reduction in the size of arrays, which makes the idea of multiple antennas at handheld terminals feasible. In fact, at any point of the electromagnetic field there exist six independent states of polarization (three magnetic and three electrical states), which can, in principle, be sampled by an electromagnetic vector sensor. However, practical utilization of more than three uncorrelated states seems unlikely in radio communications (Yong and Thompson, 2005).

Real, multiplexed, and virtual arrays

The elements of an antenna array provide spatial samples of the electromagnetic field. Measurement setups differ in the way this sampling is performed.

A straightforward, but expensive, way is to provide *one receive chain for every antenna* (see Fig. 23.4(a)). Transmission must be done such that the multiple receivers can distinguish signals from different Tx antennas, which is usually achieved by orthogonal codes in either time or frequency. While this offers truly simultaneous multiple spatial samples, the problems of balancing all receive chains in amplitude and phase are large.

Therefore, *multiplexing* the antenna signals, within short time intervals, to one and the same receiver has become standard practice (see Fig. 23.4(b)). The objection that the environment may change in the meantime is valid, but has been countered by ever faster multiplexers. A limitation that remains at the Tx side is the amount of power that can be switched in a short time. Therefore, channel sounder output power has been restricted so far to around 1 W.

Both the real and the multiplexed array types suffer from mutual coupling. In actual channel sounding, mutual coupling is taken care of by careful calibration of the channel sounder in combination with the antenna array (Fleury et al., 2002; Thomä et al., 2005).

Arbitrary array configurations become possible when using *virtual arrays* [see Fig. 23.4(c)]. A single optimized antenna is moved along a predefined trajectory, and samples of the electromagnetic field are taken ad libitum. The environment has to remain static during the measuring period, but this scheme solves the problem of mutual coupling for good.

23.2.3 Channel sounders

The antennas themselves are only one aspect of probing the electromagnetic environment. Inspired by the benefits that smart antennas and MIMO techniques can offer, the technology of channel sounders has taken enormous strides ahead in the recent past. Many institutions have designed and built their own channel sounding equipment (Correia, 2001), but some instruments are now commercially available. The fields of plain measurements and of direction estimation (Chapter 13) have merged, through pre- and postprocessing, into the intricate area of *multidimensional channel characterization*. Besides space/angle and polarization properties at receiver and transmitter, the dimensions of interest are frequency/delay and time/Doppler.

In analogy to the spatial domain in the previous section, channel sounders differ in their method of sampling the frequency band of interest. A simple way to probe the channel is by a vector network analyzer. The radio channel is the circuit under test. By stepping a sinusoidal signal through a (large) number of frequencies that span the frequency region of interest, the transfer function of the channel is obtained. A Fourier transform yields the channel's impulse response. When combined with antenna arrays, the (double-)directional impulse response characterizes the channel both in delay and angle. Again, the environment has to remain static.

Alternatively, all frequencies can be excited simultaneously. Differentiated by the postprocessing of the received signals, two main methods are applied: in frequency-domain correlation methods, temporally orthogonal codes are transmitted on individual frequency tones (Thomä et al., 2005). The transmitter separates the codes and yields the channel transfer function in the frequency domain. The temporal correlation method utilizes a Tx sequence with good autocorrelation properties (Ylitalo et al., 2004). The receiver correlates the Rx signal with the same sequence and yields the channel response in the delay domain.

For both methods, the correlation gains increase with the length of the test signal period. While the signal has to be longer than the maximum delay expected in the channel, it must be short enough not to violate the implicit stationarity assumption. That is, the channel has to remain constant, but—in contrast to the network analyzer—only for tens of microseconds. In this way, MIMO channels can be measured "on the fly".

23.2.4 Mutual coupling and radiation efficiency

The role of mutual coupling in MIMO arrays is still under investigation. Seemingly conflicting reports on mutual coupling between antenna elements increasing (Svantesson and Ranheim, 2001) or reducing ergodic capacity might not be conflicting at all. Both effects are, in principle, plausible. On the one hand, mutual coupling changes the individual antenna patterns. This creates diversity, as each antenna "sees" different portions of the surrounding scatterers ("pattern diversity"). But the effect seems to be small unless the antennas are located very close to each other. On the other hand, mutual coupling may, by reradiation of received power, result in higher spatial correlation between antenna signals, which is a possible cause for reduced capacity. Let us be clear, however: correlation is *not* solely caused by mutual coupling and depends on a number of other factors as well.

Antennas in close proximity to each other and to lossy material, like human tissue, also suffer from reduced *radiation efficiency*. Though not necessarily so (Dossche *et al.*, 2004), mutual coupling may reduce radiation efficiency. This is an intricate effect to measure, but it becomes extremely important if we want to compare the performance of an antenna array with that of a single antenna. One way to tackle this problem is the method of so-called "embedded patterns" (Kildal and Rosengren, 2004), which involves full-wave electromagnetic computation of actual antennas and the creation of a well-defined multipath environment. Current findings indicate that spatial correlation has little effect as compared with radiation efficiency. So, after all, we do encourage multiantenna measurements including the antennas—to study near-field effects of specific antenna structures in specified field configurations.

23.2.5 Spatial stationarity

Although radio channels are often assumed to be stationary within the temporal and spatial region under consideration, this assumption is rarely tested.

However, knowing the limits of the stationarity region is important for the operation of MIMO systems and analysis of channel measurements alike. Concentrating on the spatial properties of MIMO channels, Herdin and Bonek (2004) introduced the *correlation matrix distance* (CMD),

$$d_{\mathrm{CMD}} = 1 - \frac{\mathrm{tr}\{\boldsymbol{R}_1 \boldsymbol{R}_2\}}{\|\boldsymbol{R}_1\|_{\mathrm{F}} \, \|\boldsymbol{R}_2\|_{\mathrm{F}}},$$

where $\boldsymbol{R}_1$ and $\boldsymbol{R}_2$ are two correlation matrices taken, e.g., at different positions or time instants. The CMD characterizes the difference between the spatial structure of two MIMO channels. Changes in the spatial structure leading to large values of the CMD (≥ 0.2) also show up as a significant performance reduction in MIMO transmission schemes using outdated CSI.

Concerning measurement analysis, a particularly challenging problem related to the question of stationarity is the proper size of the measured statistical sample in space, time, and frequency. The quality of statistical estimates depends on the number of measured channel realizations—the more the better. On the one hand, these realizations should be independent, i.e., their distance in at least one domain has to be approximately equal to or longer than the coherence distance in the respective domain. Additionally, hardware restrictions limit minimum sampling distances in all domains. On the other hand, the measured sample must not span a region that exceeds the stationarity region of the statistics under consideration because this would mean averaging over different statistics. Taken together, both minimum sampling distance and maximum sampling region pose an upper limit on the number of realizations that can be taken, and, thus, on the quality of statistical estimates. Both coherence and stationarity regions are statistical metaparameters that are a-priori unknown and have to be estimated as well (Matz, 2003).

In measurement analysis it is customary to plot the amplitude envelope of MIMO channel coefficients as normalized probability density functions (pdfs) and see whether they are Rayleigh, Rice, double-Rayleigh (Erceg *et al.*, 1997) or otherwise distributed. However, conclusions from pdf agreement— obtained by visual inspection—should be drawn with caution. Along the lines of the previous paragraph, great care has to be taken to make sure that the pdf under investigation represents data produced by one and the same statistics. In fact, quite arbitrary distributions may happen to be generated by superimposing scaled versions of different Rayleigh or Rice distributions. The crucial point is the proper extent of the statistical sample.

23.2.6 Normalization

The received power of measured MIMO channels is the result of the transmitted power and the power budget of the measurement equipment (antennas, filters, etc.) on the one hand, and the physical *path loss* on the other hand. To make the analysis of measured channels independent of the measurement equipment, it is common to normalize the channel matrices to a predefined average power value. There are at least three different normalization regimes: first, if one normalizes each measured channel matrix separately, the entire information about path loss and fast or slow fading is lost. The only information left is the spatial structure of the MIMO channel matrices. Second, normalizing all matrices of a single measured location with the same factor preserves the path loss information at this position. This method allows for a full analysis of this specific location, but it prevents a comparison of different locations. Such a comparison can only be done by means of the third normalization approach, which normalizes all matrices of all locations of an entire measurement campaign with one and the same factor. Ignoring the path loss by means of normalization corresponds to a constant receive SNR, which is often taken as a parameter in theoretical analysis, but will be difficult to achieve in operation without sophisticated power control at the transmitter.

Especially when it comes to analyzing channel capacity, the proper normalization method is crucial. The discovery of spatial multiplexing has led to a concentration of research efforts on the spatial channel structure, cumulating in the belief that rich scattering (without a line-of-sight component) provides "good" MIMO channels. However, rich scattering is often associated with a high path loss. Measurements at the Institut für Nachrichtentechnik und Hochfrequenztechnik, Technische Universität Wien, Vienna (Herdin *et al.*, 2002), suggested that, given that the environment provides some multipaths, channel capacity is dominated by the path loss. Spatial structure was only of secondary importance. Fig. 23.5 shows a scatter plot of the mean mutual information of 72 locations, where all matrices of the entire measurement campaign have been normalized with the same factor. The normalization was performed such that the Rx SNR of a reference location (Rx26D3) was set to 20 dB. The mutual information values are clustered around the solid curve, corresponding to the mean capacity of the reference location for varying SNR.

Great care has to be taken when comparing the performance of MIMO systems with various array sizes with each other and with the performance of SISO or SIMO systems. The average channel coefficient power as well as the total transmit power have to be the same for each system.

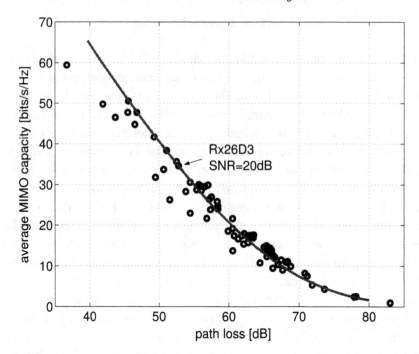

Fig. 23.5. Mean mutual information of 72 measured locations. The solid curve corresponds to the mean capacity of the reference location (Rx26D3) for varying SNR. From Bonek *et al.* (2003), reprinted with permission.

23.3 What we can derive from multiantenna measurements

What do we do with the information about the MIMO channel obtained from measurements? We have to distinguish between different purposes for which we use this information: MIMO system *design*, the *deployment* of such a system, or *operation* of an already deployed system. For the latter purpose, we need on-line information, in essence channel state information (CSI), which is treated in Chapters 6 and 15. The off-line information that we gather with measurements about the propagation environment provides the basis for MIMO channel modeling and for MIMO model validation. This is the classical domain of multiantenna measurements.

23.3.1 Scattering clusters

Electromagnetic waves propagate via scattering objects (here "scattering" is used as a generic term for all kinds of propagation mechanisms, e.g., specular reflection, diffraction, and diffuse scattering from rough surfaces). Many publications assume the propagation environment to be spatially white, i.e., multipath components (MPCs) arrive at the array from all directions. This

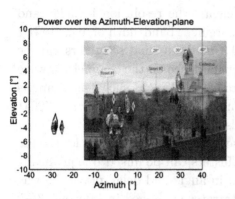

Above -80 [dB] ○   -80 to -90 [dB] □ -91 to -100 [dB] ○
-101 to -110 [dB] ×   Below -110 [dB] ✳

(a) 2 GHz "photograph" of incoming waves, superimposed on a photograph at visible wavelengths, taken from the same position.
Courtesy: K. Kalliola, Helsinki University of Technology, now with Nokia Research Center, Helsinki.

(b) Multipath clusters measured with a UWB MIMO technique and resolved in delay and angle.
From Tsuchiya *et al.* (2004), reprinted with permission.

Fig. 23.6.

is rarely the case; MPCs tend to come in bundles or *"clusters"*. The temporal clustering approach, proposed in Saleh and Valenzuela (1987), was extended to the spatial domain in Spencer *et al.* (2000). The effect of clustering on the capacity of MIMO systems is considerable (Li *et al.*, 2002). Clusters can be understood as MPCs that have similar AoAs (or AoDs), delays, and, possibly, polarizations; they are distinct in these properties from other MPCs. The degree of clustering resulting from a specific environment can be assessed by the angular delay power spectrum (ADPS). Among other things, it shows how AoAs and AoDs are linked via scattering objects in the environment.

How well we can resolve the scattering environment is ruled by the method we use to determine directions and delays (see Chapter 13), by the number of antennas, and by the measurement bandwidth. Data postprocessing by sophisticated superresolution techniques has become an integral part of the measurement procedure (Fleury *et al.*, 2002; Thomä *et al.*, 2005) and has produced spectacular results. An exemplary result of such measurements is given in Fig. 23.6(a); it shows a "photograph" of the incoming waves at the measurement wavelength. The clustering of MPCs originating from scatterers is clearly visible. Another example, Fig. 23.3(b), reveals strong high-elevation MPCs in an urban environment. This result challenged the general assumption of scatterers being positioned only in the horizontal plane.[†]

[†]This assumption, together with a vertical linear dipole antenna, is essential for the omnipresent bathtub-shaped Jakes Doppler spectrum.

By means of *ultra-wideband* measurements, the resolution in delay and distance can be further enhanced; a measurement bandwidth of, e.g., 1 GHz corresponds to a spatial resolution of 0.3 m. Individual scatterers can be precisely identified and located, and their influence on the channel via wave propagation can be studied in great detail. Fig. 23.6(b) shows an example of a measurement campaign in a Japanese home (Tsuchiya *et al.*, 2004). Clusters of MPCs that come from walls, ceiling, floor, window and door frames, and pieces of furniture can be observed. While the figure is for the Rx, very similar patterns occur at the Tx. Clusters can be linked unequivocally to AoDs and AoAs. This high resolution in angle and delay paves the way for MIMO channel modeling for the data rates envisaged in "Beyond 3G", say 100 to 1000 Mbit/s.

As the above examples show, the ADPS cannot, in general, be factorized into an angular power spectrum (APS) and a power delay profile (PDP). Also, it is pointless to speak of the second-order moment of the APS as *an* angular spread in such environments. Rather, each cluster of MPCs has a specific *cluster angular spread*. It is evident that the number of clusters increases and the cluster angular spread becomes smaller as the tools to resolve clusters evolve. For instance, Spencer *et al.* (2000), Chong *et al.* (2003) found cluster angular spreads in office environments around 20°, whereas Czink *et al.* (2004) report cluster angular spreads of approx. 4°.

23.3.2 Spatial correlation

Spatial correlation of antenna signals has been identified as a problem area of MIMO systems rather early (Shiu *et al.*, 2000). To determine the spatial correlation in a specific environment, we have to measure the full MIMO system with specific Tx and Rx arrays in place.

In the current literature, it is common to characterize the MIMO correlation properties by separate receive, $\boldsymbol{R}_{\mathrm{Rx}} = \mathrm{E}\{\boldsymbol{H}\boldsymbol{H}^{\mathrm{H}}\}$, and transmit, $\boldsymbol{R}_{\mathrm{Tx}} = \mathrm{E}\{\boldsymbol{H}^{\mathrm{T}}\boldsymbol{H}^{*}\}$, correlation matrices. However, we want to stress that these two matrices neglect the correlation terms across the link ("cross correlation", "joint correlation"), which do matter, at least in indoor scenarios (Herdin *et al.*, 2002). Another surprising result, emphasizing that separate Rx and Tx correlation matrices are not able to completely describe MIMO channels, was recently found in Oestges and Paulraj (2004): so-called "*diagonal correlations*" may boost the ergodic capacity *beyond* the previously accepted upper limit of i.i.d. random entries of $\boldsymbol{H}$. Özcelik and Oestges (2005) showed that such channels are not unrealistic for symmetric $n \times n$ MIMO systems.

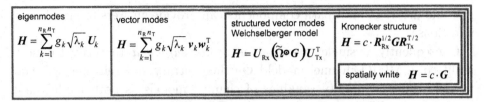

Fig. 23.7. Nested hierarchy of analytical models based on the spatial eigenstructure of MIMO channels.

To cover the spatial characteristics of a MIMO channel, one has to use the *full* correlation matrix R_H, given by $R_H = \mathrm{E}\left\{\mathrm{vec}(H)\,\mathrm{vec}(H)^{\mathrm{H}}\right\}$, where the $\mathrm{vec}(\cdot)$ operator stacks all columns of a matrix into a tall vector.

Weichselberger (2003) has recently established a hierarchy of correlation models based on the eigenstructure of MIMO channels (see Fig. 23.7), ranging from detailed to simple. The matrix-valued eigenmodes of R_H constitute the top-level of this model hierarchy. The number of real-valued parameters of each model decreases from $(n_{\mathrm{R}}n_{\mathrm{T}})^2$ (top level), to $n_{\mathrm{R}}^2 + n_{\mathrm{T}}^2 + n_{\mathrm{R}}n_{\mathrm{T}} - n_{\mathrm{R}} - n_{\mathrm{T}}$ in the model using structured vector modes (Weichselberger's model), to $n_{\mathrm{R}}^2 + n_{\mathrm{T}}^2$ in the "Kronecker" model, and, finally, to 1 in the i.i.d. model. We will come back to these models in Section 23.3.3. What matters here is that the model parameters of all hierarchy levels are easily accessible to measurements.

23.3.3 Model validation, channel assessment, and related performance figures

The ultimate test of any model is its experimental validation. In terms of MIMO channels this means comparing channel models with channel measurements. A number of metrics and performance figures have been proposed to judge the goodness of a MIMO channel model. Among these are ergodic and outage capacity, eigenvalue statistics, diversity order, correlation figures, and APS. However, none of these metrics is capable of capturing all properties of a MIMO channel. Modeling always implies a reduction of reality to some selected aspects, and so does the application of a specific metric. Significantly different propagation environments may result in a similar aggregate metric. Evaluating MIMO measurements or models only in terms of, e.g., channel capacity does not reveal which propagation mechanisms lead to these results, or how and to which extent the theoretical capacity values can be achieved in real systems. Only a full analysis with respect to angles and delays of multipath components can provide answers to all relevant questions.

Consequently, a single metric alone is *not sufficient* to verify the suitability of a MIMO channel model.[†] On the contrary, different metrics tend to yield different quality rankings of channel models as both models and metrics cover different channel aspects. We will illustrate this point by experimental validation of three current analytical models using three different metrics (Özcelik, 2004; Özcelik *et al.*, 2005). As metrics, we apply *mutual information*, essentially describing the potential spatial multiplexing gain; the *double-directional APS*, giving insight into the multipath structure, the spatial correlation, and the potential beamforming gain; and a *"diversity measure"* introduced by Ivrlac and Nossek (2003), describing the channel's diversity order. The three exemplary models are the Kronecker model, originally introduced by Shiu *et al.* (2000) and discussed in detail in Kermoal *et al.* (2002); Weichselberger's model (Weichselberger *et al.*, 2006), which relaxes the separability restriction of the Kronecker model; and the "virtual channel representation" (Sayeed, 2002) modeling the MIMO channel in beamspace, not in eigenspace.

For model validation we (i) extract model parameters from measurement; (ii) generate synthesized channels by Monte-Carlo simulations, according to the three models; (iii) compare the metrics calculated from the synthesized channels with those extracted directly from measurement. As the basis for our comparison, we took 58 locations of the extensive 5.2 GHz measurement campaign in the offices of the Institut für Nachrichtentechnik und Hochfrequenztechnik, Technische Universität Wien, Vienna (Özcelik, 2004). Figs. 23.8 and 23.9 compare the synthesized 4×4 MIMO channels with the measured ones.[‡] Fig. 23.8(a) shows a scatter plot of the models' prediction of average mutual information versus the true values. Here, the Kronecker model usually underestimates mutual information; the virtual channel representation significantly overestimates it; and Weichselberger's model fits the measurements equally well as the full R_H-model. The second metric, shown in Fig. 23.8(b), is the diversity measure of Ivrlac and Nossek (2003). Again, a scatter plot of the models' diversity metric versus the true diversity metric is shown. All three models overestimate the diversity order.

Fig. 23.9 depicts the double-directional APS of one exemplary location. In the measured channel, specific angles-of-departure (AoDs) are clearly linked to specific angles-of-arrival (AoAs), such that the joint APS is not separable. The Kronecker model introduces artifact paths at the intersections of the

[†]This is particularly true for ergodic capacity.

[‡]Results for 2×2 and 8×8 channels with variable antenna spacing are given in Özcelik (2004).

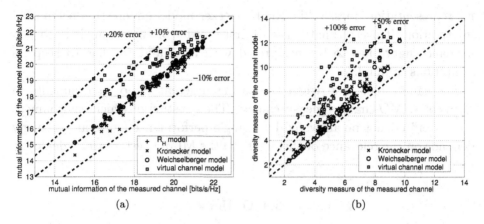

(a) (b)

Fig. 23.8. Comparison of models' predictions of (a) average mutual information and (b) "diversity measure" versus respective measured values. Each data point, for which 10,422 realizations in frequency and space were averaged, corresponds to one measurement location. Antenna spacing was 0.4λ at Rx and 0.5λ at Tx.

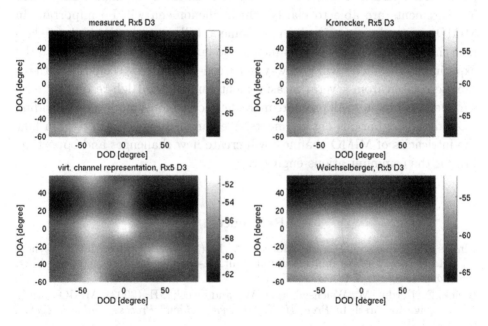

Fig. 23.9. Measured double-directional APS of an exemplary location compared with three different models.

AoA and AoD spectral peaks. The virtual channel representation shows the best fit with the measured APS.

While Weichselberger's model predicts average mutual information almost correctly, it distorts the double-directional APS more than the virtual channel representation does. While the Kronecker model underestimates mutual information, it overestimates the diversity order of the channel. Each performance figure has to be interpreted with care, and one has to keep its limitations in mind.

The same holds true when comparing different radio environments in terms of MIMO channel measurements. The question of which environment is "better" cannot be answered by a single performance figure and strongly depends on the relevance of this figure to the MIMO system to be deployed.

23.4 Outlook

Ergodic and outage MIMO capacity are elusive metrics—it is left to the engineering scientist to find ways to come close to them. Careful multi-antenna measurements, performed off-line for channel modeling and on-line for MIMO operation, are the basis for the endeavor for higher data rates. Measurements are able to clarify which phenomena will be important in MIMO system design, and which phenomena will not (e.g., the rank-one keyhole channel). "Typical" environments—so far the one that is nearest to the experimenter's office—will give way to agreed-on reference scenarios with parameters taken from several measurement campaigns. High-resolution measurement techniques will promote more detailed channel models by providing more details of various environments. We can still expect surprises in how the intricacies of MIMO channels will create new challenges and opportunities for the communications engineer.

References

Almers, P., Tufvesson, F., and Molisch, A. F. (2003). Measurement of keyhole effects in a wireless multiple-input multiple-output (MIMO) channel. *IEEE Commun. Lett.*, **7** (8), 373–375.

Bonek, E., Herdin, M., Weichselberger, W., and Özcelik, H. (2003). MIMO—study propagation first! In *Proc. IEEE Int. Symp. Signal Process. and Inf. Techn.*, 150–153.

Chong, C.-C., Tan, C.-M., Laurenson, D. I., McLaughlin, S., Beach, M. A., and Nix, A. R. (2003). A new statistical wideband spatio-temporal channel model for 5GHz band WLAN systems. *IEEE J. Sel. Areas Commun.*, **21** (3), 139–150.

Correia, L. M. (ed.) (2001). *Wireless Flexible Personalised Communications, COST 259: European Co-operation in Mobile Radio Research* (Wiley).

Czink, N., Herdin, M., Özcelik, H., and Bonek, E. (2004). Number of multipath clusters in indoor MIMO propagation environments. *Electron. Lett.*, **40** (23), 1498–1499.

Dossche, S., Blanch, S., and Romeu, J. (2004). Optimum antenna matching to minimise signal correlation in a two-port antenna diversity system. *Electron. Lett.*, **40** (19), 1164–1165.

Erceg, V., Fortune, S. J., Ling, J., Rustako, A. J., and Valenzuela, R. A. (1997). Comparisons of a computer-based propagation prediction tool with experimental data collected in urban microcellular environments. *IEEE J. Sel. Areas Commun.*, **15** (4), 677–684.

Fleury, B., Yin, X., Rohbrandt, K., Jourdan, P., and Stucki, A. (2002). Performance of a high-resolution scheme for joint estimation of delay and bidirection dispersion in the radio channel. In *Proc. IEEE Veh. Technol. Conf. (VTC Spring)*, vol. 1, 522–526.

Gans, M. J., Valenzuela, R. A., Yeh, Y.-S., Amitay, N., Sizer, T., Tran, C., Taylor, D., and Storz, R. (2002). Precise incident power density pattern measurement through antenna pattern deconvolution. *Wireless Personal Commun.*, **21** (2), 181–200.

Gesbert, D., Bölcskei, H., Gore, D. A., and Paulraj, A. J. (2000). MIMO wireless channels: Capacity and performance prediction. In *Proc. IEEE Global Telecommun. Conf.*, vol. 2, 1083–1088.

Herdin, M. and Bonek, E. (2004). A MIMO correlation matrix based metric for characterizing non-stationarity. In *Proc. IST Mobile Summit*.

Herdin, M., Özcelik, H., Hofstetter, H., and Bonek, E. (2002). Variation of measured indoor MIMO capacity with receive direction and position at 5.2 GHz. *Electron. Lett.*, **38** (21), 1283–1285.

Hottinen, A., Tirkkonen, O., and Wichman, R. (2003). *Multi-antenna Transceiver Techniques for 3G and Beyond* (John Wiley & Sons, Chichester, U.K.).

Ivrlac, M. T. and Nossek, J. A. (2003). Quantifying diversity and correlation of Rayleigh fading MIMO channels. In *Proc. IEEE Int. Symp. Signal Process. and Inf. Techn.*, 158–161.

Kalliola, K., Laitinen, H., Vainikainen, P., Toeltsch, M., Laurila, J., and Bonek, E. (2003). 3-D double-directional radio channel characterization for urban macrocellular applications. *IEEE Trans. Antennas Propagat.*, **51** (11), 3122–3133.

Kalliola, K., Laitinen, H., Vaskelainen, L. I., and Vainikainen, P. (2000). Real-time 3-D spatial-temporal dual-polarized measurement of wideband radio channel at mobile station. *IEEE Trans. Instrum. Meas.*, **49** (2), 439–448.

Kermoal, J. P., Schumacher, L., Pedersen, K. I., Mogensen, P. E., and Frederiksen, F. (2002). A stochastic MIMO radio channel model with experimental validation. *IEEE J. Sel. Areas Commun.*, **20** (6), 1211–1226.

Kildal, P.-S. and Rosengren, K. (2004). Electromagnetic characterization of MIMO antennas including coupling using classical embedded element pattern and radiation efficiency. In *Proc. IEEE Antennas and Propagat. Symp.*, vol. 2, 1259–1262.

Li, K.-H., Ingram, M. A., and Nguyen, A. V. (2002). Impact of clustering in statistical indoor propagation models on link capacity. *IEEE Trans. Commun.*, **50**, 521–523.

Matz, G. (2003). Characterization of non-WSSUS fading dispersive channels. In *Proc. IEEE Int. Conf. Commun.*, 2480–2484.

Oestges, C. and Paulraj, A. J. (2004). Beneficial impact of channel correlations on MIMO capacity. *Electron. Lett.*, **40** (10), 606–608.

Özcelik, H. (2004). *Indoor MIMO Channel Models*. Ph.D. thesis, Institut für

Nachrichtentechnik und Hochfrequenztechnik, Technische Universität Wien. URL http://www.nt.tuwien.ac.at/mobile/theses_fnished/ .

Özcelik, H., Czink, N., and Bonek, E. (2005). What makes a good MIMO channel model? In *Proc. IEEE Veh. Technol. Conf. (VTC Spring)*, vol. 1, 156–160.

Özcelik, H. and Oestges, C. (2005). Some remarkable properties of diagonally correlated MIMO channels. *IEEE Trans. Veh. Technol.*, **54** (6), 2143–2145.

Saleh, A. and Valenzuela, R. (1987). A statistical model for indoor multipath propagation. *IEEE J. Sel. Areas Commun.*, **5** (2), 128–137.

Sayeed, A. M. (2002). Deconstructing multiantenna fading channels. *IEEE Trans. Signal Process.*, **50** (10), 2563–2579.

Shiu, D.-S., Foschini, G. J., Gans, M. J., and Kahn, J. M. (2000). Fading correlation and its effect on the capacity of multielement antenna systems. *IEEE Trans. Commun.*, **48** (3), 502–513.

Spencer, Q. H., Jeffs, B. D., Jensen, M. A., and Swindlehurst, A. L. (2000). Modeling the statistical time and angle of arrival characteristics of an indoor multipath channel. *IEEE J. Sel. Areas Commun.*, **18** (3), 347–360.

Steinbauer, M., Molisch, A. F., and Bonek, E. (2001). The double-directional radio channel. *IEEE Antennas Propagat. Mag.*, **43** (4), 51–63.

Svantesson, T. and Ranheim, A. (2001). Mutual coupling effects on the capacity of multielement antenna systems. In *Proc. IEEE Int. Conf. Acoust., Speech, Signal Process.*, vol. 4, 2485–2488.

Thomä, R. S., Landmann, M., Richter, A., and Trautwein, U. (2005). Multidimensional high-resolution channel sounding measurement. In *Smart Antennas— State of the Art*, vol. 3 of *EURASIP Book Series on Signal Process. and Commun.*, 241–270.

Tsuchiya, H., Haneda, K., and Takada, J. (2004). UWB indoor double-directional channel sounding for understanding the microscopic propagation mechanisms. In *Proc. Int. Symp. Wireless Personal Multimedia Commun.*, 95–99.

Weichselberger, W. (2003). *Spatial Structure of Multiple Antenna Radio Channels— A Signal Processing Viewpoint*. Ph.D. thesis, Institut für Nachrichtentechnik und Hochfrequenztechnik, Technische Universität Wien. URL http://www.nt. tuwien.ac.at/mobile/theses_fnished/ .

Weichselberger, W., Herdin, M., Özcelik, H., and Bonek, E. (2006). A stochastic MIMO channel model with joint correlation of both link ends. *IEEE Trans. Wireless Commun.*, **5** (1), 90–100.

Ylitalo, J., Nuutinen, J.-P., Hämäläinen, J., Jämsä, T., and Hämäläinen, M. (2004). Multi-dimensional wideband radio channel characterisation for 2–6GHz band. In *Proc. 11th Wireless World Research Forum*.

Yong, S. K. and Thompson, J. S. (2005). 3-dimensional spatial fading correlation models for compact MIMO receivers. *IEEE Trans. Wireless Commun.*, **4** (6), 2856–2869.

24

Experiments in space-time modulation and demodulation

Weijun Zhu, Heechoon Lee, Daniel Liu, David Browne,
and Michael P. Fitz
University of California Los Angeles

24.1 Introduction

Over the past decade, there has been a great deal of research to improve the performance of wireless communications in fading environments by exploiting transmitter and/or receiver diversity. The pioneering work by Teletar (1995), Foschini and Gans (1998) showed that multiple antennas in a wireless communication system can greatly improve performance and spectral efficiency. For L_t transmit antennas and L_r receive antennas in Rayleigh fading, it was shown that, with spatial independence, there are essentially $L_t L_r$ levels of diversity available and there are $\min(L_t, L_r)$ independent parallel channels that can be established. These information-theoretic studies spawned two lines of work; one where the number of independent channels is large (Foschini, 1996) and one where the number of independent channels is small (Tarokh *et al.*, 1998; Guey *et al.*, 1996). With eight years of intense engineering research and development effort after these insights, Multiple antenna radio (MAR) techniques are starting to make a significant impact on how wireless services are provided. Examples include the nascent IEEE 802.11n standard and 4G mobile telecommunication systems. Progress in this area is such that researchers are calling the area mature.

The open problems in MAR communications relate to situations where more sophisticated and detailed aspects of communication systems need to be modeled and understood. For example, performance is not easily understood in channel models that are not well modeled as Gaussian/Rayleigh, or where the scattering is not rich or isotropic, or where system/channel parameters are time-varying, or how hardware nonidealities impact system performance. These problems are not well addressed by simulation or analysis. The sophistication of these types of research questions usually prohibits analysis in most cases and the utility of simulation is limited to the accuracy of the models used for simulation. Because of this trend in MAR research, wireless researchers (Batariers *et al.*, 2001; Sampath *et al.*, 2002; Lang *et al.*, 2004) have increasingly decided that experimental MIMO communications will be important in advancing the technology toward one that will achieve ubiquitous communications.

This chapter examines a small subset of the open issues in the literature and reports on experiments that attempt to resolve these issues regarding real systems and real channels. The focus here is on the following systems:

(i) *Land mobile wireless*: mobility and multipath typical of this environment will be the focus of the study presented in this chapter.

(ii) *Frequency-flat channels*: a vast majority of the work in space-time signaling has used frequency-flat models. This corresponds to relatively narrowband transmission in traditional land mobile wireless channels. It should be noted that orthogonal frequency-division multiplexing is usually designed so that each subcarrier has frequency flat fading.

(iii) *Linear modulation*: only linear modulation will be the focus of the study presented in this chapter.

(iv) *Short packet communication*: the packet lengths of the system presented in the chapter will be approximately 300 symbols. This type of system is typical of speech communication or paging systems.

Within this fairly focused area, we see the following open issues:

(i) *Trained-coherent versus noncoherent*: in general, there are three types of communications paradigms for signal design when channel state information (CSI) is not known at the transmitter: coherent, noncoherent, and differentially coherent. The coherent system assumes that the receiver perfectly knows the CSI. This paradigm implies CSI must be learned (e.g., with training symbols) before demodulation begins. Consequently, a trade-off exists between the bandwidth efficiency of coding schemes versus the performance in the presence of CSI estimation errors. This trade-off is a function of the transmission rate and the channel characteristics.

(ii) *Robustness issues*: code performance is very much a function of the channel conditions (Kose and Wesel, 2003) and channel models. Consequently, it is useful to see if any interesting characteristics are experienced in real wireless channels that impact the choice of coded modulation.

(iii) *Diversity versus multiplexing*: systems that try to achieve full diversity often need large constellations while systems that use spatial

multiplexing can achieve the same transmission rate with smaller constellations. Some sophisticated modulation schemes offer both multiplexing and diversity at the expense of a more complex demodulator.
For a given transmission rate, where is the best trade-off between
diversity, multiplexing, and complexity of the receiver? It should be
noted that this diversity/multiplexing trade-off is different than that
popularized by Zheng and Tse (2003).

This chapter is organized by having Section 24.2 overview the models, Section 24.3 present the experimental system and detail the nontrivial demodulation algorithms, Section 24.4 provide the experimental results, and Section 24.5 present the take-away conclusions.

24.2 Signal models

This chapter adopts the standard models for frequency-flat MAR communications. Assume a linear modulation is used with a Nyquist pulse shape,
$X_i(l)$ is the modulation symbol on the ith antenna at symbol l, and N_f is
the length of the frame in symbols. If the fading is slow enough, the sampled
matched-filter outputs are the sufficient statistics for the demodulation, and
these samples for the kth symbol are given as an $L_r \times 1$ vector[†]

$$\vec{Y}(k) = \mathbf{H}(k)\sqrt{E_s}\vec{X}(k) + \vec{N}(k),$$

where E_s is the energy per transmitted symbol; $H_{ji}(k)$ is the complex path
gain from transmit antenna i to receive antenna j at time kT; $\vec{X}(k)$ is
the $L_t \times 1$ vector of symbols transmitted at symbol time k; $\vec{N}(k)$ is the
additive white Gaussian noise vector of size $L_r \times 1$. The noise is modeled
as independent circularly symmetric zero-mean complex Gaussian random
variables with variance $N_0/2$ per dimension. $\mathbf{X}$, an $L_t \times N_f$ matrix, is often
denoted the space-time codeword. The codebook for space-time signaling
is denoted with $\mathcal{X}$, and the rate of the transmission scheme is defined as
$R = \log_2\left(|\mathcal{X}|\right)/N_f$.

Coherent demodulation refers to the case of finding the most likely transmitted word when the CSI is known. The optimum word demodulator is
denoted the maximum likelihood (ML) word demodulator. Denote $\vec{B}$ as the
transmitted binary word, i.e., $\mathbf{X} = a(\vec{B})$, where $a(\cdot)$ is determined by the
modulation and coding. For Nyquist linear modulation when the channel is

---

[†]This chapter uses the notation $\vec{X}$ for a vector and $\mathbf{H}$ for a matrix. When dealing with random
variables and processes, capital letters will refer to the random variable or process and lowercase
letters will indicate a sample function (i.e., an observation) of the random variable or process.

$\mathbf{H}(k) = \mathbf{h}(k)$ and observations are $\vec{Y}(k) = \vec{y}(k)$, the ML word demodulator is

$$\hat{B} = \arg\min_{\vec{B}=n} \sum_{k=1}^{N_f} (\vec{y}(k) - \vec{s}_n(k))^H (\vec{y}(k) - \vec{s}_n(k)) \qquad (24.1)$$

$$= \arg\max_{\vec{B}=n} \sum_{k=1}^{N_f} \left(2\Re \left[\vec{s}_n^H(k)\vec{y}(k) \right] - \vec{s}_n^H(k)\vec{s}_n(k) \right),$$

where $\vec{s}_n(k) = \sqrt{E_s}\,\mathbf{h}(k)\,\vec{x}^{(n)}(k)$ is the vector of noiseless received points on each of the antennas, $\vec{x}^{(n)}(k)$, an $L_t \times 1$ vector, is used to denote the transmitted codeword at the kth symbol for the transmitted bit sequence $\vec{B} = n$, and $\vec{y}(k)$ is the observed matched-filter output vector for the kth symbol corresponding to the outputs from all receive antennas. The ML demodulator essentially finds the code matrix $\mathbf{X} = \mathbf{x}_n$ that produces the minimum Euclidean square distance between the matched-filter outputs $\vec{Y}(k)$ and the channel output $\vec{s}_n(k) = \sqrt{E_s}\mathbf{h}(k)\vec{x}^{(n)}(k)$. If the modulation is defined on a trellis then the Viterbi algorithm can be used to find this minimum-distance transmitted codeword, and if the transmitted codeword is defined by a lattice then a lattice search algorithm can be used to find the best codeword.

The core idea of *differential modulation and demodulation* is that unitary group codes have characteristics that enable a simple modulator and demodulator (Hughes, 2000). A unitary group code is characterized by a set of unitary matrices $\mathbf{g}_n \in \mathcal{G}$ of size $N_b \times N_b$ that satisfy the group property for matrix multiplication. Differential encoding corresponds to generating a sequence of space-time block codes of size $L_t \times N_b$ via

$$\mathbf{x}(k) = \mathbf{x}(k-1)\mathbf{g}(k),$$

where the value of $\mathbf{g}(k) = \mathbf{g}_n$ is determined by the bits that are transmitted at the kth block time. The supremum of the rate of the code is $R < \log_2(|\mathcal{G}|)/N_b$. The simplest differentially coherent demodulator uses two consecutive blocks of matched-filter outputs to demodulate as

$$\hat{\mathbf{g}}(k) = \arg\max_{\mathbf{g}_n \in \mathcal{G}} \Re \left[\mathrm{Tr} \left(\mathbf{g}_n^H \mathbf{Y}(k-1)^H \mathbf{Y}(k) \right) \right],$$

where $\mathbf{Y}(k)$ is a block of matched-filter outputs of size $L_r \times N_b$. Some comments are noteworthy here: 1) if $\mathbf{Y}(k-1)$ is replaced by $\mathbf{H}(k)\mathbf{x}(k-1)$, this demodulator would be exactly the coherent demodulator given in (24.1) for the unitary modulation, 2) the demodulator is very simple and attractive but suffers two degradations, a) a 3 dB performance loss compared with coherent demodulation due to the noisy channel estimate $\mathbf{Y}(k-1)$, b) an error floor when the channel is time varying, since the demodulator assumes the channel is constant during two consecutive codewords.

The core idea of *noncoherent modulation and demodulation* is that a detector can be formed that is based only on the statistics of the fading channel and the observations. In the case of noncoherent detection, it was shown by Hochwald and Marzetta (2000) that a unitary constellation is capacity-achieving in the limit of large block lengths or high SNR (if $N_b \geq L_t$). A unitary noncoherent code is characterized by a set of unitary matrices $\mathbf{x}_n \in \mathcal{X}$ of size $L_t \times N_b$. The rate of the code is $R = \log_2\left(|\mathcal{X}|\right)/N_b$. The noncoherent demodulator (Hochwald *et al.*, 2000) uses the block of matched-filter outputs to demodulate the signal with a form

$$\hat{\mathbf{x}}(k) = \arg\max_{\mathbf{x}_n \in \mathcal{X}} \Re\left[\mathrm{Tr}\left(\mathbf{x}_n \mathbf{Y}(k)^H \mathbf{Y}(k) \mathbf{x}_n^H\right)\right],$$

where $\mathbf{Y}(k)$ is a kth block of matched-filter outputs of size $L_r \times N_b$. The demodulator is very simple and attractive but suffers two degradations 1) a performance loss compared with coherent demodulation 2) an error floor when the channel is time varying.

24.3 Experimental system

The experimental system that has been deployed for the work detailed in this chapter is a narrowband 3×4 MAR system. We have chosen a carrier frequency of 220 MHz and a bandwidth of around 4 kHz. All modulation types are linear, with a spectral raised-cosine pulse shape whose excess bandwidth is 0.2, and a symbol rate of 3.2 kHz. This carrier frequency and bandwidth allow us to do realistic land mobile testing and still be confident that the frequency-flat assumption is valid.

24.3.1 Radio system

The UnWiReD Laboratory narrowband testbed is a software-defined real-time multiantenna testbed. The information bits are encoded and pulse shaped by two Analog Devices (ADI) fixed-point digital signal processors (DSPs). The baseband signals are then digitally up-converted to 10 MHz IF signals. The three-TX up-converter radio further up-converts the IF signals to the 220 MHz RF, and amplifies it for transmission with a maximum transmission power of 35 dBm.

The receiver is a high-performance system for narrowband MIMO processing. The received signals are down-converted from RF to 10 MHz IF signals

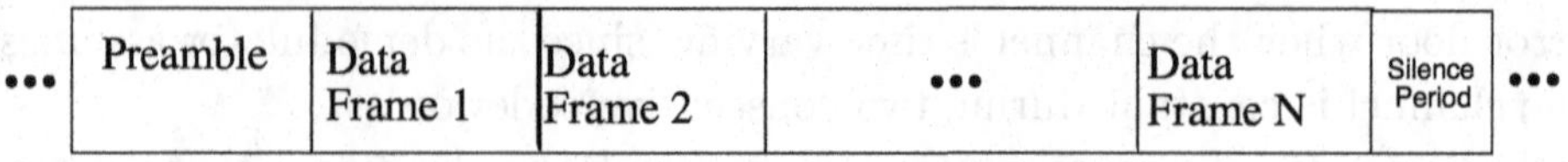

...	Preamble	Data Frame 1	Data Frame 2		...	Data Frame N	Silence Period	...

Fig. 24.1. The superframe used in the field experiments.

by a calibrated fixed gain four-channel down-converter radio, and then digitally down-converted to baseband by a four-channel digital receiver. The four-channel digital receiver oversamples the input signals at 64 MHz. The overall receiver dynamic range is greater than 80 dB. The overall error vector magnitude through both the transmit and receive chains is less than 2%. The demodulation is performed by two floating-point ADI DSPs. The demodulated data, as well as other important test information, are transferred to a laptop for recording and for displaying real-time test results. A system of this form allows a great deal of information to be extracted from the experiments.

24.3.2 Packet format

The format for the transmitted signals of the experimental system was designed to allow many modulation types to be tested in a time-interleaved fashion. The structure that allows this comparison is a superframe at the transmitter that is repeated about every four seconds. During this superframe, a preamble is generated and followed by 42 frames of potentially different space-time modulation types. The preamble is used to initially acquire accurate symbol-time estimates, a coarse frequency estimate, and frame alignment. The preamble-signal format permits high performance symbol-time estimation and frame synchronization by implementing a dotting pattern and a unique word. Each of the subsequent space-time modulated frames is 300 symbols in length (93.75 ms). Modulations are independent from frame to frame for the experiments documented in this chapter. At the end of the superframe there is a silence period of about 70 symbols. The noise power, which can vary significantly at a 220 MHz carrier frequency in various scenarios due to man-made noise, is measured every frame in the silence period with a 50 symbol average of the received energy.

24.3.3 Receiver processing overview

All of the receiver functions are implemented in real-time in a DSP. Symbol-time estimates are derived by using a nonlinear open loop timing estimator

(Oerder and Meyr, 1988). Rapid acquisition is performed during the preamble and slow tracking is accomplished during the remainder of the frame. Frequency estimation and frame synchronization are achieved during the first preamble portion of the frame during the decoding. These synchronization techniques are based on ML principles and provide near-optimum performance. Having a unique word for frame synchronization allows pilot symbols to be inserted and used for CSI estimation, and also provides block-boundary synchronization for all coded modulation types during demodulation. The details of the pilot symbol processing are given in Section 24.3.6. A wide variety of decoding algorithms are implemented in this testbed. For coherent demodulation, ML receivers based on a trellis search and on a sphere or lattice decoder have been implemented. A quick overview of the sphere or lattice decoding algorithms used in the testbed is given in Section 24.3.4. Suboptimal spatial demultiplexing algorithms are detailed in Section 24.3.5. The great flexibility of having a DSP-based solution allows many algorithms to be compared in order to understand the complexity-performance trade-offs in real implementations. One of the powerful characteristics of the implementation is that the same transmitted data can be used to compare decoding versus the number of receive antennas. In most modulation formats, decoding for any number of receive antennas (from $L_r = 1$ to $L_r = 4$) can be accomplished simultaneously in real time.

24.3.4 Coherent sphere or lattice decoding

In MIMO fading channels, ML detection is desirable to achieve high performance. As the spectral efficiency increases an exhaustive search implementation of ML decoder becomes increasingly infeasible due to a complexity that is growing exponentially with the spectral efficiency. However, if the received signal can be represented as a lattice, lattice decoding algorithms can be utilized to attain ML performance with significantly reduced complexity.

The sphere or lattice decoder was first introduced in digital communications by Viterbo and Biglieri (1993). The lattice representation of a MIMO system was first defined by Damen *et al.* (2000). The idea of sphere or lattice decoder can be described simply. The sphere decoder attempts to minimize the ML metric $||\vec{y}(k) - \mathbf{h}(k)\vec{x}_n(k)||^2$, with the constraint that the search is confined to the lattice points within a sphere of radius $\sqrt{C}$ centered around the received vector $\vec{y}(k)$. Whenever a valid point is found within the sphere, the search is restarted and looks for points at a distance less than the last identified point. The search is continued until no closer point is found. Then, the decoder outputs the last valid point as the ML solution. If the initial sphere contains no valid point, the radius of sphere can be increased and searching can be restarted. Therefore, if there is no restriction on the complexity, the sphere or lattice decoder can achieve ML performance by updating the radius as needed.

The sphere or lattice decoder is summarized in the following three steps:

(i) *Preparation*: a sphere decoder imposes a triangular channel matrix by implementing a QR factorization of the channel matrix.

(ii) *Front end (optional)*: several front-end processing techniques have been developed to expedite the searching time. These include Lenstra–Lenstra–Lovász (LLL) and Korkine–Zolotareff (KZ) reduction (Agrell *et al.*, 2002), and ZF-DFE and MMSE-DFE preprocessing (Damen *et al.*, 2003). When the channel remains fixed for a long time, these front-end processing techniques are very effective for reducing the overall complexity. However, if the channel changes significantly from block to block, front-end processing can increase the overall complexity.

(iii) *Searching*: depending on the enumeration of lattice points within the sphere, the search time can be quite different. Pohst enumeration (Pohst, 1981; Fincke and Pohst, 1985) is a natural spanning of lattice points from one end to the other. Schnorr–Euchner enumeration starts from the center point (Babai point (Agrell *et al.*, 2002)) and expands in zig-zag manner.

In the experiments reported in this chapter front-end processing was not implemented (because of the time-varying nature of the fading encountered in the experimental system) and Pohst enumeration was implemented.

24.3.5 Coherent spatial equalization techniques

Suboptimal detectors are possible for MAR demodulation. Due to the computational complexity of an ML decoding algorithm, a simple linear receiving frontend is often employed. This linear frontend is given as

$$\hat{\vec{X}} = \mathbf{W}^H(k)\mathbf{Y}(k),$$

where $\mathbf{W}^H(k)$ corresponds to the implemented linear filter. The linear frontend processing can be chosen to satisfy the zero-forcing (ZF) solution or the minimum mean square error (MMSE) solution. The MMSE solution requires knowledge or an estimate of the noise variance, while the ZF solution does not. For spatial multiplexing, linear frontend processing can be combined with interference cancellation to produce better performance. In the experiments reported in this chapter, spatial multiplexing is decoded with an MMSE linear frontend processor, and often successive interference cancellation (SIC) is used. SIC in this chapter is very similar to the V-BLAST (Foschini, 1996) architecture. MMSE filtering and SIC can produce hard or soft decisions as needed.

ZF processing can be used for Alamouti type decoding on rapidly-varying channels. Many of the experiments reported in this chapter are high-mobility experiments. In high mobility, the codeword threads of Alamouti-like signaling lose their orthogonality due to the rapidly-varying channel. This loss of orthogonality, while relatively small, can cause significant degradations in performance for large constellations. We have found it useful in cases with mobility to implement the Alamouti decoder with a ZF frontend filter. For example, in the $L_t = 2$ and $L_r = 1$ case, the Alamouti signaling and the ZF processing is characterized by

$$\begin{bmatrix} Y(1) \\ Y^*(2) \end{bmatrix} = \begin{bmatrix} h_1(1) & h_2(1) \\ -h_2^*(2) & h_1^*(2) \end{bmatrix} \begin{bmatrix} X(1) \\ X(2) \end{bmatrix} + \begin{bmatrix} N(1) \\ N^*(2) \end{bmatrix}$$

$$\begin{bmatrix} \hat{X}(1) \\ \hat{X}(2) \end{bmatrix} = \begin{bmatrix} ah_1^*(2) & -ah_2(1) \\ ah_2^*(2) & ah_1(1) \end{bmatrix} \begin{bmatrix} Y(1) \\ Y^*(2), \end{bmatrix}$$

where $1/a = h_1(1)h_1^*(2) + h_2(1)h_2^*(2)$. The complexity of the processing is increased compared with that proposed by Alamouti (1998) (especially for $L_t > 2$), but the resulting performance improvements often warrant the increased complexity.

24.3.6 Channel estimation

Accurate estimation of CSI is crucial for reliable decoding of coherent coding schemes. Pilot symbol assisted demodulation (PSAD) is employed when good performance in high-mobility situations is desired at a reasonable complexity. Pilot symbol based frame design and CSI estimation is essentially an application of sampling theory and optimal interpolation of Gaussian processes (Cavers, 1991; Kuo and Fitz, 1994; Guey *et al.*, 1996; Liu, 2001). Due to the Gaussian nature of the assumed Rayleigh fading, linear interpolation filtering is optimal. For a finite frame size, interpolation at the frame edges performs worse, and hence it is important to have more samples at the frame edges. Uniform pilot sampling in the middle of the frame is optimal as long as the sampling is above the Nyquist rate of the channel. Guey *et al.* (1996) showed that orthogonal pilot elements on each transmit antenna have many desirable characteristics. An orthogonal pilot symbol pattern maintains good performance (but not orthogonality at the receiver) even with high mobility.

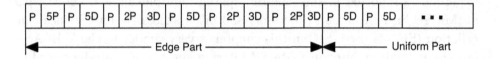

Fig. 24.2. The frame design for the pilot symbol processing for $L_t = 2$. P indicates pilot symbols, D indicates data symbols, and a number indicates the multiplicity of contiguous symbols.

The pilot symbol frame structure for the experiments reported in this chapter is optimized for the short 300-symbol frame structure and the rapid fading that is possible with high mobility. The optimal strategy is that more pilot symbols are used at the edge of the frame, while the pilot insertion pattern becomes uniform in the center of the frame. A different pilot symbol frame is produced for a different numbers of transmit antennas. For example, the two-Tx frame is shown in Fig. 24.2. In this example 72 out of 300 total symbols are used for training. Hence, to maintain a fair comparison with a modulation/demodulation not needing training for CSI estimation, a code rate increase of roughly 4/3 needs to be implemented for the coherent coding and decoding. The channel gains between any transmitter-receiver pair are assumed to be spatially independent for interpolation filter design. Also, the channel coefficients are assumed to be constant over a symbol period, but vary from symbol to symbol according to a Clarke spectrum (Clarke, 1968) modeled with $R_H(m) = J_0(2\pi f_D Tm)$, where J_0 is the zeroth order Bessel function of the first kind and f_D is the Doppler spread of the channel. An FIR Wiener filter optimized for $E_b/N_0 = 30$ dB and Doppler fading rate $f_D T = 0.01$ is used for pilot interpolation in the experiments reported in this chapter.

24.4 Experimental results

The experimentation was done on the UCLA campus and the surrounding West Los Angeles area. In all cases, the antennas were whip dipole antennas (see Fig. 24.3) that were vertically polarized and had roughly a 3 dB gain compared with isotropic antennas. The testing was broken up into three types:

(i) *Indoor stationary*: this environment has both the transmitter (TX) and the receiver (RX) radio deployed in a building on the same floor (Boelter Hall at UCLA). The tests were run during both day and night time so foot traffic was typical of a campus building. Unless

(a) Transmitter deployment.   (b) Receiver deployment.

Fig. 24.3. Example deployment for the outdoor tests.

specified otherwise, the receiver array was linear with a $\lambda/2$ spacing, and the transmitter array was triangular also with a $\lambda/2$ spacing. The antennas were either mounted on a tripod or on a table top.

(ii) *Outdoor local*: this environment has one radio (TX) deployed on the top of a five story building and one radio (RX) on a vehicle (a cart or a van). The test consisted of the receiver radio being driven around the campus area. The UCLA campus area is heavily urbanized. Unless specified otherwise, the receiver array was square with a $\lambda/2$ spacing on each side and mounted on the top of a van. The transmitter array was linear with a 2λ spacing and the mounting was on the top of a building.

(iii) *Outdoor high speed*: this environment has one radio (TX) deployed on the top of a five story building and one radio (RX) on a vehicle (a van). The test consisted of the receiver radio being driven on the 405 Freeway in West Los Angeles. The testing was done at times of relatively free flowing traffic, so speeds of 50–75 mph were maintained during the testing. The antenna geometries are the same as the outdoor local scenarios unless noted otherwise.

An example of the testbed deployment for the mobile test is shown in Fig. 24.3. It is apparent that the building where the transmitter is deployed has the same height as many of the surrounding buildings. Consequently, the deployment is not a typical tower-type deployment and should not require large antenna separation to get independent spatial channels.

A large amount of data was collected and not all of it will be reported. The point of this chapter is to answer the questions that have been posed. We will highlight the data that supports our conclusions. The rest of the data will be available upon request. The major findings reported will be the bit error rate and frame error rate versus E_b/N_0. The measured E_b/N_0 reported in the experiments are computed by the averages over the entire superframe and all the receive antennas. This measure gives something closer to the average SNR in high-mobility tests and something closer to the instantaneous SNR in static testing, but the measurement was viewed as the best compromise in reporting the data.

24.4.1 Trained versus noncoherent

In this test, simple coherent demodulation with pilot symbol based CSI estimation is compared with noncoherent modulation and demodulation techniques. Achieving the desired unitary form for the modulation placed a significant restriction on the constellations that can be used for noncoherent demodulation. Hence, we limit our study to coding schemes that have small effective coding rates. For $L_t = 2$, a rate-fair comparison that accounts for pilot symbols uses the Alamouti code (Alamouti, 1998) ($R = 2$) with QPSK, the $R = 1.5$ with QPSK differentially coherent code of Hughes (2000), the $R = 1.5$ and $N_b = 4$ noncoherent code of Kammoun and Belfiore (2003), and the $R = 1.5$ and $N_b = 4$ Alamouti noncoherent code with 8-PSK. The noncoherent Alamouti code has an identity matrix concatenated with a coherent Alamouti code. In the Alamouti-code case, 448 bits are transmitted with 112 Alamouti block codes. For the differentially coherent code, 447 bits are transmitted with 150 blocks of length-two symbols. The noncoherent codes use 75 blocks of length-four symbols to send 450 bits.

These tests and comparisons were completed both in a static and a mobile environment. The performance in the static environment is shown in Fig. 24.4. The differentially coherent and trained coherent results give about the same performance. The reason for this is that the higher rate code ($R = 2$ versus $R = 1.5$) and the imperfect CSI estimation cause enough degradation in coherent demodulation to give differentially coherent demodulation comparable performance, in spite of a theoretical 3 dB loss inherent in differentially coherent demodulation. The noncoherent modulation types have about 3–4 dB degradation compared with the others. The reason for this degradation is that the noncoherent coding schemes need to implicitly include training to demodulate in the presence of a random channel. Hence, the constellation needs to be larger to keep the same average rate. The differentially coherent and trained coherent schemes obtain this channel estimate in a more bandwidth-efficient manner. Longer block length noncoherent codes might be able to recover this inefficiency, but then the codes will be more susceptible to time-varying fading. Longer noncoherent codes were not tested. The performance in the high-mobility environment is shown in Fig. 24.5. The results follow the same general trend of Fig. 24.4, except that trained coherent demodulation has a slightly better performance than differentially coherent demodulation, and all modulation types experience an error floor at high SNR. The slight performance improvement is probably due to the fact that the pilot symbol aided CSI estimation was designed to accommodate Doppler spread, while differential demodulation was not. The error floor is expected and predicted by theory for the differential and noncoherent demodulators, but the manifestation of an error floor for coherent demodulation was unexpected. An interesting characteristic of the Kammoun code seen in Fig. 24.5 is that as the number of receive antennas grows from $L_r = 1$ to $L_r = 4$, the Kammoun code realizes greater diversity gain than does the noncoherent Alamouti scheme. Also the Kammoun code has better performance in the high SNR region for high mobility and a larger number of receive antennas than the noncoherent Alamouti code.

24.4.2 Robustness

The performance achieved by a space-time code is a function of the interaction of the code structure and the channel structure. As a first step toward quantifying this characteristic of space-time codes, an experiment was run that implemented a wide variety of powerful trellis codes and compared their performance. The code comparison that is reported here includes the $L_t = 2$ codes optimized for spatially white Rayleigh fading reported by Yan and Blum (2000), the superorthogonal (SO) codes (Siwamogsatam and Fitz, 2001, 2002; Ionescu *et al.*, 2001; Jafarkhani and Seshadri, 2002) (specifically the codes of Siwamogsatam and Fitz (2001)), and the universal trellis codes of Kose and Wesel (2003, 2004). Typical results that were obtained are shown in Fig. 24.6. This figure shows the compilation of the outdoor local test results for $L_r = 2$.

Several conclusions can be drawn about the investigated trellis codes' performance in actual channels. First, the bit error probability performance does not change dramatically between each of the powerful codes. The variations in performance with SNR are much greater than is typically observed in simulation. This characteristic is due to the great variability of the wireless channel with propagation geometry, and the difficulty in acquiring statistically significant amount of data compared with the variability in the channel. Though a full day of drive testing collected a large amount of data, the channels corresponding to this collected data are only a small sampling of all pos-

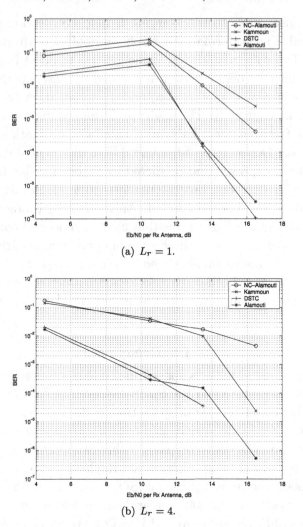

(a) $L_r = 1$.

(b) $L_r = 4$.

Fig. 24.4. Comparison of $R = 1.5$ (effective) trained and noncoherent modulation performance in stationary tests.

sible channels. Second, the powerful codes also seem to exhibit an error floor in the mobile environment. This leads to a postulation that the error floor that occurs in our experimental system is not an aspect of the modulation, but was due to an unforeseen aspect of CSI estimation. There is a wide variety of possible explanations for the error floor with high mobility (Doppler-induced intersymbol interference, frequency or timing estimation failure in high Doppler, and/or channel estimation biases due to filter mismatches). It has not been possible yet to conclusively identify the cause of the error floor, but it is important to note that even a high fidelity receiver and advanced

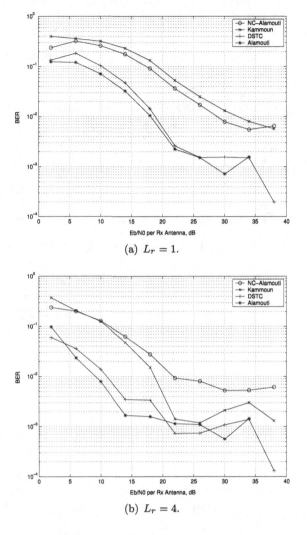

(a) $L_r = 1$.

(b) $L_r = 4$.

Fig. 24.5. Comparison of $R = 1.5$ (effective) trained and noncoherent modulation performance in high mobility tests.

processing algorithms produce this error floor. This should be a cautionary note to designers of space-time modems operating with high mobility.

An interesting characteristic of the SO codes is that they yield an increasing bit error rate with the number of states in the code (as they do in simulation). In terms of frame error rate, the SO codes hold a clear advantage. Consequently, it is postulated that for short packet delivery, an SO code might be a good choice. Alternately, if the space-time trellis code is to be used in a system that concatenates an inner code with an outer code

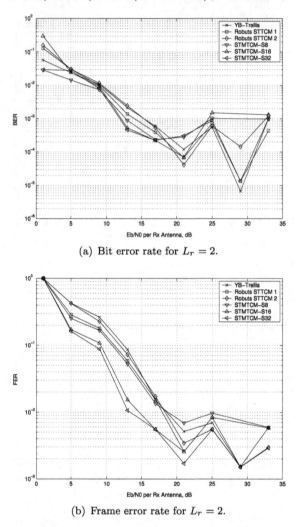

(a) Bit error rate for $L_r = 2$.

(b) Frame error rate for $L_r = 2$.

Fig. 24.6. Comparison of $R = 2$ trellis codes in mobile outdoor tests.

(like a Reed–Solomon code), the universal codes of Kose and Wesel (2003) or the optimized trellis codes of Yan and Blum (2000) might be better choices. This concatenated coding experiment will be a subject of future work to understand the best coding structures in fading channels. Finally, at least with this initial experiment, a significant variation between simulation results and experimental results was not seen, nor was there much evidence of a lack of robustness in any of the codes. A more careful future experiment is planned that specifically characterizes the performance versus the channel gain conditions, and it might generate different conclusions than this experiment, which aggregated all data obtained during a test.

24.4.3 Diversity versus multiplexing

The choice in an engineering trade-off between using diversity signaling and spatial multiplexing at a fixed rate in modems operating in realistic channels with noisy synchronization and channel estimation is an open question. To achieve full diversity at $R > 2$ typically implies that large constellations are needed. This can be done with a standard space-time code like the Alamouti block code or by using algebraic rotations of smaller constellations as proposed by Damen *et al.* (2002), El Gamal and Damen (2003). Spatial multiplexing does not worry about achieving full diversity, and hence it can achieve a given throughput with a smaller constellation. Spatial multiplexing and Alamouti coding can be decoded via simple linear receivers, while the threaded algebraic codes require the more complex sphere or lattice decoder. A good understanding of the performance complexity trade-off between these signaling options on realistic channels is an open issue and the subject of the experiments reported in this section.

For $L_t = 2$, the diversity signaling has an advantage when $L_r = 2$, but that advantage disappears with larger number of receive antennas. For example, Fig. 24.7 shows the bit error rate performance of the various proposed alternatives for $L_r = 2$ and $L_r = 4$ in the indoor stationary environment. Spatial multiplexing uses the MMSE detector with SIC. For a smaller number of receive antennas, the Alamouti code shows a clear advantage. The extra diversity of the modulation can be thought of as making up for the closer decision boundaries inherent in using a larger constellation. For a larger number of receive antennas, the extra diversity inherent in the MMSE processing for spatial multiplexing tilts the trade-off toward using smaller constellations. While this characteristic would be expected from theory, it is interesting to see where the optimal operating points are for each rate and number of transmit antennas.

Other trade-offs can be explored between diversity and multiplexing. For instance, if the system expands to three TX antennas, then diversity signaling is not worth the effort above a certain rate. Fig. 24.8(a) shows the bit error rate performance of spatial multiplexing and the super orthogonal block code (Siwamogsatam and Fitz, 2002; Lee *et al.*, 2004) with roughly comparable rates for indoor testing. It is clear from this figure that the rate loss for $L_t = 3$ or more antennas associated with orthogonal like codes makes it hard for large constellation diversity-type signaling to be competitive with smaller constellations that can be achieved with spatial multiplexing. Similarly, Fig. 24.8(b) shows the frame error rate performance of the powerful TAST codes (El Gamal and Damen, 2003) compared with diversity and

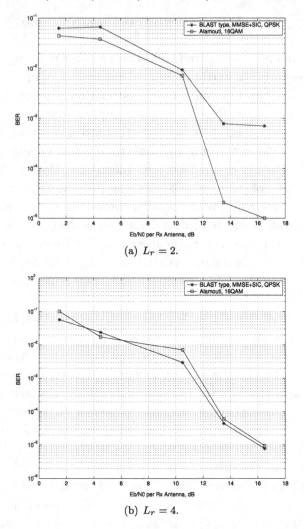

(a) $L_r = 2$.

(b) $L_r = 4$.

Fig. 24.7. Comparison of diversity signaling and spatial multiplexing during indoor stationary tests.

spatial multiplexing type signaling in outdoor testing. These codes clearly improve the performance in most cases, but come at a significant complexity increase. Our experiments also show us that the advantage of TAST is less when bit error probability is considered. Some degradations of TAST decoding are apparent at high SNR in high-mobility testing. This might be due to the assumption in the decoder that the channel is constant over the block length of the TAST decoding. Engineering trade-offs will determine the utility of using such signaling and demodulation schemes.

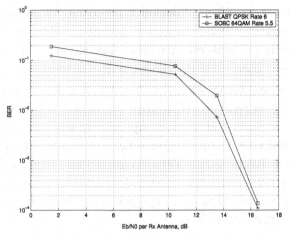

(a) $L_t = 3$ and $L_r = 3$ diversity multiplexing trade-off.

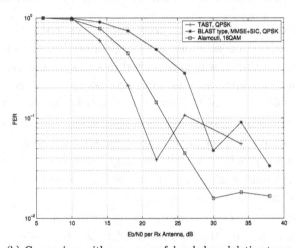

(b) Comparison with more powerful coded modulation types.

Fig. 24.8. Comparison of diversity signaling and spatial multiplexing during outdoor mobile tests.

24.5 Conclusions

The following conclusions can be obtained from the experiments reported in this chapter:

(i) *Trained coherent versus noncoherent*: the experiments presented in this chapter show that for low rates ($R \leq 2$), differentially coherent demodulation has good performance and low complexity. The loss in rate due to training and noisy CSI estimates in trained coherent system cause a 3 dB degradation, so that performance is comparable.

Error floors for all modulation types occur at high mobility. Short block length noncoherent codes do not provide competitive performance, but perhaps longer codes would give better performance.

(ii) *Robustness issues*: for the reported first order tests, there appears to be no obvious issues with the robustness of space-time codes.

(iii) **Diversity versus multiplexing**: the choice between low-complexity diversity signaling and low-complexity spatial multiplexing depends on the desired rate (higher rates favor multiplexing) and on the number of receive antennas (lower number of antennas favor diversity signaling). Powerful threaded codes can offer the advantages of both diversity and multiplexing type signaling, but require more complex demodulation algorithms.

The following open issues are still worth exploring in follow-on work:

(i) *Error floor in high mobility*: error floors occurred with all modulation and demodulation schemes with high mobility. This error floor was surprising in that it was unexpected in an experimental system with a well designed radio and baseband processing architecture. Identifying the possible causes of this error floor will be important.

(ii) *Performance of more powerful coding schemes*: the experiments reported in this chapter were mostly for space-time constellations and spatial multiplexing. The interesting question is if the conclusions would still hold with powerful concatenated coding as are typical in a normal modern communication system.

(iii) *More focused tests to quantify robustness*: robustness is a function of the channel and the code. More focused tests are needed that appropriately parameterize the channel and the performance of each of the coded modulation types as opposed to the results in this chapter, which report the performance aggregated over many channels.

Acknowledgments

The National Science Foundation has provided significant support for this effort, which had a time constant of not months but years. The vision of the program managers and reviewers that enabled this effort is acknowledged. Analog Devices provided hardware and software support in the construction of this testbed. Since the author list was limited, a great deal of intellectual capital that has gone into this project needs to be acknowledged. Past UnWiReD Laboratory members who participated in the project are Parul

Gupta, Shingwa Wong, and Sunder Venkateswaran. A portion of the software used in the experiments was the result of class work in the Wireless Communication Theory course that is taught by Prof. Fitz at UCLA, and people whose projects contributed to the results reported in this chapter are Cenk Kose, Maryam Owrang, and Scott Enserink. Prof. Belfiore also gave us the unpublished noncoherent $R = 1.5$ code based on the results in Kammoun and Belfiore (2003).

References

Agrell, E., Eriksson, T., Vardy, A., and Zeger, K. (2002). Closest point search in lattices. *IEEE Trans. Inf. Theory*, **48** (8), 2201–2214.

Alamouti, S. M. (1998). A simple transmit diversity technique for wireless communications. *IEEE J. Sel. Areas Commun.*, **16** (8), 1451–1458.

Batariers *et al.* (2001). An experimental OFDM system for broadband mobile communications. In *Proc. IEEE Veh. Technol. Conf., Fall*, vol. 4, 1947–1951.

Cavers, J. K. (1991). An analysis of pilot symbol assisted modulation for Rayleigh faded channels. *IEEE Trans. Veh. Technol.*, **40** (4), 686–693.

Clarke, R. H. (1968). A statistical theory for mobile radio reception. *Bell Syst. Tech. J.*, **47** (4), 957–1000.

Damen, M. O., Chkeif, A., and Belfiore, J.-C. (2000). Lattice codes decoder for space-time codes. *IEEE Commun. Lett.*, **4** (5), 161–163.

Damen, M. O., Gamal, H. E., and Caire, G. (2003). On maximum-likelihood detection and the search for the closest lattice point. *IEEE Trans. Inf. Theory*, **49** (10), 2389–2402.

Damen, M. O., Tewfik, A., and Belfiore, J.-C. (2002). A construction of a space-time code based on number theory. *IEEE Trans. Inf. Theory*, **48** (3), 753–760.

El Gamal, H. and Damen, M. O. (2003). Universal space-time coding. *IEEE Trans. Inf. Theory*, **49** (5), 1097–1119.

Fincke, U. and Pohst, M. (1985). Improved methods for calculating vectors of short length in a lattice, including a complexity analysis. *Math. Comp.*, **44** (170), 463–471.

Foschini, G. J. (1996). Layered space-time architecture for wireless communications in a fading environment when using multiple antennass. *Bell Labs Tech. J.*, **1** (2), 41–59.

Foschini, G. J. and Gans, M. (1998). On the limits of wireless communication in a fading environment. *Wireless Personal Commun.*, **6** (3), 311–355.

Guey, J.-C., Fitz, M. P., Bell, M. R., and Kuo, W.-Y. (1996). Signal design for transmitter diversity wireless communication systems over Rayleigh fading channels. In *Proc. IEEE Veh. Technol. Conf.*, 136–140.

Hochwald, B. M. and Marzetta, T. L. (2000). Unitary space-time modulation for multiple-antenna communications in Rayleigh flat fading. *IEEE Trans. Inf. Theory*, **46** (2), 543–564.

Hochwald, B. M., Marzetta, T. L., Richardson, T. J., Sweldens, W., and Urbanke, R. (2000). Systematic design of unitary space-time constellations. *IEEE Trans. Inf. Theory*, **46** (6), 1962–1973.

Hughes, B. L. (2000). Differential space-time modulation. *IEEE Trans. Inf. Theory*, **46** (7), 2567–2578.

Ionescu, M., Mukkavilli, K. K., Yan, Z., and Lilleberg, J. (2001). Improved 8- and 16-state space-time codes for 4PSK with two transmit antennas. *IEEE Commun. Lett.*, **5** (7), 301–303.

Jafarkhani, H. and Seshadri, N. (2002). Super orthogonal space-time trellis codes. In *Proc. IEEE Int. Conf. Commun.*, vol. 3, 1439–1443.

Kammoun, I. and Belfiore, J.-C. (2003). A new family of Grassmann space-time codes for non-coherent MIMO systems. *IEEE Commun. Lett.*, **7** (11), 528–530.

Kose, C. and Wesel, R. D. (2003). Universal space-time trellis codes. *IEEE Trans. Inf. Theory*, **49** (10), 2717–2727.

—— (2004). Universal space-time codes from two-dimensional trellis codes. In *IEEE Global Commun. Conf.*, 391–395.

Kuo, W. Y. and Fitz, M. P. (1994). User slot design and performance analysis for burst mode communication with fading and frequency uncertainty. *Int. J. Wireless Inf. Netw.*, **1** (10), 239–252.

Lang, S., Rao, R. M., and Daneshrad, B. (2004). Design and development of a 5.25 GHz, software defined wireless OFDM communication platform. *IEEE Commun. Mag., Radio Commun. Supplement*, **42** (6), S6–12.

Lee, H., Siti, M., Zhu, W., and Fitz, M. P. (2004). Super-orthogonal space-time block code using a unitary expansion. In *Proc. IEEE Veh. Technol. Conf., 2004-Fall*, 2513–2517.

Liu, Z. (2001). *Implementation of Transmit Antenna Diversity in Wireless Communication Systems*. Master's thesis, The Ohio State University.

Oerder, M. and Meyr, H. (1988). Digital filter and square timing recovery. *IEEE Trans. Commun.*, **36** (5), 605–612.

Pohst, M. (1981). On the computation of lattice vectors of minimal length, successive minima and reduced basis with applications. *ACM SIGSAM Bulletin*, **15** (1), 37–44.

Sampath, H., Talwar, S., Tellado, J., Erceg, V., and Paulraj, A. (2002). A fourth-generation MIMO-OFDM broadband wireless system: Design, performance, and field trial results. *IEEE Commun. Mag.*, **40** (9), 143–149.

Siwamogsatam, S. and Fitz, M. P. (2001). Improved high rate space–time TCM via orthogonality and set partitioning. In *Proc. Int. Symp. Wireless Personal Multimedia Commun.*

—— (2002). Improved high rate space–time TCM via concatenation of expanded orthogonal block codes and MTCM. In *Proc. IEEE Int. Conf. Commun.*, 636–640.

Tarokh, V., Seshadri, N., and Calderbank, A. R. (1998). Space-time codes for high data rate wireless communication: Performance criterion and code construction. *IEEE Trans. Inf. Theory*, **44** (2), 744–765.

Teletar, E. (1995). Capacity of multi-antenna Gaussian channels. Tech. rep., AT&T–Bell Labs.

Viterbo, E. and Biglieri, E. (1993). A universal decoding algorithm for lattice codes. In *GRETSI 14-ème Colloque*, 611–614.

Yan, Q. and Blum, R. S. (2000). Optimum space-time convolutional codes. In *IEEE Wireless Commun. and Netw. Confernce*, 1351–1355.

Zheng, L. and Tse, D. (2003). Diversity and multiplexing: a fundamental tradeoff in multiple-antenna channels. *IEEE Trans. Inf. Theory*, **49** (5), 1073–1096.

25

Multiple antenna techniques in 3G wireless systems

Robert A. Soni

Lucent Technologies

R. Michael Buehrer

Virginia Tech

25.1 Introduction

Multiple antenna techniques (beyond two receive antennas at the base station) have now achieved a level of technical maturity that allows their implementation in commercial cellular systems. Specifically, multiple antenna technologies have been integrated into Third Generation (3G) cellular systems, and will soon be part of the 802.11n standard. In this chapter, we examine commercial implementations of multiple antenna techniques. While multiple antennas can be used at either the transmitter or the receiver, commercial standard specifications primarily focus on application at the transmitter. Multiple antenna techniques that are applied at the receiver are not specified by the standard and are vendor specific. As a result, while some of the techniques discussed in this chapter (e.g., transmit diversity) are defined by the 3G standards, others (e.g., receive beamforming) can be used in 3G systems, but are not specifically defined by standards-based technical specifications. Further, there are techniques that are under investigation in 3G standards bodies, e.g., spatial multiplexing, which requires multiple antennas at both the receiver and transmitter, often referred to as multiple input and multiple output (MIMO). We will describe techniques that fall into these categories, and will be careful to distinguish those techniques that are specified by the standard, that are under investigation by a specific standards group, or that are allowed by the standard. Section 25.2 presents the system model used throughout the chapter. Transmit diversity techniques are specified by both of the major 3G standards, and are discussed in detail in Section 25.3. The impact of transmit beamforming in 3G is described in Section 25.4, while receiver-oriented techniques are discussed in Section 25.5. Finally, applications of spatial multiplexing are investigated in Section 25.6.

25.2 System model

Cellular 3G systems predominantly use code division multiple access
(CDMA). Consider a CDMA system that uses orthogonal Walsh functions
to separate users on the downlink along with a base-station-specific covering
code. For a system with K mobiles, receiving signals from a common base
station, the baseband (i.e., before upconversion) transmitted chip-rate
signal on a single antenna is

$$x(t) = \left(\sum_{i=1}^{K} \sqrt{P_i} s_i(t) w_i(t) + \sqrt{P_p} w_0(t) \right) p(t),$$

where P_i is the power transmitted to the ith mobile, $s_i(t)$ and $w_i(t)$ are the
data signal and unique Walsh function intended for the ith mobile, respectively, P_p is the power of the pilot signal, which uses Walsh function 0, and
$p(t)$ is the long pseudorandom covering code for the base station of interest.
The symbol (i.e., chip) rate of $x(t)$ is equal to the symbol rate of $s_i(t)$ times
the length N of the Walsh function $w_i(t)$. That is, $T_c = T_s/N$, where T_c is
the chip duration and T_s is the data symbol duration. Further, the Walsh
functions are real-valued, orthogonal, and periodic with every data symbol
period T_s, i.e.,

$$\frac{1}{T_s} \int_0^{T_s} w_i(t) w_j(t) dt = \begin{cases} 1 & i = j \\ 0 & i \neq j. \end{cases}$$

At the mobile, the following signal is received on a single antenna:

$$y(t) = h(t) x(t) + n(t),$$

where $h(t)$ is the complex multiplicative distortion (i.e., time-varying, flat
fading) caused by the wireless channel, and $n(t)$ is a complex Gaussian
random process that represents thermal noise and all out-of-cell interference.
Mobile i correlates the received signal with the ith Walsh function during
the kth symbol interval after uncovering (i.e., removing $p(t)$) to obtain the
decision statistic $z_i[k]$:

$$z_i[k] = \int_{(k-1)T_s}^{kT_s} y(t) p^*(t) w_i(t) dt$$

$$= \sqrt{P_i} h[k] s_i[k] + n[k],$$

where $h[k]$ represents the cumulative effect of the channel $h(t)$ over the
kth symbol interval, $s_i[k]$ is the kth transmitted symbol for the ith mobile,
and $(\cdot)^*$ represents the complex conjugate operation.

The transmitted symbol can be recovered by using an estimate of the channel distortion $\hat{h}[k]$ obtainable from the pilot channel, i.e.,

$$\hat{s}_i[k] = f\left(z_i[k]\hat{h}^*[k]\right),$$

where $f(\cdot)$ is an appropriate estimator. Alternatively, in a coded system $z_i[k]\hat{h}^*[k]$ may be used to compute the branch metric in the Viterbi decoder. Assuming a flat, slow-fading Rayleigh channel, the resulting performance of the link will be rather poor in the absence of fast, accurate power control, due to the lack of diversity.[†] As a result, it is desirable to have a second antenna at the receiver to provide diversity reception, which improves performance considerably. However, mobile handsets do not easily allow a second antenna to be added.

In the uplink of 3G cellular systems, the signal is slightly different. The received baseband signal from K mobile units is modeled as

$$r(t) = \sum_{i=1}^{K} h_i\left(t - \tau_i\right) x_i\left(t - \tau_i\right) + n(t),$$

where τ_i is the delay of the ith signal, and $x_i(t)$ is the received signal from the ith mobile defined as

$$x_i(t) = \left(\sqrt{P_d}s_i(t)w_d(t) + \sqrt{P_p}w_p(t)\right) p_i(t).$$

P_d and $w_d(t)$ are the power and Walsh code dedicated to the data signal, P_p and $w_p(t)$ are the power and Walsh code dedicated to the pilot signal, and $p_i(t)$ is the pseudorandom spreading code for the ith mobile. Data demodulation is accomplished in a similar manner as in the downlink.

On both the uplink and downlink channels, multipath is likely to be resolvable. Resolvable multipath is typically handled using a rake receiver, which coherently combines the resolvable multipath components. Multiple antenna techniques can be extended to the case of resolvable multipaths. We will describe the extension explicitly only when necessary.

25.3 Transmit diversity

The performance of wireless systems is directly related to the statistics of the received signal. In Rayleigh fading environments, some form of diversity (spatial, temporal, or frequency) is needed to improve the received signal statistics and thus the performance. The worst performance in the downlink

---

[†]In a convolutionally coded system, the performance can be acceptable in fast fading due to the diversity in the path metric arising from appropriate interleaving.

of cellular CDMA systems occurs in slow, flat fading, when the mobile is not in soft hand-off nor in line-of-sight to the base station. In such a case there is no diversity present in the received signal in either time, frequency or space during a frame duration, resulting in degraded performance. As the mobile speed increases, the increased fading rate combined with the interleaver will provide sufficient temporal diversity in the Viterbi path metrics (assuming convolutional coding) to provide good performance. Similarly, if the mobile is in soft hand-off (i.e., communicating with two base stations simultaneously), the channel provides resolvable multipath. Further, if the mobile has line-of-sight propagation from the base station, the received signal will also show less variance. Thus, the goal of transmit diversity is to improve the performance in situations where neither time, frequency, nor space diversity is present, while maintaining the needed complexity at the base station, where it is more easily afforded (Thompson et al., 2000). Note that 3G standards for shared packet data pipes have other forms of diversity, which mitigate the gains due to explicit transmit diversity. These include physical-layer-based retransmission and multiuser scheduling diversity, which is discussed later in this chapter.

The current *cdma2000* 3G standard provides two means of transmit diversity: space-time spreading (TR45.5, 1999; Papadias et al., 1999; Soni et al., 1999; Hochwald et al., 2001; Soni and Buehrer, 2004) and orthogonal transmit diversity (TR45.5, 1999; Rohani, 2000). We will compare and contrast the two schemes in this section, as well as investigate their performance characteristics. Phase-sweep transmit diversity (PSTD) is also possible in 3G systems, since it is not standard-dependent. PSTD will not be examined here, but the interested reader can find more details in related publications, like the ones by Thompson et al. (2000), Buehrer et al. (2004).

One method to obtain downlink diversity takes advantage of the decoding process when forward error correction codes are used. Orthogonal transmit diversity (OTD) achieves diversity in the Viterbi decoder path metrics rather than in the individual symbols (Rohani, 2000). OTD transmits alternate bits over spatially separated (or orthogonally polarized) antennas. Even bits are transmitted on antenna 1 using one Walsh code, and odd bits are transmitted on antenna 2 using a second Walsh code. In the standard, these codes are closely related. For example, if a user i is assigned a Walsh code $w_i^N(t)$ of length N in non–diversity mode, then the same user i would be assigned two extended codes, which are formed from $w_i^N(t)$, in the optional OTD mode.

By transmitting the even- and odd- numbered data bits on different antennas, a form of diversity gain is obtained. This gain is achievable because the

Viterbi decoder creates path metrics that are based on several consecutive bits after deinterleaving. Since alternate bits are transmitted from different antennas, the path metrics will inherently contain diversity. By practical arguments, it is easy to observe that the diversity gain is a function of the strength of the code. The more powerful the code, the closer the performance will be to a scheme that explicitly obtains diversity on each symbol. At low Doppler frequencies, and for powerful codes (R = 1/4 convolutional), the diversity gains over schemes without diversity can be dramatic. However, the scheme is code-dependent, and lower rate codes (R = 1/2 convolutional) do not benefit as much. Further, if one antenna path becomes disabled, the mobile receiver will no longer be able to recover the frame.

The second form of transmit diversity in *cdma2000* is termed space-time spreading (STS) and is an extension of the famous space-time Alamouti code (Alamouti, 1998; Tarokh *et al.*, 1998). The Alamouti code was originally applied across two simultaneous time slots occurring on physically separate antennas. This idea was extended to Walsh codes as follows (Papadias *et al.*, 1999; Soni *et al.*, 1999; Hochwald *et al.*, 2001; Soni and Buehrer, 2004). With space-time spreading, we transmit the following signal on the first antenna

$$x_1(t) = \left(\sqrt{\frac{P_1}{2}} s_e(t) w_1^{2N}(t) - \sqrt{\frac{P_1}{2}} s_o^*(t) w_{1+N}^{2N}(t) + \sqrt{P_{p1}} w_{p1} \right) p(t), \quad (25.1)$$

and on the second antenna we transmit

$$x_2(t) = \left(\sqrt{\frac{P_1}{2}} s_e^*(t) w_{1+N}^{2N}(t) + \sqrt{\frac{P_1}{2}} s_o(t) w_1^{2N}(t) + \sqrt{P_{p2}} w_{p2} \right) p(t). \quad (25.2)$$

To decode the STS signal at the mobile receiver, we again uncover and correlate with the two extended Walsh codes to obtain (dropping the dependence on symbol interval)

$$z_1 = \sqrt{\frac{P_1}{2}} h_1 s_e + \sqrt{\frac{P_1}{2}} h_2 s_o + n_1,$$

$$z_2 = \sqrt{\frac{P_1}{2}} h_2 s_e^* - \sqrt{\frac{P_1}{2}} h_1 s_o^* + n_2,$$

which obviously introduces interference terms in the decision statistics. However, if we have estimates $\hat{h}_1$ and $\hat{h}_2$ of the channel h_1 and h_2, respectively, from pilot signals 1 and 2, we can obtain a signal estimate for the even data by

$$\hat{s}_e = f\left\{\hat{h}_1^* z_1 + \hat{h}_2 z_2^*\right\}$$

$$= f\left\{\left(\sqrt{\frac{P_1}{2}}|h_1|^2 + \sqrt{\frac{P_1}{2}}|h_2|^2\right)s_e + h_1^* n_1 + h_2 n_2^*\right\},\qquad (25.3)$$

where we have assumed that the channel estimation is exact, $\hat{h}_1 = h_1$ and $\hat{h}_2 = h_2$.

It can be easily shown that this is identical to the decision statistic of two-antenna diversity reception without the 3 dB aperture gain (Proakis, 1995). Thus, we achieve diversity gain at the coded symbol level (i.e., before decoding). The difference between STS and OTD is clearly visible in the final state metrics of the Viterbi decoder (Soni and Buehrer, 2004). At slow speeds (3 km/h) a single antenna receiver sees large variations in the metrics and has high probability of obtaining very low values. However, at high speeds (100 km/h), the Viterbi path metrics show significantly less variability due to interleaving, and there is a much lower probability of obtaining a small value. With OTD there is no symbol-level diversity; the interleaving across antennas allows for diversity in the path metrics even at low speeds. However, this diversity advantage is not as great as that achieved with STS, which obtains diversity at the symbol level.

The performance of OTD and STS was examined for three different types of channels associated with the forward link of the *cdma2000* system (Kogiantis *et al.*, 1999; Lucent Technologies, 1999; Soni and Buehrer, 2004). The performance of OTD and STS are presented for one- and two-path Rayleigh fading channels for voice applications (i.e., the fundamental channel) in Fig. 25.1 (a) and (b). The transmit power fraction requirements (E_c/I_{or}) for full-rate voice using radio configurations RC3 and RC4 assuming a geometry of $I_{or}/I_{oc} = 6$dB [†] were derived from simulations. We plot the fraction of base station transmit power required to achieve a 1% frame error rate versus mobile speed. RC3 and RC4 are 1/4 rate and 1/2 rate convolutional codes, respectively. Both configurations are associated with a family of data rates that are multiples of 9.6 kbps. The effects of power control, puncturing, and coding using the interleaver specified in the *cdma2000* standard (TR45.5, 1999) were included.

As seen in Figs. 25.1(a) and (b), STS and OTD offer significant perfor-

---

[†]In keeping with the terminology frequently used within the standards bodies, E_c is the transmit energy per chip, and I_{or} is the total full load transmit power, both referenced to the mobile, and E_c/I_{or} is the fraction of the total transmit power. Further, I_{oc} is the out-of-cell interference at the mobile, including thermal noise.

mance advantage by reducing the amount of power required to achieve the target frame error rate, particularly at low speeds and in flat fading. The largest improvement is obtained when no diversity is present in either time or space. Since RC3 and RC4 use 1/4 rate and 1/2 rate convolutional codes, respectively, RC3 uses length 64 Walsh codes while RC4 uses length 128 Walsh codes. Thus, while RC3 offers better performance, RC4 is less likely to experience a capacity limit due to a Walsh code limitation, and may be preferable in situations where the number of Walsh codes is a concern. The results for a two-path (equal-power) channel are shown in Fig. 25.1(b). The gains for transmit diversity are not as substantial when diversity is present in the channel itself (as expected). Note that the gains of STS over OTD are more substantial in RC4 in flat fading, since it uses a higher rate convolutional code. As mentioned previously, this is an expected result since OTD relies on the decoding process to obtain diversity gain, whereas STS does not.

The other major 3G standard, known as the Universal Mobile Telephony Standard (UMTS) or Wideband CDMA (WCDMA), developed by the third generation partnership project (3GPP), also provides transmit diversity options. Specifically, two options are available: space-time transmit diversity (STTD) and closed loop transmit diversity (CLTD). STTD is a time domain two-dimensional space-time block code. Thus, its performance is similar to space-time spreading, and will not be discussed further. It does offer some performance benefit in terms of peak to average ratio requirements when compared with STS, as it does not increase the spreading factor associated with data (Dabak *et al.*, 2000). CLTD, however, is substantially different, and involves using feedback from the mobile. It is similar to transmit beamforming, as discussed in the next section. Based on feedback from the mobile, The weights are chosen such that the signals from two antennas arrive coherently at the receiver. The scheme provides both two-fold diversity, as well as a 3 dB power gain due to coherent combining.

25.4 Transmit beamforming

Transmit beamforming can be used to improve the distribution of signal-to-interference ratios (SIR) throughout the cell. When multiple antenna elements at the base station are closely spaced, the received signals from a particular mobile will be highly correlated. Using the uplink signals, the base station can estimate long-term statistics of the downlink channel, i.e., second order moments such as the correlation matrix of the channel. Using this partial channel information, the base station can perform transmit beam-

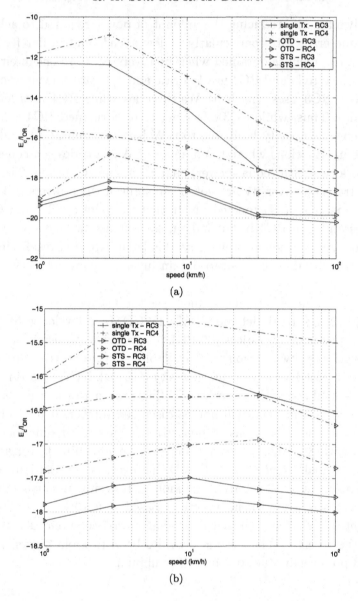

Fig. 25.1. Transmit power fraction, E_c/I_{or}, versus speed for a fundamental channel (9.6 kbps) in RC3 and RC4. (a) flat fading, (b) two equal power paths, $I_{or}/I_{oc} = 6$ dB.

forming, or optimal energy steering to a mobile. The performance benefit of this directive energy steering will be a function of the channel propagation environment.

Most beamsteering algorithms in 3G systems calculate steering weights, which are a function of the arrival direction of the received signals in the uplink. These weights are designed to satisfy an optimality criterion, e.g., minimizing the uncoded symbol error rate at the mobile, or maximizing the signal-to-interference ratio at the mobile. When these weights premultiply the transmitted signal, the signals arrive cophased at the mobile location, resulting in an immediate SIR improvement. The SIR improvement translates into reduced requirements on the transmit power at the base station. This reduction in required power will result in a larger overall system capacity. In the case of code multiplexed shared channels for packet data, which typically use all the available power, the SIR benefits result in improved throughput and reduced delay.

The possible performance enhancement of a transmit beamforming algorithm is determined in large part by the pilot, or training, sequence structure associated with the air interface. Both the *cdma2000* and UMTS standards require that base stations transmit a common pilot training sequence using the same Walsh code in all sectors. The mobile receiver uses this common pilot training sequence for many purposes. Initially, the mobile searches for the presence of this training sequence to determine if the mobile station is in range of a compatible base station. Once a mobile is synchronized and in communication with this base station, it continues to monitor the pilot to maintain synchronization. The pilot is also very important for the demodulation process. Both systems use coherent quadrature phase-shift-keyed (QPSK) or binary phase-shift-keyed (BPSK) modulation, and require a coherent reference of the underlying fading channel to reliably demodulate the data.

The UMTS standard supports two other types of pilot channels. Dedicated pilot channel bits are included in a dedicated control channel. These bits are useful for SIR estimation and possibly beamforming, and are time-multiplexed on the physical layer-based dedicated per-user control channel (DPCCH). Further, the UMTS standard provides for secondary common pilot channels (S-CPICH), which can be useful for a group of users, i.e., using a predefined set of fixed beams. These other pilots can also be used for coherent reference for the fading channel.

A group of N cophased antennas will achieve a link level improvement of $10 \log N \, \mathrm{dB}$ aperture gain in an ideal environment. In environments with small angle spread, this gain is achievable. In environments with large angle spread, aperture gains will be reduced in common pilot schemes. This is due to a decreased coherence (attributed to angle spread) between the steered traffic and the common pilot channel. Use of a dedicated pilot can alleviate this problem, but requires additional transmit power. Secondary pilots can also reduce sensitivity with less power than a dedicated pilot reference (Conner *et al.*, 2004). The paper by Kuzminskiy *et al.* (2005) describes an approach that attempts to mitigate some of the loss in beamforming gain via semiblind signal processing at the receiver.

We can detect the presence of large angle spread by measuring the magnitude of the correlation of uplink signals (presumably pilots) between closely spaced pairs of antennas. Using knowledge of the correlation, we can tune the beams to be optimal for a particular angle spread. For complex Gaussian channels (i.e., traditional Rayleigh or Ricean fading), the magnitude of the correlation $|\rho|$ will decrease with angle spread. For perfect line-of-sight channels with no scattering, $|\rho| = 1$.

Studies of the performance of adaptive algorithms for transmit beamforming in cellular systems are separated into two parts. To obtain intuition and for detailed algorithm development, algorithms are evaluated with simulations that focus on link level (physical layer) effects for the benefit of a single user. Algorithms are then mapped to system level simulations, and their operation is abstracted in terms of their overall impact on the SIR of individual users, and the impact as interference to other users. With system level simulations, it becomes possible to judge the overall impact on capacity and throughput of the system. As an example of link level (physical layer) analysis, we will evaluate the performance of transmit beamforming for *cdma2000* when combined with transmit diversity.

Fig. 25.2 displays the transmit power requirements for a scheme that combines transmit beamforming and transmit diversity in flat fading for a range of speeds from 1 km/h to 100 km/h. The scheme is labeled as SSTS (steered-STS) in Fig. 25.2. The addition of beam steering for each transmit diversity path can improve performance further. In this configuration, we use a two-element array with widely-separated antenna elements for both the base station transmitter and receiver to support a combination of beamforming and diversity. Beamforming provides close to an additional 3 dB reduction across the entire range of speeds for an angle spread of two degrees (i.e., fairly narrow angle spread). In this simulation, angle spread is described by the standard deviation of the angle-of-arrival distribution at the base station. Performance gains are reduced by a few tenths of a dB at an angle spread of four degrees. At 8 degrees rms angle spread, this gain is anywhere from 1–2 dB. Even at 16 degrees rms angle spread, 1 dB of beamforming gain can be achieved at high speed. This gain for large angle spreads can be attributed to the fact that the beamforming algorithm widens the beam as the angle spread increases to reduce pilot/data mismatch, but still provides some directivity (Soni *et al.*, 2002).

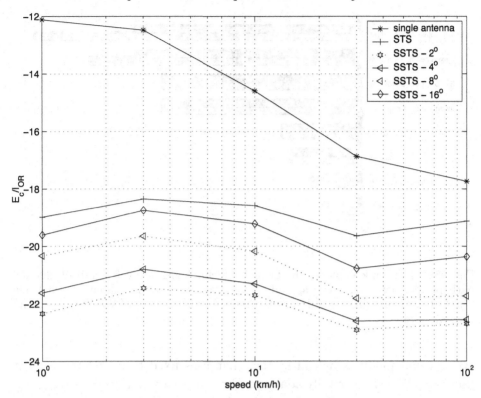

Fig. 25.2. Transmit power requirements for transmit beamsteering and transmit diversity in a single path Rayleigh faded channel with various angle spreads, $I_{or}/I_{oc} = 6$dB.

Predictions for system-level voice user capacity enhancements follow for the UMTS standard. In this case, a few candidate antenna architectures are evaluated. For comparison purposes, both fully adaptive transmit beamforming solutions and fixed beamforming solutions are studied. It is not insightful to study the performance of fixed beamforming solutions with link level analysis, and thus no link level results are provided for fixed beamforming.

For the system-level simulation, evaluation of the capacity limits for the downlink is based on a combination of three different parameters: (1) transmit power limits, (2) transmit power overload fractions, and (3) quality of service outage (Conner *et al.*, 2004). Fig. 25.3 illustrates the downlink capacity improvement over a single transmit antenna for a 3-sector system when using transmit diversity (both space-time and closed-loop), a combination of transmit diversity and beamforming (two pairs of diversity antennas can be used to achieve this), beamsteering alone with four antennas, a fixed beam

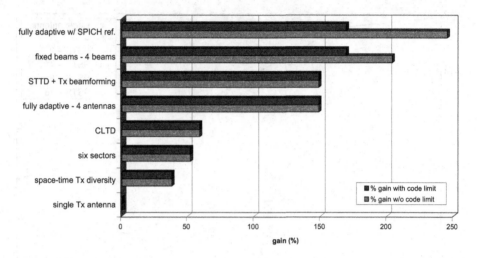

Fig. 25.3. Downlink capacity limits of IA technologies for voice, mixed channel path profile, per path angle spread = 5 degrees with and without Walsh Code Limit of 128 simultaneous codes. Note that the fixed number of Walsh codes limits the potential gain.

system, and finally a hybrid system that uses fixed beam references, but users are allowed to be fully adaptive. For comparison purposes, a six-sector system is also shown. Beamforming techniques can provide substantial capacity benefits, particularly when compared with the standard six-sector solution (Conner et al., 2004).

25.5 Multiantenna receiver techniques

Current third generation wireless systems do not standardize or specify the use of multiple receive antennas at either the base station or the mobile receiver. However, as in any wireless system, multiple receive antennas can be used to achieve performance and capacity gains. Since the use is not standard specific but rather specific to the type of system, the gains achieved in UMTS and cdma2000 are similar.

Increasing the number of antennas at the receiver in 3G wireless systems provides three possible benefits: aperture gain, diversity gain, and interference rejection. The translation of those link-based gains into system gains depends on the type of system being discussed. Specifically, we will examine 3G voice systems, mixed voice and data systems, and packet-based data-only systems. In general, the use of power or rate control means that increasing the number of antennas will result in any of the following gains: increased capacity in terms of simultaneous voice users, reduced transmit power requirements, or increased throughput.

One of the primary benefits to service providers (and actually one of the original drivers for 3G) is an increase in voice capacity. A primary gain provided by multiple receive antennas on either up - or downlink is simple aperture gain. Aperture gain is the collection of additional energy due to the coherent combining of multiple antenna outputs. In general the signal-to-noise ratio after combining is increased by 3 dB for every doubling of the number of antennas. However, since 3G systems are (predominantly) CDMA-based systems, we must determine whether this gain is still relevant. With CDMA, the system is interference limited rather than noise limited. The gain is still relevant, provided that the interference can be modeled as "wide-sense" stationary white Gaussian noise. It is well documented that for a modest number of signals, the interference in the uplink of a CDMA system, as seen at the detector (i.e., after despreading), is temporally white and well modeled by a Gaussian distribution (Pursley, 1977; Viterbi, 1995). If the users are uniformly distributed and equal in received power, the interference can also be shown to be spatially white. The interference signal shows little correlation from one antenna to another, especially if the antennas are spaced far apart. This is a reasonable approximation for sectored systems.

The resulting SINR improvement can directly benefit the system in two ways. First, 3G systems incorporate fast methods of power control, which attempt to achieve a desired SINR at the base station receiver. The additional receive antennas provide an aperture gain resulting in a reduction of the mobile transmit power for the same target SINR. This reduction of the transmit power can result in increased battery life. The second direct system benefit is to allow more voice users in the system. The SINR improvement due to coherent combining can be directly transferred to an increased number of users so that the original SINR target is achieved without a change in transmit power.

In addition to increasing the capacity via aperture gain, increasing the number of antennas can increase the capacity via additional diversity. If the receive antennas are spaced sufficiently far apart (the exact required spacing depends on the angular spread of the received signal, as discussed previously), increasing the number of antennas increases the diversity order of the received signal. The diversity increase depends on the other diversity mechanisms available. The diversity gain reduces the required SIR, which is converted directly into capacity improvement in a voice-only CDMA system by increasing the number of allowable users.

As an example, consider a four-antenna architecture for *cdma2000* (Soni et al., 2002) that uses two pairs of closely-spaced antennas at the base station. Typical base stations have two diversity antennas. For this four-antenna architecture, the pairs are separated by ten wavelengths. The widely separated pairs provide diversity, while the closely spaced antennas provide additional aperture gain. Fig. 25.4(a) provides the performance of the four-antenna architecture and a traditional two-antenna diversity receiver. The four-antenna system uses MMSE combining whereas the two antenna system uses maximal ratio combining. For this example, the interference is spatially white, and the difference between MRC and MMSE combining is minimal. The four antenna receiver provides approximately 2 dB improvement in performance. This improvement results in 2 dB less E_b/N_t required per antenna and thus 2 dB less transmit power or a 2 dB increase in capacity. While doubling the number of antennas theoretically reduces the required SINR by a factor of two, the improvement is approximately 1 dB less. The main reason for the reduction is the degradation in channel estimation because each channel has a lower SINR.

While 3G systems provide increased capacity for voice-only networks, the true distinction between 2G and 3G systems are the dedicated channels for data applications. The high data rate channels will generate high-powered interference on the uplink that is not spatially white. Thus, multiple antennas with simple coherent combining are not particularly effective. However, adaptive interference rejection techniques can provide substantial improvement in SINR, which again leads to capacity improvement (Soni et al., 2002). As an example, consider the four-antenna system discussed above (Soni et al., 2002) in a mixed voice and data network, such as cdma2000. The performance of the four-antenna system is compared with the traditional two-branch diversity system in Fig. 25.4(b). For a 1% FER requirement, the four-antenna system requires 4.25 dB less SINR, resulting in a capacity increase of over two times.

Having considered the impact of additional receive antennas at the base station beyond the normal two antennas, we will now focus on adding antennas to the mobile receiver. Third generation packet data systems such as cdma2000 3G-1x EV-DO (EVolution-Data Only) transmit from the base station to a single user at a time, relying on feedback information to determine both which mobile should be served and at what data rate. Thus, these systems are fundamentally different from the voice-oriented systems discussed previously. By allowing a variable delay, the system takes advantage of what is termed multiuser diversity. Multiuser diversity increases the overall throughput of the system, but may not provide fairness.

As mentioned before, multiple receive antennas increase the SIR at the

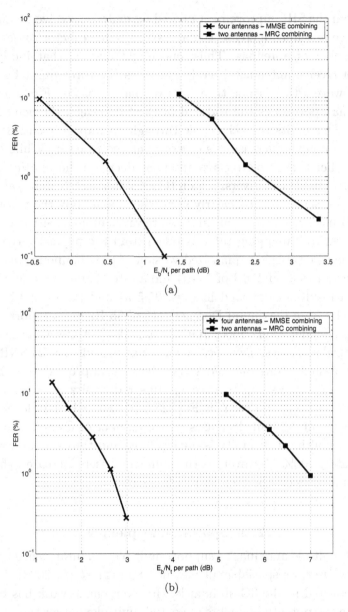

Fig. 25.4. Uplink performance of *cdma2000* with traditional two-branch receive diversity and four-branch intelligent antenna (MMSE processing) (a) voice signals only, (b) voice with high-speed data user present.

receiver. This holds equally for mobile receivers and base station receivers. The increase in SIR results in a higher rate in the downlink, and thus in an increase in the system throughput. Thus, data-only systems can directly benefit from the SIR improvement offered by multiple receive antennas. Multiple receive antennas increase the average SIR by a factor directly related to the number of antennas. Thus, one might assume that increasing the number of receive antennas directly increases the throughput by the same factor. However, this is not true for two main reasons. First, as the SIR per antenna approaches the top end of the SIR range, the achievable rate saturates. This limits the achievable throughput improvement. Second, the additional antennas increase the diversity at the receiver. While in general this is beneficial, multiple receive antennas decrease the *relative* (although not absolute) multiuser diversity gain (Gozali *et al.*, 2003; Jiang *et al.*, 2004).

These two factors can be clearly seen when examining the throughput of an EV-DO-like system with one and two receive antennas at the mobile unit. The assumed rate mapping table is taken from the paper by Bender *et al.* (2000). The rate table is a quantized version of the Shannon capacity minus 5 dB. It saturates at an SNR of 10 dB and a rate of approximately 2.4 Mbps. Simulation results are plotted in Figs. 25.5(a) and (b) for a Rayleigh fading channel with a maximum Doppler rate of 55 Hz and proportionally fair scheduling (Bender *et al.*, 2000; Jiang *et al.*, 2004). All users are assumed to have identical statistics. Two receive antennas provide a 3 dB SNR improvement over a single antenna and some diversity improvement. The spatial diversity gain reduces the *relative* multiuser diversity gain achieved from scheduling. This can be seen in Fig. 25.5(b). For all SNR levels, the improvement in throughput decreases as the number of users increases. For a single user, multiuser diversity is nonexistent, and thus the improvement is approximately 100%. However, as the number of users increases, the relative increase is less, as can be seen in Fig. 25.5(b).

25.6 Spatial multiplexing

The use of multiple antennas to support multiple, simultaneous data streams for a single user, or spatial multiplexing (Paulraj *et al.*, 2003), is not currently supported in the 3G standards. However, much work has been done to examine the potential benefits of spatial multiplexing for the packet data portions of the 3G standards, and in particular in the shared packet channel.

There are several methods of accomplishing spatial multiplexing. In 3G packet data systems, capacity needs are primarily anticipated on the downlink. This is based on an assumption that web traffic will dominate packet data systems, and that web traffic is primarily from the content provider to the end user. Spatial multiplexing can be accomplished through layered receiver-centric techniques such as V-BLAST, or through transmitter-centric

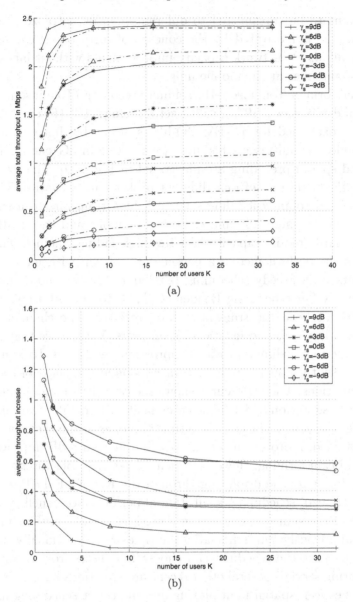

Fig. 25.5. Single sector (a) throughput and (b) gain versus the number of users for one (solid curves) and two (dashed curves) receive antennas at various average SNR values (Rayleigh fading 55 Hz max Doppler) (Jiang *et al.*, 2004).

techniques such as spatial precoding. Spatial Tomlinson–Harashima precoding (Fischer and et. al., 2002; Jiang *et al.*, 2003) is a distributed MIMO scheme that is the precoding analogy to V-BLAST, which allows multiple data streams to be transmitted to different (i.e., decentralized) end users. This technique is motivated by the concept of dirty paper coding (Costa, 1983) and allows for network capacity improvements without increasing the number of antennas at the mobile unit.

A spatial form of Tomlinson–Harashima precoding (Tomlinson, 1971; Harashima and Miyakawa, 1972), called zero-forcing THP (Fischer and et. al., 2002), was examined for 1X-EV/DO-like systems by Jiang et $al.$ (2003). The synergies of multiuser scheduling (as is done in 3G packet data standards) and spatial precoding are illustrated in Fig. 25.6. The figure shows the theoretical sum rate of four different packet data scenarios. The first is a simple round robin scheduling scheme, which transmits to a single user at a time, with a single antenna at both the mobile and base station. This is the standard ergodic capacity of a Rayleigh fading channel. The second scenario is the performance of a packet data system with multiuser diversity, obtained via greedy scheduling (similar results can be obtained with proportionally fair scheduling Bender et $al.$, 2000), $K{=}20$ users, and a single antenna at both the transmitter and receiver. The channel statistics are assumed Rayleigh and identical for all users. We can see that multiuser diversity due to scheduling provides approximately 2.5 b/s/Hz sum rate improvement, as it is able to schedule transmission at the peaks of the channel gain distributions. The third curve shows the capacity of a four antenna system at the base station and a single antenna at each mobile with round robin scheduling (one user). The use of multiple antennas at the base station again provides a fixed improvement of 2.5 b/s/Hz due to a 6 dB improvement in SNR from beam-forming. However, when we use multiple antennas (four in this case) for spatial multiplexing through the use of ZF-THP, along with multiuser ($K{=}20$) scheduling, a substantial performance benefit is obtained. The slope of the sum rate curve is improved, since we can schedule to four users simultaneously through spatial multiplexing. Additionally, we obtain a benefit because the scheduler can choose four users from among twenty users awaiting service. Several other recent investigations have examined the synergies between spatial multiplexing and multiuser scheduling as is done in 3G packet data systems, including Le and Hossain (2004) and Airy et $al.$ (2003).

MIMO is not currently included in the 3GPP UTRA standard. However, it is actively investigated for inclusion in Release 7. Several MIMO options for UTRA are under discussion, using one, two, or four antennas at the Node B and at the UE (Group, 2004). The first proposed option (termed "Proposal 1" in the 3GPP technical report, Group, 2004) allows spatial multiplexing using T transmit antennas supporting up to T simultaneous data streams.

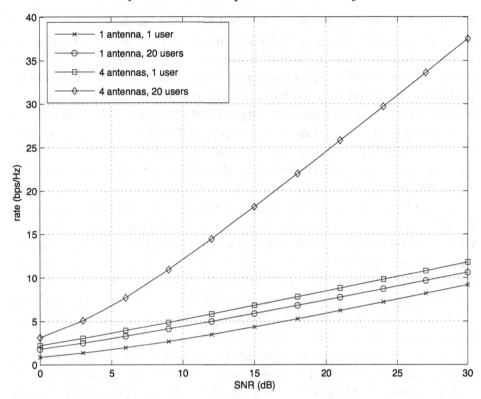

Fig. 25.6. Sum capacity of packet data system with one and four transmit antennas with and without proportionally fair multiuser diversity (fading = 5.6 Hz Rayleigh, slot duration 8.3 ms).

Each stream can have a different data rate through different modulation and coding schemes, motivating the name per antenna rate control (PARC). Although receiver structures are not typically specified by the standard, it is assumed that an MMSE or successive interference cancellation (SIC) receiver, which is common to standard V-BLAST processing, is used. The second and third architectures described by the Group (2004) are called rate control multipath diversity and double space-time transmit diversity subgroup rate control. Both schemes combine transmit diversity and spatial multiplexing. Transmit diversity is achieved in UMTS through space-time transmit diversity (STTD). Five other MIMO proposals are included, which use various forms of spatial multiplexing and transmit beamforming.

Several recent publications have examined the performance impact of MIMO techniques on HSDPA (e.g., Pandey *et al.* (2003)). Pandey *et al.* (2003) examine the impact of two-stream spatial multiplexing with proportionally fair scheduling compared with a single antenna configuration. They also examine the interaction of CLTD and scheduling. It is found that in a slow-fading, flat fading channel a 2x2 dual-stream spatial multiplexing scheme provides 50% improvement over a single antenna scheme with proportionally fair scheduling. Further, it is found that the relative gains due to multiuser scheduling are less with larger numbers of antennas, especially if some antennas are used for diversity purposes. This partially results because systems with a higher number of antennas are likely to saturate the adaptive rate table, and because a larger number of antennas typically allows for larger diversity gains without multiuser scheduling.

As an alternative to multiplexing streams to the same user through a diversity antenna array, spatial multiplexing with transmit beamforming with a phased array can also result in significant improvements for packet data systems. Although the SIR improvement due to beamforming results in throughput enhancement, a more dramatic benefit comes from the fact that beamforming enables simultaneously scheduled users, separated spatially using beams, to share the same bandwidth (or codes), and split the available power. This results because throughput increases only linearly with bandwidth, while it increases logarithmically with SIR. The simultaneous reuse of resources across spatially distributed users is referred to as spatial division multiple access (SDMA) (Pedersen et al., 2003; Conner et al., 2004). Considering complexity and ease of implementation, the most appropriate beamforming technique is fixed beamforming. The benefits of SDMA on HSDPA (UMTS Release 5) are shown by Conner et al. (2004), and for UMTS Release 4 by Pedersen et al. (2003). Both papers claim that SDMA improves overall throughputs and individual user throughputs by 80–100%.

References

Airy, M., Shakkottai, S., and Heath, R. W. (2003). Spatially greedy scheduling in multi-user MIMO wireless systems. In *Proc. Asilomar 2003*, 982–986.

Alamouti, S. M. (1998). A simple transmitter diversity scheme for wireless communications. *IEEE J. Sel. Areas Commun.*, **16** (8), 1451–1458.

Bender, P., Black, P., Grob, M., Padovani, R., Sindushayana, N., and Viterbi, A. (2000). CDMA/HDR: A bandwidth-efficent high-speed wireless data service for nomadic users. *IEEE Commun. Mag.*, **38** (7), 70–77.

Buehrer, R. M., Soni, R. A., and Benning, R. D. (2004). Transmit diversity for combined 2G and 3G systems. *IEEE Trans. Commun.*, **52** (10), 1648–1653.

Conner, K., Gollamudi, S., Lee, J., Nagaraj, S., Moustakis, A., Monogioudis, P., Rao, A., Soni, R. A., and Yuan, Y. (2004). Intelligent antenna solutions for UMTS: Algorithms and simulations results. *IEEE Commun. Mag.*, **42** (10), 28–39.

Costa, M. H. N. (1983). Writing on dirty paper. *IEEE Trans. Inf. Theory*, **29** (3), 439–441.

Dabak, A., Hosur, S., Schmidl, T., and Sengupta, C. (2000). A comparison of the open loop transmit diversity schemes for third generation wireless systems. In *Proc. IEEE Wireless Commun. Netw. Conf.*, vol. 1, 437–442.

Fischer, R. F. H. and et. al. (2002). MIMO precoding for decentralized receivers. In *Proc. IEEE Int. Symp. Inf. Theory*, 496.

Gozali, R., Buehrer, R. M., and Woerner, B. D. (2003). The impact of multiuser diversity on space-time block coding. *IEEE Commun. Lett.*, **7** (5), 1–3.

Group, G. T. R. (2004). *Multiple-Input Multiple Output in UTRA* (3GPP TR 25.876 v1.7.0).

Harashima, H. and Miyakawa, H. (1972). Matched-transmission technique for channels with intersymbol interference. *IEEE Trans. Commun.*, **20** (4), 774–780.

Hochwald, B., Marzetta, T., and Papadias, C. B. (2001). A transmitter diversity scheme for wideband CDMA based on space-time spreading. *IEEE J. Sel. Areas Commun.*, **49** (3), 48–60.

Jiang, J., Buehrer, R. M., and Tranter, W. H. (2003). Spatial T-H precoding for packet data systems with scheduling. In *Proc. IEEE Veh. Technol. Conf. Fall 2003*, vol. 1, 537–541.

—— (2004). Antenna diversity in multiuser data networks. *IEEE Trans. Commun.*, **52** (3), 490–497.

Kogiantis, A., Soni, R. A., Hochwald, B., and Papadias, C. B. (1999). Downlink diversity improvements through space-time spreading. Lucent Technologies proposal 3GPP2-C30-19990817-014 to the cdma2000 standard.

Kuzminskiy, A., Mullany, F., and Papadias, C. B. (2005). Semi-blind channel estimation at the receiver for steered-STS transmit antenna architecture in cdma2000. In *Proc. IEEE Int. Conf. Acoust., Speech, Signal Process.*

Le, L. B. and Hossain, E. (2004). On the performance of spatial multiplexing MIMO cellular systems with adaptive modulation and scheduling. In *Proc. IEEE Wireless Commun. Netw. Conf.*, 1282–1286.

Lucent Technologies (1999). Performance of Space Time Spreading (STS) for IS-2000. Contribution 3GPP2-C30-19990914-013.

Pandey, A., Emeott, S., Pautler, J., and Rohani, K. (2003). Application of MIMO and proportional fair scheduling to CDMA downlink data channels. In *Proc. IEEE Veh. Technol. Conf. Fall 2002*, 1046–1050.

Papadias, C. B., Hochwald, B., Marzetta, T., Buehrer, R. M., and Soni, R. A. (1999). Space-time spreading for CDMA systems. In *Stanford Sixth Workshop Smart Antennas for Mobile Commun.*

Paulraj, A., Nabar, R., and Gore, D. (2003). *Introduction to Space-Time Wireless Communications* (Cambridge Univ. Press).

Pedersen, K., Mogensen, P., and Ramiro-Moreno, J. (2003). Application and performance of downlink beamforming techniques in UMTS. *IEEE Commun. Mag.*, **41** (10), 134–143.

Proakis, J. G. (1995). *Digital Communications* (McGraw-Hill, New York, NY), third edn.

Pursley, M. B. (1977). Performance evaluation for phase-coded spread-spectrum multiple-access communication—Part I: System analysis. *IEEE Trans. Commun.*, **25** (8), 795–799.

Rohani, K. (2000). Open-loop transmit diversity for CDMA forward link. In *Proc. Emerging Technologies Symp.: Broadband, Wireless Internet Access.*

Soni, R. A. and Buehrer, R. M. (2004). On the performance of open loop transmit diversity systems for IS-2000 systems: A comparative study. *IEEE Trans.
Wireless Commun.*, **3** (5), 1602–1615.

Soni, R. A., Buehrer, R. M., and Benning, R. D. (2002). An intelligent antenna
system for cdma2000. *IEEE Signal Process. Mag.*, **19** (4), 54–67.

Soni, R. A., Buehrer, R. M., and Tsai, J.-A. (1999). Open-loop transmit diversity
methods in IS-2000 systems. In *Proc. Asilomar Conf. Signals, Syst., Comput.*,
vol. 1, 654–658.

Tarokh, V., Seshadri, N., and Calderbank, A. R. (1998). Space-time codes for high
data rate wireless communication: Performance criterion and code construction.
IEEE Trans. Inf. Theory, **44** (2), 744–765.

Thompson, J. S., Grant, P. M., and Mulgrew, B. (2000). Downlink transmit diversity schemes for CDMA networks. *IEE Proc. - Commun.*, **147** (6), 371–380.

Tomlinson, M. (1971). New automatic equalizer employing modulo arithmetic. *Electron. Lett.*, **7**, 138–139.

TR45.5 (1999). *Physical Layer Standard for cdma2000 Spread Spectrum Systems
(TIA/EIA/IS-2000.2)*. (Ballot Version).

Viterbi, A. J. (1995). *CDMA: Principles of Spread Spectrum Communication*
(Addison-Wesley, Reading, MA).

26

MIMO wireless local area networks

Karine Gosse, Marc de Courville, Markus Muck,
Stéphanie Rouquette, and Sébastien Simoens
Motorola Labs

26.1 Introduction: high throughput short range systems—from SISO solutions to new MIMO standards

Until recently, wireless local area network (WLAN) devices, heavily constrained in cost and size, have fulfilled short range communication needs with very simple spatial diversity schemes at the access points (AP) realized, in general, with two antennas. Now WLAN AP products embedding phased array antenna designs have been designed with the objective of extending the range of the WLAN coverage, especially outdoors. However, with the demand for increasing bit rate, and the limited amount of spectrum available, the application of multiple-input multiple-output (MIMO) techniques to WLAN has been identified as a key enabler for high throughput WLAN; as a consequence, for instance, all the proposals made to the IEEE 802.11n task group, created in 2004 with the objective of devising next generation WLAN, include MIMO processing of the data.

The objective of this chapter is to present a design methodology for WLAN air interfaces relying on multiantenna signal processing solutions, and meeting feasibility requirements in terms of bit rate, bandwidth, and complexity. Here, we limit ourselves to a design compatible with the requirements set by the IEEE 802.11n task group: the aim is to provide a peak throughput of at least 100 Mbps at the medium access control (MAC) data service access point (SAP). Thus, with an overall MAC efficiency of 80%, thanks to a carefully optimized MAC, the peak physical (PHY) data rate of the designed WLAN systems should be at least of 125 Mbps. We assume for backward compatibility reasons to operate over a 20 MHz bandwidth.

Note that reference to IEEE 802.11n is only made here to provide a framework for comparison. This chapter is not intended to detail or comment on any IEEE 802.11n solution but rather to propose a generic MIMO WLAN design methodology and to present the resulting system performance.

In addition to the throughput versus bandwidth requirements presented above, the design process for a MIMO WLAN interface needs to take into account specific constraints. The first constraint is related to the nature and the level of backward compatibility desired with respect to existing single antenna devices. For instance, it is expected that for a given bit rate, by using multiple antennas, the robustness of the radio link should be improved and the range covered by the multiantenna device should be expanded compared with legacy devices. For sake of future reference, Fig. 26.9(b) presents the performance of IEEE 802.11a transmission modes in terms of packet error rate (PER) versus range.

The fact that a wireless system is deployed in a given spectrum region sets physical constraints on the wave propagation and on the corresponding transmission channel seen by the wireless transceiver. In Section 26.2, we present the underlying assumptions to be made in the case of a multiantenna transmission in the 5 GHz spectrum band. Such assumptions include the selection of an appropriate channel model suited for performance assessments using software simulation tools.

Afterwards, given the propagation encountered and the bit rates envisaged in the given bandwidth, modulation parameters are set. Here, the backward compatibility assumption with legacy WLAN devices relies on the ability for legacy single-antenna devices to connect to the updated multiantenna AP. This sets constraints on the preambles to be transmitted, which in turn, for the sake of hardware reuse, points toward the preferred usage of an orthogonal frequency-division multiplexing (OFDM) modulator as in IEEE 802.11a/g. The tuning of the OFDM parameters, as well as related implementation constraints, is also addressed in Section 26.2.

In this framework, we then describe appropriate multiantenna techniques, relying on MIMO concepts that are candidates for next generation WLAN air interface enhancements (26.3.1), and propose a complete transmission chain based on these approaches in Section 26.3. The core design of the first MIMO WLAN devices will probably rely on simple, robust, and proven techniques. We focus here on multiantenna techniques that do not rely on any channel state information (CSI) knowledge at the transmitter (so-called open-loop schemes). Receiver techniques are discussed in Section 26.3.2, and overall expected system performance in terms of bit error rate (BER) or PER is presented in Section 26.3.3, thus validating the selection of the proposed techniques for meeting the MIMO WLAN design requirements. As part of the overall system design, preamble design and related support functions such as gain control and synchronization are discussed, and an example of preambles suited for MIMO next generation WLAN is given in Section 26.4.

It has to be noted that the result of the design cannot be a single set of parameters and link definition, since the WLAN radio interface has to support not only different user requirements in terms of bit rate, distance to the AP, and speeds, but also various propagation environments ranging from indoor office transmissions to indoor-to-outdoor hot spot scenarios.

Thus, it is crucial to introduce in the system design the ability to adapt transmission parameters to optimize the link with respect to user and environment specificities; these parameters, determining the bit rate of the radio link, are the modulation and coding scheme (constellation order, coding rate) and the number of parallel data streams processed by the MIMO transceivers (potentially the way these streams are transmitted). However, depending on the propagation characteristics, some combinations of modulation/coding/MIMO schemes will not enable the target PER for the given application (typically 10^{-2} or 10^{-3}) to be attained. Alternatively, even if the link quality requirements are met, other criteria may be used to select the link parameters, such as the power consumption of the mobile device. In all cases, due to the variability of wireless channel, the link adaptation will be a dynamic process selecting a transmission mode defined by its modulation constellation, coding rate, and parallel data streams transmitted. In turn, the final selection of the MIMO technique to be applied will depend on the number of transmit and receive antennas available. So, finally, Section 26.5 gives an overview on how the WLAN system exploits the radio interface modes designed in the previous steps by addressing coverage issues as well as the link adaptation procedure.

26.2 System design methodology: derivation of multicarrier parameters

To properly design a new wireless system, it is important to adapt its main parameters to the characteristics of the environment considered for operation, namely the propagation conditions and mobility of the devices and surroundings. Therefore, the first challenge is to negotiate with regulation agencies both a maximum allowed transmit power and a spectrum niche suited for the target application.

The goal is to obtain a total transmit power and a carrier frequency compatible with the expected range of coverage and, based on this information, a bandwidth large enough to enable a full scale deployment yielding a large enough frequency reuse pattern to grant a level of interference compliant with the expected system performance.

For next generation WLAN, the target is to use the same bands as for IEEE 802.11a/g and, particularly in the United States, the license exempt unlicensed national information infrastructure (U-NII) band, which is composed of three parts: the low band, 5.15–5.25 GHz reserved for very low power (50 mW equivalent isotropic radiated power, EIRP) and mostly intended for indoor applications, the middle band, 5.25–5.35 GHz allowing higher power operation (250 mW EIRP) and finally the high band 5.725–5.825 GHz for outdoor, larger range operations (1 W EIRP). In total, around 300 MHz of spectrum is available, representing a total of twelve channels of 20 MHz each. This figure is to be compared with the 83.5 MHz available in the ISM 2.4–2.4835 GHz band for IEEE 802.11b/g systems.

Then, maximum throughput required at the application level, targeted mobility capturing both the terminal velocity and environment fluctuations, and expected range of coverage are used as inputs to the system specification methodology.

The overall goal of this section is to link all these constraints to the parameters of a MIMO OFDM system and derive i) the carrier frequency, the sampling rate, and the useful bandwidth; ii) the modulation scheme parameters: constellation used, the block size N_c representing the number of carriers, and the size of the cyclic prefix D in number of samples; iii) the maximum burst length ensuring compatibility with the propagation channel models at the carrier frequency of choice, the delay spread, the Doppler spread, and the path loss model; iv) the required antenna configurations and the multi-antenna signal processing to be supported. It is important to note that in the overall parameter design the propagation plays a major role.

In total, five typical single-input single-output (SISO) channel models, resulting from real measurement campaigns (performed in the framework of the ETSI BRAN HIPERLAN/2 standardization), were identified for indoor and limited outdoor scenarios with line-of-sight (LOS) and non-LOS (NLOS) conditions (802.11n Task Group, 2004), which are illustrated in Table 26.1.

The MIMO extensions of these SISO channels are based on a cluster modeling approach where uniform distributions are used for the relative cluster mean angle of arrival and departure (AoA/AoD). The delay spread (DS) and angular spread (AS) are modeled as correlated log normal random variables. Moreover, a two-slope path loss model using a path loss exponent of 2 before the break point distance and 3.5 after the break point distance is associated with each channel model. Various Doppler spectra are also specified for each environment under consideration; for example, for outdoor operation (chan-

Table 26.1. *IEEE 802.11n channel model parameters*

Channel model	Environment	RMS delay spread (ns)
A	AWGN, flat fading channel for calibration and benchmarking purposes	0
B LOS & NLOS	residential environment	15
C LOS & NLOS	residential/small office environments	30
D LOS & NLOS	office environments	50
E LOS & NLOS	large open spaces and office environments	100
F LOS & NLOS	large open spaces for indoor and outdoor	150

nel F), a Doppler component from a moving vehicle is introduced and for the office environment (channel D-E), the presence of fluorescent lights is modeled.

The design methodology of a single antenna indoor OFDM WLAN can be outlined by the following steps:

(i) determine the targeted system or (layer 3) application data rate (for 802.11a, $R_{\text{appl}} = 40$ Mbps); estimate the inherent signaling overhead of the MAC considered (typically $\textit{eff}_{\text{MAC}} = 75\%$) and translate these figures into the required data rate on top of the PHY (total PHY rate $R_{\text{PHY}} \approx 53$ Mbps for legacy 802.11a);

(ii) convert in a useful bandwidth B given a feasible spectrum efficiency (if $s_{\text{eff}} = 3$ bit/s/Hz for a SISO system: $B \approx R_{\text{PHY}}/s_{\text{eff}} \approx 17$ MHz < 20 MHz); derive the resulting critical sampling frequency $f_e = B$ and the maximum sample duration: $T = 1/B = 50$ ns;

(iii) operate in appropriate spectrum bands, as the U-NII band described above, around 5.2 GHz and perform propagation measurements at the selected carrier frequency $f_c = 5.2$ GHz;

(iv) specify the average root mean squared (RMS) delay spread, which is typically $\tau_{\text{rms}} = 100$ ns for the frequencies at hand in indoor environments. The resulting overall channel model duration, including transmit and receive filters, is on the order of $T_m \approx 8\tau_{\text{rms}} = 800$ ns. The OFDM guard interval (i.e., cyclic prefix) consisting of D samples has to absorb the resulting intersymbol interference (ISI), $D \approx T_m/T = 16$ samples for allowing a low complexity frequency-domain equalization;

(v) set the total number of carriers such that the insertion of the guard interval represents an acceptable overhead. The block size needs to be large enough to mitigate this overhead, and ideally $N_c \to \infty$. The limiting factors for the size of the OFDM block are:

- the compromise with implementation complexity and latency of the overall system, classically limiting the number of carriers N_c to $D/N_c \approx 25\%$ ($N_c \approx 4D = 64$ in 802.11a);
- the constraint of experiencing a constant channel over an OFDM baud, that is to say $(N_c + D)T \ll \Delta T_c$ with $\Delta T_c = 1/B_d$. For a station velocity of $v = 3$ m/s, with $c = 3 \times 10^8$ m/s standing for the speed of light, we have to cope with a maximum Doppler shift of $f_d = f_c v/c = 52$ Hz, representing a Doppler spread of $B_d = 104$ Hz, or, equivalently, a maximum channel coherence time of $\Delta T_c = 1/B_d \approx 10^{-2}$ s. In this example, $\Delta T_c/T \approx 2 \times 10^5$, and the channel varies very slowly compared with the duration of an OFDM symbol with 64 carriers;
- an acceptable intercarrier interference level (in terms of matching the specification and of cost) resulting from oscillators phase noise. This condition is verified with an intercarrier spacing of $B/N_c = 312.5$ kHz and a typical 5.2 GHz oscillator phase noise of -70 dBc/Hz at 10 kHz, -90 dBc/Hz at 100 kHz and -110 dBc/Hz at 1 MHz generating through inter carrier interference (ICI) a noise of -37 dB compatible with achieving 10^{-4} BER with a 64-quadrature amplitude modulation (QAM).

(vi) set a burst/frame duration much smaller than the channel coherence time to ensure that the identification of the CSI at the start of the frame remains valid long enough for its full duration. In IEEE 802.11a, a complete change of CSI occurs after $\Delta T_c/((N_c + D)T) \approx 2400$ OFDM symbols suggesting a maximum limitation of burst/frame size to 500 OFDM symbols.

Note that the figures derived using the proposed methodology closely match the definition of IEEE 802.11a OFDM PHY layer: 64 carriers over 20 MHz bandwidth with 52 useful carriers representing a total of 16.25 MHz of spectrum used and 800 ns cyclic prefix. Considering the IEEE 802.11a forward error correction (FEC) constraint length $K = 7$ convolutional code (CC) (o171/o133) applied before bit interleaving over a single OFDM block followed by QAM constellation mapping from binary phase shift keying (BPSK) to 64-QAM, we obtain, on top of the PHY and using various puncturing patterns, data rates ranging from 6 Mbps up to 54 Mbps.

To adapt this SISO methodology to systems using multiple antennas for transmission, the overall spectral efficiency and data rates need to be scaled appropriately by the number of spatial streams (SS) to be sent over the air. For classical space division multiplexing (SDM), the number of SS is equal to the number of antennas and for space-time block coding (STBC) the rate of the code considered has to be taken into account. Thus, to comply with the bit-rate requirements of 100 Mbps at the SAP, translated into a 120 Mbps PHY rate with an improved 80% efficiency MAC, at least two transmit antennas will be required using SDM (reaching 108 Mbps), and potentially a compromise will have to be found either with an increased number of useful carriers, or with a higher coding rate, or with higher-order constellations. This results in a spectral efficiency of 6 bits/s/Hz in a 20 MHz band.

26.3 A robust MIMO transceiver to reach high data rates

26.3.1 *MIMO modulator without channel state information*

The challenge for designing a MIMO WLAN system is to find the proper physical layer modes that will allow us to adapt the range of coverage versus throughput/PER trade-offs to accommodate the QoS requirements of the user or application targeted under propagation condition constraints.

In single antenna systems, such as IEEE 802.11a/b/g, these trade-offs are achieved by adapting FEC encoding (e.g., using different puncturing patterns) and the carrier symbol constellations. MIMO systems also rely on these techniques, but additionally offer a new degree of freedom inherent to the space-time transmission scheme. For instance, classical STBCs (Tarokh *et al.*, 1998; Alamouti, 1998) seek to improve link robustness or range for low to medium data rates (suited to small packet size, e.g., VoIP), and SDM is a means to increase the spectrum efficiency and the peak data rates.

In this section we focus on specific open-loop multiple transmit schemes that rely on hybrid combinations of SDM and STBC, enabling a trade-off between robustness provided by the STBC Alamouti encoder (Alamouti, 1998) and data rate increase of pure SDM approaches. The first solution consists in transmitting two parallel Alamouti codes on four antennas (Texas Instruments, 2001; Zhuang *et al.*, 2003), and the second solution consists of transmitting one Alamouti coded stream in parallel with uncoded streams on three or four antennas.

The corresponding architectures for up to four transmit antennas supporting from one to three parallel streams are detailed in Figs. 26.1, 26.2, and 26.3. As a summary, the studied space time transmit schemes are: i) Alamouti encoder for two transmit antennas, ii) space division multiplexing for two and three transmit antennas, iii) hybrid combinations of SDM and

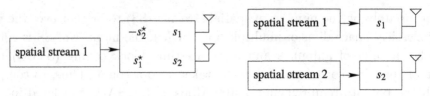

Fig. 26.1. Alamouti and SDM encoders to transmit one and two SS respectively on two antennas.

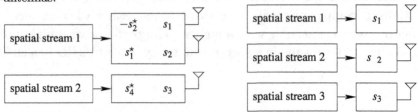

Fig. 26.2. Hybrid Alamouti/SDM and SDM schemes to transmit two and three parallel streams respectively on three antennas.

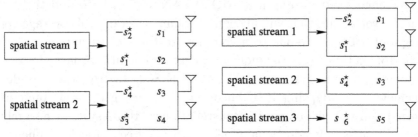

Fig. 26.3. Hybrid Alamouti/SDM schemes to transmit two or three parallel streams on four antennas.

STBC transmitting two parallel streams on three transmit antennas, and two or three parallel streams on four transmit antennas. Note that prior to this modulator, a spatial interleaver is usually applied, to distribute the symbols over the parallel streams with cyclic permutation. This interleaver is specifically designed taking into account the parameters of the frequency interleaver to ensure that adjacent bits are assigned to different streams.

The advantages of such space-time transmission schemes are manifold. First, they enable us to tune the range versus data throughput/PER trade-offs with increased granularity. Second, since the decoding of n data streams requires n or more receive antennas, these transmission schemes can be applied to asymmetrical antenna configurations, where the number of transmit antennas N_{TX} is greater than the number of receive antennas N_{RX}. This feature is especially desirable for accommodating a wide range of device sizes, from laptops to handheld devices such as PDAs or phones having constraints on the number of antennas that can be physically implemented, or for granting a smooth evolution path by allowing independent station (STA) or AP antenna configuration upgrades. Note that the number of receive antennas determines the maximum number of SS that can be transmitted but additional antennas can be used at the receiver for achieving a greater diversity gain. This way, for a given number of SS supported, several solutions are provided for achieving the same data rate.

As presented in Fig. 26.3, the rate-2 hybrid scheme, for instance, consists in transmitting four complex symbols s_1, s_2, s_3, and s_4 according to the following code matrix

$$C = \begin{bmatrix} s_1 & s_2 & s_3 & s_4 \\ -s_2^* & s_1^* & -s_4^* & s_3^* \end{bmatrix},$$

where the nth column and ith line element of C stands for the symbol sent on the nth antenna at time i.

The design philosophy is to propose STBC codes with rates lower than the number of transmit antennas without trying to achieve the highest diversity gain and limiting the STBC code to span only two OFDM symbols. The main motivation is to decrease the overall decoding latency and limit the size of the matrices in the ZF/MMSE equalizers for complexity reduction's sake.

26.3.2 MIMO receivers

When the hybrid transmission scheme is applied to each data subcarrier of an OFDM-based system, the received signal on a given subcarrier can be expressed as:

$$\underbrace{\begin{bmatrix} \mathbf{y}_1 \\ \mathbf{y}_2^* \end{bmatrix}}_{\mathbf{y}} = \underbrace{\begin{bmatrix} \mathbf{h}_1 & \mathbf{h}_2 & \mathbf{h}_3 & \mathbf{h}_4 \\ \mathbf{h}_2^* & -\mathbf{h}_1^* & \mathbf{h}_4^* & -\mathbf{h}_3^* \end{bmatrix}}_{[H_1|H_2]} \times \underbrace{\begin{bmatrix} s_1 \\ s_2 \\ s_3 \\ s_4 \end{bmatrix}}_{\mathbf{s}} + \underbrace{\begin{bmatrix} \mathbf{n}_1 \\ \mathbf{n}_2^* \end{bmatrix}}_{\mathbf{n}}. \tag{26.1}$$

Here $\mathbf{h}_n$ represents the $N_{\mathrm{RX}} \times 1$ vector gathering the channel coefficients between the transmit antenna n and the N_{RX} receive antennas. The subcarrier index is omitted for sake of notation simplicity. The vectors $\mathbf{y}_1$ and $\mathbf{y}_2$ denote the $N_{\mathrm{RX}} \times 1$ signals received on the first and second time slots, and $\mathbf{n}_1$ and $\mathbf{n}_2$ denote the corresponding additive noise components.

The channel matrix $\mathbf{H} = [\mathbf{H}_1|\mathbf{H}_2]$ is the concatenation of two orthogonal submatrices $\mathbf{H}_1$ and $\mathbf{H}_2$, for which the following relations hold:

$$\mathbf{H}_1^H \mathbf{H}_1 = \left(\|\mathbf{h}_1\|^2 + \|\mathbf{h}_2\|^2\right) \mathbf{I}_2 \doteq c_1 \mathbf{I}_2$$
$$\mathbf{H}_2^H \mathbf{H}_2 = \left(\|\mathbf{h}_3\|^2 + \|\mathbf{h}_4\|^2\right) \mathbf{I}_2 \doteq c_2 \mathbf{I}_2.$$

Then, various strategies can be applied to recover the four transmitted symbols, from optimal maximum likelihood to zero-forcing (ZF) or minimum mean squared error (MMSE) equalization, while exploiting these orthogonality properties. In the following, we focus on standard ZF- or MMSE-based receivers in order to derive a scheme that is suited for real-time implementation on current FPGA-based prototyping hardware. Here, for any ZF or MMSE based receiver, the minimum number of receive antennas required for decoding the hybrid SDM and STBC combination modes is equal to $N_{\mathrm{TX}}/2$ instead of N_{TX}. Moreover, the hybrid scheme also benefits from the orthogonal design and inherent full diversity of each Alamouti code.

Zero-forcing receiver To recover the four transmitted symbols s_1, s_2, s_3, and s_4, ZF equalization can be applied to the hybrid system (26.1) by applying the Moore–Penrose pseudoinverse matrix of $\mathbf{H}$, defined as $\mathbf{H}^\# = \left(\mathbf{H}^H \mathbf{H}\right)^{-1} \mathbf{H}^H$, to the received signal $\mathbf{y}$. Let us introduce the orthogonal matrix $\mathbf{P}$ defined as $\mathbf{P} = \mathbf{H}_1^H \mathbf{H}_2$, which satisfies $\mathbf{PP}^H = \mathbf{P}^H \mathbf{P} = c_3 \mathbf{I}_2$, where $c_3 = \left|\mathbf{h}_1^H \mathbf{h}_3 + \mathbf{h}_2^T \mathbf{h}_4^*\right|^2 + \left|\mathbf{h}_1^H \mathbf{h}_4 - \mathbf{h}_2^T \mathbf{h}_3^*\right|^2$.

An analytic form of the pseudoinverse of the channel matrix can therefore be derived as

$$\mathbf{H}^\# = \frac{1}{c_1 c_2 - c_3} \underbrace{\begin{bmatrix} c_2 \mathbf{I}_2 & -\mathbf{P} \\ -\mathbf{P}^H & c_1 \mathbf{I}_2 \end{bmatrix}}_{W_2^{\mathrm{ZF}}} \underbrace{\begin{bmatrix} \mathbf{H}_1^H \\ \mathbf{H}_2^H \end{bmatrix}}_{W_1^{\mathrm{ZF}}},$$

such that no matrix inversion is actually required.

The ZF receiver can then be decomposed into two stages. First, the received signal $\mathbf{y}$ is filtered by $\mathbf{W}_1^{\mathrm{ZF}}$ (matched filter) to remove the contribution from the other signal involved in each Alamouti code. Secondly, the filter $\mathbf{W}_2^{\mathrm{ZF}}$ is applied to the resulting signal to recover the two transmitted streams. These two stages lead to the same signal-to-noise ratio (SNR) for each of the Alamouti codes, i.e.,

$$\mathrm{SNR}_1 = c_1 - c_3/c_2 \quad \text{for the first stream, and}$$
$$\mathrm{SNR}_2 = c_2 - c_3/c_1 \quad \text{for the second one.} \tag{26.2}$$

The second term in the expression of the SNR (26.2) corresponds to the cost paid by employing spatial multiplexing compared with the SNR achieved by the a pure classical Alamouti scheme.

Minimum mean squared error receiver In the previous subsection, an analytic expression of the ZF receiver has been derived. Similarly, the MMSE receiver represented by the matrix $\mathbf{W}$ to be applied onto the received signal vector $\mathbf{y}$, given by (26.1), is

$$\mathbf{W} = \left(\mathbf{H}^H \mathbf{H} + \sigma_n^2 \mathbf{I}_4\right)^{-1} \mathbf{H}^H,$$

where σ_n^2 denotes the variance of the additive white Gaussian noise, and the data symbols are assumed to be normalized to unit norm. $\mathbf{W}$ can be calculated without matrix inversion as

$$\mathbf{W} = \frac{1}{d_1 d_2 - c_3} \underbrace{\begin{bmatrix} d_2 \mathbf{I}_2 & -\mathbf{P} \\ -\mathbf{P}^H & d_1 \mathbf{I}_2 \end{bmatrix}}_{W_2^{\mathrm{MMSE}}} \underbrace{\begin{bmatrix} \mathbf{H}_1^H \\ \mathbf{H}_2^H \end{bmatrix}}_{W_1^{\mathrm{MMSE}}},$$

with

$$d_1 = \|\mathbf{h}_1\|^2 + \|\mathbf{h}_2\|^2 + \sigma_n^2 = c_1 + \sigma_n^2\ ,$$
$$d_2 = \|\mathbf{h}_3\|^2 + \|\mathbf{h}_4\|^2 + \sigma_n^2 = c_2 + \sigma_n^2\ .$$

The first stage of filtering, using $\mathbf{W}_1^{\mathrm{MMSE}}$, is performed in the same way as for the ZF detection to benefit from the orthogonality design of each Alamouti code. The second stage, based on the $\mathbf{W}_2^{\mathrm{MMSE}}$ filter, then takes into account the variance of the additive noise to achieve a better detection of the two transmitted streams and is specific to the MMSE equalizer.

26.3.3 Performance of the hybrid schemes

To illustrate the performance of the hybrid STBC/SDM codes proposed, the following parameters and assumptions were used for the simulations: 20 MHz bandwidth, IEEE 802.11a convolutional code and puncturing, 48 data subcarriers, channel TGn D NLOS, packet size: 1000 bytes, MMSE MIMO detection and perfect channel estimation.

Here the performance of the hybrid STBC/SDM schemes is compared with that of classical SDM. The PER versus SNR curves presented in Fig. 26.4 show that a significant gain is achieved by the combination of SDM and STBC for low to medium data rates. In particular, the gains observed at PER = 10^{-2} for the three comparison scenarios are the following:

- $2 \times 2 \rightarrow 3 \times 2$: 2.7 dB to 5.1 dB gain for 72 Mbps, 96 Mbps and 120 Mbps,
- $2 \times 2 \rightarrow 4 \times 2$: 4.7 dB to 8.1 dB gain for 72 Mbps, 96 Mbps and 120 Mbps,
- $3 \times 3 \rightarrow 4 \times 3$: 2.2 dB to 4.8 dB gain for 144 Mbps, 162 Mbps and 180 Mbps.

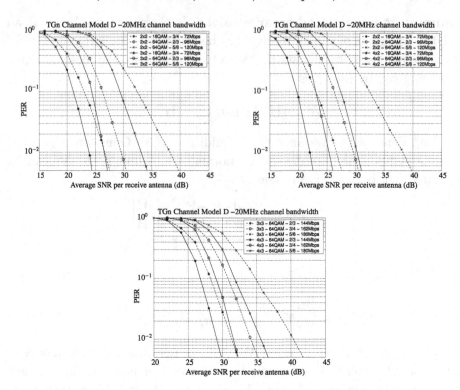

Fig. 26.4. Performance comparison of (a) 2×2 vs. 3×2 and (b) 2×2 vs. 4×2 antenna configurations with two SS and (c) 3×3 vs. 4×3 antenna configurations with three SS.

Corresponding PER vs. SNR curves that take into account receiver impairments as well as channel estimation and synchronization are given in Fig. 26.5.

26.4 Framing: preamble design for efficient automatic gain control, synchronization and CSI estimation

In the same way as for the single antenna context, to enable the transmission of the useful payload over the air, several tasks need to be performed first by the transceiver:

- *Automatic gain control (AGC) adaptation:* the power level of the received signal is estimated and the amplifier gains in the radio frequency (RF) front end are adjusted to align the signal to the voltage dynamic of the analog-to-digital converters (ADC) to exploit its resolution in an optimum manner.

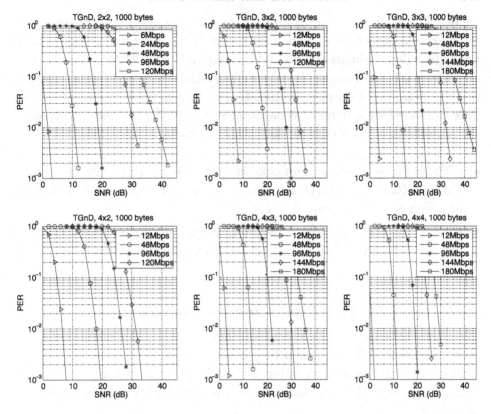

Fig. 26.5. Simulation results for TGn channel model D.

- *Frame detection and rough time synchronization:* the start of a new burst (frame) is detected and its starting point is roughly estimated.
- *Fine time and frequency offset synchronization:* based on the initial rough estimates, the sample time synchronization is refined and the carrier frequency offset is estimated. Frequency offsets occur due to impairments in the local oscillators.
- *CSI estimation:* the channel impulse responses between all transmit and receive antennas are estimated.

For that purpose, data is sent in bursts or frames, each beginning with a known reference sequence called preamble, specifically designed for the above steps.

This section focuses on detailing a possible preamble design building on the current IEEE 802.11a standard to cope with multiple transmit, multiple receive systems. The traditional approach is to split the preambles into two separate subsequences: the *short training symbols (STS)* part for AGC adaptation, rough time synchronization, and frequency offset estimation and the *long training symbols (LTS)* part for fine time synchronization, refined frequency offset estimation, and CSI estimation. In the MIMO context, the difficulty lies in performing all the required estimations based on a superposition of preamble sequences emitted by the various transmit antennas and thus convolved by different propagation channels. To overcome this issue, the advantages of the following design are discussed in what follows:

- nSTS: use a code overlay time-domain sequence design relying on a finite alphabet $\{0, \pm 1, \pm j\}$ targeting a simple cross correlator implementation for achieving time synchronization;
- nLTS: a frequency domain orthogonal design known to provide an optimum MMSE on the channel estimation using Walsh–Hadamard weighting.

26.4.1 Short training symbols

The main challenge of STS design is to find an acceptable compromise between accuracy of the AGC and time offset estimation.

Classically, time synchronization in OFDM is achieved using a cross- or autocorrelation of the received time-domain signal with the preamble template or itself.

For the AGC, the main goal is to avoid destructive or constructive recombination effects on the received antennas of the preambles sent on the various transmit antennas that could lead to an erroneous received power estimation. For that purpose, artificial decorrelation of the transmitted signals is achieved by sending cyclically shifted versions of the same frequency-domain sequence. This approach has the disadvantages to yield a complex cross correlator architecture for hardware implementation and to create multiple peaks at the output of this correlator, which are difficult to discriminate for achieving time synchronization.

To avoid this design trade-off, we propose to change paradigm and assimilate the various transmitting antennas to several users in a synchronous code division multiple access (CDMA) system and adopt a new STS design aligned with this philosophy using different preambles/codes for each antenna composed from a constant amplitude alphabet. For a system using up to four antennas, the sequence design procedure consists in finding, for each antenna, one time-domain STS preamble of length 16 formed of the constant amplitude symbols $\{1, -1, j, -j\}$, which allows for an extremely simple correlator implementation. This is achieved by:

Table 26.2. *Definition of short training symbol subsequences*

STS1	1	1	−1	1	−1	−1	1	−1	−1	0	1	−1	−1	−1	1	1
STS2	0	−1	−1	1	−1	−1	1	−1	1	1	1	1	−1	−1	−1	1
STS3	−j	−j	−j	−j	−j	j	1	−1	−1	1	0	1	−1	−1	1	j
STS4	j	j	j	j	j	−j	1	−1	−1	1	0	1	−1	−1	1	j

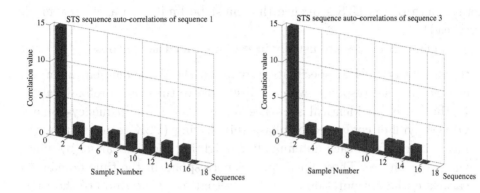

Fig. 26.6. Autocorrelation properties of sequences STS1 and STS4.

- sorting all possible sets of symbols according to their spectrum properties; that is, yielding a low out-of-band radiation and in-band flatness taking into account the null side carriers since it is of no use to put energy there that would be removed afterwards by transmit/receive filters and would alter the correlation properties (sorting is based on corresponding cost functions);

- removing all sets with bad autocorrelation properties (above a threshold);

- selecting the subset of the best remaining ones to consider for cross correlation optimization and isolating the four best sequences among the considered subsets yielding the lower cross correlation properties;

- performing all possible cyclic delays and weighting by $\{1, -1, j, -j\}$ such that the sum of all four sequences has maximum power (i.e., avoid inherent destructive interference in the presence of correlated channels).

An exhaustive search based on the above heuristics provides the desired STS as detailed in Table 26.2. The correlation properties of the resulting STS are illustrated Fig. 26.6.

26.4.2 Long training symbols

The LTS are mainly dedicated to initial channel estimation. Compared with single antenna systems, the challenge in LTS design in a MIMO context is to find the right sequence that yields a CSI estimator with good minimum squared error (MSE) for all the channels between each pair of transmit and receive antennas. When the number of transmit antennas grows, one has to find an adequate compromise between accuracy of the estimator and the total length of the LTS sequence that must be limited to avoid too much overhead.

Two main strategies are usually investigated for that purpose:

- to exploit the OFDM specificity, relying on the time confinement of the channel impulse response, which is restricted to the cyclic prefix duration. In that case, an identical sequence represented by a frequency-domain OFDM symbol composed of phase shift keying (PSK) constellation elements is sent from each antenna after weighting each subcarrier by different linear phases. These linear phases correspond in the time domain to various cyclic permutations adjusted so that all cyclic shifts of the time-domain channels are nonoverlapping. Thus, at the reception, separation of all components of the MIMO channel can be achieved as follows: i) de-modulation by the fast Fourier transform (FFT) ii) cancellation of the initial OFDM sequence by multiplication in the frequency domain by its conjugate and iii) mapping back to the time domain. With this method it is possible to estimate all the transmit channels with only one OFDM symbol as long as the "nonoverlapping criterion" is fulfilled: zero forcing channel invertibility is granted by exploiting the time-domain constraints. Note that a lot of details have been omitted here for the sake of clarity of explanation. In practice, the channel estimators rely on pseudoinverses.

- to adopt an orthogonal frequency-domain design, which is known (Suh *et al.*, 2003) to provide the lowest MSE. This design also avoids the explosion of the MSE resulting from a bad conditioning of the matrices involved with the previous method when considering the usual presence of null side carriers (for spectrum shaping) arising for bad configurations of the number of channel taps and carriers. The price paid for this more accurate estimator is the number of symbols used: one OFDM symbol per transmit antenna is required. With this approach, zero-forcing channel estimation is directly possible in the frequency domain and better performance through denoising can be further reached taking into account the time-domain confinement of the channel.

Here, a possible orthogonal frequency-domain LTS sequence design is il-

Table 26.3. *LTS frequency-domain definition for carriers* $-28, \ldots, +28$

subLTS$_{-28\ldots+28}$	$-1,\quad 1, -1,\quad 1,\quad 1,\quad 1,\quad 1, -1, -1,\quad 1,\quad 1,\quad 1, -1,\quad 1,$ $1, -1, -1, -1, -1,\quad 1,\quad 1, -1,\quad 1,\quad 1, -1,\quad 1, -1, -1,$ $0, -1, -1, -1,\quad 1, -1,\quad 1, -1, -1, -1,\quad 1,\quad 1,\quad 1,\quad 1,$ $-1, -1,\quad 1,\quad 1,\quad 1,\quad 1,\quad 1,\quad 1,\quad 1,\quad 1,\quad 1, -1, -1,\quad 1, -1$

+ subLTS	+ subLTS	+ subLTS	+ subLTS	
+ subLTS (1600ns CS)	− subLTS (1600ns CS)	+ subLTS (1600ns CS)	− subLTS (1600ns CS)	
+ subLTS (100ns CS)	+ subLTS (100ns CS)	− subLTS (100ns CS)	− subLTS (100ns CS)	
+ subLTS (1700ns CS)	− subLTS (1700ns CS)	− subLTS (1700ns CS)	+ subLTS (1700ns CS)	

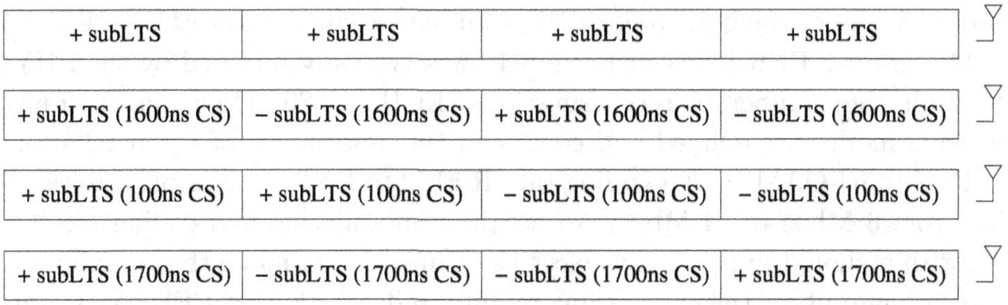

Fig. 26.7. Entire LTS sequence defined in the time domain.

lustrated. The proposed LTS is defined as a concatenation of a weighted and cyclically shifted subsequence subLTS given by Table 26.3. This subsequence is defined in the frequency domain, where each nonzero carrier from index -28 to $+28$ is chosen from the limited alphabet $\{\pm 1\}$ and has been selected for its low peak to average power ratio. The weighting of the subLTS in the LTS is performed across time and space according to a Walsh–Hadamard structure to benefit from an orthogonal design (see Fig. 26.7).

The introduction of a cyclic shift in the transmitted subsequences is made to reduce the dynamic of the total received power by limiting constructive or destructive recombination on a receive antenna between the various weighted subLTS sent over the transmit antennas.

This particular choice enables the frequency-domain estimation of the channel coefficients using a simple ZF approach and minimizes the mean squared error on the channel estimates. The corresponding time-domain sequence is obtained by a 64-point FFT, with undefined carriers set to zero. In addition, more accurate channel estimates can be obtained by including a time confinement constraint into the estimator.

26.5 Throughput with adaptive modulation and coding

Due to the fluctuating nature of the channel, it is inefficient to target a very low PHY layer PER in WLAN systems. Therefore, a fast retransmission mechanism is implemented close to the PHY, in addition to the one provided by higher layer protocols such as TCP. This mechanism is known as automatic repeat request (ARQ) and is enabled by the cyclic redundancy check (CRC) sequence appended to each packet. In a WLAN, the delays resulting from the ARQ are low with respect to those tolerated by higher layer protocols. Therefore, a large number of retransmissions is possible. Under unlimited retransmissions and neglecting the overhead introduced by the MAC protocol, the throughput ρ for a PHY bit rate r equals: $\rho = r\,(1 - \mathrm{PER})$.

Moreover, a PER between 1% and 10% is typically targeted by the PHY, to limit the number of retransmissions. In IEEE 802.11a/g, eight transmission modes are defined and consist in the association of a constellation (BPSK to 64-QAM) and a coding rate R=1/2 to R=3/4, for a bit rate ranging from 6 Mbps to 54 Mbps. An adaptive modulation and coding (AMC) algorithm should dynamically select the transmission mode that maximizes the throughput for the current link quality under the target PER constraint. In the open-loop MIMO WLAN PHY layer that we described previously, the number of dimensions that the AMC has to deal with increases, as the AMC must also select among the multiple-antenna techniques (SDM, STBC,...) available for $N_{\mathrm{TX}} \times N_{\mathrm{RX}}$ antennas.

Ideally, assuming that the instant of the next packet transmission is known, the AMC must base its decision on the best PER prediction for the various modes. A simple approach is to estimate the PER by dividing the number of erroneous packets by the total number of received packets during a given observation window. However, such an estimator takes too many packets to converge, especially at low PER values. Faster convergence can be expected by predicting the future CSI and deriving a PER estimate from the predicted CSI. In this section, we do not focus on CSI prediction but on the derivation of PER from the CSI.

In a noise-limited environment, the PER is a function of the transmission mode, packet size, noise variance, and of the complex channel matrix. The number of dimensions of this function is huge, and many attempts have been made to map these parameters onto a single link quality metric, which could be associated with the PER by means of a look-up table. For instance, the instantaneous SNR was used in Simoens and Bartolomé (2001). In a single antenna OFDM system, it depends on the average of the squared modulus of the complex channel coefficients on every subcarrier. However, it was shown in Lampe *et al.* (2002) that the PER vs. instantaneous SNR performance of a given transmission mode could still differ significantly from one multipath channel realization to another. Therefore, there are some "bad" channels for which an AMC relying on instantaneous SNR may select a bit rate that exhibits a PER much higher than the target PER. Since the channel coherence time in a WLAN can be very long (hundreds of ms), the link could be practically broken. A solution to this problem can be to conservatively shift the SNR thresholds used by the AMC by a few dBs. To reduce this margin, Lampe *et al.* (2002) define an "effective SNR", which depends not only on the average, but also on the variance of the channel coefficients magnitude. In Nanda and Rege (1998), the effective SNR is derived from the union bound applied to a convolutionally-coded system. Among the link quality metrics taking into account FEC effects, the exponential effective SNR mapping (Ericsson, 2003) represents a good performance/complexity trade-off.

In this section, we show how Shannon capacity can also be used as a link quality metric, as proposed in FITNESS (2003). To improve the PER prediction, we compute the capacity of a channel comprising not only the noisy multipath channel but also the OFDM modulator/demodulator and the space-time encoder/decoder. The formula thus differs for the various modes (SDM, STBC, ...) and corresponding space-time receivers (ZF, MMSE, ...). We denote by $H_{m,n}$ the elements of the $N_{RX} \times N_{TX}$ channel matrix $\mathbf{H}$ on a given OFDM subcarrier. The noise variance per receive antenna is denoted by σ^2. The signal variance per antenna is P_{TX}/N_{TX}, where P_{TX} is the total transmitted power. For notational simplicity, we neglect the subcarrier index but remind the reader that the capacity shall be ultimately averaged over the subcarriers. For Alamouti STBC, the capacity equals

$$
C = \log\left(1 + P_{TX} \sum_{n=1}^{2} \sum_{m=1}^{N_{RX}} \frac{|H_{m,n}|^2}{2\sigma^2}\right).
$$

For the $N_{TX} = 2$, $N_{RX} = 2$ SDM scheme with linear MMSE processing, the length-2 received vector is multiplied by the matrix W defined by:

$$
W = H^H \left(HH^H + \sigma^2 \frac{N_{TX}}{P_{TX}} I_{N_{RX}}\right)^{-1}.
$$

The output of the linear MMSE estimator is affected by colored noise and interstream interference. Modeling the latter by a Gaussian random variable, the overall noise plus interstream interference covariance matrix equals:

$$
R = \frac{P_{TX}}{N_{TX}} (WH - \mathrm{diag}(WH))(WH - \mathrm{diag}(WH))^H + \sigma^2 WW^H.
$$

We assume that the effect of nondiagonal terms of R on capacity is negligible

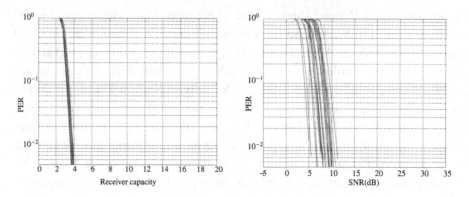

Fig. 26.8. PER prediction using (a) Shannon capacity; (b) instantaneous SNR.

thanks to bit interleaving, yielding

$$C \approx \log\left[\det\left(I_{N_{\mathrm{TX}}} + \frac{P_{\mathrm{TX}}}{N_{\mathrm{TX}}} \mathrm{diag}(WH)(\mathrm{diag}(R))^{-1}(\mathrm{diag}(WH))^{H}\right)\right].$$

The same reasoning holds for STBC modes, when H is replaced with an equivalent channel matrix. In Fig. 26.8, the PER vs. capacity performance of the $N_{\mathrm{TX}} = 3$, $N_{\mathrm{RX}} = 2$, QPSK, R=1/2, SDM-STBC transmission mode is plotted for 40 independent 802.11n channel realizations. Clearly, simulations confirm that the capacity provides a much better prediction than the instantaneous SNR. For each transmission mode, it is possible to define an outage probability P_{out} and a target capacity value C_{target} such that in a random channel realization with capacity higher than the target capacity, there is a probability $1 - P_{\mathrm{out}}$ that the PER is below the target PER. Note that the target capacity should not be mistaken for the well-known outage capacity concept.

To quantify the gain achievable by AMC, the average throughput is plotted versus average SNR in Fig. 26.9(a) for an $N_{\mathrm{TX}} = 2$, $N_{\mathrm{RX}} = 2$ system for a typical indoor channel. The five low bit rate modes rely on Alamouti STBC, whereas the four higher bit rate modes rely on SDM. The target PER and P_{out} both equal 5%. Since no MAC overhead is assumed, the curves represent an upper bound on the achievable MAC throughput on a single link. The throughput obtained with a perfect PER knowledge for every channel realization is plotted as a dotted line and represents an upper bound on the performance of any AMC algorithm. The thin curves represent the average throughput of each mode but without any constraint on target PER or outage. The better PER prediction with capacity yields up to 3 dB gain

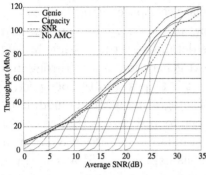
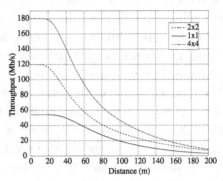

(a) Impact of the PER prediction quality on the AMC throughput performance.

(b) Improvement of WLAN peak throughput and range by multiple antenna techniques.

Fig. 26.9.

over instantaneous SNR (dashed vs. solid line), but the prediction remains imperfect, because (among other reasons) the real system including FEC does not operate at capacity.

To evaluate the impact of multiple antennas on a typical indoor WLAN deployment, the average throughput versus distance is plotted in Fig. 26.9(b) assuming a 100 mW total RF output power and a −90 dBm noise floor at the receiver. For fast WLAN deployment evaluation, a log-distance path loss model and log-normal shadowing are assumed, as described in 802.11n Task Group (2004). The single antenna system corresponds to 802.11a/g. The 2×2 and 4×4 systems double and triple the peak throughput, respectively, and also significantly extend the cell range for a given throughput.

26.6 Conclusion

In this contribution, we presented a design methodology for devising a new MIMO short range WLAN system addressing requirements that are typical for a standardization process, such as the one currently conducted by IEEE 802.11n. This methodology relies on an evolutionary approach with the goal of extending the current IEEE 802.11a solution to address higher throughput while preserving backward compatibility. The emphasis has been on detailing new OFDM MIMO modes with the constraints to i) handle asymmetric transmit/receive antenna configurations with one, two, or three parallel spatial streams and ii) focus on open-loop schemes for stability, avoiding calibration circuit or feedback signaling. A possible solution addressing these constraints, which we propose, is based on the exploitation of a hybrid combination of spatial division multiplexing to increase spectrum efficiency and peak data rates and classical space time block coding to improve link robustness or range for low to medium data rates (suited to small packet size, e.g., VoIP). As well as detailing all the possible MIMO solutions suited for consideration for the next generation of high throughput WLAN systems, one of the goals of this part is to put into the perspective of a complete system design the problem of integrating a MIMO modulator. Although we have restricted ourselves to open loop solutions that will most probably form the basis of future WLAN standards, i.e., IEEE 802.11n, it is worth mentioning that alternative solutions exploiting CSI at the transmitter, referred to as advanced beamforming, are likely to be introduced in medium-term products.

Acknowledgment

The authors would like to thank AAU-CSys and FUNDP-INFO for providing us with MATLAB™ packages of the IEEE 802.11 TGn channel models, and the IST project IST-2000-30148 I-METRA (http://www.ist-imetra.org/).

References

802.11n Task Group (2004). Tgn channel models, 802.11-03/940r4. Tech. rep., IEEE P802.11.

Alamouti, S. M. (1998). A simple transmit diversity technique for wireless communications. *IEEE J. Sel. Areas Commun.*, **16** (8), 1451–1458.

Ericsson (2003). System-level evaluation of OFDM—further considerations. Tech. rep., 3GPP TSG-RAN WG1.

FITNESS (2003). D3.3.1—MTMR baseband transceivers needs for intra-system and inter-system (UMTS/WLAN) reconfigurability.

Lampe, M., Rohling, H., and Zirwas, W. (2002). Misunderstandings about link adaptation for frequency selective fading channels. In *IEEE Personal Indoor Mobile Radio Commun. Conf.*, vol. 2, 710–714.

Nanda, S. and Rege, K. (1998). Frame error rates for convolutional codes on fading channels and the concept of effective E_b/N_0. *IEEE Trans. Veh. Technol.*, **47** (4), 1245–1250.

Simoens, S. and Bartolomé, D. (2001). Optimum performance of link adaptation in HIPERLAN/2 networks. In *Proc. IEEE Veh. Technol. Conf.*, vol. 2, 1129–1133.

Suh, C., Hwang, C.-S., and Choi, H. (2003). Preamble design for channel estimation in MIMO-OFDM systems. In *Proc. IEEE Global Telecommun. Conf.*, vol. 1, 317–321.

Tarokh, V., Seshadri, N., and Calderbank, A. R. (1998). Space-time codes for high data rate wireless communication: Performance criterion and code construction. *IEEE Trans. Inf. Theory*, **44** (2), 744–765.

Texas Instruments (2001). Double-STTD scheme for HSDPA systems with four transmit antennas: Link level simulation results. Tech. rep., TSG-R WG1 document, TSGR1#20(01)0458.

Zhuang, X., Vook, F. W., Rouquette-Léveil, S., and Gosse, K. (2003). Transmit diversity and spatial multiplexing in four-transmit-antenna OFDM. In *Proc. IEEE Int. Conf. Commun.*, vol. 4, 2316–2320.

27

VLSI implementation of MIMO detection

Andreas Burg
ETH Zurich

David Garrett
Beceem Communications, Inc.

27.1 Introduction

MIMO techniques play an important role in numerous wireless standards, such as the HSDPA extension of WCDMA, IEEE 802.11n wireless LAN, and IEEE 802.16 wireless MAN. Their success will critically depend on the availability of high-performance, low complexity receivers, which requires careful study of the implementation aspects.

Initially, most efforts toward optimizing MIMO detection for implementations were concerned with highly suboptimal linear and successive interference cancellation (SIC) techniques, since they are associated with the lowest order of complexity. However, as demonstrated recently in a number of application-specific IC (ASIC) implementations, the combination of advances in silicon technology, innovative VLSI architectures, and low complexity algorithms have enabled the implementation of better performing MIMO detection schemes that come closer to realizing the full channel capacity. The highest performing detectors for MIMO systems employ a full maximum likelihood (ML) search of the transmit constellation space. The exhaustive ML approach has been readily demonstrated in implementations for rates up to 8 bits per channel use (bpcu). The problem is that for higher rates, the exhaustive-search ML solution far exceeds current silicon capabilities. Sphere decoding algorithms have emerged as the most promising decoding strategies, providing full, or close to ML, bit error rate (BER) performance at reasonable cost for transmission rates that are relevant in practical systems.

In this chapter, the VLSI implementation aspects of MIMO detection are discussed. First, a general review of the capabilities and limitations of today's silicon technology is presented. Next, the basic tools and concepts are introduced that enable us to estimate the true implementation complexity

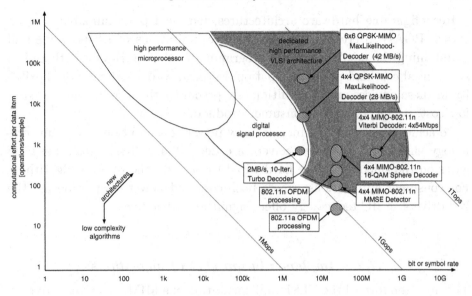

Fig. 27.1. Comparison of implementation architectures and processing requirements
of wireless communication algorithms.

of an algorithm, and to design new VLSI architectures and low complexity MIMO detection schemes. The most important algorithmic choices for MIMO detection are then summarized and compared with respect to implementation examples from the literature. Finally, the corresponding specific hardware architectures, optimizations and implementation trade-offs are described in more detail.

27.2 Considerations for implementation

To put the complexity challenge of MIMO into perspective, consider the chart in Fig. 27.1. On the horizontal axis, the chart shows the number of data items to be processed per second. A data item may be a symbol, a vector, or an individual bit, since algorithms operate on different atomic data types. The vertical axis gives the number of operations per data item. On the log-log scale, the diagonal lines represent constant numbers of operations per second. Various implementation technologies are overlaid on the plot to show their respective regions of operation.

Microprocessors and digital signal processors (DSPs) provide a high degree of flexibility and are straightforward to program. However, most of the examples shown in the figure lie on the edge or exceed the capabilities of even the highest performance software-programmable devices.

Reconfigurable hardware architectures, like field programmable gate arrays (FPGAs), overcome the performance limitations of DSPs, while still maintaining the flexibility of a reprogrammable device. However, there is a substantial cost and power dissipation overhead associated with the flexibility in these devices, and this ultimately precludes them from being used in low-cost, battery-powered consumer products.

To achieve high performance and low power consumption, the main technology choice for wireless receivers are dedicated ASICs, where the logic gates on the silicon chip are hard-wired to perform the receiver algorithms. For consumer applications, where both cost and power dissipation are the key drivers, ASIC technology offers significant advantages.

27.2.1 Implementation cost of algorithms

The first step toward the VLSI implementation of a MIMO receiver is to evaluate the total number of operations required by the candidate algorithms for a given set of system parameters, such as data rate, number of antennas, and the modulation scheme. In combination with an overall system performance metric, such as packet error rate (PER) or system throughput, a designer can start to understand and balance the implementation trade-offs. Once initial requirements are met, an algorithm must be evaluated using four additional factors, which have significant impact on its true silicon implementation complexity:

- The *relative VLSI complexity* of arithmetic operations associates different operations with their hardware cost and with the time needed for their execution. Additions, for example, are simple to implement, while multiplications and divisions are slower and require more silicon area.

- *Data dependencies* determine the degree of dependence between steps in the algorithm, and consequently limit the number of operations that can be completed in parallel to increase throughput.

- *Numerical precision* is concerned with the number of significant digits in the number representation, and directly affects the required accuracy of arithmetic operations in an algorithm. Reduced numerical precision and approximations of complex expressions reduce area, but introduce quantization noise. The goal is to reduce complexity with minimal implementation loss of system performance.

- *Memory* requirements often show up as *latency* and are determined by the amount of information that needs to be retained during the execution of an algorithm. Examples in communication systems that introduce significant memory are bit-interleaving or block operations, such as FFTs.

27.2.2 Architectural transformations

VLSI design provides many degrees of freedom to trade area for throughput using *architectural transformations*. The most relevant are briefly outlined:

- *Parallel processing* replicates processing elements, such as arithmetic units, to increase throughput at the expense of substantial additional area.
- *Iterative decomposition and time sharing* schedule multiple operations to a single resource to save area at the expense of throughput.
- *Pipelining* breaks up a complex single-cycle operation into a sequence of shorter operations (pipeline stages), which are separated by memory elements and allow for a higher clock rate. In each cycle, the stages pass their intermediate results to the next stage and accept a new data item. Effectively, multiple data items are processed in parallel without additional combinatorial hardware. The drawback is the introduction of latency, which adds overhead for storage and can cause problems due to data dependencies in iterative algorithms, such as SIC or sphere decoding.

27.2.3 Assessing VLSI circuit complexity

Once the number and type of operations, the numerical requirements, and the necessary degree of parallelism are known, a rough estimate of the circuit complexity can be obtained from the cost (area and delay) of the individual VLSI building blocks. To allow for a fair comparison of implementations across different silicon technologies, area is measured in gate equivalents (GE) instead of square millimeters. One GE corresponds to the area of a two-input NAND gate in the respective technology. Unfortunately, accurately predicting the delay of a circuit for different technologies is much more difficult, especially with modern deep-submicron processes. As a rule of thumb, one can expect about 30% delay reduction with each new generation of silicon technology. Table 27.1 summarizes these trends (ITRS, 2004).

The logic complexity of digital signal processing circuits is often dominated by the arithmetic units in the data path. To obtain an understanding of the basic implementation trade-offs in the most frequently used arithmetic operations, consider Fig. 27.2. The chart compares addition, multiplication of two complex numbers, multiplication of a complex number with

Table 27.1. *Scaling of VLSI process technology (ITRS, 2004)*

Process (nm)	Area of NAND2 (μm^2)	kGE/mm^2	Normalized delay	Area of 1-bit register (μm^2)	Area of 1-bit SRAM cell (μm^2)
350	54	19	1.5	260	20
250	24	40	1.0	150	10
180	12	80	0.67	75	5
130	5	200	0.44	36	2.5

a limited set of constellation points (i.e., canonical signed digit representation of 16-QAM constellation points), and Givens rotations (implemented as CORDICs) with respect to area and delay for a single-cycle operation implemented in a 0.25 μm technology. It highlights the large disparity in area and speed of addition and multiplication and shows that significant savings can be obtained from the use of optimized multipliers when, for example, one of the operands is only chosen from a small set of constellation points. The graph also illustrates how the complexity of additions grows linearly with the number precision W, while multiplications grow as $\mathcal{O}(W^2)$. Finally, it can be seen that the area for silicon implementations of arithmetic components grows quickly when they are driven toward their highest speed.

Synchronous digital design, in addition to the arithmetic combinatorial logic, requires *memory storage* elements, which often account for more than 50% of the area. Register-based memories are used in small, local memories and pipeline stages, consuming 4–6 GEs per bit. Area-optimized on-chip random access memories (RAMs) concentrate the storage in large arrays, so that area requirements can be driven down to 0.5 GE/bit. However, a constant area overhead is involved, which reduces area efficiency when RAMs are small. Moreover, RAMs only allow access to a few words at a time, while registers can be integrated directly within logic for full parallel access. A chip design will almost always be a mix of registers and RAMs mainly because of access requirements.

27.3 MIMO algorithm choices

We begin with a high-level discussion of the merits and costs of different MIMO detection algorithms. The goal is to review the available algorithm choices for MIMO detection, and to study their silicon complexity and ar-

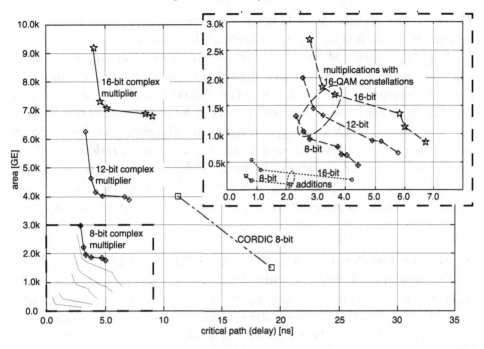

Fig. 27.2. Comparison of VLSI complexity of arithmetic operations.

Table 27.2. *ASIC implementations of MIMO detection*

System configuration	Algorithm	Area (kGE)	Throughput (Mbps)	Reference
4 × 4 (any)	V-BLAST	190	—	(Guo and Nilsson, 2003)
4 × 4 16-QAM	MMSE	68	320	(Haene et al., 2004)
4 × 4 QPSK	ML	40	50	(Burg et al., 2003)
4 × 4 QPSK	ML-APP	140	19	(Garrett et al., 2004)
4 × 4 16-QAM	K-best	91	52	(Guo and Nilsson, 2004)
4 × 4 16-QAM	sphere	50	169[a]	(Burg et al., 2005)

[a] Throughput depends on the SNR and on the channel realization. The average throughput at 20 dB SNR is given when the entries of **H** are i.i.d. Gaussian distributed.

eas of application. A survey of the various implementations is shown in Table 27.2. See Chapter 15 for more details on the algorithms used.

Linear detectors have the lowest BER performance but are also associated with the lowest complexity order. Zero-forcing (ZF) or minimum mean squared error (MMSE) detectors first obtain a set of nulling vectors from the channel matrix **H**, for example according to $\mathbf{G} = (\mathbf{H}^H \mathbf{H})^{-1} \mathbf{H}^H$ in the

ZF case. The computation of $\mathbf{G}$ is often referred to as *preprocessing*, as it is only carried out when the channel changes. The associated complexity order is $\mathcal{O}(M_T^2 M_R)$, where M_T and M_R are the number of transmit and receive antennas, respectively. The actual detection itself is performed with a simple matrix-vector multiplication, $\hat{\mathbf{x}} = \mathbf{G}\mathbf{y}$, followed by slicing of $\hat{\mathbf{x}}$, where $\mathbf{y}$ denotes the received vector. The complexity order of the symbol rate processing is therefore only $\mathcal{O}(M_T M_R)$, and is almost independent of the employed modulation scheme. However, despite the low complexity order, the silicon complexity of linear receivers can be rather high. The reason for this are the stringent numerical requirements of the matrix inversion algorithms, which lead to arithmetic components with high circuit complexity (Haene *et al.*, 2004).

Successive interference cancellation (SIC) outperforms linear detection in terms of BER. SIC algorithms detect the parallel streams sequentially, make a decision, remodulate the signal, and subtract it from the received vector before detecting the next stream. As opposed to linear schemes, an additional overhead in the preprocessing is associated with finding a suitable detection order. The design by Guo and Nilsson (2003), for example, implements the V-BLAST scheme (Foschini and Gans, 1998) according to Hassibi (2000), which guarantees an optimum ordering. The actual detection procedure for SIC is only slightly more complex than for linear receivers, as the additional cancellation stage can be optimized significantly.

Exhaustive-search maximum likelihood detection achieves optimum BER performance by finding $\hat{\mathbf{s}} = \arg\min\|\mathbf{y} - \mathbf{H}\mathbf{s}\|^2$ through an exhaustive search over all possible candidate vector symbols $\mathbf{s}$. Unfortunately, the number of candidates (i.e., the complexity) increases exponentially with the rate. However, the implementations by Burg *et al.* (2003) and Garrett *et al.* (2004) show that an exhaustive-search detector is feasible *and* efficient for rates up to 8 bpcu, and can have an even lower silicon complexity than linear or SIC schemes. An additional advantage of exhaustive-search ML is the ability to provide soft-information (Garrett *et al.*, 2005) to improve the performance of a subsequent channel decoder with only little additional hardware effort.

K-best and sphere decoding techniques also search through the set of possible vector symbols. However, complexity is greatly reduced by limiting the search space. As a result, rates of up to 16–24 bpcu become feasible with low hardware complexity and high throughput. The *K-best algorithm* features close to ML performance with a deterministic throughput and an architecture that is very suitable for VLSI implementation. The first ASIC realization of the algorithm was reported by Wong *et al.* (2002) and was later improved by Guo and Nilsson (2004). *Sequential sphere decoding* (SD), on the other hand, can achieve ML performance, but the iterative approach has a nondeterministic throughput. A guaranteed throughput can be enforced with early termination at the expense of a BER performance penalty. The implementations by Burg *et al.* (2005), however, demonstrate that SD (even without early termination) outperforms K-best decoders in most scenarios, providing higher throughput with lower silicon complexity.

When comparing the K-best and the sphere decoder to other detector types in Table 27.2, it is important to note that lattice decoding also requires a preprocessing stage for QR-decomposition, which is not included in the area of the reported ASIC implementations. Its complexity is similar to the preprocessing required for a linear detector.

27.4 VLSI architectures for MIMO detection algorithms

27.4.1 Linear and SIC detectors

For the sake of analyzing the implementation of linear and SIC detectors, four types of receivers are considered in this section: *zero-forcing* (ZF), *MMSE*, *suboptimal ordered SIC*, and *V-BLAST*. While all four approaches differ in their BER performance and in their implementation complexity, the underlying algorithms and architectures are very similar.

Implementation of preprocessing for linear and SIC detectors

Preprocessing for linear detectors revolves around the decomposition and inversion of the dense, unstructured channel matrix $\mathbf{H}$. From an implementation point of view, the numerous algorithms that are available to solve this problem (Golub and Van Loan, 1996; Press *et al.*, 1992) can be partitioned into four categories:

Direct methods are based, for example, on *LU-decomposition*, *Cholesky factorization*, or the *Riccati recursion*. The drawback of these techniques is that they usually operate on the covariance matrix $(\mathbf{H}^H\mathbf{H} + \sigma^2\mathbf{I})$, whose eigenvalue spread is squared compared with $\mathbf{H}$. This aggravates numerical problems and increases the dynamic range of the intermediate results. Consequently, arithmetic components become more costly in terms of hardware due to the increased numerical precision. On the positive side, the total number of operations is typically lower than for other algorithms. The main applications for these techniques are designs that target floating-point DSPs or modern FPGAs, which already have dedicated built-in multipliers with high number precision, so that the bare number of operations becomes the main concern.

Unitary transformations lead to numerically more stable algorithms, such as *QR-decomposition* for MMSE, or ZF detection, or the *square-root algorithm* for V-BLAST (Hassibi, 2000). Efficient implementations of these algorithms are mostly based on *Givens rotations*. These can be implemented efficiently using *CORDIC* circuits (Volder, 1959; Parhami, 2000), which carry out vector rotation, and can also determine the phase of a vector through a series of simple shift-and-add operations. Multiple CORDICs can be instantiated in parallel to meet throughput requirements (Lightbody *et al.*, 2003). However, for the needs of today's wireless systems, moderately parallel architectures are typically sufficient. The V-BLAST implementation by Guo and Nilsson (2003), for example, implements the *square-root algorithm* with three CORDICs combined in a supercell for a complex-valued Givens rotation.

Iterative methods start from an estimate of the channel matrix and use algorithms such as *gradient descent, Schultz–Hotelling*, or *Gauss–Seidel* to find the pseudoinverse. These schemes are rarely efficient, as many iterations are needed to converge, and a single iteration often has almost the complexity of other schemes that obtain the inverse immediately.

Adaptive approaches are based for example on LMS. The main advantages are that no explicit channel estimation has to be performed, and that numerical requirements in the adaptation loop are often very low, so that approximations can be applied that reduce circuit area. An example for the realization of such an adaptive scheme has been presented by Garrett *et al.* (2005), where a low-complexity NLMS algorithm is used for the implementation of a space-time equalizer for a MIMO-HSDPA receiver.

Relative implementation complexity

The complexity of the preprocessing for linear and SIC MIMO receivers is governed by numerical requirements that depend on the applied algorithms, the system (antenna) configuration, and the SNR operating range. The most relevant linear and SIC algorithms are therefore considered *only relative to each other*, assuming implementations based on unitary transformations. This approach applies to all considered detection algorithms in a

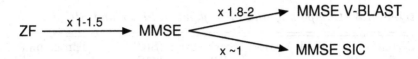

Fig. 27.3. Relative preprocessing complexity increase for linear and SIC-based MIMO detection schemes.

very similar way and is therefore a good basis for comparisons. The results are summarized in Fig. 27.3 and are detailed as follows:

Zero-forcing has the lowest implementation complexity if the number of operations is taken into account only. However, the algorithm leads to a potentially unlimited dynamic range, and is sensitive to quantization effects in fixed-point arithmetic. In *MMSE detection*, the noise variance term can be set to an implementation-dependent minimum value at high-SNR to serve as a *regularization parameter*, which limits the dynamic range of the result. MMSE is therefore numerically much more convenient than *ZF*. In terms of the number of operations, computing the MMSE detector is roughly 50% more complex than computing the ZF detector, assuming a QR-based approach. This results, because the decomposition now needs to be carried out on the larger $(M_T + M_R) \times M_T$ dimensional matrix $[\mathbf{H}^T \quad \sigma\mathbf{I}]^T$. However, the sparse nature of the augmented channel matrix keeps the overhead low, and the possible reduction of word length often more than compensates for the additional operations.

Compared with MMSE detection, the implementation of the better performing V-BLAST algorithm requires an approximately 1.8–2 times higher number of operations, while the favorable numerical properties of MMSE are preserved. An alternative approach is the use of suboptimal low-complexity ordering criteria that can be determined directly on the channel matrix $\mathbf{H}$ (e.g., column-norm ordering). Then, the nulling vectors can be computed using straightforward MMSE after rearranging the columns of $\mathbf{H}$ appropriately. While the performance of these algorithms is between MMSE without ordering and V-BLAST, their complexity is almost identical to MMSE.

Detection stage

The complexity of the detection stage in a linear detector is governed by the multiply-accumulate (MAC) operation (4.5–7 kGE) for computing $\hat{\mathbf{x}} = \mathbf{Gy}$, where $\mathbf{G}$ is the matrix of nulling vectors, and $\mathbf{y}$ is the received vector. With a single MAC unit, $M_T M_R$ cycles are required for each vector symbol. To increase throughput, multiple MAC units can be instantiated at the expense of area. The step from linear to SIC detection mainly involves augmenting

Table 27.3. *Complexity and throughput of linear and SIC detectors*

System configuration	# of units	Area (linear) (kGE)	Area (SIC) (kGE)	Throughput[a] [vector symbols/s]
2×2	1	4.5	5.5	25 M
4×4	1	4.5	5.5	6.25 M
6×6	1	4.5	5.5	2.7 M
2×2	2	9	11	50 M
4×4	4	18	22	25 M
6×6	6	27	33	16.6 M

[a] A clock rate of 100 MHz is assumed.

the basic MAC unit with an interference cancellation stage. As the latter only involves multiplications of the channel coefficients with constellation points (here 16-QAM), the complexity overhead is rather low (1–2 kGE). The corresponding trade-offs are illustrated in Table 27.3.

27.4.2 Exhaustive-search maximum-likelihood (ML) detection

Despite the exponential growth of complexity with rate, exhaustive-search ML detection has a number of advantages over linear or SIC detectors:

(i) No costly operations, such as divisions, square roots, or CORDICs are required.

(ii) The dynamic range of all intermediate results is limited and small.

(iii) The algorithm has no recursive component, which allows for the full range of architectural transformations (e.g., pipelining).

(iv) Algorithmic transformations can reduce the number of costly operations significantly.

These favorable properties, if exploited properly, push the implementation limit to higher rates than one might expect.

Reducing exhaustive-search circuit complexity

The circuit complexity of an *exhaustive-search* ML detector is dominated by the repeated evaluation of the cost function $\|\mathbf{y} - \mathbf{Hs}\|^2$ for all possible candidate vector symbols $\mathbf{s}$ to find the minimum among the results. The following optimizations can help to greatly reduce the implementation complexity of this type of problems:

Preprocessing extracts common terms from a series of computations to ensure they are evaluated only once. An obvious application to the implementation of ML detection is to precompute all possible values of **Hs** *only when the channel changes*, and to store the results in a cache, ready to be used for the detection of many subsequent symbols (Garrett *et al.*, 2004).

Algorithmic transformations can often increase the gains achieved from preprocessing. The goal is both to maximize the number of terms that can be precomputed *and* to reduce complexity by precomputing the costly operations at lower rate, even if this may come at the expense of additional, less complex operations. Application to ML detection is demonstrated by Burg *et al.* (2003). Therein, $\|\mathbf{y} - \mathbf{Hs}\|^2$ is simply written as $\|\mathbf{Hs}\|^2 - 2\Re\{(\mathbf{y}^H\mathbf{H})\mathbf{s}\} + c$, where c is independent of **s**, and can therefore be omitted. As a result of this factorization, the costly squaring operation for each candidate vector symbol can be precomputed and has been removed from the detection process, at the cost of few additional but much cheaper multiplications with constellation points. In addition to the reduced logic complexity, a significant reduction in cache size is achieved, as now only a single real-valued number $\|\mathbf{Hs}\|^2$ needs to be stored for each candidate vector symbol.

Implementation of ML-APP detection

Providing a-posteriori probability (APP) information about the individual bits from the MIMO detector to a soft-input forward error correction (FEC) decoder can significantly improve BER performance. APPs are usually expressed as *log-likelihood ratios* (LLRs), which can be approximated with various degrees of complexity based on the *"Jacobian Logarithm"* (Erfanian *et al.*, 1994). The simplest, yet quite accurate, form is the max-log approximation, which yields

$$LLR_j \approx \min_{b_j(\mathbf{s})=0} \|\mathbf{Hs} - \mathbf{y}\|^2 - \min_{b_j(\mathbf{s})=1} \|\mathbf{Hs} - \mathbf{y}\|^2, \qquad (27.1)$$

where $b_j(\mathbf{s})$ denotes the jth bit in the candidate vector symbol **s**.

From (27.1) another advantage of exhaustive-search detection becomes apparent: in computing the ML cost function $\|\mathbf{y} - \mathbf{Hs}\|^2$ for the possible transmit candidates, all the information required to compute the LLRs for each bit is available (Garrett *et al.*, 2005). Essentially, the hardware only needs to keep track of the minimum of the cost function for each bit in the symbol vector being *zero* and *one* individually. Once all candidate symbols have been examined, the LLRs can be computed.

Complexity and scaling of exhaustive-search ML

The downside of exhaustive-search algorithms is that they suffer from an exponential growth in complexity. To illustrate the limits of ML detection, the basic building blocks from the ML detector designs by Burg *et al.* (2003) and Garrett *et al.* (2004) have been scaled up to larger constellation sizes and to higher number of transmit antennas. Table 27.4 shows the expected complexities.

The ML detector by Burg *et al.* (2003) is designed specifically for QPSK modulation; therefore, only the number of parallel streams is scaled. With each additional antenna, the area of the cache that stores the precomputed $\|\mathbf{Hs}\|^2$ grows by a factor of four, and the number of vector symbols that can be processed per second reduces to one fourth. To partially regain the original throughput, the number of detection units can be increased to process multiple candidates in parallel at the expense of GEs. Due to extensive *sharing of common terms* in the computation, however, the area penalty is quite low as can be seen from lines 2 and 3 in Table 27.4. Configurations up to 4×4 (8 bpcu) are therefore efficient and provide high throughput.

The ML-APP is shown for a fixed MIMO symbol rate of 2.4M vectors/s. At higher rates, more candidate symbols must be considered in parallel. As expected, the ML-APP detector shows the same exponential scaling behavior as the hard-decision ML detector. However, the computation of the LLRs and the support for 16-QAM modulation consume additional area. An extrapolation of the circuit to the case of 16-QAM with four parallel streams at a constant vector symbol rate clearly illustrates the limits of the exhaustive-search approach. The 22M GE correspond to around 117 mm^2 of silicon area in a 130 nm process, which is significantly larger than entire advanced microprocessor designs. The conclusions is that exhaustive-search ML detection becomes computationally prohibitive beyond 8 bpcu.

27.4.3 K-best and sphere decoding

K-best lattice decoding and *sphere decoding* have been proposed to achieve close to ML and full ML performance, respectively, with a complexity that is well below an exhaustive search. Despite their fundamental differences, both algorithms share an important concept: they start from a triangularized form of the ML detection cost function $d(\mathbf{s}) = \|\hat{\mathbf{y}} - \mathbf{Rs}\|^2$, where $\mathbf{Q}$ is unitary and $\mathbf{R}$ is upper triangular, so that $\mathbf{H} = \mathbf{QR}$ and $\hat{\mathbf{y}} = \mathbf{Q}^H \mathbf{y}$. As a result, $d(\mathbf{s})$ can be written as a monotonically increasing series of *partial Euclidean*

Table 27.4. *Complexity of exhaustive-search ML (Burg et al., 2003) and ML-APP detection (Garrett et al., 2004)*

System configuration	Rate (bpcu)	Detector type	Throughput at 100 MHz clock (Mbps)	Area (kGE)	# of **s** considered in parallel
2 × 2 QPSK	4	ML	100.0	13	4
4 × 4 QPSK	8	ML	12.5	40	4
4 × 4 QPSK	8	ML	50.0	42	16
6 × 6 QPSK	12	ML	18.8	160	64
2 × 2 QPSK	4	ML-APP	9.6	93	4
4 × 4 QPSK	8	ML-APP	19.2	140	8
4 × 4 8-PSK	12	ML-APP	28.8	1,500	120
4 × 4 16-QAM	16	ML-APP	38.4	22,000	1920

distances (PEDs)

$$T_i(\mathbf{s}^{(i)}) = T_{i+1}(\mathbf{s}^{(i+1)}) + \left| e(\mathbf{s}^{(i+1)}) - R_{ii}s_i \right|^2 \qquad (27.2)$$

with $e(\mathbf{s}^{(i+1)}) = y_i - \sum_{j=i+1}^{M_T} R_{ij}s_j$. One can visualize this recursion as traversing a tree, where the nodes on level i are associated with a partial vector symbol $\mathbf{s}^{(i)} = [s_i \ \ s_{i+1} \ \ \cdots \ \ s_{M_T}]^T$ and with the corresponding PEDs $T_i(\mathbf{s}^{(i)})$. The root of the tree is in $i = M_T + 1$, and the metric at the leaves finally corresponds to $d(\mathbf{s}) = T_1(\mathbf{s}^{(1)})$.

Breadth-first tree traversal with the K-best algorithm

The *K-best algorithm* (Wong et al., 2002) traverses the tree *breadth-first*, meaning that the decoder starts from the root of the tree and proceeds in forward direction only. Unfortunately, the number of nodes increases exponentially from one level to the next. Complexity reduction is achieved by keeping only those K nodes that are associated with the smallest PEDs when proceeding to the next level. All other nodes are pruned from the tree, together with all their children. The design parameter K determines the trade-off between the complexity of the algorithm and the BER performance. It was shown by Guo and Nilsson (2004) that for a 4×4 system with 16-QAM modulation $K = 5$ is sufficient for closce-to-ML performance, provided that a suitable detection order has been chosen.

A VLSI architecture for the K-best algorithm can be designed based on a pipelined linear array of processing elements (PEs) (Wong et al., 2002; Guo and Nilsson, 2004). Each PE considers one level of the tree and examines the children of the K parent nodes that it receives from the previous level (i.e., from the preceding PE in the pipeline). As multiple PEs are needed, their individual circuit complexity has to be kept low. Thereto, each of them computes the PEDs of the children sequentially, over multiple cycles (*time sharing*). A bubble sort is used to maintain a list of the K preferred candidates to be passed on to the next stage. For a given K, the degree of *time sharing* within each stage can be adjusted to increase the decoding throughput at the expense of area.

Efficient parallel PED computation for multiple children of a node is a key to achieving high throughput with still reasonably low area. Similar to the implementation of the exhaustive-search ML detector, this can be achieved through *algorithmic transformations* (Burg et al., 2005), which factor the expression for the PEDs in (27.2) in such a way that most terms that require costly operations become independent of s_i and can therefore be shared (precomputed).

Depth-first tree traversal with sphere decoding (SD)

Sequential *sphere decoding* (Pohst, 1981; Damen et al., 2003) is a recursive procedure, which traverses the tree in forward and backward direction following a *depth-first* approach. As in the K-best algorithm, *tree pruning* is the key to reducing complexity. However, SD takes a less heuristic approach: the *sphere constraint* (SC) restricts the search to only those candidate symbols **Rs** that lie inside a hypersphere with radius r around $\hat{y}$. As PEDs can only increase from one level to the next, a node $s^{(i)}$ can be pruned from the tree with all its children if $T_i(s^{(i)}) \geq r^2$. Whenever the decoder reaches a leaf $s_k^{(1)}$, the sphere is shrunk by setting $r^2 = d(s_k^{(1)})$. This immediately removes more nodes from the tree. Radius updating is most efficient when the depth-first tree traversal is supplemented with a metric-first strategy, which, if multiple children of a node meet the SC, chooses the one with the smallest metric as the preferred child (Schnorr and Euchner, 1994). Radius updating is key to achieving high throughput, and allows, in general, to start with an infinite initial radius.

The VLSI architecture for the depth-first tree traversal is designed to *visit a new node in each cycle, ensuring that no node is ever visited twice* (Burg et al., 2005). To this end, two independent units are employed, as shown in Fig. 27.4. The metric computation unit (MCU) starts from a parent node at its input and finds the preferred child according to Schnorr and

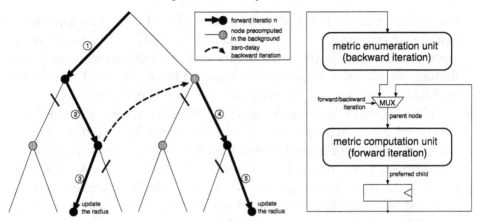

Fig. 27.4. Depth-first tree traversal example (left) and *one-node-per-cycle* VLSI architecture (right).

Euchner (1994). If it meets the SC, the decoder proceeds to the next level, and the preferred child becomes the new parent node (cycles 1–3 and 5 in the example in Fig. 27.4). The metric enumeration unit (MEU) runs in the background and considers the siblings of the preferred children. When a leaf is reached, or when no child meets the SC, the MEU can provide a new parent node to the MCU, from which the search can continue (cycle 4).

The throughput of such an implementation in terms of vector symbols per second is determined by the *cycle time* and, on the algorithmic level, by the *number of visited nodes*, which depends on the efficiency of the tree pruning.

Finding the preferred child for the forward iteration according to the metric-first paradigm is the most challenging part in the implementation. Three approaches have been proposed to solve this problem:

- *Real-lattice decomposition* first breaks down the M_T-dimensional complex-valued problem into a $2M_T$-dimensional real-valued problem. From (27.2), one can then obtain an *admissible interval* for s_i, whose center point minimizes the PED. However, real-lattice decomposition also doubles the depth of the tree, i.e., the number of visited nodes, and the computation of the admissible interval is costly in terms of hardware. Therefore, the approach is not suited for VLSI implementation. Also, it is only applicable to QAM modulation.

- An *exhaustive search* can be performed by computing the PEDs of all children of a node, as discussed earlier for the K-best implementation. Finding

the minimum among the results yields the preferred child. The approach is
efficient for moderate constellation sizes (up to 16-QAM) when costly operations are shared. Beyond that, hardware complexity grows quickly, and
finding the minimum becomes rather slow. The advantage of the scheme
is that it can operate on arbitrary complex-valued constellations.

• A *hybrid approach* between computing *admissible intervals* and *exhaustive search* has been proposed by Hochwald and ten Brink (2003) and was
refined for hardware implementation by Burg *et al.* (2005). It starts by
decomposing QAM constellations into PSK constellations. Within each
of them, the constellation point with the smallest PED can be easily determined based on simple relations between the real and imaginary parts
of $e(\mathbf{s}^{(i+1)})$, so that only a single PED needs to be computed for each
PSK subset. These PEDs are then compared to find the minimum, i.e.,
the preferred child across the 3, 9, or 32 subsets in the case of 16-QAM,
64-QAM and 256-QAM, respectively.

Infinity-norm sphere decoding (Burg *et al.*, 2005) increases throughput
and reduces circuit complexity. The algorithm starts by taking the square-
root of the SC and replaces the resulting ℓ^2-norm with the ℓ^∞-norm, so that
one obtains the modified PED update equation

$$X_i(\mathbf{s}^{(i)}) = \max\left(X_{i+1}(\mathbf{s}^{(i+1)}), \left|e(\mathbf{s}^{(i+1)}) - R_{ii}s_i\right|_\infty\right),$$

where $|\cdot|_\infty$ denotes the ℓ^∞-norm of a complex number. Circuit complexity
reduction results from replacing the squaring operation in the original PED
computation with finding the maximum, which reduces the contribution to
the chip area from close to multiplication to below an addition. The increase
in throughput comes from both a *significant reduction in the number of vis-
ited nodes* (as $|\cdot|_\infty \leq |\cdot|_2$), and from the *reduced delay of the PED computa-
tion*. The improvement can be seen in Fig. 27.5 for the case of a 4×4 system
with 16-QAM modulation. The savings are expected to grow with increasing
dimension of the problem. On the negative side, however, the use of the ℓ^∞-
norm incurs a small but constant loss in SNR, compared with the ℓ^2-norm
SD (in the case of a 4×4 system with 16-QAM modulation roughly 1.4 dB).

Comparison and scaling of lattice decoders

Fig. 27.5 plots the throughput for different lattice decoder implementations
and summarizes their technical details. The K-best strategy guarantees constant deterministic throughput, while the average throughput (over different
channel realizations) of *sequential SD* varies as a function of the SNR. However, a direct comparison shows that the average throughput of the ℓ^∞-norm

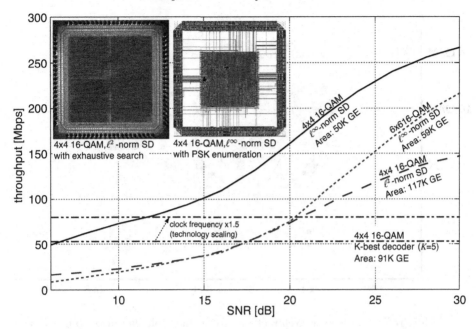

Fig. 27.5. Comparison of throughput and area of lattice decoder ASICs based on depth-first sequential SD and on the K-best strategy.

SD outperforms the constant throughput of the K-best decoder over a wide range of SNRs, making it the more attractive choice. Also, early termination can be used to enforce a guaranteed minimum throughput. In terms of BER performance, the K-best and the ℓ^∞-norm SD are close to ML, while the ℓ^2-norm SD can achieve full ML performance *if no early termination is employed.*

As the number of antennas is increased, the average throughput of the *sequential SD* drops, but the area increases only slightly (see annotations in Fig. 27.5). In a pipelined K-best implementation, the throughput remains constant when K remains unchanged, but the number of stages (i.e., the silicon area) grows proportional to the number of antennas. In addition to that, K also needs to be increased to maintain close to ML performance. This decreases throughput, or further increases area.

27.5 Conclusions

The implementation of MIMO algorithms has come a long way from initially only considering the least complex, but suboptimal, algorithms to recent silicon implementations of (close to) optimal detection schemes. This has been

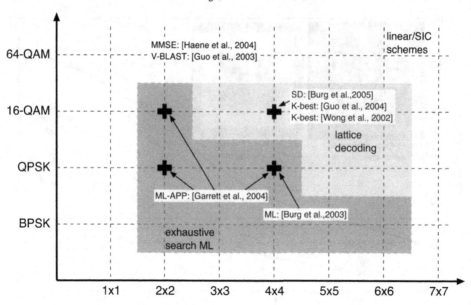

Fig. 27.6. Preferred receiver architectures from an implementation point-of-view for different antenna configurations and modulation schemes.

enabled by new *reduced complexity algorithms, optimized VLSI architectures,* and *advances in silicon process technology.*

For MIMO systems with less than 8 bpcu, exhaustive-search ML detection techniques are the best approach in performance, and are feasible in silicon chips. When the spectral efficiency exceeds 8 bpcu, sequential sphere decoding with radius reduction appears to be the most promising technique. Only for rates in excess of 16–24 bpcu, or for very high symbol rates suboptimal MIMO detection techniques must be employed. These results are summarized in Fig. 27.6.

Acknowledgments

The first author expresses his gratitude to his colleagues at ETH: S. Haene, D. Perels, M. Borgmann, and P. Luethi for their important contributions and to N. Felber, W. Fichtner, and H. Bölcskei for invaluable discussions and continuous support. Furthermore, the authors would like to acknowledge the contributions from past and present colleagues at Lucent Bell Laboratories, L. Davis, G. Woodward, C. Nicol, S. ten Brink, B. Hochwald, S. Venkatesan, L. Mailaender, D. Samaradzija, and F. Mullany among many others. Finally, the authors would like to acknowledge B. Lorenz for his feedback on this chapter.

References

Burg, A., Borgmann, M., Wenk, M., Zellweger, M., Fichtner, W., and Bölcskei, H. (2005). VLSI implementation of MIMO detection using the sphere decoder algorithm. *IEEE J. Solid-State Circuits*, **40** (7), 1566–1577.

Burg, A., Felber, N., and Fichtner, W. (2003). A 50 Mbps 4×4 maximum likelihood decoder for multiple-input multiple-output systems with QPSK modulation. In *Proc. IEEE Int. Conf. Electron., Circuits, Syst.*, vol. 1, 332–335.

Damen, M. O., El Gamal, H., and Caire, G. (2003). On maximum-likelihood detection and the search for the closest lattice point. *IEEE Trans. Inf. Theory*, **49** (10), 2389–2402.

Erfanian, J., Pasupathy, S., and Gulak, G. (1994). Reduced complexity symbol detectors with parallel structures for ISI channels. *IEEE Trans. Commun.*, **42** (234), 1661–1671.

Foschini, G. and Gans, M. (1998). On limits of wireless communications in a fading environment when using multiple antennas. *Wireless Personal Commun.*, **6** (3), 311–334.

Garrett, D., Davis, L., ten Brink, S., Hochwald, B., and Knagge, G. (2004). Silicon complexity for maximum likelihood MIMO detection using spherical decoding. *IEEE J. Solid-State Circuits*, **39** (9), 1544–1552.

Garrett, D., Woodward, G. K., Davis, L., and Nicol, C. (2005). A 28.8 Mb/s 4 × 4 MIMO 3G CDMA receiver for frequency selective channels. *IEEE J. Solid-State Circuits*, **40** (1), 320–330.

Golub, G. H. and Van Loan, C. F. (1996). *Matrix Computations* (John Hopkins Univ. Press).

Guo, Z. and Nilsson, P. (2003). A VLSI implementation of MIMO detection for future wireless communications. In *Proc. IEEE Int. Symp. Personal Indoor Mobile Radio Commun.*, vol. 3, 2852–2856.

—— (2004). A VLSI architecture of the Schnorr-Euchner decoder for MIMO systems. In *Proc. IEEE CAS Symp. Emerging Technologies*, 65–68.

Haene, S., Perels, D., Baum, D. S., Borgmann, M., Burg, A., Felber, N., Fichtner, W., and Bölcskei, H. (2004). Implementation aspects of a real-time multiterminal MIMO-OFDM testbed. In *IEEE Radio and Wireless Conf.* URL http://www.nari.ee.ethz.ch/commth/pubs/p/RAWCON2004.

Hassibi, B. (2000). An efficient square-root algorithm for BLAST. In *Proc. IEEE Int. Conf. Acoust., Speech, Signal Process.*, vol. 2, 737–740.

Hochwald, B. M. and ten Brink, S. (2003). Achieving near-capacity on a multiple-antenna channel. *IEEE Trans. Commun.*, **51** (3), 389–399.

ITRS (2004). International technology roadmap for semiconductors. URL http://public.itrs.net.

Lightbody, G., Woods, R., and Walke, R. (2003). Design of a parameterized silicon intellectual property core for QR-based RLS filtering. *IEEE Trans. VLSI Syst.*, **11** (4), 659–678.

Parhami, B. (2000). *Computer Arithmetic, Algorithms and Hardware Design* (Oxford Univ. Press).

Pohst, M. (1981). On the computation of lattice vectors of minimal length, successive minima and reduced bases with applications. *SIGSAM Bulletin*, **15** (1), 37–44.

Press, W. H., Teukolsky, S. A., Vetterling, W. T., and Flannery, B. P. (1992). *Numerical Receipes in C* (Cambridge Univ. Press).

Schnorr, C. P. and Euchner, M. (1994). Lattice basis reduction: Improved practical algorithms and solving subset sum problems. *Math. Program.*, **66** (2), 181–191.

Volder, J. (1959). The CORDIC trigonometric computing technique. *IRE Trans. Electron. Comput.*, **8** (3), 330–334.

Wong, K., Tsui, C., Cheng, R. S.-K., and Mow, W. (2002). A VLSI architecture of a K-best lattice decoding algorithm for MIMO channels. In *Proc. IEEE Int. Symp. Circuits and Syst.*, vol. 3, 273–276.

Index